DATE DUE

VLSI TECHNOLOGY

VLSI
TECHNOLOGY

Second Edition

Edited by
S. M. Sze

AT&T Bell Laboratories
Murray Hill, New Jersey

McGraw-Hill Book Company

New York   St. Louis   San Francisco   Auckland   Bogotá   Caracas   Colorado Springs
Hamburg   Lisbon   London   Madrid   Mexico   Milan   Montreal   New Delhi
Oklahoma City   Panama   Paris   San Juan   São Paulo   Singapore   Sydney   Tokyo   Toronto

This book was set in Times Roman by Publication Services.
The editor was Alar E. Elken;
the designer was Albert M. Cetta;
the production supervisor was Salvador Gonzales.
Drawings were done by Bell Laboratories, Inc.
Project supervision was done by Publication Services.
Arcata Graphics/Halliday was printer and binder.

VLSI TECHNOLOGY

1 2 3 4 5 6 7 8 9 0 HALHAL 8 9 3 2 1 0 9 8

ISBN 0-07-062735-5

Library of Congress Cataloging-in-Publication Data
VLSI technology.

   (McGraw-Hill series in electrical engineering.
Electronics and electronic circuits)
   Includes index.
   1. Integrated circuits—Very large scale integration.
I. Sze, S. M., 1936–   . II. Series.
TK7874.V566   1988   621.381'73   87-22803

CONTENTS

LIST OF CONTRIBUTORS

A. C. ADAMS
AT&T Bell Laboratories
Murray Hill, New Jersey

W. J. BERTRAM
AT&T Bell Laboratories
Murray Hill, New Jersey

J. B. BINDELL
AT&T Bell Laboratories
Allentown, Pennsylvania

W. FICHTNER*
Swiss Federal Institute of Technology
Zurich, Switzerland

M. D. GILES
AT&T Bell Laboratories
Murray Hill, New Jersey

S. J. HILLENIUS
AT&T Bell Laboratories
Murray Hill, New Jersey

L. E. KATZ
AT&T Bell Laboratories
Allentown, Pennsylvania

S. P. MURARKA*
Rensselaer Polytechnic Institute
Troy, New York

C. W. PEARCE
AT&T Network Systems
Allentown, Pennsylvania

R. J. SCHUTZ
AT&T Bell Laboratories
Murray Hill, New Jersey

K. M. STRINY
AT&T Bell Laboratories
Holmdel, New Jersey

J. C. C. TSAI
AT&T Bell Laboratories
Reading, Pennsylvania

R. K. WATTS
AT&T Bell Laboratories
Murray Hill, New Jersey

*Consultant to AT&T Bell Laboratories, Murray Hill, New Jersey

PREFACE

Because of the rapid progress of the fabrication technology in recent years, the first edition of *VLSI Technology* clearly needed substantial revision if it was to continue to serve its purpose. Like the first edition, the new edition describes the theoretical and practical aspects of the most advanced state of electronics technology—very-large-scale integration (VLSI). From crystal growth to reliability testing, the reader is presented with all the major steps in the fabrication of VLSI circuits. In addition many broader topics, such as process simulation and analytical techniques, are considered in detail. Each chapter describes one aspect of VLSI processing. The chapter's introduction provides a general discussion of the topic, and subsequent sections present the basic science underlying individual process steps, the necessity for particular steps in achieving required parameters, and the tradeoffs in optimizing device performance and manufacturability. The problems at the end of each chapter form an integral part of the development of the topic.

The book is intended as a textbook for senior undergraduate or first-year graduate students in applied physics, electrical engineering, and materials science; it assumes that the reader has already acquired an introductory understanding of the physics and technology of semiconductor devices. Because it elaborates on integrated circuit (IC) processing technology in a detailed and comprehensive manner, it can also serve as a reference for those actively involved in integrated circuit fabrication and process development.

The first edition of *VLSI Technology* was published in 1983, and the book was subsequently translated into Italian, Japanese, and Russian. The text of the first edition was derived from a set of lecture notes prepared for an in-hours continuing education course at AT&T Bell Laboratories. The course, called "Silicon Integrated Circuit Processing," has been given to hundreds of engineers and scientists engaged in research, development, fabrication, and application work of ICs. In the second edition of *VLSI Technology*, 50 percent of the material has been revised or updated to include the most advanced and important topics in VLSI processing. About 900 references have been cited, of which 40 percent were published in the past five years, and over 400 technical illustrations are included, of which 55 percent are new.

In the course of writing the second edition of *VLSI Technology,* many people have assisted us and offered their support. We would first like to express our appreciation to the management of AT&T Bell Laboratories and AT&T Network Systems for providing the environment in which we worked on the book. We wish to thank Drs. D. B. Fraser, R. B. Marcus, D. A. McGillis, C. J. Mogab, L. C. Parrillo, T. E. Seidel, and C. A. Steidel, each of whom authored one chapter for the first edition of *VLSI Technology;* we have benefited from their style of presentation and have adopted many background materials from their chapters. We have also benefited from suggestions made by the reviewers: Drs. J. N. Arnold, D. K. Atwood, W. A. Baker, K. E. Benson, G. K. Celler, D. P. Chesire, J. T. Clemens, W. T. Cochran, L. C. Feldman, P. D. Foo, G. W. Hills, R. W. Hurlbrink, T. Kook, P. H. Langer, H. J. Levinstein, R. A. Levy, A. M. R. Lin, W. T. Lynch, M. A. Menendez, D. W. Peters, M. Pinto, J. M. Poate, V. Pol, A. G. Rane, D. Rehrig, T. T. Sheng, K. Short, A. K. Sinha, R. K. Smith, G. R. Weber, D. S. Williams, and A. W. Yanof of AT&T Bell Laboratories; Prof. S. Bibyk of Ohio State University, Prof. R. W. Dutton of Stanford University, Prof. W. Guy and Prof. T. Riley of Lafayette College, Prof. R. Jaccodine of Lehigh University, Prof. C. Muckenfuss of Rensselaer Polytechnic Institute, Prof. D. Neikirk of University of Texas, Prof. A. Neureuther of University of California at Berkeley, Prof. K. Scoles of Drexel University, Prof. P. Sukanek of Clarkson University, and Prof. H. E. Talley of University of Kansas.

We are further indebted to Mr. E. Labate and Dr. R. A. Matula for their literature searches; Mr. N. Erdos, Mr. E. V. Bacher, Mr. J. A. Elkins, and Ms. A. C. Johnson for technical editing of the manuscript, and Ms. A. M. Jackson and M. L. McSpiritt for providing thousands of technical papers on IC processing cataloged at the Murray Hill Library of AT&T Bell Laboratories. Finally, we wish to acknowledge Ms. E. C. Hung, Ms. J. Marotta, Ms. J. Maye, Ms. A. M. McDonough, and the members of the Text Processing Centers who typed the initial drafts and the final manuscript; Mr. D. A. Spranza and Ms. C. A. Vella and members of the Art Studio who furnished the hundreds of technical illustrations used in the book; and Mrs. T. W. Sze, who prepared the Appendixes and Index.

S. M. Sze

INTRODUCTION

GROWTH OF THE INDUSTRY

The electronics industry in the United States has grown rapidly in recent years, with factory sales increasing by a factor of 15 since the early 1960s [see Fig. 1, curve (a)[1,2]]. Electronics sales, which were $210 billion in 1986, are projected to increase at an average annual rate of 12% and finally reach $1 trillion by the year 2000. The integrated circuit (IC) market has increased at an even higher rate than electronic sales [see Fig. 1, curve (b)]. IC sales in the United States were $11 billion in 1986 and are expected to grow by 16% annually, reaching $90 billion by the year 2000. The main impetuses for such phenomenal market growth are the intrinsic pervasiveness of electronic products and the continued technological breakthroughs in integrated circuits. The world market of electronics (about twice the size of the US market) will grow at a comparable rate.[3] By the year 2000, it will surpass the automobile, chemical, and steel industries in sales volume.

Figure 2 shows the sales of major IC groups and how sales have changed in recent years.[1] In the 1960s the IC market was broadly based on bipolar transistors. Since 1975, however, digital MOS ICs have prevailed. At present, even the intrinsic speed advantage of bipolar transistors is being challenged by MOSFETs. Because of the advantages in device miniaturization, low power dissipation, and high yield, by 2000 digital MOS ICs will dominate the IC market and capture a major market share of all semiconductor devices sold. This book, therefore, emphasizes MOS-related VLSI technology.

1

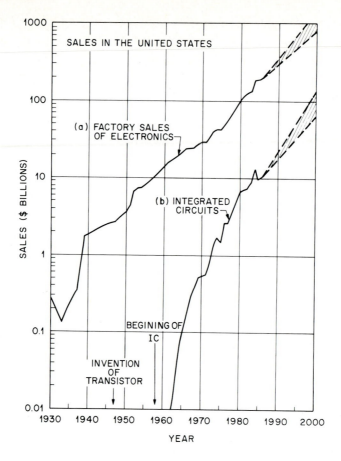

FIGURE 1

(a) Factory sales of electronics in the United States for the 57 years between 1930 and 1986 and projected to 2000. (b) Integrated circuit market in the United States for the 25 years between 1962 and 1986 and projected to 2000. (*After Refs. 1 and 2.*)

DEVICE MINIATURIZATION

Figure 3, curve (a), shows the exponential growth of the number of components per IC chip.[4,5] Note that IC complexity has advanced from small-scale integration (SSI) to medium-scale integration (MSI), to large-scale integration (LSI), and finally to very-large-scale integration (VLSI), which has 10^5 or more components per chip. Although the rate of growth has slowed in recent years because of difficulties in defining, designing, and processing complicated chips, about 100 million devices per chip will be available before 2000.

The most important factor in achieving such complexity is the continued reduction of the minimum device dimension [see Fig. 3, curve (b)]. Since 1960, the annual rate of reduction has been 13%; at that rate, the minimum feature length will shrink from its present length of 1 μm to 0.1 μm in the year 2000.

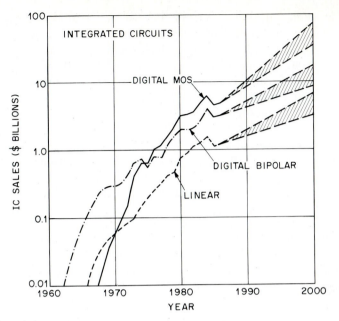

FIGURE 2

Sales of major IC groups in the United States. (*After Ref. 1.*)

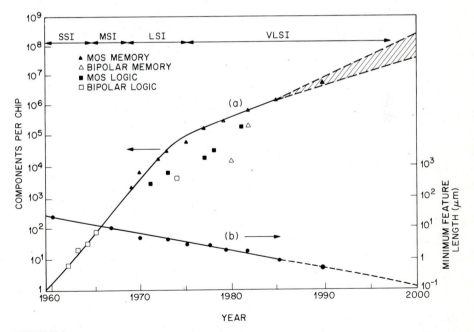

FIGURE 3

(a) Exponential growth of the number of components per IC chip. (*After Moore, Ref. 4, and Bezold and Olsen, Ref. 5.*) (b) Exponential decrease of the minimum device dimensions.

Device miniaturization results in reduced unit cost per function and in improved performance. Figure 4, curve (a), gives an example of the cost reduction. The cost per bit of memory chips has halved every 2 years for successive generations of random-access memories.[6] By 2000 the cost per bit is expected to be less than 0.1 millicent for a 64-megabit memory chip. Similar cost reductions are expected for logic ICs.

As device dimension decreases, the intrinsic switching time in MOSFETs decreases linearly. (The intrinsic delay is given approximately by the channel length divided by the carrier velocity.) The device speed has improved by four orders of magnitude since 1960. Higher speeds lead to expanded IC functional throughput rates. In the future, digital ICs will be able to perform data processing, numerical computation, and signal conditioning at 10 and higher gigabit-per-second rates. Another benefit of miniaturization is the reduction of power consumption. As the device becomes smaller, it consumes less power. Therefore, device miniaturization also reduces the energy used for each switching operation. Figure 4, curve (b), shows the trend of this energy consumption, called the power-delay product.[7] The energy dissipated per logic gate has decreased by over five orders of magnitude since 1960.

INFORMATION AGE

Figure 5 shows four periods of change in the electronics industry in the United States. Each period exhibits normal life-cycle characteristics[8] (i.e., from inception to rapid growth, to saturation, and finally to decline). The development of the

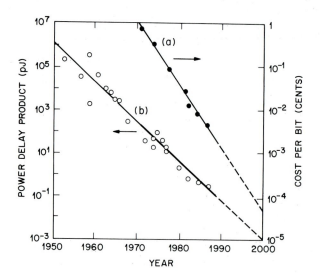

FIGURE 4
(a) Reduction of cost per bit of RAM chips. (*After Noyce, Ref. 6.*) (b) Power-delay product per logic gate versus year. (*After Keyes, Ref. 7.*)

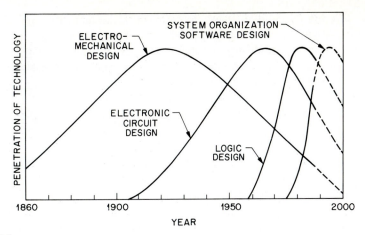

FIGURE 5
Penetration of technology into the industrial output versus year for four periods of change in the US electronic industry. (*After Connell, Ref. 8.*)

vacuum tube in 1906 and the invention of transistors[9] in 1947 opened the field of electronic circuit designs. The development of integrated circuits[10] in 1958 led to a new generation of logic families. Since 1975, the beginning of VLSI, the frontier has moved to system organization of ICs and the associated software designs.

Many system-oriented VLSI chips, such as speech analysis/recognition and storage circuits, will be built in response to the enormous market demand for sophisticated electronic systems to handle the growing complexities of the Information Age.[11,12] In this age a major portion of our work force can be called "information workers"; they are involved in gathering, creating, processing, disseminating, and using information. Figure 6 shows the changing composition of

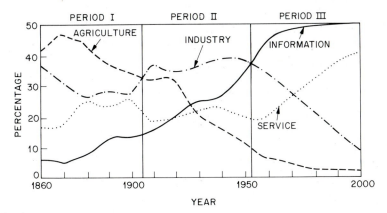

FIGURE 6
Changing composition of work force in the United States. (*After Robinson, Ref.11.*)

the work force in the United States. Prior to 1906, the largest single group was involved in agriculture. In the next period, until the mid 1950s, the predominant group was involved in industry. Currently, the predominant group consists of information workers; about 50% of the total work force is in this category. In Europe and Japan, information workers now constitute about 40 to 45% of the work force, which is also expected to reach 50% before the end of the century.[13] Advances in VLSI will have a profound effect on the world economy, because VLSI is the key technology for the Information Age.

ORGANIZATION OF THE BOOK

Figure 7 shows how the 14 chapters of this book are organized. Chapter 1 considers crystal growth and wafer preparation. VLSI technology is synonymous with *silicon* VLSI technology. The unique combination of silicon's adequate bandgap, stable oxide, and abundance in nature ensures that in the foreseeable future, no other semiconductor will seriously challenge its preeminent position in VLSI applications. (Some important properties of silicon are listed in Appendix A.) Once the silicon wafer is prepared, we enter into the wafer-processing sequence, described in Chapters 2 through 9, and depicted in the wafer-shaped central circle of Fig. 7. Each of these chapters considers a specific processing step. Of course, many processing steps are repeated many times in IC fabrication; for example, lithography and reactive plasma etching steps may be repeated 5 to 15 times.

Chapter 10 considers process simulation of all the major steps covered in Chapters 2 through 9. Process simulation is emerging as an elegant aid to process

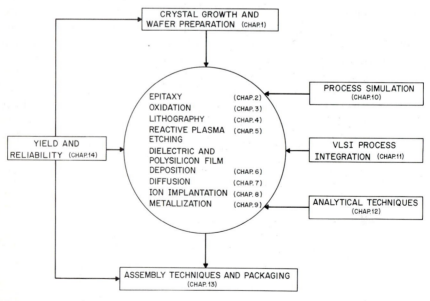

FIGURE 7
Organization of this book.

development. This approach is attractive because of its rapid turn-around time and lower cost when compared to the experimental approach. Process simulation coupled with device and circuit simulations can provide a total design system that allows on-line process design and simulation to predict desired device and circuit parametric sensitivities and to facilitate circuit design and layout.

The individual processing steps described in Chapters 2 through 9 are combined in Chapter 11 to form devices and logic circuits. Chapter 11 considers the three most important IC families: the NMOS (n-channel MOSFET) ICs, the CMOS (complementary MOSFET) ICs, and the bipolar ICs. As the device dimension decreases and circuit complexity increases, sophisticated tools are needed for process analysis. Chapter 12 covers many advanced analytical techniques, such as scanning and transmission electron microscopy for morphology determination, Auger electron spectroscopy for chemical analysis, and X-ray diffraction for structural analysis.

After completely processed wafers are tested, those chips that pass the tests are ready to be packaged. Chapter 13 describes the assembly and packaging of VLSI chips. Chapter 14 describes the yield at every step of the processing and the reliability of the packaged ICs. As device dimensions move to the submicron regime, VLSI processing becomes more automated, resulting in tighter control of all processing parameters. At every step of production, from crystal growth to device packaging, numerous refinements are being made to improve the yield and reliability.

To keep the notation simple in this book, we sometimes found it necessary to use a simple symbol more than once, with different meanings. For example, in Chapter 1 S means 4-point probe spacing, in Chapter 4 it means a measure of coherence, and in Chapter 14 it means slope of a failure plot. Within each chapter, however, a symbol has only one meaning and is defined the first time it appears. Many symbols do have the same or similar meanings consistently throughout this book; they are summarized in Appendix B.

At present, VLSI technology is moving at a rapid pace. The number of VLSI publications (i.e., papers with the acronym "VLSI" in the title or abstract) has grown from virtually zero in 1975 to over 2500 in 1986 with an average annual growth rate of over 200%! Note that many topics, such as lithography and process simulation, are still under intensive study. Their ultimate capabilities are still not fully understood. The material presented in this book is intended to serve as a foundation. The references listed at the end of each chapter can supply more information.

REFERENCES

1 *1987 Electronic Market Data Book*, Electronic Industries Association, Washington, D.C., 1987.

2 "1987 U.S. Market Report," *Electronics,* **60**, No.1, 51 (1987).

3 *Mackintosh Year Book, Electronic Data*, Vol.2, Benn Electronics Publication, 1987.

4 G. Moore, "VLSI, What Does the Future Hold," *Electron Aust.*, **42**, 14 (1980).

5 C. Bezold and R. Olsen, *The Information Millennium: Alternative Futures*, Information Industry Association, Washington, D.C., 1986.

6 R. N. Noyce, "Microelectronics," in T. Forester, Ed., *The Microelectronics Revolution*, MIT Press, Cambridge, Mass., 1981, p.29.

7 R. W. Keyes, "Limitations of Small Devices and Large Systems," in N. G. Einspruch, Ed., *VLSI Electronics*, Academic, New York, 1981, Vol. 1, p. 186.

8 J. M. Connell, "Forecasting a New Generation of Electronic Components," *Digest IEEE Spring Compcon.*, **81**, 14 (1981).

9 W. Shockley, "The Path to the Conception of the Junction Transistor," *IEEE Trans. Electron Devices*, **ED-31**, 1523 (1984).

10 J. S. Kilby, "Invention of the Integrated Circuits," *IEEE Trans. Electron Devices*, **ED-23**, 648 (1976).

11 A. L. Robinson, "Electronics and Employment: Displacement Effects," in T. Forester, Ed., *The Microelectronics Revolution*, MIT Press, Cambridge, Mass., 1981, p. 318.

12 J. S. Mayo, "Technology Requirements of the Information Age," *Bell Lab. Rec.*, **60**, 55 (1982).

13 D. Kimbel, *Microelectronics, Productivity and Employment*, Organization for Economic Cooperational Development, Paris, 1981, p.15.

CHAPTER

1

CRYSTAL GROWTH AND WAFER PREPARATION

C. W. PEARCE

1.1 INTRODUCTION

Silicon, occurring naturally in the form of silica and silicates, is the most important semiconductor for the electronics industry. At present, silicon-based devices constitute over 95% of all semiconductor devices sold worldwide. Silicon is one of the most studied elements in the periodic table. A literature search on published papers using silicon as a search word yields over 25,000 references. Silicon has been well studied and many of its physical properties have been measured. Appendix A is a compilation of some useful constants. [1,2]

The advent of solid-state electronics dates from the invention of the bipolar transistor effect by Bardeen, Brattain, and Shockley. [3] The technology progressed during the early 1950s, using germanium as the semiconductor material. However, germanium proved unsuitable in certain applications because of its propensity to exhibit high junction leakage currents. These currents result from germanium's relatively narrow bandgap (0.66 eV). For this reason, silicon (1.1 eV) became a practical substitute and has almost fully supplanted germanium as a material for solid-state device fabrication. Silicon devices can operate up to 150°C versus 100°C for germanium.

In retrospect, other reasons could have ultimately led to the same material substitution. Planar processing technology derives its success from the properties of thermally grown silicon dioxide. Germanium oxide is unsuited for device applications. The intrinsic (undoped) resistivity of germanium is about 47 Ω-cm, which would have precluded the fabrication of rectifying devices with high break-down voltages. In contrast, the intrinsic resistivity of silicon is about 230,000 Ω-cm. Thus, high-voltage rectifying devices and certain infrared sensing devices are practical with silicon. Finally, there is an economic consideration—electronic-grade germanium is now more costly than silicon.

Similar problems impeded the widespread use of compound semiconductors. For example, it is difficult to grow a high-quality oxide on GaAs. One element oxidizes more readily than the other, leaving a metallic phase at the interface. Such material is difficult to dope and obtain in large diameters with high crystal perfection. In fact the technology of Group III–V compounds has advanced partly because of the advances in silicon technology.

This chapter provides an overview of silicon wafers. It starts with the production of the raw material and discusses the major steps required to produce wafers that can be used to manufacture devices. It will also emphasize those attributes of the wafers that influence device properties.

1.2   ELECTRONIC-GRADE SILICON

Electronic-grade silicon (EGS), a polycrystalline material of high purity, is the raw material for the preparation of single-crystal silicon. EGS is undoubtedly

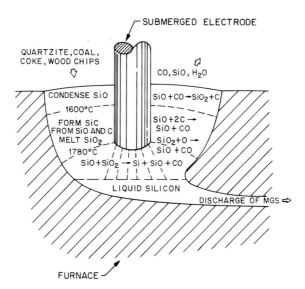

FIGURE 1
Schematic of a submerged-electrode arc furnace for the production of metallurgical-grade silicon.
(*After Crossman and Baker, Ref. 4.*)

one of the purest materials routinely available. The major impurities are boron, carbon, and residual donors. Pure EGS generally requires that doping elements be in the parts per billion (ppb) range, and carbon be less than 2 parts per million (ppm).[4] These properties are usually evaluated on test ingots rather than measured on the material itself.[5] In the case of the doping elements, ppb levels are below the capabilities of most laboratory methods, so the doping level can be inferred from resistivity measurements on the test ingot. Alternatively, low temperature infrared absorption measurements or photoluminence can be used.

To obtain EGS requires a multistep process.[4] First, metallurgical-grade silicon (MGS) is produced in a submerged-electrode arc furnace, as shown in Fig. 1. The furnace is charged with quartzite, a relatively pure form of sand (SiO_2), and carbon in the form of coal, coke, and wood chips. In the furnace a number of reactions take place, the overall reaction being

$$SiC \text{ (solid)} + SiO_2\text{(solid)} \rightarrow Si \text{ (liquid)} + SiO \text{ (gas)} + CO \text{ (gas)} \qquad (1)$$

The process is power intensive, requiring 13 kWh/kg, and MGS is drawn off and solidified at a purity of 98%. Table 1 shows typical purities of various

TABLE 1
Comparison of typical impurity contents in various materials (values in ppm except as noted)

Impurity	Quartzite	Carbon	MGS*	EGS†	Crucible quartz
Al	620	5500	1570	. . .	. . .
B	8	40	44	<1 ppb	. . .
Cu	<5	14	. . .	0.4	0.23
Au	. . .	. . .	. . .	0.07 ppb	. . .
Fe	75	1700	2070	4	5.9
P	10	140	28	<2 ppb	. . .
Ca	. . .	. . .	. . .	. . .	. . .
Cr	. . .	. . .	137	1	0.02
Co	. . .	. . .	. . .	0.2	0.01
Mn	. . .	. . .	70	0.7	. . .
Sb	. . .	. . .	. . .	0.001	0.003
Ni	. . .	. . .	4	6	0.9
As	. . .	. . .	. . .	0.01	0.005
Ti	. . .	. . .	163	. . .	. . .
La	. . .	. . .	. . .	1 ppb	. . .
V	. . .	. . .	100	. . .	. . .
Mo	. . .	. . .	. . .	1.0	5.1
C	. . .	. . .	80	0.6	. . .
W	. . .	. . .	. . .	0.02	0.048
O	. . .	. . .	. . .	. . .	. . .
Na	. . .	. . .	. . .	0.2	3.7

* Metallurgical-grade silicon.

†Electronic-grade silicon.

materials used in the arc furnace. The MGS used in the making of metal alloys is not sufficiently pure to use in the manufacture of solid-state devices.

The next process step is to pulverize the silicon mechanically and react it with anhydrous hydrogen chloride to form trichlorosilane ($SiHCl_3$), according to the reaction:

$$Si(solid) + 3HCl(gas) \rightarrow SiHCl_3(gas) + H_2(gas) + heat \qquad (2)$$

The reaction takes place in a fluidized bed at a nominal temperature of 300°C using a catalylst. Here silicon tetrachloride and the chlorides of impurities are formed. At this point the purification process occurs. Trichlorosilane is a liquid at room temperature (boiling point 32°C), as are many of the unwanted chlorides. Purification is therefore done by fractional distillation.

EGS is prepared from the purified $SiHCl_3$ in a chemical vapor deposition (CVD) process similar to the epitaxial CVD process that is presented in Chapter 2. The chemical reaction is a hydrogen reduction of trichlorosilane.

$$2SiHCl_3(gas) + 2H_2(gas) \rightarrow 2Si(solid) + 6HCl(gas) \qquad (3)$$

This reaction is conducted in the type of system shown in Fig. 2. A resistance-heated rod of silicon, called a "slim rod," serves as the nucleation point for the deposition of silicon. A complete process cycle takes many hours and results in rods of EGS, which are polycrystalline in structure, up to 0.2 m (or more) in diamater and several meters in length. EGS can be cut from these rods as single chunks or crushed into nuggets (Fig. 3). Figure 4 illustrates the entire EGS preparation process starting with MGS and including the feedback or recycling of reaction byproducts to achieve high overall process efficiency. An alternate

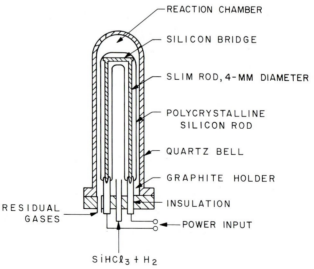

FIGURE 2
Schematic of a CVD reactor used for EGS production. (*After Crossman and Baker, Ref. 4.*)

FIGURE 3
EGS in chunk form loaded into a 30 cm diameter silica crucible.

process for the production of EGS that is starting to receive commercial attention is the pyrolysis of silane.[6] In this process the CVD reactor of Fig. 2 is operated at approximately 900°C and supplied with silane instead of trichlorosilane. The overall reaction is

$$SiH_4 \text{ (gas)} + \text{ heat } \rightarrow Si \text{ (solid)} + 2H_2(\text{gas}) \quad (4)$$

The advantages of producing EGS from silane are potentially lower cost and less harmful reaction byproducts. At present the worldwide consumption[6] of electronic-grade polysilicon is approximately 5×10^6 kg per year.

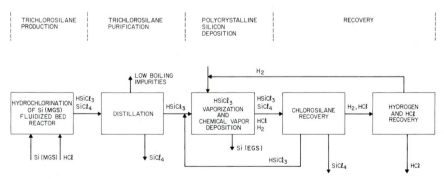

FIGURE 4
Schematic diagram of the production of Electronic-Grade Silicon from the hydrogen reduction of trichlorosilane. (*After McCormick, Ref. 6.*)

1.3 CZOCHRALSKI CRYSTAL GROWING

A substantial percentage (80 to 90%) of the silicon crystals prepared for the semiconductor industry are prepared by the Czochralski (CZ) technique.[7] Virtually all the silicon used for integrated circuit fabrication is prepared by this technique.

1.3.1 Crystal Structure

The processing characteristics and some material properties of silicon wafers depend on its orientation. The {111} planes have the highest density of atoms on the surface,[8] so crystals grow most easily on these planes. Mechanical properties such as tensile strength are highest for <111> directions. The moduli of elasticity also show an orientation dependence (Appendix A). Processing characteristics such as oxidation are similarly orientation dependent. For example, {111} planes oxidize faster than {100} planes, because they have more atoms per unit surface area available for the oxidation reaction to occur. The choice of crystal orientation, therefore, is generally not left to the discretion of the crystal grower, but is a device design consideration. Historically, bipolar circuits have preferred <111> oriented material and MOS devices <100>. There are, of course, exceptions. Growth on other orientations such as <110> has been demonstrated, but is more difficult to achieve routinely.[9]

A real crystal, as represented by a silicon wafer, differs from a mathematically ideal crystal in several respects. It is finite, not infinite; thus, surface atoms are incompletely bonded. The atoms are displaced from their ideal locations by thermal agitation. Most importantly, real crystals have the following classifications of defects: (1) point defect, (2) line defect (dislocation), (3) area or planar defect, and (4) volume defect.[10,11] Defects influence the optical, electrical, and mechanical properties of silicon.

Point defects. Point defects take several forms as shown in Fig. 5. Any nonsilicon atom incorporated into the lattice at either a substitutional (i.e., replacing a host silicon atom) or interstitial (i.e., between silicon atoms) site is considered

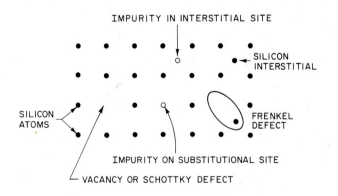

FIGURE 5
The location and types of point defects in a simple lattice.

a point defect. This is true whether the atom is an intentional dopant or unintentional impurity. Missing atoms create a vacancy in the lattice called a "Schottky defect," which is also considered a point defect. A silicon atom in an interstitial lattice site with an associated vacancy is called a "Frenkel defect." Vacancies and interstitials have equilibrium concentrations that depend on temperature. From thermodynamic principles the concentration as a function of temperature can be derived and has the following relation:[11]

$$N_d = A \exp{(-E_a/kT)} \tag{5}$$

where N_d is the concentration of the point defect, A is a constant, E_a is the activation energy (2.6 eV for vacancies and 4.5 eV for interstitials), T is the absolute temperature, and k is Boltzmann's constant.

Point defects are important in the kinetics of diffusion and oxidation. The diffusion of many impurities depends on the vacancy concentration, as does the oxidation rate of silicon. Vacancies and interstitials are also associated with defect formation in processing.[10]

To be electrically active, atoms must usually be located on substitutional sites.[12] When in such sites, they introduce an energy level in the bandgap. Shallow levels are characteristic of efficient donor and acceptor dopants. Midgap levels act as centers for the generation and recombination of carriers to and from the conduction and valence bands. Some impurities are entirely substitutional or interstitial in behavior, but others can exist in either lattice position.

Dislocations. Dislocations form the second class of defects. Two general categories of dislocations are screw and line (edge), the terms being aptly descriptive of their shape. Figure 6 is a schematic representation of a line dislocation in a cubic lattice; it can be seen as an extra plane of atoms AB inserted into the lattice.

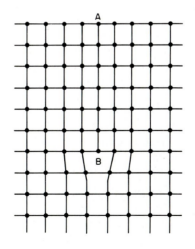

FIGURE 6
An edge dislocation in a cubic lattice created by an extra plane of atoms. The line of the dislocation is perpendicular to the page.

The line of dislocation would be perpendicular to the plane of the page. Dislocations in a lattice are dynamic defects; that is, they can move under applied stress, disassociate into two or more dislocations, or combine with other dislocations. A vector notation developed by Burgers characterizes dislocations in the crystal as to their length and direction.[13] The vector notation is also used to describe dislocation interactions.

Crystals for IC usage are generally grown free of edge dislocations,[10] but may contain small dislocation loops from excess point-defect condensation.[14] These defects act as nuclei for the precipitation of impurities such as oxygen and are responsible for a swirl pattern seen in wafers.[10] Dislocations (edge type) are also introduced by thermal stress on the wafer during processing[15,16] or by the introduction of an excessive concentration of an impurity atom. Substitutional impurities with covalent radii larger or smaller than silicon compress or expand the lattice accordingly. The stress S (in dynes per cm^2) induced depends on the size of the impurity and its concentration:[10]

$$S = \frac{\beta C E}{1 - V} \tag{6}$$

where β is the lattice contraction constant reflecting the degree of distortion introduced by the impurity ($\beta = 8 \times 10^{-24}$ cm^3/atom for boron), C is the impurity concentration; E is Young's modulus, and V is Poisson's ratio (see Appendix A). Dislocations in devices are generally undesirable, because they act as sinks for metallic impurities and alter diffusion profiles. Dislocations can be revealed by preferential etching (see Section 1.3.5).

Area (planar) defects. Two area defects are twins and grain boundaries. Twinning represents a change in the crystal orientation across a twin plane, such that a certain symmetry (like a mirror image) exists across that plane. In silicon, the twin plane is {111}. A grain boundary represents a transition between crystals having no particular orientation relationship to one another. Grain boundaries are more disordered than twins and separate grains of single crystals in polycrystalline silicon. Area defects, such as twins or grain boundaries, represent a large area discontinuity in the lattice. The crystal on either side of the discontinuity may be otherwise perfect. The defects appear during crystal growth, but crystals having such defects are not considered usable for IC manufacture and are discarded.

Volume defects. Precipitates of impurity or dopant atoms constitute the fourth class of defects. Every impurity introduced into the lattice has a solubility; that is, a concentration that the host lattice can accept in a solid solution of itself and the impurity. Figure 7 illustrates the solubility versus temperature behavior for a variety of elements in silicon. Most impurities have a retrograde solubility, which is defined as a solubility that decreases with decreasing temperature. Thus, if an impurity is introduced (at a temperature T_1) at the maximum concentration allowed by its solubility, and the crystal is then cooled to a lower temperature T_2, a supersaturated condition is said to exist (also see Fig. 16). The degree of supersaturation is expressed as the ratio of the concentration introduced at T_1 to

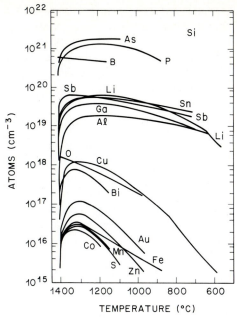

FIGURE 7
Solid solubilities of impurity elements in silicon. (*After Milnes, Ref. 12.*)

the solubility at T_2. The crystal achieves an equilibrium state by precipitating the impurity atoms in excess of the solubility level as a second phase, the second phase being a material of different composition and structure. For instance, excess metallic impurities can react with silicon and form silicides within the host lattice. The kinetics of precipitation depend on the degree of supersaturation, time, and nucleating sites where the precipitates form.

Precipitates are generally undesirable because they act as sites for dislocation generation. Dislocations result from the volume mismatch between the precipitate and the lattice, inducing a strain that is relieved by the formation of dislocations. Precipitation in silicon processing has been observed for dopants such as boron, oxygen, and metallic impurities.[17,18,19]

1.3.2 Crystal Growing Theory

Growing crystals, in the most general sense, involves a phase change from solid, liquid, or gas phases to crystalline solid phase. Czochralski growth, named for the inventor, is the process used to grow most of the crystals from which silicon wafers are produced. This process can be characterized, as applied to silicon, to be a liquid–solid monocomponent growth system. This section discusses some elements of this process as they relate to the understanding of the properties of the grown crystals. For a more complete treatment of crystal growth, refer to the many excellent books devoted to the subject.[10,20]

The growth of a Czochralski (CZ) crystal involves the solidification of atoms from a liquid phase at an interface. The speed of growth is determined by the number of sites on the face of the crystal and the specifics of heat transfer at the interface. Figure 8 schematically represents the transport process and temperature gradients involved. Macroscopically, the heat transfer conditions about the interface can be modeled by the following equation[21]

$$L\frac{dm}{dt} + k_l\frac{dT}{dx_1}A_1 = k_s\frac{dT}{dx_2}A_2 \tag{7}$$

where L is the latent heat of fusion, dm/dt is the mass solidification rate, T is the temperature, k_l and k_s are the thermal conductivities of the liquid and solid, respectively, dT/dx_1 and dT/dx_2 are the thermal gradients at points 1 and 2 (near the interface in the liquid and solid, respectively), and A_1 and A_2 are the areas of the isotherms at positions 1 and 2, respectively. Computer simulation of the growth process for large diameter silicon crystals has been reported.[22]

From Eq. 7 the maximum pull rate of a crystal under the condition of zero thermal gradient in the melt (i.e., $dT/dx_1 = 0$) can be deduced. Converting the mass solidification rate to a growth rate using density and area yields

$$V_{max} = \frac{k_s}{Ld}\frac{dT}{dx} \tag{8}$$

where V_{max} is the maximum pull rate (or pull speed) and d is the density of solid silicon.[7,23] Figure 9 is an experimentally determined temperature variation along a crystal.

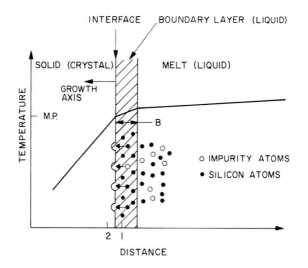

FIGURE 8

Temperature gradients, solidification, and transport phenomena involved in Czochralski growth. Positions 1 and 2 represent the location of isotherms associated with Eq. 7 and the crystal solidification at the interface. Impurity atoms are transported across the boundary layer (B) and incorporated into the growing crystal interface. M.P. is the melting point.

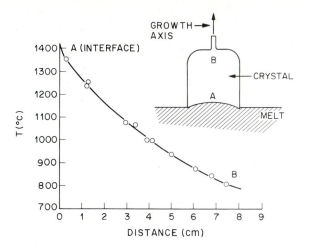

FIGURE 9
Experimentally determined temperature gradient in a silicon crystal as referenced to insert showing a growing crystal. (*After deKock and van de Wijgert, Ref. 14.*)

The pull rate influences the incorporation of impurities into the crystal and is a factor in defect generation. Generally, when the temperature gradient in the melt is small, the heat transferred to the crystal is the latent heat of fusion. As a result, the pull rate generally varies inversely with the diameter[7,23] (Fig. 10). The pull rates obtained in practice are 30 to 50% slower than the maximum values suggested by theoretical considerations.[7]

The growth rate (or growth velocity) of the crystal, actually distinct from the pull rate, is perhaps the most important growth parameter. Pull rate is the macroscopic indication of net solidification rate, whereas growth rate is the instantaneous solidification rate. The two differ because of temperature fluctuations near the interface. The growth rate can exceed the pull rate or even be negative at a given time. When the growth rate is negative, remelting is said to occur. That is, the crystal dissolves back into the melt. The growth rate influences the defect structure and dopant distribution in the crystal on a microscopic scale.

Pull rate affects the defect properties of CZ crystals in the following way. The condensation of thermal point defects (i.e., vacancies and silicon self-interstitials) into small dislocation loops (termed microdefects) occurs as the crystal cools from the solidification temperature. This process occurs above 950° C. The number of defects depends on the cooling rate, which is a function of pull rate and diameter, at temperatures above 950° C. A pull rate of 2 mm/min eliminates defect formation by quenching the point defects in the lattice before they can agglomerate. We find from Fig. 10 that large diameters preclude this pull rate for crystal diameters above 75 mm. Large defects such as edge dislocations are not influenced by the point defect condensation process. A related phenomenon is the remelting of the crystal that occurs because of temperature instabilities in the melt caused by thermal convection. This condition can also be suppressed by attaining a pull rate of 2.7 mm/min,[24] which is half the maximum attainable pull

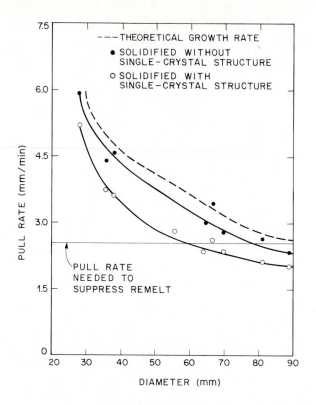

FIGURE 10

Theoretical and experimental pull rates for Czochralski-grown crystals. The dashed line is the theoretical growth rate according to Rea (Ref. 7). (*After Digges, Jr. and Shima, Ref. 23.*)

rate (Eq. 8). Crystals in which remelt has not been suppressed exhibit impurity striation and defect swirls.[25,26] Elimination of remelt results in more uniformly doped crystals, discussed below, but will not necessarily eliminate dopant striation if the growth velocity still varies on a microscopic level.

As mentioned earlier, every impurity has a solid solubility in silicon. The impurity has a different equilibrium solubility in the melt. For dilute solutions commonly encountered in silicon growth, an equilibrium segregation coefficient k_0 may be defined as

$$k_0 = \frac{C_s}{C_l} \tag{9}$$

where C_s and C_l are the equilibrium concentrations of the impurity in the solid and liquid near the interface, respectively.

Table 2 lists the equilibrium segregation coefficients for common impurity and dopant atoms. Note that all are below unity, so that during growth, the impurities at the interface are left in the liquid (melt). Thus, as the crystal grows, the melt becomes progressively enriched with the impurity.

The distribution of an impurity in the grown crystal can be described mathematically by the normal freezing relation:[8,21]

TABLE 2
Segregation coefficients for common impurites in silicon

Impurity	Al	As	B	C	Cu	Fe	O	P	Sb
k_0	0.002	0.3	0.8	0.07	4×10^{-6}	8×10^{-6}	0.25	0.35	0.023

$$C_s = k_0 C_0 (1 - X)^{k_0 - 1} \tag{10}$$

where X is the fraction of the melt solidified, C_0 is the initial melt concentration, C_s is the solid concentration, and k_0 is the segregation coefficient.

Figure 11 illustrates the segregation behavior for several segregation coefficients. It is found experimentally that segregation coefficients differ from equilibrium values and it is necessary to define an effective segregation coefficient k_e:

$$k_e = \frac{k_0}{k_0 + (1 - k_0)\exp(-VB/D)} \tag{11}$$

where V is the growth velocity (or pull rate), D is the diffusion coefficient of dopant in melt, and B is the boundary, or stagnant, layer thickness.[8,27]

The boundary layer thickness is a function of the convection conditions in the melt. Rotation of a crystal in a melt (forced convection) produces a boundary

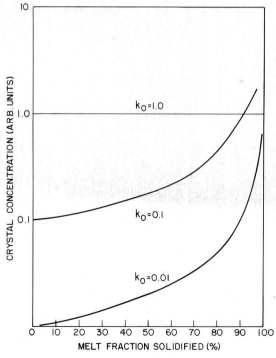

FIGURE 11
Impurity concentration profiles for various k_0 with $C_0 = 1$.

layer B defined by the approximate expression

$$B \simeq 1.8D^{1/3}V^{1/6}W^{-1/2} \tag{12}$$

where W is the rotational velocity.[28]

Our presentation so far represents a first-order approach to the problem. In large melts the convection forced by rotation is often secondary to the thermal convection caused by temperature gradients in the crucible.[28] Because the thermal convection is a random process, the thickness of the boundary layer fluctuates with time, resulting in a variable value for B.

The net result of thermal convection effects is an inhomogeneous distribution of dopant in the crystal on a microscale (Fig. 12). The boundary layer thickness also varies radially across the face of the crystal, resulting in a radial distribution of dopant. Generally, less dopant is incorporated at the edges.

Another effect occuring in heavily doped melts is constitutional supercooling.[29] This effect, particularly prevalent with Sb, occurs when the concentration in the melt ahead of the growing interface is sufficient to depress the solidification temperature (freezing point). When this occurs, the planar liquid–

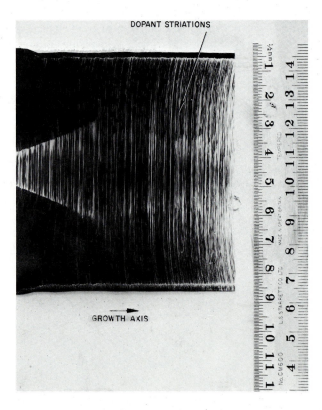

FIGURE 12
Dopant striations in an Sb-doped ingot revealed by preferentially etching a longitudinal section from the seed end of an ingot.

solid interface breaks down, causing the crystal to solidify irregularly and dislocations to appear. Constitutional supercooling limits the ultimate dopant incorporation for certain impurities.

The pull speed is also a factor in determining the shape of the growing interface, as are the melt radial temperature gradient and the crystal surface cooling conditions.[30] A proper choice of these conditions is needed to ensure the stability of the growth process.

Problem
Using Eq. 10 and the segregation coefficient for oxygen from Table 2, predict the concentration of oxygen in the crystal at a fraction solidified of 0.4 in Fig. 15.

Solution
First, find the concentration of oxygen in the melt at the top of the crystal ($X = 0.05$) using Eq. 10.

$$1.3 \times 10^{18} = 0.25 \times C_0(1 - 0.05)^{.25-1}$$

$$C_0 = 5.0 \times 10^{18} \text{atoms/cm}^3$$

Second, calculate the expected value at a fraction solidified of 0.4.

$$C_s = 0.25 \times 5.0 \times 10^{18}(1 - 0.4)^{0.25-1}$$

$$C_s = 1.83 \times 10^{18} \text{atoms/cm}^3$$

Since this is higher than the observed value of 1.0×10^{18} atoms/cm^3, it indicates a loss of oxygen from the melt during growth and shows that Eq. 10 is not strictly applicable for this case.

1.3.3 Crystal Growing Practice

A Czochralski crystal growth apparatus, also called a "puller," is shown in Fig. 13. The one pictured weighs 17,600 kg and is 6.5 m tall. This puller can be configured to hold a melt charge of 60 kg of silicon, which can be transformed into a crystal of 100 mm diameter and 3.0 m length. The puller has four subsystems as follows (Fig. 14):

1. *Furnace*: crucible, susceptor (crucible support) and rotation mechanism, heating element and power supply, and chamber.
2. *Crystal-pulling mechanism*: seed shaft or chain, rotation mechanism, and seed chuck.
3. *Ambient control*: gas source, flow control, purge tube, and exhaust or vacuum system.
4. *Control system*: microprocessor, sensors, and outputs.

Furnace. Perhaps the most important component of the growing system is the crucible (Fig. 3). Since it contains the melt, the crucible material should be chemically unreactive with molten silicon. This is a major consideration, because the electrical properties of silicon are sensitive to even ppb levels of impurities.

FIGURE 13
An industrial-sized Czochralski grower. Numbers relate to the four basic parts of the growers.

Other desirable characteristics for crucible material are a high melting point, thermal stability, and hardness. Additionally, the crucible should be inexpensive or reusable. Unfortunately, molten silicon can dissolve virtually all commonly used high-temperature materials, such as refractory carbides (TiC or TaC), thus introducing unacceptable levels of the metallic species into the crystal.[31]

Carbon or silicon carbide crucibles are also unsuitable. Although carbon is electrically inactive in silicon, high-quality crystals cannot be grown with carbon-saturated melts.[32] During growth in a carbon or silicon carbide crucible, a two-phase solidification occurs once the solid solubility has been exceeded. The second phase is SiC, which is responsible for dislocation generation and loss of single-crystal structure. The remaining choices for a crucible are silicon nitride (Si_3N_4) and fused silica (SiO_2), the latter of which is in exclusive use today.

Fused silica, however, reacts with silicon, releasing silicon and oxygen

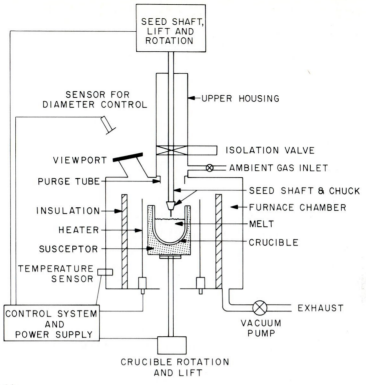

FIGURE 14
Schematic representation of a crystal grower.

into the melt. The dissolution rate is quite substantial,[33] being in the range 8–25 × 10^7 g/cm²-s. The actual rate of erosion is a function of temperature and the convection conditions, either forced or thermal, in the melt.[28] Most of the oxygen in the melt escapes by the formation of gaseous silicon monoxide. The SiO condenses on the inside of the furnace chamber, creating a cleanliness problem in the puller. Crystals grown with these crucibles also contain substantial amounts of interstitial oxygen that can be either beneficial or detrimental, as will be discussed later. The purity of the silica itself (Table 1) also affects the silicon purity, because the silica can contain sufficient acceptor impurities to limit the upper values of resistivity of the silicon that is being grown. The presence of carbon in the melt also accelerates the dissolution rate up to twofold.[34] One possibles reaction is

$$C + SiO_2 \rightarrow SiO + CO \tag{13}$$

Crucibles for large CZ pullers have a diameter-to-height ratio of approximately 1 or slightly greater; common diameters are 25, 30, and 35 cm for charge sizes of 12, 20, and 30 kg, respectively. A 45cm, 60kg configuration has even been proposed. Wall thicknesses of 0.25 cm are used, but the silica is sufficiently soft to require the use of a susceptor for mechanical support. Upon cooling, the thermal mismatch between residual silicon and silica usually results in the fracture of the crucible.

The feasibility of using silicon nitride as a crucible material has been demonstrated using CVD-deposited nitride.[35] Such an approach is attractive as a means of eliminating oxygen from crucible-grown crystals. However, even the nitride is eroded, resulting in a doping of the crystal with nitrogen, a weak donor. CVD nitride is the only form of nitride with sufficient purity for crucible use. However, this method needs further development before it becomes practical.

The susceptor, as mentioned previously, is used to support the silica crucible. It also provides for better thermal conditions. Graphite, because of its high-temperature properties, is the material of choice for the susceptor. A high-purity, nuclear-grade graphite is usually specified. This high purity is necessary to prevent contamination of the crystal from impurities that would be volatilized from the graphite at the temperature involved. Besides the susceptor, other graphite parts in the hot zone of the furnace require high purity. The susceptor rests on a pedestal whose shaft is connected to a motor that provides rotation. The whole assembly can usually be raised and lowered to keep the melt level equidistant from a fixed reference point, which is needed for automatic diameter control.

The chamber housing the furnace must meet several criteria. It should provide easy access to the furnace components to facilitate maintenance and cleaning. The furnace structure must be airtight to prevent contamination from the atmosphere, and have a specific design that does not allow any part of the chamber to become so hot that its vapor pressure would be a factor in contaminating the crystal. As a rule, the hottest parts of the puller are water cooled. Insulation is usually provided between the heater and the chamber wall.

To melt the charge, radio frequency (induction heating) or resistance heating have been used. Induction heating is useful for small melt sizes, but resistance heating is used exclusively in large pullers. Resistance heaters, at the power levels involved (tens of kilowatts), are generally smaller, cheaper, easier to instrument, and more efficient. Typically, a graphite heater is connected to a dc power supply.

Crystal-pulling mechanism. The crystal-pulling mechanism must, with minimal vibration and great precision, control two parameters of the growth process: the pull rate and crystal rotation. Seed crystals, for example, are prepared to precise orientation tolerances, and the seed holder and pulling mechanism must maintain this precise orientation perpendicular to the melt surface. Lead screws are often used to withdraw and rotate the crystal. This method unambiguously centers the crystal relative to the crucible, but may require an excessively tall apparatus if the grower is to produce long crystals. Since precise mechanical tolerance is difficult to maintain over a long shaft, pulling with a cable may be necessary. Centering the crystal and crucible is more difficult when using cable. Furthermore, although the cable provides a smooth pulling action, it is prone to pendulum effects. However, since the cable can be wound on a drum, the machine can be shorter than a similar lead-screw puller. The crystal leaves the furnace through a purge tube, where (if present) ambient gas is directed along the surface of the crystal to cool it. From the purge tube, the crystal enters an upper chamber, which is usually separated from the furnace by an isolation valve.

Ambient control. Czochralski growth of silicon must be conducted in an inert gas or vacuum, because (1) the hot graphite parts must be protected from oxygen to prevent erosion and (2) the gas around the process should not react with the molten silicon. Growth in a vacuum meets these requirements; it also has the advantage of removing silicon monoxide from the system, thus preventing its buildup inside the furnace chamber. Growth in a gaseous atmosphere, generally used on large growers, must use an inert gas such as helium or argon. The inert gas may be at atmospheric pressure or at reduced pressure. Growing operations on an industrial scale use argon because of its lower cost. A typical consumption is 1500 L/kg of silicon grown. The argon is supplied from a liquid source by evaporation, and must meet purity requirements for amount of moisture, hydrocarbon content, and so on.

Control system. The control system can take many forms, and it provides control of process parameters such as temperature, crystal diameter, pull rate, and rotation speeds. This control may be closed-loop (i.e., with feedback control) or open-loop. Parameters, including pull speed and rotation, with a high response speed are most amenable to closed-loop control. The large thermal mass of the melt generally precludes any short-term control of the process based on temperature. For example, to control the diameter, an infrared temperature sensor can be focused on the melt–crystal interface and used to detect changes in the meniscus temperature. The sensor output is linked to the pulling mechanism and controls the diameter by varying the pull rate. The trend in control systems is to use digital, microprocessor-based systems. These rely less on operator intervention and have many parts of the process preprogrammed.

1.3.4 Oxygen and Carbon in Silicon

Oxygen in silicon is an unintentional impurity arising from the dissolution of the crucible during growth. Typical values[36] range from 5×10^{17} to 10^{18} atoms/cm^3. A reported segregation coefficient[37] for oxygen is 0.25; however, the axial distribution of impurities often reflects the specifics of the puller and process parameters in use, because they influence crucible erosion and evaporation of oxygen from the melt. For example, less dissolution of the crucible occurs as the melt level is lowered in the crucible, and thus less oxygen impurity is available for incorporation.[38] Rotation speeds, ambient partial pressure, and free melt surface area are all factors that determine the level and distribution of oxygen in the crystal.[30] Figure 15 shows a diagram of oxygen concentration versus fraction solidified for one set of growing conditions.

A novel method to reduce crucible erosion is to suppress thermal convection currents by applying a magnetic field to the melt.[39] Such an approach also reduces

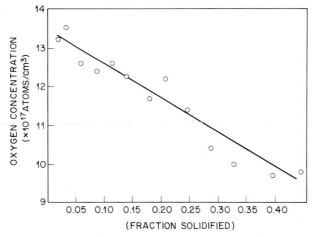

FIGURE 15
The axial distribution of oxygen in a Czochralski ingot. (*After Liaw, Ref. 36.*)

the thermal fluctuations in the boundary, or stagnant, layer, resulting in a more homogeneous distribution of dopant atoms. The magnetic fields at which the suppression occurs were found to be 4.0 kG for an axially applied field[40] and 2.0 kG for transverse fields.[41]

The level of oxygen in the melt is also affected by high levels of impurity-doping. It has generally been found that high levels of boron-doping tend to enhance the dissolution rate of the silica crucible.[42] At doping levels above 1×10^{19} atoms/cm^3 of boron, the oxygen content of a crystal may be up to 40% higher than a lightly doped crystal grown under otherwise identical conditions. An opposite effect occurs for antimony at levels above 1×10^{18} atoms/cm,[3] at which the oxygen level in the crystal decreases.[43]

As an impurity, oxygen has three effects: donor formation, yield strength improvement, and defect generation by oxygen precipitation.[44] In the crystal as grown, over 95% of the oxygen atoms occupy interstitial lattice sites. Oxygen in this state can be detected using an infrared absorption line[45] at 1106 cm^{-1}. The remainder of the oxygen polymerizes into complexes, such as SiO_4. This configuration acts as a donor and changes the resistivity of the crystal causesd by intentional doping. These complexes form rapidly in the 400 to 500° C temperature range, with a rate proportional to the oxygen content to the fourth power. The complex formation occurs as the crystal cools. Crystals of larger diameter cool more slowly and form more complexes. Fortunately, these complexes are unstable above 500°C, so crystals or wafers are commonly heated to between 600 and 700°C to dissolve the complexes. During cooling, following the 600 to 700°C treatment, the donor complexes can reform in crystals. Because wafers cool rapidly enough to circumvent this problem, treatment of wafers is preferred for large diameter material. Common treatment times are tens of minutes for wafers and about an hour for crystals. The longer time for crystals is needed to bring the center up to temperature. A typical dissolution rate for these complexes is 5 $\times$ 10^{15} donors/cm^3-h at 700°C.

Oxygen will also combine with acceptor elements to create a second type of donor complex. These complexes form more slowly, 2×10^{14} donors/cm^3-h at 700°C, allowing for a net reduction of oxygen complexes using the stabilization treatment previously described. The acceptor-element complexes are more resistant to dissolution even with high-temperature processing. From a device viewpoint, it is important that material be resistivity stabilized by a suitable heat treatment prior to processing. It is also important to avoid prolonged exposure during processing to the temperature ranges we have discussed. The trend toward low-temperature processing poses a dilemma, because complex formation could occur during the device processing.

Oxygen in interstitial lattice sites also acts to increase the yield strength of silicon[46,47] through the mechanism of solution hardening. Improvements of 25% over oxygen-free silicon have been reported. This beneficial effect increases with concentration until the oxygen begins to precipitate. Oxygen at the concentration levels mentioned earlier represents a supersaturated condition at most common processing temperatures and will precipitate during processing, given a sufficient supersaturation ratio. Precipitation usually occurs when the oxygen concentration exceeds a threshold value[48] of about 6.4×10^{17} atoms/cm^3. The precipitation may proceed homogeneously, but native defects, usually in the crystal due to growth, allow heterogeneous precipitation to dominate the kinetics. Figure 16 details the

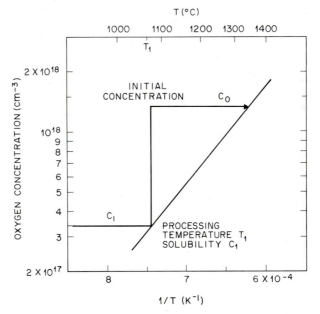

FIGURE 16
The solid solubility of oxygen in silicon. (*After Patel, Ref. 44.*)

solubility of oxygen in silicon and illustrates the supersaturation effect. A wafer containing an initial concentration of oxygen C_0 when processed at temperature T_1 results in a supersaturated condition. The supersaturation C_0/C_1 results because the oxygen solubility at T_1 is only C_1, which is less than C_0.

The precipitation kinetics are also a function of the doping level of the crystal.[49,50] At sufficient doping levels the precipitation of oxygen is suppressed or retarded as shown in Fig. 17. The doping level for the effect in n-type material is nominally 1×10^{18} atoms/cm^3 and 5×10^{19} atoms/cm^3 for p-type silicon. This apparent Fermi level dependency is tentatively explained as follows. Point defects (interstitials or vacancies) are involved in the nucleation process of the oxygen precipitate. In the case of vacancies, the charge state is strongly Fermi level dependent (see Chapter 7). For heavily doped n-silicon the dominant vacancies are V^- and $V^=$, and in heavily doped p-silicon, they are V^+. Thus, at sufficient doping levels there are few neutral vacancies available for nucleating the precipitate. The charged vacancies are postulated to pair with ionized impurity atoms as, for example, B^-V^+ or P^+V^-. In this state they are unavailable for nucleation processes.

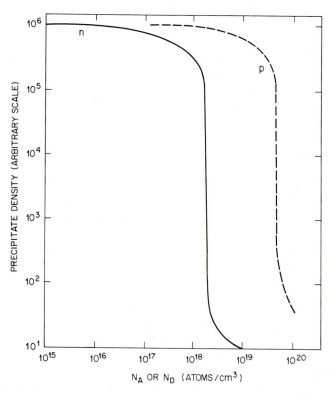

FIGURE 17
Precipitate density versus doping level, showing the reduction in density at high doping levels for n and p silicon.

The precipitates represent an SiO_2 phase. A volume mismatch occurs as the precipitates grow in size, representing a compressive strain on the lattice that is relieved by the punching out of prismatic dislocation loops. Actually, a variety of defects, including stacking faults (a type of dislocation), are associated with precipitate formation. These defects attract fast-diffusing metallic species, which give rise to large junction leakage currents.

The ability of defects to capture harmful impurities (called "gettering") can be used beneficially. Defects formed in the interior of a wafer getter impurities from the wafer surface where device junctions are located (Fig. 18). Gettering is explained more fully in Section 1.5.

Carbon is another unintended impurity in the polysilicon, and is transported to the melt from graphite parts in the furnace.[51] Carbon in silicon occupies a substitutional lattice site and is conveniently measured using infrared transmission measurements of an absorption line at 603 cm^{-1}. Because carbon's segregation coefficient is small (0.07), its axial variation is large. Typical seed-end concentrations range from 10^{16} atoms/cm^3 and lower. For butt ends at a high percentage of melt solidification, values range[36] up to 5×10^{17} atoms/cm^3. At these levels carbon does not precipitate like oxygen, nor is it electrically active. Carbon has been linked to the precipitation kinetics of oxygen and point defects.[52] In this regard, its presence is undesirable because it aids the formation of defects.

1.3.5   Characteristics and Evaluation of Crystals

Routine evaluation of crystals (also called ingots or boules) involves testing their resistivity, evaluating their crystal perfection, and examining their mechanical properties, such as size and mass. Other less routine evaluations include measuring the crystals' oxygen, carbon, and heavy metal content. The evaluations of heavy metal content are made by minority carrier measurements or neutron activation analysis.

After growth, the crystal is usually weighed, then inspected visually. Gross

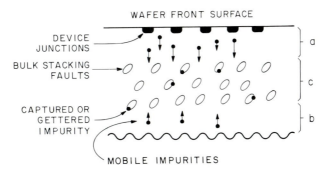

FIGURE 18
Schematic of a denuded zone in a wafer cross section and gettering sites. *a* and *b* are zones denuded of defects, and *c* represents the region of intrinsic gettering.

crystalline imperfections such as twinning are apparent to the unaided eye. Sections of the ingot containing such defects are cut from the boule, as are sections of the boule that are irregularly shaped or undersized. Total silicon loss can equal 50% at this step. Next the butt (or tang) end of the ingot, or a slice cut from that position, is preferentially etched to reveal defects such as dislocations. A common etchant is Sirtl's etch, which is a one/one mixture of HF acid (49%) and five-molar chromic acid.[53] This same etch can be used on polish and processed wafers to delineate other types of microdefects or impurity precipitates. Cracks can be detected by a method that uses ultrasonic waves.[54]

Resistivity measurements are made on the flat ends of the crystal by the four-point probe technique (Fig. 19). Current I (mA) is passed through the outer probes and voltage V (mV) measured between the inner probes. The measure resistance (V/I) is converted to resistivity (Ω-cm) using the formula

$$\rho = (V/I)2\pi S \tag{14}$$

where S is the probe spacing in centimeters. Measurements can be reproduced to $\pm 2\%$ if care is taken in selecting instrumentation, probe pressure, and current levels.[55,56] For example, current levels are raised as the resistivity of the material is lowered, such that the measured voltage is maintained between 2 and 20 mV. The variation of resistivity with ambient temperature is a significant effect — approximately 1%/°C at 23°C for 10 Ω-cm material. The resistivity of the material is related to the doping density through the carrier mobility.[57] Figure 20 shows the relationship for boron- and phosphorus-doped samples.

Boron-doped CZ silicon is available in resistivity from 0.0005 to 50 Ω-cm, with radial uniformities of 5% or better. Arsenic- and phosphorus-doped silicon is available in the range 0.005 to 40 Ω-cm, with arsenic being the preferred dopant in the lower resistivity ranges. Antimony is also used to dope crystals in the

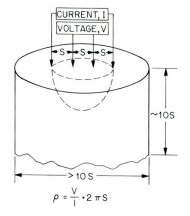

FIGURE 19
Four-point probe measurement on crystal end. (*Courtesy P. H. Langer*).

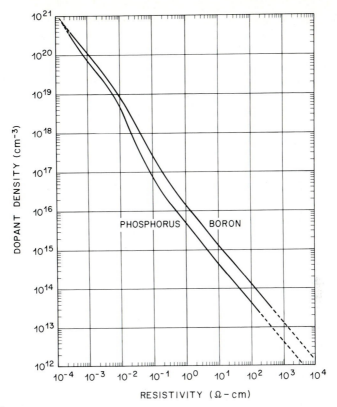

FIGURE 20
Conversion between resistivity and dopant density in silicon. (*After Thurber, Mattis, and Liu, Ref. 57.*)

0.01 Ω-cm range. Antimony-doped substrates are preferred as epitaxial substrates because autodoping effects are minimized (Chapter 2). Radial resistivity uniformities for n-doped material range from 10 to 50% depending on the diameter, dopant, crystal orientation, and process conditions.[58]

Problem
A boron-doped crystal is measured at its seed end with a four-point probe of spacing 1 mm. The (*V/I*) reading is 10 ohms. What is the seed end doping and the expected reading at 0.95 fraction solidified?

Solution
First use Eq. 14 to find the seed end resistivity.

$$\rho = 10 \times 2 \times \pi \times 0.1$$

$$\rho = 6.3 \ \Omega\text{-cm}$$

Figure 20 shows that this equates to a doping of 2×10^{15} atoms/cm^3.

Next use Eq. 10 and Table 2 to find melt concentration.

$$2 \times 10^{15} = 0.8 \, C_0 (1 - 0)^{0.8-1}$$
$$C_0 = 2.5 \times 10^{15} \, \text{atoms/cm}^3$$

At 0.95 fraction solidified the solid concentration is

$$C_s = 0.8 \times 2.5 \times 10^{15} (1 - 0.95)^{0.8-1}$$
$$C_s = 3.6 \times 10^{15} \, \text{atoms/cm}^3$$

or about 5 Ω-cm.

1.4   SILICON SHAPING

Silicon is a hard, brittle material. The most suitable material for shaping and cutting silicon is industrial-grade diamond, although SiC and Al_2O_3 have also been used. This section highlights the major shaping methods, but in some cases alternatives do exist. This section also discusses the relationship of these operations to the device processing needs required of silicon slices.

Conversion of silicon ingots into polished wafers requires nominally six machining operations, two chemical operations, and one or two polishing operations.[54,59] Additionally, assorted inspections and evaluations are performed between the major process steps. A finished wafer is subject to a number of dimensional tolerances, dictated by the needs of the device fabrication technology. As shown in Table 3, these tolerances are somewhat loose compared to metal machining capabilities. The existence of organizations to standardize these factors[60] and their measurements[61] proves the maturity of the silicon materials industry. The motivation for standards is twofold: (1) they help to standardize wafer production, resulting in efficiency and cost savings, and (2) producers of process equipment and fixtures benefit from knowing the wafer dimensions when designing equipment.

1.4.1   Shaping Operations

The first shaping operation removes the seed and tang ends from the ingot. Portions of the ingot that fail the resistivity and perfection evaluations previously mentioned are also cut away. The cuttings are sufficiently pure to be recycled, after cleaning, in the growing operation. The rejected ingot pieces can also be sold as metallurgical-grade silicon. The cutting is conveniently done as a manual operation using a circular saw.

The next operation is a surface grinding, and is the step that defines the diameter of the material. Silicon ingots are grown slightly oversized because the automatic diameter control in crystal growing cannot maintain the needed diameter tolerance, and crystals cannot be grown perfectly round. Figure 21

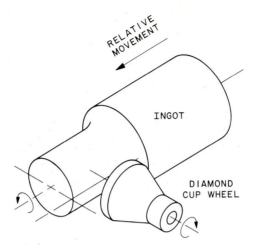

FIGURE 21
Schematic of grinding process. (*After Bonora, Ref. 54.*)

shows schematically the lathelike machine tool used to grind the ingot to diameter. A rotating cutting tool makes multiple passes down a rotating ingot until the chosen diameter is attained. Precise diameter control is required for many kinds of processing equipment, and is a consideration in the design of processing and furnace racks.

Following diameter grinding, one or more flats are ground along the length of the ingot. The largest flat, called the "major" or "primary" flat, is usually relative to a specific crystal direction. The flat is located by an X-ray technique. The primary flat serves several purposes. It is used as a mechanical locator in automated processing equipment to position the wafer, and also serves to orient the IC device relative to the crystal. Other smaller flats are called "secondary flats," and serve to identify the orientation and conductivity type of the material (Fig. 22). Secondary flats provide a means of quickly sorting and identifying wafers should mixing occur.

Once these operations have been completed, the ingot is ready to be sliced into wafers. Slicing is important because it determines four wafer parameters: surface orientation, thickness, taper, and bow. The surface orientation is determined by cutting several wafers, measuring the orientation by an X-ray method, and then resetting the saw until the correct orientation is achieved. Wafers of $<100>$ orientation are usually cut "on orientation" (Table 3). The tolerances allowed for orientation do not adversely affect MOS device characteristics such as interface-trap density. The other common orientation, $<111>$, is usually cut "off orientation" (by about 3°), as required for epitaxial processing (Chapter 2). Routine manufacturing tolerances are also acceptable here.

The wafer thicknesss is essentially fixed by slicing, although the final value depends on subsequent shaping operations. Thicker wafers are better able to withstand the stresses of subsequent thermal processes (epitaxy, oxidation, and diffusion), and as a result exhibit less tendency to deform plastically or elastically

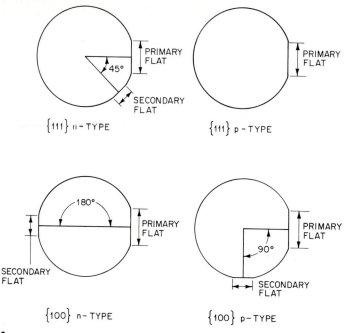

FIGURE 22
Identifying flats on silicon wafer. (*SEMI standard, used by permission.*)

in such processing. A major concern in slicing is the blade's continued ability to cut wafers from the crystal in very flat planes. If the blade deflects during slicing, this will not be achieved. By positioning a capacitive sensing device near the blade, the blade position and vibration in the blade can be monitored, and higher-quality cutting achieved. If a wafer is sliced with excessive curvature (bow), subsequent lapping operations may not be able to correct it, and surface flatness requirements cannot be obtained by polishing.

Inner diameter (ID) slicing is the most common mode of slicing. ID slicing uses a saw blade whose cutting edge is on the interior of an annulus. Figure 23 shows a schematic of this process. The ingot is prepared for slicing by mounting

TABLE 3
Typical specifications for 100, 125, and 150 mm diameter wafers

Diameter(mm)	100±0.5	125±0.5	150±0.5
Primary flat(mm)	30–35	40–45	55–60
Secondary flat (mm)	16–20	25–30	35–40
Thickness (mm)	0.50–0.55	0.60–0.65	0.65–0.7
Bow (μm)	40	25	60
Total Thickness Variation (μm)	25	20	50
Surface Orientation	$<100>\pm1°$	Same	Same
	$<111>$ off orientation	Same	Same

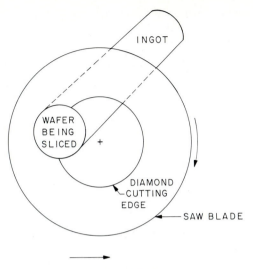

FIGURE 23
Schematic of ID slicing process.

it in wax or epoxy on a support, and then positioning the support on the saw. This procedure ensures that the ingot is held rigid for the slicing process. Some success has been obtained mounting ingots in a fixture using hydraulic pressure. The saw blade is a thin sheet of stainless steel (325 μm), with diamond bonded on the inner rim. This blade is tensioned in a collar and then mounted on a drum that rotates at high speed (2000 r/min) on the saw. Saw blades up to 58 cm in diameter with a 20 cm opening are available. These blades have a slicing capability up to nearly the ID opening of 20 cm. The blade is moved relative to the stationary ingot. The cutting process is water cooled. The kerf loss (loss due to blade width) at slicing is 325 μm, which means that approximately one-third of the crystal is lost as sawdust. Cutting speeds are nominally 0.05 cm/s, which, considering that wafers are sliced sequentially, is a rather slow process. Another shortcoming is the drum's finite depth, which limits the length of the ingot section that can be cut into wafers. Another style of ID saw has the blade mounted on an air-bearing and is rotated by means of a belt drive, an arrangement allowing any length of ingot to be sliced. After the ingot is sliced, individual wafers are recovered opposite the feed side and placed in a cassette. This type of saw, which hydraulically mounts the ingot, represents a highly automatic approach to sawing.

The wafer as cut varies enough in thickness to warrant an additional operation if the wafers are intended for VLSI application. A mechanical, two-sided lapping operation (Fig. 24), performed under pressure using a mixture of Al_2O_3 and glycerine, produces a wafer with flatness uniform to within 2 μm. This process helps ensure that surface flatness requirements for photolithography can be achieved in the subsequent polishing steps. Approximately 20 μm per side is removed.

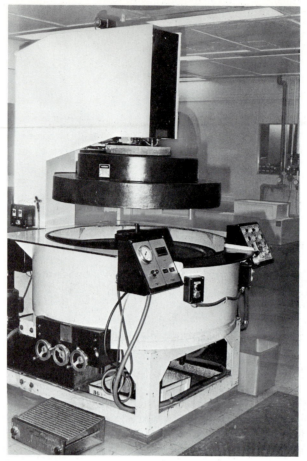

FIGURE 24
Double-sided lapping machine.

A final shaping step is edge contouring, where a radius is ground on the rim of the wafer (Fig. 25). This process is usually done in cassette-fed, high-speed equipment. Edge-rounded wafers develop fewer edge chips during device fabrication and aid in controlling the buildup of photoresist (Chapter 4) at the wafer edge. Chipped edges act as places where dislocations can be introduced during thermal cycles and as places where wafer fracture can be initiated. The silicon particles originating from the chipped edges can, if present on the wafer surface, add to the D_0 (defect density) of the IC process reducing yield (Chapter 14).

1.4.2   Etching

The previously described shaping operations leave the surface and edges of the wafer damaged and contaminated, with the depth of work damage depending

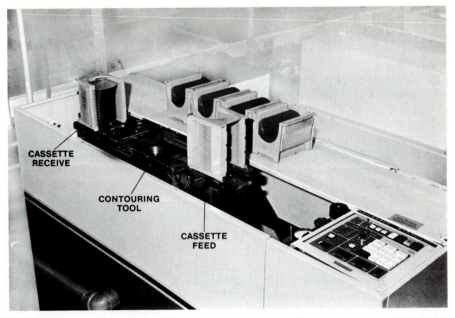

FIGURE 25
(a) Cassette feed edge-contouring tool.

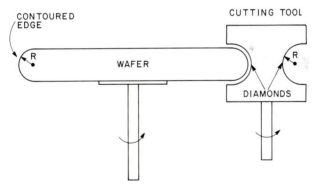

FIGURE 25
(b) Schematic of edge-containing process.

on the specifics of the machining operations. The damaged and contaminated regions are on the order of 10 μm deep and can be removed by chemical etching. Historically, mixtures of hydrofluoric, nitric, and acetic acids have been used, but alkaline etching, using potassium or sodium hydroxide, is also in common use.

The process equipment includes an acid sink, which contains a tank to hold the etching solution, and one or more positions for rinsing the wafers in water. The process is batch in nature, involving tens of wafers. The best process equipment provides a means of rotating the wafer during acid etching to maintain uniformity. Processing is usually performed with a substantial overetch to assure all damage is removed. Removing 20 μm per side is typical. The etching process is checked frequently by gauging wafers for thickness before and after etching. Etch times are usually on the order of several minutes per batch.

The etching involves oxidation–reduction processes, followed by a dissolution of an oxidation product. In the hydrofluoric, nitric, and acetic acid etching system[62,63] nitric acid is the oxidant and hydrofluoric acid dissolves the oxidized products according to the following reaction:

$$3Si + 4HNO_3 + 18HF \rightarrow 3H_2SiF_6 + 4NO + 8H_2O \qquad (15)$$

Acetic acid dilutes the system so that etching can be better controlled. Water can also be used, but acetic acid is preferable because water is a by-product of the reaction. The etching can be isotropic or anisotropic, according to the acid mixture or temperature. In HF-rich solutions, the reaction is limited by the oxidation step. This regime of etching is anisotropic, and the oxidation reaction is sensitive to doping, orientation, and defect structure of the crystal (where the oxidation occurs preferentially). The use of HNO_3-rich mixtures produces a condition of isotropic etching, and the dissolution process is then rate limiting. Over the range 30 to 50°C, the etching kinetics of an HNO_3-rich solution have been found to be diffusion controlled rather than reaction rate limited (Fig. 26). Thus, transport (diffusion) of reactant products to the wafer surface across a stagnant boundary layer is the controlling mechanism. For these reasons, the HNO_3-rich solutions are preferred for removing work damage. Rotating the wafers in the solution controls the boundary layer thickness and thereby effects dimensional control of the wafer. The isotropic character of the etch produces a smooth, somewhat specular surface. A common etch formulation is a 4:1:3 mixture of HNO_3 (79% by weight), HF (49% by weight) and CH_3COOH acids.

Unfortunately, the dimensional uniformity introduced by the lapping step is not maintained across larger diameter wafers ($>$ 75 mm) to a degree compatible with maintaining surface flatness in polishing. The hydrodynamics of rotating a large diameter wafer in solution do not allow for a uniform boundary layer, so a taper is introduced into the wafer. Projection lithography places demands on surface flatness that necessitate the use of alkaline etching. Alkaline etching is by nature anisotropic; thus, the etch rate depends on the surface orientation. The reaction is apparently dominated by the number of dangling bonds present on the surface. The reaction is generally reaction rate limited, and wafers do not have to be rotated in the solution. Since boundary layer transport is not a factor, excellent uniformity can be achieved. As in acid etching, the reaction is twofold when a mixture of KOH/H_2O or $NaOH/H_2O$ is used.[64] A typical formulation uses KOH and H_2O in a 45% by weight solution (i.e. 45% KOH and 55% H_2O) at 90°C to achieve an etch rate of 25 μm/min for {100} surfaces.

An occasional problem with the damage-removal process is insufficient

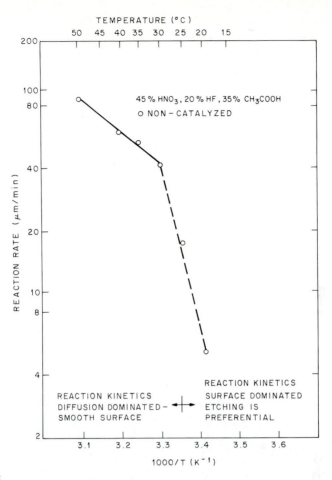

FIGURE 26
Typical etch rate versus temperature curve for one mixture of HF, HNO$_3$ and CH$_3$COOH. (*After Robbins and Schwartz, Ref. 62.*)

etching, which can lead to the generation of dislocations in subsequent treatments because of residual damage. Residual damage can be evaluated by measuring the fracture strength of the material after etching or by thermal-wave mapping.[65] Before etching, a laser is commonly used to engrave an alphanumeric identification mark on each wafer. Laser marking will replace the secondary flat identification method for larger diameter wafers ($\geq$ 150 mm).

1.4.3 Polishing

Polishing is the final step. Its purpose is to provide a smooth, specular surface on which device features can be photoengraved. A main VLSI concern is to produce a surface with a high degree of surface flatness and minimum local slope to meet the requirements of optical projection lithography.[66] Values between 5 and 10 μm are typical surface flatness specifications. The surface is also required to be free from contamination and damage.

Figure 27 shows a typical polishing machine and a schematic of the process. The process requires considerable operator attention for loading and unloading. It can be conducted as a single-wafer or batch-wafer process depending on the equipment. Economics determines the choice of single or batch processing; single-wafer processing is preferred for large wafers. Single-wafer processing also offers a better means of achieving surface flatness goals. In both single and batch processing, the process involves a polishing pad made of artificial fabric, such as a polyester felt, polyurethane laminate. Wafers are mounted on a fixture, pressed against the pad under high pressure, and rotated relative to the pad. A mixture of polishing slurry and water, dripped onto the pad, does the polishing (which is both a chemical and mechanical process). The porosity of the pad is a factor in carrying slurry to the wafer for polishing. The slurry is a colloidal suspension of fine SiO_2 particles (100Å diameter) in an aqueous solution of sodium hydroxide. Under the heat generated by friction, the sodium hydroxide oxidizes the silicon with the OH^- radical. This is the chemical step. In the mechanical step the silica

WAFER HOLDER   SLURRY FEED   POLISHING HEAD

FIGURE 27
(a) Photograph of polishing machine.

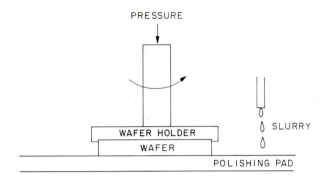

FIGURE 27
(b) Schematic of polishing process.

particles in the slurry abrade the oxidized silicon away. Polishing rate and surface finish are complex functions of pressure, pad properties, rotation speed, slurry composition, and pH of the polishing solution. Typical processes remove 25 μm of silicon. In a batch process involving tens of wafers, silicon removal can take 30 to 60 min; single-wafer processing can be accomplished in 5 min. Single-wafer processes use higher pressures than the batch processes.

The method of mounting wafers for polishing also deserves attention. Historically, wafers were waxed onto a metal plate. This method is costly and may not yield the best surface flatness. An alternative (Fig. 28) is a waxless technique where wafers are applied to a conformal pad, typically a two-layer vinyl.[67] This method is cost-effective and eliminates the influence of rear surface particles on front surface flatness. After polishing, wafers are chemically cleaned with acid, base, and/or solvent mixtures to remove slurry residue (and wax), and readied for inspection. Polished wafers are subjected to a number of measurements that are concerned with cosmetic, crystal perfection, mechanical, and electrical attributes.

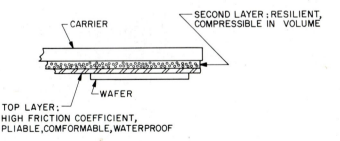

FIGURE 28
Schematic of Flex-Mount[TM] (Flex-Mount is a trademark of Siltec Corp., Menlo Park, California) polishing process. (*After Bonara, Ref. 67.*)

Figure 29 shows how the industry has used wafers of increasingly larger diameter motivated in part by the trend to larger chip areas.

1.5   PROCESSING CONSIDERATIONS

In the IC processing of silicon wafers it is usually necessary to maintain the purity and perfection of the material.

1.5.1   Chemical Cleaning

Prior to use, silicon wafers are usually cleaned chemically to remove organic films, heavy metals, and particulates. Commonly used are aqueous mixtures of NH_4OH–H_2O_2, HCl–H_2O_2 and H_2SO_4–H_2O_2.[68,69] All of the solutions are efficient in removing metallic impurities, but the HCl–H_2O_2 mixture is the best. The ammonium hydroxide and sulfuric acid based mixtures will also remove

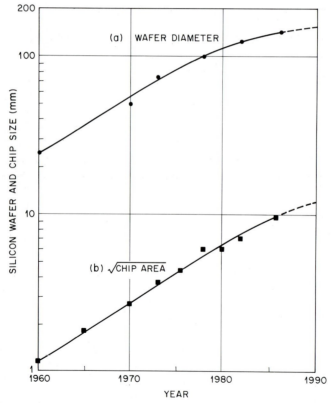

FIGURE 29
Diameter and chip area progression for silicon wafers by year.

organic contaminants, but the latter is better in this regard. These cleaning solutions leave the surface of a wafer in a hydrophilic state due to the oxidizing nature of the peroxide. In a hydrophilic state, water will wet the wafer surface (i.e., will be retained by surface tension). Since the chemically grown oxide can contain impurities from the chemicals, it is usually removed by a short immersion in dilute hydrofluoric acid. A typical cleaning sequence would be a sulfuric acid–hydrogen peroxide clean followed by the hydrofluoric acid dip, with deionized water rinses following each acid step.

l.5.2 Gettering Treatments

Many VLSI circuits (e.g., dynamic random access memories) require low junction leakage currents. Narrow-base bipolar transistors are sensitive to conductive impurity precipitates, which act like shorts between the emitter and collector (the pipe effect). Metallic impurities, such as transition group elements, are responsible for these effects. These elements are located at interstitial or substitutional lattice sites and are generation-recombination centers for carriers. The precipitated forms of these impurities are usually silicides, which are electrically conductive. To remove impurities from devices, a variety of processing techniques, termed "gettering" treatments, are available.[48] Gettering is a general term taken to mean a process that removes harmful impurities or defects from the regions in a wafer where devices are fabricated. Among these techniques are ways to pretreat (i.e., pregetter) silicon wafers prior to IC processing. Pregettering provides a wafer with sinks that can absorb impurities as they are introduced during device processing.

One technique of removing impurities involves intentionally damaging the back surface of the wafer. Mechanical abrasion methods such as lapping or sand blasting have been used for this purpose. A more controllable process uses damage created by a focused heat beam.[70] This process requires a threshold energy density of 5 J/cm^2. One configuration of this technique involves using a Q-pulsed, Nd:YAG laser. The laser beam is rastered along the back surface to create an array of micromachined spots. Depending on the energy density and proximity of the spots, the silicon lattice is damaged and/or strained by the high-energy pulse. During thermal processing, dislocations emanate from the spots. If the stresses placed on the wafer during furnace processing are low, the dislocations remain localized on the back surface. The dislocations represent favorable trapping sites for fast-diffusing species. When trapped on the rear surface, these impurities are innocuous.

Another series of methods uses the defects associated with oxygen precipitation for trapping sites. These methods use one or more thermal cycles to produce the desired result.[71] They usually involve a high-temperature cycle (over 1050°C in nitrogen), which removes oxygen from the surface of the wafer by evaporation. This lowers the oxygen content near the surface so that precipitation does not occur because the supersaturated condition has been removed. The depth of the oxygen-poor region is a function of time and temperature, and depends on the diffusivity of oxygen in silicon (Fig.30). The region represents a defect-free zone (denuded zone) for device fabrication. Additional thermal cycles can be added to promote the formation of precipitates and defects in the interior of the wafer. This approach is called "intrinsic gettering" because the oxygen is native to the wafer. Intrinsic gettering is attractive because it fills the volume of the wafer with trapping sites. Otherwise, the bulk of the wafer serves no useful function beyond mechanically supporting the thin layer where the device is formed.

Although gettering has generally been associated with improvements in junction leakage currents, recent results have shown that a lack of gettering can lead to degraded gate-oxide quality in MOSFET devices in the presence of excessive metallic impurities.[72] When impurity levels exceed their solid solubility, precipitation occurs similar to the oxygen precipitation process described earlier. Metallic

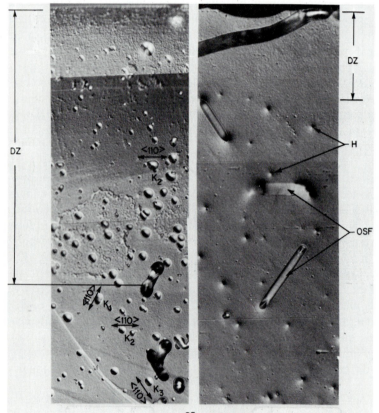

FIGURE 30

(a) The left-hand photograph shows a denuded zone (DZ) in a wafer cross section after a preferential etch. K_1, K_2, and K_3 are small stacking fault defects. The right-hand photograph is another wafer cross section showing stacking faults (OSF) and precipitate features (H) below the DZ. (*Courtesy G. A. Rozgonyi*).

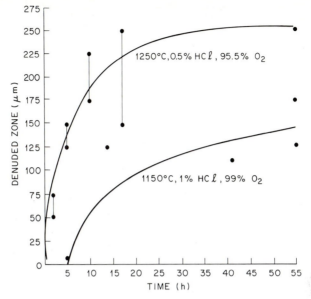

FIGURE 30
(b) Denuded zone width for two sets of processing conditions.

precipitates formed at the wafer surface disrupt the growth of thin oxide layers and act as points of localized breakdown; Figure 31 illustrates the influence of gettering on oxide quality. Another commercially available gettering technique is to deposit approximately 1 μm of polysilicon, using the methods described in Chapter 6, after the chemical etching of the wafer prior to polishing. After polishing, the polysilicon is resident on the rear surface. The grain boundaries in the polysilicon readily retain process-induced metallic contamination. This method of gettering is very advantageous for n$^+$ wafers where intrinsic gettering is inadequate or nonexistent. All of these methods have been successfully employed in circuit fabrication processes to improve junction leakage currents.[73,74]

1.5.3 Thermal Stress Factors

We generally want to maintain the crystal perfection of wafers throughout the device fabrication process, and to keep wafers mechanically undeformed. Wafers are typically processed in furnaces using racks with a high wafer-diameter-to-spacing ratio. When the wafer is removed from a high-temperature furnace, the wafer edges cool rapidly by radiation to the surroundings, but the wafer centers remain relatively hot.[15] The resultant temperature gradient creates a thermal stress S that can be estimated as

$$S = aE \, \Delta T \tag{16}$$

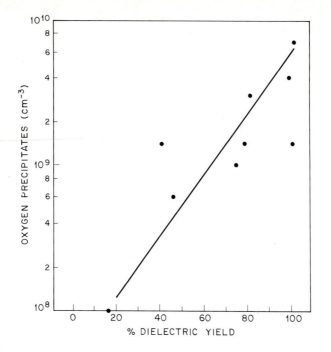

FIGURE 31
Correlation between oxygen precipitate density in the substrate (i.e., intrinsic gettering level) and the yield of oxide test capacitors. A good capacitor sustains a leakage current of $\leq 1\ \mu A$ at 10 V.

where a is the coefficient of thermal expansion, E is Young's modulus, and ΔT is the temperature difference across the wafer.

If these stresses exceed the yield strength (the maximum, or critical shear, stress the material will accommodate without irreversible deformation) of the material, dislocations will form. Stresses are usually kept to acceptable levels by withdrawing wafers from the furnace slowly to minimize the temperature gradient, or by lowering the furnace temperature (prior to removing the wafers) to the point where the yield strength at the removal temperature exceeds the stresses imposed (Fig. 32).[16]

Material parameters must also be considered. Oxygen precipitates (useful for gettering) can reduce the yield strength (critical shear stress) up to fivefold (Fig. 32).[75]

Problem
A silicon wafer having an oxygen content of 6.5×10^{17} atoms/cm^3 is processed at 1000°C for a time sufficient to precipitate all excess oxygen. Upon removal from the furnace, the wafer has a center-to-edge temperature difference of 5°C. Will dislocations result?

Solution
Calculate the resultant stress using Eq. 16.

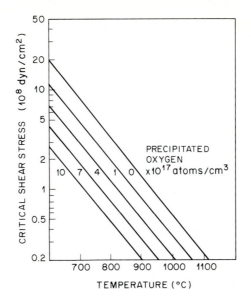

FIGURE 32
Yield strength of silicon showing the influence of oxygen precipitates. (*After Leroy and Plougoven, Ref. 76.*)

$$S = 2.6 \times 10^{-6} \times 1.09 \times 10^{12} \times 5$$

$$S = 1.4 \times 10^{7} \text{dyn/cm}^2$$

According to Fig. 16, the equilibrium solubility of oxygen in silicon at 1000° C is 2.5×10^{17} atoms/cm^3. Thus, the precipitated oxygen is 4.0×10^{17} atoms/cm^3. According to Fig. 32, the critical stress at 1000° C would be 2×10^7 dyn/cm^2. Thus, dislocations would not occur.

1.6   SUMMARY AND FUTURE TRENDS

Silicon wafers have been and will continue to be the predominant material for solid-state device manufacture. In the VLSI device arena, some other material technologies will also become common. The two main contenders are silicon-on-insulator (e.g., Si on SiO$_2$) and compound semiconductors (notably GaAs). These technologies will be chosen when high-speed circuitry or the need to optimize other circuit parameters are the deciding factors.

The specifications placed on wafers will become more stringent for VLSI applications. Unintentional impurities that are now ignored in specifications will need maximum allowable levels placed upon them. This would also be true for carbon and metallic species.[76] Oxygen is already subject to such specifications, but additional control over oxygen precipitation as it relates to the growing process and thermal cycling will probably be forthcoming. Mechanical dimensions will continue to be driven by equipment and processing needs. In particular, lithographic evolution will require flatter wafers. Surface cleanliness and other surface characteristics, because they influence oxide–silicon interface trap density, may represent a new class of specification and a new area of study.

Large diameter wafers (> 150 mm) are feasible, and 350 mm wafers have been produced. Practical implementation is awaiting the need for further productivity improvements and improved circuit fabrication capability, particularly in the lithographic area. Larger diameters ingots will be grown in large pullers. The slower cooling rates experienced by these ingots may alter the properties of the material, particularly the point-defect kinetics. This topic provides an area for continued research, because properties of the material are related to defects formed in device processing, and are thus related to IC yield.

PROBLEMS

1 Iron is an impurity in quartz crucibles. Using a concentration value of $2 \times 10^{18}/cm^3$ in the crucible, assume 300 cm^3 of the crucible is dissolved into a 6500 g melt, all at the beginning of the cycle. Calculate the seed (0% solidified) and tang end (90% solidified) iron concentration in the ingot.

2 Using the equation governing crystal growth (Eq. 7), derive an expression relating growth rate inversely to crystal diameter. Assume no temperature gradient in the melt. Since the heat flow down the crystal is small, assume heat flow from the crystal is predominantly from radiation.

3 Calculate the number of gallons of HF and HNO_3 acid needed to remove the work damage from 5000 wafers of 100 mm diameter.

4 The seed crystal used in CZ growing is usually "necked down" to a small diameter (3 mm) as a means to initiate dislocation-free growth. Using the yield strength of silicon, calculate the maximum mass of silicon that could be supported by such a seed. Convert this to a length for 100 and 125 mm diameter cystals.

5 Large growers, such as those pictured in the chapter, require 120 kWh to convert a kilogram of polysilicon into a crystal. Account for this energy in terms of the energy needed to melt the silicon, the radiation loss from the melt surface, and conduction down the crystal. Is all the energy accounted for? Assume that a 10 kg charge of polysilicon is used to grow a 10 cm diameter crystal from a 25 cm diameter crucible at a rate of 0.0025 cm/s.

6 Solar cells have been suggested as an alternative energy source. Conduct the following feasibility calculations: How much polysilicon would be required to supply all the United States' electrical needs from 100 mm diameter Czochralski-grown silicon wafers, and how much land area would this require? Compare the silicon used to current consumption. Use the following data:

1. The average US weekly power consumption is 42×10^9 kWh.
2. Assume that each gram of silicon in the finished cell requires 5 g of polysilicon.
3. The average solar energy falling on the earth's surface is 1340 W/m^2; assume 50 h of daylight per week.
4. The cell will convert 8% of all incident energy to electrical power.

7 There are several economic motivations to scale up the melt sizes of industrial crystal growers. Larger melt sizes increase the time a machine is actually growing a crystal,

thus making it more productive. Calculate the minimum crucible wall thickness needed under the following conditions: given a charge size of l00 kg, a crucible with a 25% larger volume than that of the silicon volume, a l2.5 cm diameter crystal growing at a rate of 0.002 cm/s, and a crucible erosion rate of 2×10^{-7} g/cm²-s. Assume a unity aspect ratio for the crucible. Also calculate the energy loss in kilowatt-hours from the surface of the melt using a temperature of 1450°C.

8 Using the gradient of Fig. 9 and Eq. 8, calculate a maximum pull speed. Assume the latent heat of fusion to be 264 cal/g and the solid thermal conductivity to be 0.05 cal/s-cm-°C. Compare the result to Fig. l0. What do you conclude?

9 Calculate the boron concentration in the crystal that would lead to misfit dislocation formation at a temperature of 1000°C.

10 A Czochralski melt is simultaneously doped with boron to a level of 10^{17} atoms/cm³ and phosphorous to a level of 8×10^{16} atoms/cm³. Does a pn junction form during growth? If so, at what fraction solidified?

REFERENCES

1 C. L. Yaws, R. Lutwack, L. Dickens, and G. Hsiu, "Semiconductor Industry Silicon: Physical and Thermodynamic Properties," *Solid State Technol.*, **24**, 87 (1981).
2 W. E. Beadle, J. C. C. Tsai, and R. D. Plummer, Eds, *Quick Reference Manual for Silicon Integrated Circuit Technology*, Wiley, New York, 1985.
3 W. Shockley, "The Theory of p-n Junctions in Semiconductor and p-n Junction Transistors," *Bell Syst. Tech. J.*, **28**, 435 (1949).
4 L. D. Crossman and J. A. Baker, "Polysilicon Technology," *Semiconductor Silicon 1977*. Electrochem. Soc., Pennington, New Jersey, 1977, p. 18.
5 Am. Soc. Test. Mater., ASTM Standard, F574, Part 43.
6 J. R. McCormick, "Polycrystalline Silicon—1986," *Semiconductor Silicon 1986*. Electrochem. Soc., Pennington, New Jersey, 1986, p. 43.
7 S. N. Rea, "Czochralski Silicon Pull Rate Limits," *J. Cryst. Growth*, **54**, 267 (1981).
8 S. M. Sze, *Semiconductor Devices, Physics and Technology*, Wiley, New York, 1985.
9 L. D. Dyer, "Dislocation-Free Czochralski Growth of <110> Silicon Crystals," *J. Cryst. Growth*, **47**, 533 (1979).
10 K. V. Ravi, *Imperfections and Impurities in Semiconductor Silicon*, Wiley, New York, 1981.
11 R. K. Watts, *Point Defects in Crystals*, Wiley, 1977.
12 A. G. Milnes, *Deep Levels in Semiconductors*, Wiley, New York, 1973.
13 J. Friedel, *Dislocations*, Pergamon Press, New York, 1964.
14 A. J. R. deKock and W. M. van de Wijgert, "The Effect of Doping on the Formation of Swirl Defects in Dislocation-Free Czochralski-Grown Silicon," *J. Cryst. Growth*, **49**, 718 (1980).
15 S. M. Hu, "Temperature Distribution and Stresses in Circular Wafers in a Row during Radiative Cooling," *J. Appl. Phys.*, **40**, 4413 (1969).
16 K. G. Moerschel, C. W. Pearce, and R. E. Reusser, "A Study of the Effects of Oxygen Content, Initial Bow and Furnace Processing on Warpage of Three-Inch Diameter Silicon Wafers," *Semiconductor Silicon 1977*, Electrochem. Soc., Pennington, New Jersey, 1977, p. 170.
17 S. Kishino, Y. Matsushita, and M. Kanamori, "Carbon and Oxygen Role for Thermally Induced Microdefect Formation in Silicon," *Appl. Phys. Lett.*, **35**, 213 (1979).
18 A. Armigliato, D. Nobili, P. Ostoja, M. Servidori, and S. Solmi, "Solubility and Precipitation of Boron in Silicon," *Semiconductor Silicon 1977*, Electrochem. Soc., Pennington, New Jersey, 1977, p. 638.
19 W. T. Stacy, D. F. Allison, and T. C. Wu, "The Role of Metallic Impurities in the Formation of Haze Defects," *Semiconductor Silicon 1981*, Electrochem. Soc., Pennington, New Jersey, 1981, p. 344.
20 J. C. Brice, *Crystal Growth Processes*, Wiley, New York, 1986.

21 W. R. Runyan, *Silicon Semiconductor Technology*, McGraw-Hill, New York, 1965.

22 K. M. Kim, A. Dran, P. Smetana, and G. H. Schwutte, "Computer Simulation and Controlled Growth of Large Diameter Czochralski Silicon Crystals," *J. Electrochem. Soc.*, **130**, 1156 (1983).

23 T. G. Digges, Jr., and R. Shima, "The Effect of Growth Rate, Diameter and Impurity Concentration on Structure," *J. Cryst. Growth*, **50**, 865 (1980).

24 S. M. J. G. Van Run, "A Critical Pulling Rate for Remelt Suppression in Silicon Crystal Growth," *J. Cryst. Growth*, **53**, 441 (1981).

25 H. Kolker, "The Behavior of Nonrotational Striations in Silicon," *J. Cryst. Growth*, **50**, 852 (1980).

26 J. Chikawa and S. Yoshikawa, "Swirl Defect in Silicon Single Crsytals," *Solid State Technol.*, **23**, 65 (1980).

27 J. A. Burton, R. C. Prim, and P. Slichter, "The Distribution of Solute in Crystal Grown from the Melt. Part I. Theoretical," *J. Chem. Phys.*, **21**, 1987 (1953).

28 J. R. Carruthers, A. F. Witt, and R. E. Reusser, "Czochralski Growth of Large Diameter Silicon Crystals—Convection and Segregation," *Semiconductor Silicon 1977*, Electrochem. Soc., Pennington, New Jersey, 1977, p. 61.

29 K. M. Kim, "Interface Morphological Instability in Czochralski Silicon Crystal Growth from Heavily Sb-Doped Melt," *J. Electrochem. Soc.*, **126**, 875 (1979).

30 K. E. Benson, W. Lin, and E. P. Martin, "Fundamental Aspects of Czochralski Silicon-Crystal Growth for VLSI," *Semiconductor Silicon 1981*, Electrochem. Soc., Pennington, New Jersey, 1981, p. 33.

31 M. H. Liepold, T. P. O'Donnell, and M. A. Hagan, "Materials of Construction for Silicon Crystal Growth," *J. Cryst. Growth*, **40**, 366 (1980).

32 F. A. Voltmer and F. A. Padovani, "The Carbon-Silicon Phase Diagram for Dilute Carbon Concentration," *Semiconductor Silicon 1973*, Electrochem. Soc., Pennington, New Jersey, 1973, p. 75.

33 H. Hirata and K. Hoshikawa, "The Dissolution Rate of Silica in Molten Silicon," *Jpn. J. Appl. Phys.*, **19**, 1573 (1980).

34 B. Bathey, H. E. Bates, and M. Cretella, "Effect of Carbon on the Dissolution of Fused Silica in Liquid Silicon," *J. Electrochem. Soc.*, 128, 771 (1980).

35 M. Watanabe, T. Usami, H. Muroaka, S. Matsuo, Y. Imanishi, and H. Nagashima, "Oxygen-Free Silicon Single Crystal from Silicon-Nitride Crucible," *Semiconductor Silicon 1981*, The Electrochem. Soc., Pennington, New Jersey, 1981, p. 126.

36 H. M. Liaw, "Oxygen and Carbon in Czochralski-Grown Silicon," *Semicon. Inst.*, 2, 71 (1979).

37 W. Lin and D. W. Hill, "Oxygen Segregation in Czochralski Silicon Growth," *J. Appl. Phys.*, **54**, 1082 (1983).

38 T. Carlberg, T. B. King, and A. F. Witt, "Dynamic Oxygen Equilibrium in Silicon Melts during Crystal Growth," *J. Electrochem. Soc.*, 127, **189** (1981).

39 T. Suzuki, N. Isawa, Y. Okubo, and K. Hoshi, "CZ Silicon Growth in a Transverse Magnetic Field," *Semiconductor Silicon 1981*, Electrochem. Soc., Pennington, New Jersey, 1981, p. 90.

40 K. M Kim and P. Smetana, "Striations in CZ Silicon Crystals Grown Under Various Axial Magnetic Field Strengths," *J. Appl. Phys.*, **58**, 2731 (1985).

41 K. Hoshi, N. Isawa, T. Suzuki, and Y. Ohkuho, "Czochralski Silicon Crystals Grown in a Transverse Magnetic Field," *J. Electrochem. Soc.*, **132**, 693 (1985).

42 C. W. Pearce, R. J. Jaccodine, A. J. Filo, and W. Lin, "Oxygen Content of Heavily Doped Silicon," *Appl. Phys. Lett.*, **46**, 887 (1985).

43 H. Walitzki, H. J. Rath, J. Reffle, S. Pahlke, and M. Blatte, "Control of Oxygen and Precipitation Behavior of Heavily Doped Silicon Substrate Materials," *Semiconductor Silicon 1986*, Electrochem. Soc., Pennington, New Jersey, 1986, p. 86.

44 J. R. Patel, "Oxygen in Silicon," *Semiconductor Silicon 1977*, Electrochem. Soc., Pennington, New Jersey, 1977, p. 521.

45 Am. Soc. Test. Mater., ASTM Standard, F121-76, Part 43.

46 J. Doerschel and F. G. Kirscht, "Differences in Plastic Deformation Behavior of CZ and FZ Grown Silicon Crystals," *Phys. Status Solidi A*, **64**, K85 (1981).

47 K. Sumino, H. Harada, and I. Yonenaga, "The Origin of the Difference in the Mechanical Strengths of Czochralski Silicon," *Jpn. J. Appl. Phys.*, **19**, L49 (1980).

48 C. W. Pearce, L. E. Katz, and T. E. Seidel, "Considerations Regarding Gettering in Integrated Circuits," *Semiconductor Silicon 1981*, Electrochem. Soc., Pennington, New Jersey, 1981, p. 705.

49 H. Tsuya, "Behavior of Oxygen and Dopants in Heavily Doped Silicon Crystals," *Semiconductor Silicon 1986*, Electrochem. Soc., Pennington, New Jersey, 1986, p. 849.

50 C. W. Pearce, T. Kook, and R. J. Jaccodine, "Effect of Heavy Doping on the Nucleation and Growth of Bulk Stacking Faults in Silicon," *Impurity Diffusion and Gettering in Silicon*, Vol. 36, Materials Research Society, Pittsburgh, Pennsylvania, 1985, p. 231.

51 T. Nozaki, "Concentration and Behavior of Carbon in Semiconductor Silicon," *J. Electrochem. Soc.*, **117**, 1566 (1970).

52 Y. Matsushita, S. Kishino, and M. Kanamori, "A Study of Thermally Induced Microdefects in Czochralski-Grown Silicon Crystals: Dependence on Annealing Temperature and Starting Material," *Jpn. J. Appl. Phys.*, **19**, L101 (1980).

53 D. G. Schimmel, "A Comparison of Chemical Etches for Revealing <100> Silicon Crystal Defects," *J. Electrochem. Soc.*, **123**, 734 (1976).

54 A. C. Bonora, "Silicon Wafer Process Technology: Slicing, Etching, Polishing," *Semiconductor Silicon 1977*, Electrochem. Soc., Pennington, New Jersey, 1977, p. 154.

55 Am. Soc. Test. Mater., ASTM Standard, F84, Part 43.

56 Am. Soc. Test. Mater., ASTM Standard, F723, Part 43.

57 W. R. Thurber, R. L. Mattis, and Y. M. Liu, "Resistivity Dopant Density Relationship for Silicon," *Semiconductor Characterization Techniques*, Electrochem. Soc., Pennington, New Jersey, 1978, p. 81.

58 S. E. Bradshaw and J. Goorissin, "Silicon for Electronic Devices," *J. Cryst. Growth*, **48**, 514 (1980).

59 R. B. Herring, "Silicon Wafer Technology—State of the Art 1976," *Solid State Technol.*, **19**, 37 (1976).

60 Semiconductor Equipment and Materials Institute (SEMI), Mountain View, California.

61 The American Society for Testing and Materials (ASTM), Committee F-1 on Electronics, Philadelphia, Pennsylvania.

62 H. Robbins and B. Schwartz, "Chemical Etching of Silicon," *J. Electrochem. Soc.*, **106**, 505 (1959); **107**, 108 (1960); **108**, 365 (1961); and **123**, 1909 (1976).

63 W. Kern, "The Chemical Etching of Semiconductors," *RCA Rev.*, **39**, 278 (1978).

64 I. Barycka, H. Teterycz, and Z. Znarmirowski, "Sodium Hydroxide Solution Shows Selective Etching of Boron-Doped Silicon," *J. Electrochem. Soc.*, **126**, 345 (1979).

65 W. L. Smith, S. Hahn, and M. Arst, "Polishing Damage Characterization of Silicon Wafers Using Thermal Wave Mapping," *Semiconductor Silicon 1986*, Electrochem. Soc., Pennington, New Jersey, 1986, p. 206.

66 W. A. Baylies, "A Review of Flatness Effects in Microlithographic Technology," *Solid State Technol.*, **24**, 132 (1981).

67 A. C. Bonara, "Flex-Mount Polishing of Silicon Wafers," *Solid State Technol.*, **20**, 55 (1977).

68 W. Kern and D. A. Puotinen, "Cleaning Solutions Based on Hydrogen Peroxide for Use in Silicon Semiconductor Technology," *RCA Rev.*, **31**, 187 (1970).

69 W. Kern, "Purifying Si and SiO_2 Surfaces with Hydrogen Peroxide," *Semicond. Inter.*, **6**, 94 (1984).

70 C. W. Pearce and V. J. Zaleckas, "A New Approach to Lattice Damage Gettering," *J. Electrochem. Soc.*, **126**, 1436 (1979).

71 K. Yamamoto, S. Kishino, Y. Matsushita, and T. Lizuka, "Lifetime Improvement in Czochralski-Grown Silicon Wafers by the Use of a Two Step Annealing," *Appl. Phys. Lett.*, **36**, 195 (1980).

72 C. W. Pearce, "The Effect of Intrinsic Gettering on Dielectric Yield," *Reduced Temperature*

Processing for VLSI, Electrochem. Soc., Pennington, New Jersey, 1985, p. 400.

73 H. R. Huff, H. F. Schaake, J. T. Robinson, S. C. Baber, and D. Wong, "Some Observations on Oxygen Precipitation Gettering in Device Processed Czochralski Silicon," *J. Electrochem. Soc.*, **130**, 1551 (1983).

74 J. W. Medernach, "An Evaluation of Extrinsic Gettering Techniques," *Semiconductor Silicon 1986*, Electrochem. Soc., Pennington, New Jersey, 1986, p. 915.

75 B. Leroy and C. Plougoven, "Warpage of Silicon Wafers," *J. Electrochem. Soc.*, **127**, 961 (1980).

76 P. F. Schmidt and C. W. Pearce, "A Neutron Activation Analysis Study of the Sources of Transition Group Metal Contamination in the Silicon Device Manufacturing Process," *J. Electrochem. Soc.*, **128**, 630 (1981).

CHAPTER

2

EPITAXY

C. W. PEARCE

2.1 INTRODUCTION

Epitaxy, a transliteration of two Greek words *epi*, meaning "upon," and *taxis*, meaning "ordered," is a term applied to processes used to grow a thin crystalline layer on a crystalline substrate. In the epitaxial process the substrate wafer acts as a seed crystal. Epitaxial processes are differentiated from the Czochralski process described in Chapter 1 in that the crystal can be grown below the melting point. Most epitaxial processes use chemical vapor deposition (CVD) techniques. A different approach is molecular beam epitaxy (MBE), which uses an evaporation method. These processes will be described in Sections 2.2 and 2.3, respectively. When a material is grown epitaxially on a substrate of the same material, such as silicon grown on silicon, the process is termed homoepitaxy. If the layer and substrate are of different materials, such as $Al_xGa_{1-x}As$ on GaAs, the process is termed heteroepitaxy. However, in heteroepitaxy the crystal structures of the layer and the substrate should be similiar if crystalline growth is to be obtained.

Silicon epitaxy was developed to enhance the performance of discrete bipolar transistors.[1] These transistors were first fabricated in bulk wafers using the wafer's resistivity to determine the breakdown voltage of the collector. However, high breakdown voltages need high-resistivity material. This requirement, coupled with the thickness of the wafer, results in excessive collector resistance that limits high-frequency response and increases power dissipation. Epitaxial growth of a high-resistivity layer on a low-resistivity substrate solved this problem (Fig. 1). Bipolar integrated circuits (Fig. 2) utilize epitaxial structures in much the same way discrete transistors utilize them. The substrate and epitaxial layer have opposite doping types to provide isolation, and a heavily doped diffusion layer (buried layer) serves as a low-resistance collector contact. Unipolar devices such

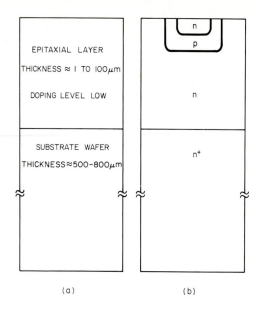

(a)   (b)

FIGURE 1
Cross-sectional schematic of a typical epitaxial layer and substrate (a), and a wafer with an npn discrete transistor fabricated in an epitaxial wafer (b).

as the junction field-effect transistor (JFET) employ an epitaxial wafer as does the VMOS tecnology.[2]

Epitaxial structures have also been used to improve the performance of dynamic random-access memory devices (RAMs) and CMOS ICs.[3] In JFETs and VMOS circuits, the doping profile provided by the epitaxial process is integral to the device structure. In the dynamic RAMs and CMOS circuits, devices could be fabricated in bulk wafers, but certain circuit parameters are optimized using epitaxial material.

Epitaxial wafers have two fundamental advantages over bulk wafers. First, epitaxial layers (one or more) on a substrate, often containing one or more buried

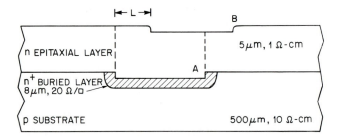

FIGURE 2
Cross-sectional schematic of an epitaxial wafer used for bipolar integrated circuit fabrication. Pattern A is present on the substrate prior to epitaxy, and its apparent location is shifted by L to B by the epitaxial process.

layers, offer the device designer a means of controlling the doping profile in a device structure beyond that available with diffusion or ion implantation. Second, the physical properties of the epitaxial layer differ from those of bulk material. For example, epitaxial layers are generally oxygen- and carbon-free, a situation not obtained with the melt-grown silicon discussed in Chapter 1.

2.2 VAPOR-PHASE EPITAXY

This section is concerned with several aspects of silicon vapor-phase epitaxy, such as process chemistry, aspects of process hardware, and current capabilities. The CVD of single-crystal silicon is usually performed in a reactor consisting of, in elemental form, a quartz reaction chamber into which a susceptor is placed. The susceptor provides physical support for the substrate wafers and provides a more uniform thermal environment. Deposition occurs at a high temperature at which several chemical reactions take place when process gases flow into the chamber.

2.2.1 Basic Transport Processes and Reaction Kinetics

A thorough study of the deposition process involves examining the thermodynamics and kinetics of the chemical reactions and the fluid mechanics of the gas flows in the reactor.[4,5]

As a starting point for discussing the fluid mechanics of the gas flow, let us consider the Reynolds number R_e, a dimensionless parameter that characterizes the type of fluid flow in the reactor:

$$R_e = \frac{D_r v \rho}{\mu} \tag{1}$$

where D_r is the diameter of the reaction tube, v is the gas velocity, ρ is the gas density, and μ is the gas viscosity.

Values of D_r and v are generally several centimeters and tens of cm/s, respectively, for industrial processes. The carrier gas is usually hydrogen, and using its typical values for ρ and μ in Eq. 1 results in values for R_e of about 100. These parameters result in gas flow in the laminar regime,[6] since R_e is less than 2000. Accordingly, a boundary layer of reduced gas velocity will form above the susceptor and at the walls of the reaction chamber. The thickness of the boundary layer y is defined as

$$y = \left[\frac{D_r x}{R_e}\right]^{1/2} \tag{2}$$

where x is distance along the reaction chamber.

Figure 3 shows that the boundary layer forms at the inlet to the reaction chamber and increases until the flow is fully established. Although fully established flows are not always encountered in the short lengths of typical reactors, it is across this boundary layer that reactants are transported to the surface. The

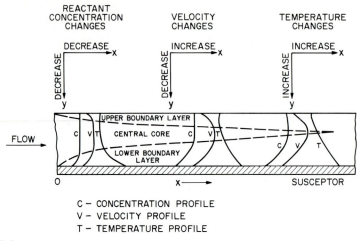

C — CONCENTRATION PROFILE
V — VELOCITY PROFILE
T — TEMPERATURE PROFILE

FIGURE 3
Boundary layer formation in a horizontal reactor. (*After Ban, Ref. 4.*)

reaction by-products diffuse back across the boundary layer and are removed by the main gas stream. The fluxes of species going to and coming from the wafer surface are complex functions of several variables, including temperature, system pressure, reactant concentration, and layer thickness. By convention, the flux is defined as

$$J = D\frac{dn}{dy} \tag{3}$$

and approximated as

$$J = \frac{D(n_g - n_s)}{y} \tag{4}$$

where n_g and n_s are the gas stream and surface reactant concentrations, respectively, D is the gas-phase diffusivity, which is a function of pressure and temperature, y is the boundary layer thickness, and J is the reactant flux of molecules per unit area per unit time.

Under steady-state conditions the flux of reactant across the boundary layer is equaled by a chemical reaction at the silicon surface where the film is growing. In approximate terms

$$J = k_s n_s \tag{5}$$

where k_s is a chemical reaction rate constant representing the surface reaction. Equating Eqs. 4 and 5 leads to an expression which determines the surface reactant concentration n_s in terms of n_g:

$$n_s = \frac{n_g}{1 + \dfrac{k_s y}{D}} \tag{6}$$

The quantity D/y is often called the gas-phase mass-transfer coefficient h_g. Equation 6 indicates that the surface concentration of reactant, which determines the layer growth rate, is a function both of the reaction chemistry and gas-phase transport processes. For example, as y approaches zero, $n_s \approx n_g$ and the growth process will be dominated by the surface chemical reaction. If k_s is very large relative to h_g, the surface concentration approaches zero, and the overall process is limited by the transport of reactant across the boundary layer.

The first-order effect of the boundary layer thickness y on the transport process must therefore be taken into consideration when designing the reactor and evaluating the operating conditions. The boundary layer must be adjusted relative to variation in temperature and reactant concentration within the reactor if uniformity of deposition is to be achieved. Figure 4 shows the sensitivity of layer growth rate to the Reynolds number. For a given reactor and set of process conditions R_e is varied by changing the gas flow (velocity), so the boundary layer y can be varied to achieve growth rate uniformity by varying the gas flow (i.e., R_e) in the reactor. This is a first-order approach to the problem.

A rigorous analysis of transport phenomena in a vertical cylinder reactor has been done.[7] This analysis is a numerical solution of the defining equations subject to appropriate boundary conditions. The temperature dependence of the various parameters has been included in this analysis. For example, D is a function of temperature T with a functionality of approximately $T^{3/2}$. Figure 5 shows a substantial temperature gradient normal to the susceptor surface. The steep temperature gradient also complicates the fluid flow, because it creates some turbulence in the gas stream. The importance of this effect relative to the laminar flow is described by the ratio of the Grashof number (Gr) to the square of the Reynolds number.[4,6] The Grashof number is a dimensionless parameter describing the effect of thermal convection in fluid flow. For Gr/R_e^2 greater than 0.5, the convection effects are significant and can be seen as oscillations in the temperature above the susceptor.

Four silicon sources have been used for growing epitaxial silicon. These are silicon tetrachloride ($SiCl_4$), dichlorosilane (SiH_2Cl_2), trichlorosilane ($SiHCl_3$), and silane[8] (SiH_4). Silicon tetrachloride has been the most studied and has seen the

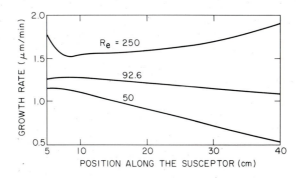

FIGURE 4
The influence of R_e number on deposition uniformity. (*After Manke and Donaghey, Ref. 7.*)

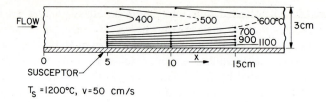

FIGURE 5
Isotherms in a horizontal reactor. (*After Ban, Ref. 4.*)

widest industrial use. It will be discussed here to exemplify the reaction chemistry. The outline of the discussion is applicable to the other halide compounds. The overall reaction can be classed as a hydrogen reduction of a gas.

$$SiCl_4(gas) + 2H_2(gas) \rightarrow Si(solid) + 4HCl(gas) \qquad (7)$$

However, a number of intermediate and competing reactions must be considered to understand the reaction fully. A starting point in the analysis is to determine for the Si—Cl—H system the equilibrium constant for each possible reaction and the partial pressure of each gaseous species at the temperature of interest. Equilibrium calculations reveal fourteen species to be in equilibrium with solid silicon.[9] In practice many of the species can be ignored because their partial pressures are less than 10^{-6} atm. Figure 6 shows the important species in the temperature range of interest. The plot is for a particular Cl/H ratio (0.01), which is representative of the ratios that occur in epitaxial deposition. Note that this ratio is constant in the reactor as neither chlorine nor hydrogen is incorporated into the layer.

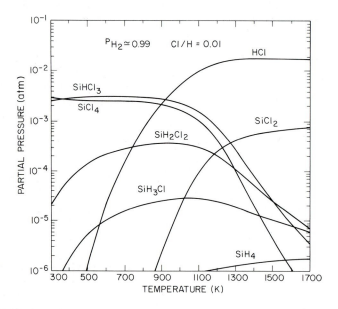

FIGURE 6
Temperature variation of the equilibrium gas phase composition at 1 atm. (*After Sirtl, Hunt, and Sawyer, Ref. 9.*)

The epitaxial process is not necessarily an equilibrium reaction. Thus, equilibrium thermodynamic calculations may not present the total picture, but relate only to the most probable reactions. In-situ measurements of the reaction process have been made by infrared spectroscopy, mass spectroscopy, and Raman spectroscopy to determine which species are actually present in the reaction. Four species in a $SiCl_4 + H_2$ reaction at 1200°C were detected.[10] Figure 7 illustrates the concentrations of each species at different positions along a horizontal reactor. Notice that the $SiCl_4$ concentration decreases while the other three constituents increase; thus the overall reaction is postulated as

$$SiCl_4 + H_2 \rightleftharpoons SiHCl_3 + HCl \tag{8}$$

$$SiHCl_3 + H_2 \rightleftharpoons SiH_2Cl_2 + HCl \tag{9}$$

$$SiH_2Cl_2 \rightleftharpoons SiCl_2 + H_2 \tag{10}$$

$$SiHCl_3 \rightleftharpoons SiCl_2 + HCl \tag{11}$$

$$SiCl_2 + H_2 \rightleftharpoons Si + 2HCl \tag{12}$$

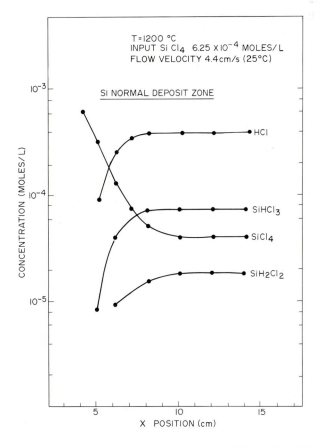

FIGURE 7

Species detected by IR spectroscopy in a horizontal reactor using $SiCl_4 + H_2$. (*After Nishizawa and Saito, Ref. 10.*)

This sequence of reactions is interesting from several viewpoints. The species $SiHCl_3$ and SiH_2Cl_2 are seen as intermediates to the overall reaction. Thus, growth with these halides would start at Eq. 9, 10, or 11. Accordingly, growth has the highest reported activation energies (1.6 to 1.7 eV) with $SiCl_4$, decreasing in turn for $SiHCl_3$ (0.8 to 1.0 eV) and SiH_2Cl_2 (0.3 to 0.6 eV). The reactions are also reversible, and under the appropriate conditions the deposition rate can become negative, causing the etching process to begin. This observation leads to a more general question about how growth rate varies with temperature.

Figure 8 shows an Arrhenius plot of growth rate, illustrating the overall reaction process.[11] In region A the process can be characterized as reaction rate or kinetic limited, that is, one of the chemical reactions is the rate-limiting step, and is even reversible. Region B represents the situation in which the transport processes are rate limiting, that is, where the growth rate is limited either by the amount of reactant reaching the wafer surface or by the reaction products diffusing away. This regime is termed mass transport or diffusion limited, and the growth rate is linearly related to the partial pressure of the silicon reactant in the carrier gas (Eq. 6). The slight increase of the growth rate at higher temperature in region B is due to the increased diffusivity of the species with temperature in the gas phase. Industrial processes at atmospheric pressure are usually operated in region B to minimize the influence of temperature variations.

Problem

Use the approximate temperature dependence of the gas-phase diffusivity to control the growth rate in a reactor where the boundary layer thickness varies per Eq. 2.

Solution

1 In Eq. 2 the boundary layer thickness y would be proportional to $x^{1/2}$; thus the gas-phase diffusivity should have a similar functionality with x to negate the boundary layer variation.

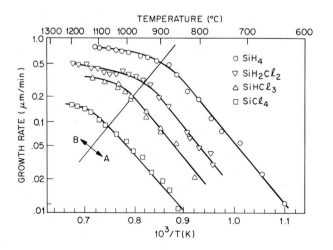

FIGURE 8
Temperature dependence of growth rate for assorted silicon sources. (*After Eversteyn, Ref. 11.*)

2 Since $D\alpha T^{3/2}$ a temperature dependence with x of $x^{1/3}$ would produce $D(x)\alpha x^{1/2}$.

3 Thus, a temperature profile with a positional dependence of $x^{1/3}$ could be used to achieve a uniform growth rate.

2.2.2 Doping and Autodoping

Incorporating dopant atoms into the epitaxial layer involves the same considerations as the growth process requires: for example, mass transport and chemical reactions.[12] Typically, hydrides of the impurity atoms are used as the source of dopant. We might expect that these compounds would decompose spontaneously, but they do not. Thermodynamic calculations indicate the hydrides are relatively stable because of the large volume of hydrogen present in the reaction. Typical of the dopant chemistry is the reaction for arsine, which is depicted with the deposition process in Fig. 9, which shows arsine being absorbed on the surface, decomposing, and being incorporated into the growing layer.

$$2AsH_3(gas) \rightarrow 2As(solid) + 3H_2(gas) \tag{13}$$

$$\rightarrow 2As(solid) \rightarrow 2As^+(solid) + 2e$$

Interactions also take place between the doping process and the growth process. First, in the case of boron and arsenic the formation of chlorides of these species is a competing reaction.[12,13] Second, the growth rate of the film influences the amount of dopant incorporated in the silicon, as shown for arsenic in Fig. 10. At low growth rates an equilibrium is established between the solid and the gas phases, which is not achieved at higher growth rates.[14]

In addition to intentional dopants incorporated into the layer, unintentional dopants are introduced from the substrate. The effect, shown in Fig. 11, is termed autodoping.[15] Dopant is released from the substrate through solid-state diffusion and evaporation. This dopant is reincorporated into the growing layer either by diffusion through the interface or through the gas phase. Autodoping is manifested

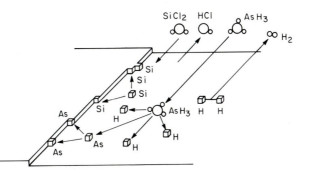

FIGURE 9
Schematic representation of arsine doping and growth processes. (*After Reif, Kamins, and Saraswat, Ref. 14.*)

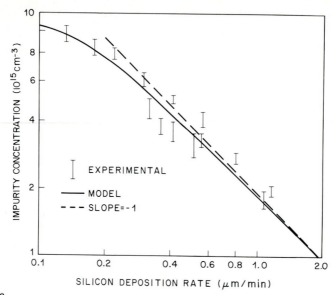

FIGURE 10
The influence of growth rate on layer concentration for arsenic doping at a fixed arsenic-to-silicon ratio entering the reactor. (*After Reif, Kamins, and Saraswat, Ref. 14.*)

as an enhanced transition region between the layer and the substrate (see Fig. 12). The shape of the doping profile, close to the substrate, is dominated by solid-state diffusion from the substrate (see Chapter 7) and is a complementary error function[16] if

$$\nu < 2(D/t)^{1/2} \tag{14}$$

where ν is the growth velocity, D is the substrate dopant diffusion constant, and t is the deposition time. The solid-state outdiffusion aspect of autodoping is easy to visualize; it determines the shape of region A in Fig. 12.

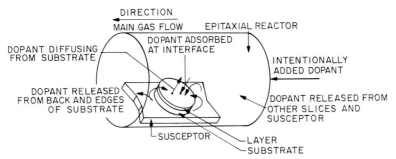

FIGURE 11
Sources of dopant for the epitaxial layer, schematically shown in a horizontal reactor. (*After Langer and Goldstein, Ref. 18.*)

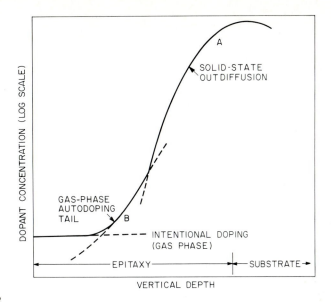

FIGURE 12
Generalized doping profile of an epitaxial layer detailing various regions of autodoping. (*After Srinivasan, Ref. 17.*)

Problem
At a temperature of 1200°C, what growth velocity would be required to exceed the solid-state diffusion of a boron-doped substrate in a time of one minute?

Solution
1 From Chapter 7 the diffusion constant of boron at 1200°C is found to be 10^{-12} cm/s.

2 $v < 2(10^{-12}/60)^{1/2}$
$v < 2.6 \times 10^{-7}$ cm/s or 0.15 microns/min

3 Growth rates exceeding this value are easily obtainable, so the condition is easily met (see Table 1).

Because the growth velocity easily outpaces the diffusion of dopant, the

TABLE 1
Epitaxial growth of silicon in hydrogen atmosphere

Chemical deposition	Nominal growth rate(μm/min)	Temperature range(°C)	Allowed oxidizer level (ppm)
$SiCl_4$	0.4–1.5	1150–1250	5-10
$SiHCl_3$	0.4–2.0	1100–1200	5-10
SiH_2Cl_2	0.2–1	1050–1120	<5
SiH_4	0.2–0.3	950–1250	<2

doping profile in region B is dominated by dopant introduced from the gas phase. If the dopant evaporated from the substrate exceeds the intentional dopant, an autodoping tail develops. Autodoping is a time-dependent phenomenon. The dopant evaporating from the wafer surface is supplied from the wafer interior by solid-state diffusion. Thus, the flux of dopant from an exposed surface is not constant, but decreases with time.

Once the autodoping is diminished, the intentional doping predominates, and the profile becomes flat. The extent of the autodoping tail is a function of the substrate dopant species and reaction parameters such as temperature and growth rate. Autodoping limits the minimum layer thickness that can be grown with controlled doping as well as the minimum doping level. Because of the technological importance of autodoping, it has been the subject of many studies.[17-21]

The discussion thus far has centered on equilibrium or at least steady-state reactions. If the dopant flow in the reactor is abruptly altered, it does not result in a rapid change in the doping profile.[22] Molecular beam epitaxy (discussed in Section 2.3) does not have this constraint.

Prior to the actual epitaxy of the substrate, autodoping can also be influenced by the pre-epitaxial processing that occurs within the reactor. One common process is in situ vapor-phase etching of the substrate with anhydrous HCl at a nominal temperature of 1200°C preceding the deposition. The reactions involved are

$$2HCl + Si \rightarrow SiCl_2 + H_2 \tag{15}$$

$$4HCl + Si \rightarrow SiCl_4 + 2H_2 \tag{16}$$

2.2.3   Equipment, Installation, and Safety Considerations

Typical reactors (Fig. 13) weigh 2000 kg and occupy 2 m^2 or more of floor space.[23] Several safety considerations must be addressed in the operation of the reactor. The reactor itself is usually designed with sufficient interlocks to prevent accidents. However, the user must remove and treat reaction by-products safely and arrange for proper delivery of process gases to the reactor. In fact, several distinct hazards require consideration: the explosion and fire potential of hydrogen, the corrosive nature of HCl, and the highly toxic nature of the doping and deposition gases. The last are particularly dangerous. Arsine, for example, is instantly lethal if a concentration of 250 ppm is inhaled, and prolonged exposure at lower levels (35 ppm) can also be lethal, depending on the length of exposure. Environmental considerations usually require a water-mist fume scrubber to remove most of the unreacted and reaction products from the carrier-gas stream.

Susceptors in epitaxial reactors are the analogs of crucibles in the crystal growing process. They provide mechanical support for the wafers and are the source of thermal energy for the reaction in induction-heated reactors. The geometric shape or configuration of the susceptor usually provides the name for

FIGURE 13
A radiant-heated barrel reactor.

reactor. Figure 14 shows three common susceptor shapes—horizontal, pancake, and barrel—which will be discussed in more detail later. Like crucibles the susceptor must be mechanically strong and noncontaminating to the process. Additionally, the susceptor must not react with the process reactants and by-products. Induction-heated reactors require a material that will couple to the rf field. The preferred material is graphite, but graphite susceptors require a coating because they are relatively impure and soft. A carbon blank is shaped to the required dimensions before the coating is applied. A coating of longstanding use is 50 to $500\mu m$ of silicon carbide applied by a CVD process similar to the silicon CVD process.

In most reactors, the reaction tube is relatively cool during operation, that is, they are operated "cold wall." Forced-air cooling carries away waste heat. Induction coils and other metal parts are water cooled. By way of comparison, the usual process for the CVD of polysilicon is a hot-wall operation (Chapter 6) resulting in a coating of silicon on the reactor tube itself.

Historically, energy for the reaction has been supplied by heating the susceptor inductively. The energy is then transported to the wafer by conduction and radiation. Motor-generator sets at 10 kHz or self-excited rf oscillators at 500 kHz are used for heating. A water-cooled coil is placed close to the susceptor so coupling can occur. The coil can be inside or outside the reaction chamber depending on the design of the reactor. Radiant heating, a newer way of supplying energy to the reaction chamber, provides more uniform heating than inductive

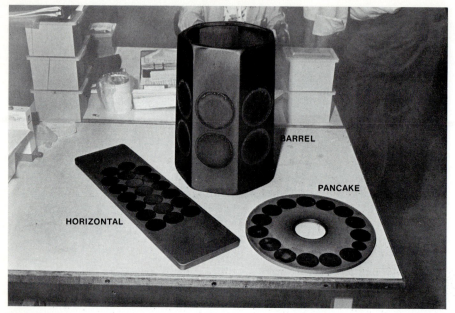

FIGURE 14
(a) Three common susceptor shapes: horizontal, pancake (vertical), and barrel.

heating provides.[23] The energy is supplied by banks of quartz halogen lamps. In most cases, process control involves maintaining gas flows and temperatures at the desired values. In modern equipment, the process cycle is generally microprocessor controlled, and the operator has only to bring wafers to the reactor and take the finished wafers away.

Three basic reactor configurations—horizontal, pancake or vertical, and barrel—(Fig. 14) have found widespread use. Each design has its relative advantages and disadvantages.[24] Horizontal reactors offer lower cost construction; however, controlling the deposition process over the entire susceptor length is a problem. Pancake reactors are capable of very uniform deposition, but suffer from mechanical complexity. Radiant-heated barrel reactors are also capable of uniform deposition, but are not suited for extended operation at temperatures above 1200°C.

A typical process for any configuration includes several steps. First, a hydrogen carrier gas purges the reactor of air. The reactor is then heated to temperature. After thermal equilibrium is established in the chamber, an HCl etch takes place at a temperature between 1150 and 1200°C for 3 min nominally. The temperature is then reduced to the growth temperature with time allowed for stabilizing the temperature and flushing the HCl. Next, the silicon source and dopant flows are turned on and growth proceeds at a rate of 0.2 to 2.0 μm/min. Once growth is complete, the dopant and silicon flows are eliminated and the temperature reduced, usually by shutting off power. As the reactor cools toward

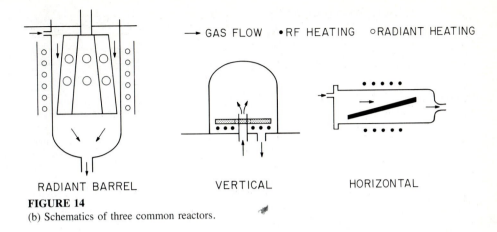

→ GAS FLOW   •RF HEATING   ◦RADIANT HEATING

RADIANT BARREL   VERTICAL   HORIZONTAL

FIGURE 14
(b) Schematics of three common reactors.

ambient temperature, the hydrogen flow is replaced by a nitrogen flow so that the reactor may be opened safely. Depending on wafer diameter and reactor type, capacities range from 10 to 50 wafers per batch. Process cycle times are about 1 h.

2.2.4. Process Selection and Capabilities

Epitaxial layers are rarely doped in excess of 10^{17} atoms/cm^3. This concentration is used in a bipolar technology where the epitaxial layer forms the transistor base.[25] The technical feasibility of doping to higher levels, approaching solid solubility, was demonstrated for phosphorus.[26,27]

The majority of applications require dopings of 10^{14} to 10^{17} atoms/cm^3. Lower doping levels, in the 10^{12} to 10^{14} atoms/cm^3 region, are used for certain types of high voltage and detector devices. These lower values are obtainable if the reactor is clean and the source is pure. Silicon sources with an equivalent purity of less than 10^{13} atoms/cm^3 in the deposited film are commercially available. Rear surface autodoping is often controlled by sealing the rear surface with an oxide or nitride layer. Uniformities of ± 10% within wafers are routinely obtained, and uniformities of ± 5% are possible in some cases. Variations within a run (batch) and from run to run are on the order of 20% or less, depending on the process and reactor.

The practical upper limit of epitaxial thickness is reached just before the layer overgrows the substrate and the film becomes contiguous with the silicon deposited on the susceptor. If the layer overgrows the substrate, the wafers become hard to separate from the susceptor, and the wafer usually cracks when removal is attempted. However, film thicknesses of several hundred micrometers are grown routinely for some power devices. As mentioned previously, thin layers are constrained by autodoping considerations, but layers as thin as .05 to 0.5 μm have been produced.[28] Layers with uniformities of ±5% are produced routinely with variations between runs of ±5% and better.

As in the case of crystal growing, the choice of dopants for epitaxial processes is limited. Boron is used for p-type doping, and arsenic or phosphorous for n-type doping. The usual dopant sources are hydrides of the dopants (PH_3, B_2H_6, or AsH_3).

The choice of a silicon source is based on several considerations. Table 1 lists each source currently in use along with characteristic growth rates and temperature ranges.[23] Silane (SiH_4) is usually chosen when a low deposition temperature is needed to minimize boron autodoping and outdiffusion. (Arsenic autodoping increases with lower temperatures.) Silane processes are prone to gas-phase nucleation (the formation of silicon particles in the gas stream above the wafer), which leads to poor film quality. Gas-phase nucleation can be suppressed[8] by adding HCl. Another disadvantage is that silane tends to coat the reactor chamber rapidly, requiring frequent cleaning. It also presents a production hazard because it is pyrophoric in concentrations above about 2%.

Dichlorosilane is a popular choice in many applications.[29] It offers high growth rates at relatively low temperatures. Although a liquid at room temperature, dichlorosilane has a high vapor pressure (< 1 atm), so it can be metered directly from a cylinder. Trichlorosilane is used for the production of electronic polysilicon, as mentioned in Chapter 1. It offers higher growth rates at lower temperatures, higher purity, and lower defect densities than silicon tetrachloride, which is the least expensive and most used of all the silicon sources. It is also a liquid at room temperature, but its low vapor pressure requires a bubbler tank to help vaporization. The high deposition temperatures required of silicon tetrachloride make it less sensitive than the other sources to oxidizers in the carrier gas and to the defects they cause.

Epitaxial reactors can generally operate at temperatures between 900 and 1250°C. Selecting the processing temperatures as well as the flow and growth rates is a complex decision based on the film thickness uniformity, doping uniformity, and doping level required, and on the defect levels, pattern shift, and distortion allowed. This chapter explains the process piecemeal, but does not tell how to design a process to meet all the objectives, many of which are in fact contradictory. For example, higher temperatures to reduce pattern shift (see Section 2.2.5) increase autodoping. A systematic approach to process design uses a factorial-design experimental approach to determine the optimum process condition for up to six variables including process temperature.[28,30] After the factorial design has been used to determine the best operating conditions, a silicon source can be chosen intelligently. Computer programs to simulate the epitaxial process are available and are constantly being refined.[22] Computer simulations are considered in detail in Chapter 10. Such programs are a useful adjunct to the factorial-design experimental method in setting up a CVD process.

Historically, the silicon CVD process has been performed at atmospheric pressure (760 Torr). However, operation in the range from 50 to 100 Torr has several advantages.[31] First, vertical and lateral autodoping effects (Fig. 15) are

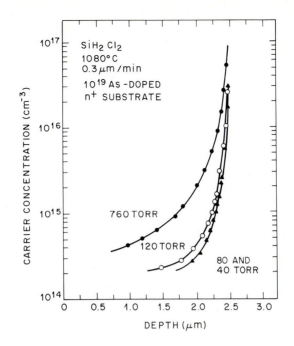

FIGURE 15
Doping profiles obtained over an arsenic-doped substrate for various reactor pressures. (*After Herring, Ref. 32.*)

significantly reduced for As and Sb (but are increased for boron). Second, pattern shift is also substantially reduced.[32]

2.2.5. Buried Layers

To fabricate bipolar ICs, usually one or more diffusions are applied to the substrate to create the necessary isolation, collector, emitter, or base functions (Fig. 2). These diffusions are applied to the substrate prior to epitaxy using the lithographic, oxidation, diffusion, or ion-implantation processes discussed in other chapters. The diffusions are called buried layers or diffusions under film. The presence of a buried layer complicates the epitaxial process because of its effect on autodoping (vertical and lateral), defects, pattern shifting, and pattern distortion.

The pre-epitaxial process leaves a step of 500 to 1000 Å around the perimeter of the buried layer that marks its location (A of Fig. 2). Subsequent masking levels must be properly aligned with the buried layer pattern. Unfortunately, the deposition process shifts the pattern on <111> oriented substrates (B of Fig. 2). Lithographic masks must compensate for the amount of the shift (L of Fig. 2) as the deposition process is not vertical but proceeds by lateral additions to steps as shown in Fig. 9. A separate but related effect is pattern distortion or washout, which alters the shape of the feature in the layer. The pattern in the epitaxial layer is thus misplaced and misshaped relative to its original configuration in the

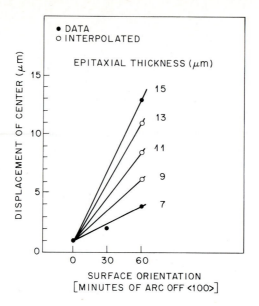

FIGURE 16

Pattern shift for a <100> orientation with various amounts of misorientation. (*After Drum and Clark, Ref. 34.*)

substrate. These effects place limitations on the design of high-density circuits and are a complicated function of substrate orientation, growth rate and temperature, and silicon source.[33]

The crystal orientation has a profound effect on pattern shift due to the dependence of growth rate on orientation.[34] Since the layer does not grow normal to the substrate but rather by addition to microsteps (Fig. 9), the step marking the diffusion is shifted. Current practice is to misorient <111> wafers by 2 to 5° towards the nearest <110> direction and to orient <100> wafers exactly on the orientation. Figure 16 illustrates the <100> case; note that the pattern shift changes with epitaxial thickness. As shown in Fig. 17, pattern shift is independent of reactor design,[35] but does show a pronounced dependence on growth rate and temperature. Pattern shift increases with growth rate and reduced deposition temperature. The magnitude of the shift is largely equal for both <111> and <100> orientations. These results are for an epitaxial process under atmospheric pressure. The pattern shift is substantially reduced as the reactor pressure is lowered.[32] Pattern distortion[33] exhibits an opposite relationship to the parameters previously mentioned, as shown in Fig. 18. For example, pattern shift is reduced at higher growth temperatures, but distortion increases and is more dependent on orientation.

A complete explanation of how all the variables affect pattern shift is not available, but includes the following elements. The step face (Fig. 9) exposes a number of crystal planes, which exhibit different growth rates. The anisotropy of growth increases at lower temperatures, as does the pattern shift. The growth rate

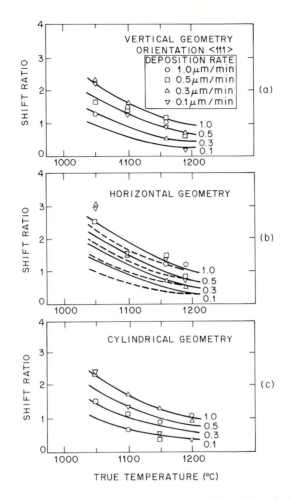

FIGURE 17

Pattern shift as a function of reactor [(a) vertical, (b) horizontal, and (c) cylindrical (or barrel)], temperature, and growth rate. (*After Lee et al., Ref. 35.*)

dependence of pattern shift is similar. The anisotropic nature of the layer growth rate increases with growth velocity. The effects of pressure and the silicon source are less clear, but are apparently interrelated. Less chlorine as a by-product in the form of HCl correlates with less distortion and shift. Reduced pressure operation would aid the escape of HCl across the boundary layer.

The discussions of Section 2.2.2 relating to autodoping also apply to the vertical autodoping profile above a buried layer. However, an effect termed lateral autodoping can be observed in such structures. Figure 19a shows that lateral autodoping is a front surface autodoping phenomenon involving the transport of dopant to regions adjacent to the diffusion prior to the onset of deposition. These regions can then diffuse as the layer grows. Figure 19b details the doping profiles on and off the buried layer. The off-profile is totally attributable to gas-phase

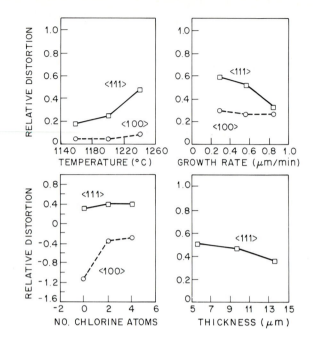

FIGURE 18
Parametric study of pattern distortion. (*After Weeks, Ref. 33.*)

autodoping. Dopant in these regions is detrimental, because if it is not eliminated by a boron isolation diffusion, it produces an electrical short circuit between the adjacent devices.[17,19] The peak concentration in the lateral autodoping profile is a function of the surface concentration in the buried layer and processing conditions such as HCl etch time, temperature, growth rate, and silicon source.

2.2.6   Epitaxial Defects

The crystal perfection of the layer never exceeds that of the substrate and is often inferior.[36,37] The crystal perfection is a function of the properties of the substrate wafer and the epitaxial process itself. Defects arising from the substrate wafer can be related to the bulk properties of the wafer or its surface finish. Item 1 in Fig. 20a is an example of an existing line dislocation continuing into the epitaxial layer. Impurity precipitates are one kind of surface defect that nucleate on an epitaxial stacking fault (item 2).[38,39,40] Process-related defects include slip (a displacement of crystal planes past each other as the result of stress). Dislocations accompany the formation of slip and of impurity precipitates from contamination (item 3). Contaminants from the susceptor and the tweezers used in handling the wafer also contaminate the layer and substrate and form precipitates that act as defect nuclei in subsequent processing.[40] These precipitates have been shown to degrade the quality of thin oxide layers used in MOSFET gates.[41] Tripyramids,

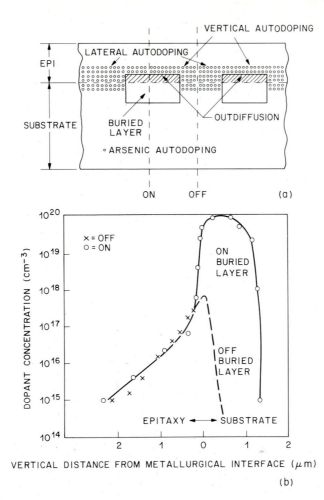

FIGURE 19
Lateral autodoping effect. (a) Cross section of epitaxial wafer showing location of lateral autodoping as adjacent to the buried layer. (b) Doping profiles above and adjacent to the buried layer. (*After Srinivasan, Ref. 19.*)

hillocks, and other growth features (item 4) can be related to the process or the surface finish of the wafer.[42] Item 5 is an example of a defect (bulk stacking fault) created in a pre-epitaxial process, such as buried layer fabrication. Such defects in turn nucleate defects in the epitaxial layer. Figure 20b is a series of photographs of defects in actual wafers. In general, the quality of the deposit is strongly related to the quality of the substrate wafer: its cleaning, layer growth rate, and temperature.[43] For example, as the deposition temperature is lowered, minor flaws in the substrate surface act as points of preferential nucleation, giving rise to stacking faults and pyramids due to lower surface mobility. Lowering the pressure can counteract the effect to some degree. Higher growth rates aggravate the problem, as discussed in the next section.

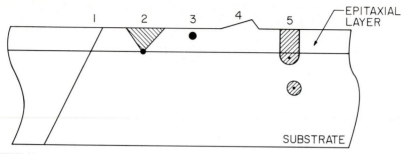

FIGURE 20

(a) Common defects occurring in epitaxial layers. Schematic representation of line (or edge) dislocation initially present in the substrate and extending into the epitaxial layer (item 1), an epitaxial stacking fault nucleated by an impurity precipitate on the substrate surface (item 2), an impurity precipitate caused by epitaxial process contamination (item 3), growth hillock (item 4), and bulk stacking faults, one of which intersects the substrate surface, thereby being extended into the layer (item 5).

A temperature gradient exists normal to the substrate in an rf-heated reactor.[44] Slip due to this gradient (during epitaxy) is produced in the following manner. Heat flow from the susceptor through the wafer equals subsequent radiation from the front surface and is given approximately by

$$\epsilon k T^4 = \frac{K dT}{dx} \tag{17}$$

where K is the thermal conductivity of silicon, dT/dx the temperature gradient normal to the wafer, ϵ the emissivity of silicon, k the radiation constant, and T the nominal (or average) wafer temperature. A front-to-rear temperature difference of only a few degrees causes a differential expansion of the wafer. In effect, the wafer curls up on the susceptor. When the wafer edge loses contact with the susceptor, the edge temperature drops, causing still further bowing. This radial temperature gradient results in sufficient stress to create dislocations in the wafer (see the section on thermal stress in Chapter 1). The inverted heat flow of the radiant-heated reactor minimizes this problem.[23]

Misfit dislocations, caused by lattice mismatch when the substrate is highly doped, constitute another class of defects.[45] The resultant strain between the layer and substrate is relieved by the formation of dislocations.

Problem

At a processing temperature of 1200°C, what wafer thickness would lead to a temperature gradient sufficient to cause slip?

Solution

1 From Chapter 1 the yield point of silicon at 1200°C in the oxygen-free case is 2×10^7 dyn/cm².

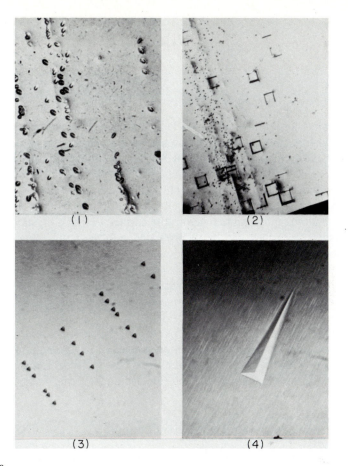

FIGURE 20
(b) Photographs of defects in actual wafers. Dislocations revealed as circular etch pits by Secco etching in a region of slip on a <100> wafer (item 1), epitaxial stacking faults on a <100> wafer (item 2), dislocations revealed by Sirtl etching in a <111> wafer (item 3), and a growth hillock on a <111> wafer, visible without etching (item 4).

2 Also from Chapter 1 the allowable temperature difference is estimated as

$$S = aE\Delta T$$
$$2.0 \times 10^7 = 2.6 \times 10^{-6} \times 1.09 \times 10^{12}\Delta T$$
$$\Delta T = 7.2°C$$

3 Using Eq. 17,

$$\epsilon kT^4 = \frac{KdT}{dx}$$
$$0.6 \times 5.7 \times 10^{-12} \times (1473)^4 = 1.5 \times 7.2/dx$$
$$dx = 0.67 \text{ cm}$$

where

$$\epsilon = 0.6$$
$$k = 5.7 \times 10^{-12} \text{watts/cm}^2$$

2.2.7 Microscopic Growth Processes

The final points to consider in the CVD process are the conditions under which single-crystal films are obtained and the mechanisms of their growth.[46] Figure 21 illustrates the maximum attainable growth rates for atmospheric pressure epitaxy. The activation energy obtained from that Arrhenius plot is 5 eV, which is equal to that of silicon self-diffusion. The physical explanation is that silicon atoms are adsorbed on the surface of the substrate after a chemical reaction takes place. These atoms must migrate across the surface to find a crystallographically favorable site where they can be incorporated into the lattice (Fig. 9). At high growth rates, insufficient time is allowed for surface migration, resulting in polycrystalline growth. The favorable sites are located at the leading edge of monolayer high steps. Thus, the growth is not vertical, but lateral. This effect accounts for the variation in growth rate with surface orientation, because the availability and movement of steps is orientation dependent.[42] The adsorbed silicon atoms compete with dopant atoms, hydrogen, chlorine, and foreign atoms for these sites. Dopant atom concentration is usually low enough to be ignored, but impurities such as carbon (initially present on the surface) affect the movement of silicon atoms on the surface and may nucleate a stacking fault or tripyramid

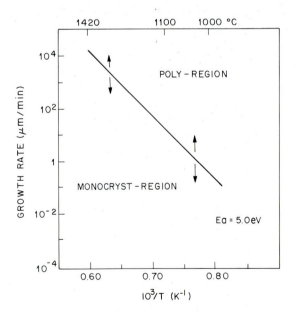

FIGURE 21
Maximum growth rate for which monocrystalline silicon can be obtained as a function of temperature. (*After Bloem, Ref. 46.*)

defect. This lateral growth mechanism accounts for the effects that were discussed under pattern shift and distortion, and is an additional reason to misorient <111> wafers. Growth of epitaxial layers exactly on <111> results in mounds being formed.[42]

2.2.8   Advanced CVD Methods

In addition to the basic CVD process, there are other ways of modifying the process to produce a technologically important result as well as a scientifically interesting process.

One of these is selective epitaxy.[47] As shown in Figure 22, it is a means of growing single-crystal silicon on a substrate patterned with oxide or nitride, which allows lateral isolation with closer packing than standard local oxidation techniques. To achieve selective epitaxial growth, it is necessary to suppress the nucleation of silicon on the dielectric film. This is accomplished by lowering the partial pressure of the reactant in the system, which results in nucleation only on the exposed silicon surfaces.

A conventional epitaxial growth process is controlled by starting and stopping gas flows, but a novel approach termed "Limited Reaction Processing" controls the process by controlling temperature.[48] In this technique radiant energy is supplied by lamps turned off and on by a microprocessor. The speed by which growth is initiated and terminated allows for the deposition of very thin layers with a minimum of autodoping whose characteristics approach those of layers prepared by molecular beam epitaxy.

A third modification to the basic CVD process is the of use plasma excitation to achieve growth at temperatures below that practical with conventional methods of supplying energy to the reaction.[49]

2.3   MOLECULAR BEAM EPITAXY

Molecular beam epitaxy (MBE) is a non-CVD epitaxial process that uses an evaporation method. Although the method has been known since the early 1960s, it has only recently been seriously considered a suitable technology for silicon-device fabrication. The two major reasons why MBE was not used were that the quality was not commensurate with device needs and that no industrial equipment existed. Equipment is now commercially available, but the process has low

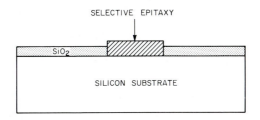

FIGURE 22
Cross-sectional schematic of a selective epitaxy process using an oxide mask.

throughput and is expensive. MBE, however, does have a number of inherent advantages over CVD techniques. Its main advantage for VLSI use is low-temperature processing. Low-temperature processing minimizes outdiffusion and autodoping, a limitation in thin layers prepared by conventional CVD. Another advantage is the precise control of doping that MBE allows. Because doping in MBE is not affected by time-constant considerations as is CVD epitaxy, complicated doping profiles can be generated. Presently, these advantages are not being exploited for IC fabrication, but they have found application in discrete microwave and photonic devices. For example, the C-V characteristic of a diode with uniform doping is nonlinear with respect to reverse bias. Varactor diodes used as FM modulators could advantageously employ a linear dependence of capacitance on voltage. This linear voltage–capacitance relationship can be achieved with a linear doping profile, which is easily obtained with MBE.

In contrast to CVD processes, MBE is not complicated by boundary-layer transport effects, nor are there chemical reactions to consider. The essence of the process is an evaporation of silicon and one or more dopants, as depicted in Fig. 23. The evaporated species are transported at a relatively high velocity in a vacuum to the substrate. The relatively low vapor pressure of silicon and the dopants ensures condensation on a low-temperature substrate. Usually, silicon MBE is performed under ultra-high vacuum (UHV) conditions of 10^{-8} to 10^{-10} Torr, where the mean free path of the atoms[50,51] is given by

$$L = 5 \times 10^{-3}/P \tag{18}$$

where L is the mean free path in cm, and P is the system pressure in Torr. At a system pressure of 10^{-9} Torr, L would be 5×10^6 cm.

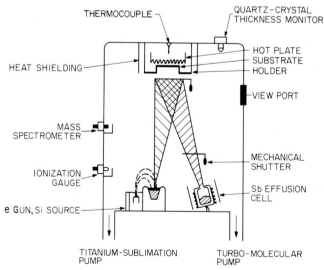

FIGURE 23
Schematic of MBE growth system. (*After Konig, Kibbel, and Kasper, Ref. 54.*)

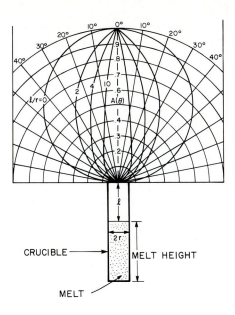

FIGURE 24

Angular distribution of flux from a crucible of radius r and melt height l referenced from the top of the crucible. (*After Luscher and Collins, Ref. 52.*)

Because collisions between atoms are unimportant in a high vacuum, transport velocity is controlled more by thermal energy effects than by diffusion effects, and deposition and its uniformity can be controlled by the source characteristics.[52] Accordingly, evaporation from a crucible produces a flux of material varying with time and angle, as shown in Fig. 24. The lack of intermediate reactions and diffusion effects, along with relatively high thermal velocities, results in film properties changing rapidly with any change at the source.

A conventional temperature range for MBE is from 400 to 800°C. Higher-temperature processes are technically feasible, but the advantages of reduced outdiffusion and autodoping are lost. Growth rates in the range 0.01 to 0.3 μm/min have been reported.[50] The higher value is comparable to those obtained in CVD epitaxy.

In situ cleaning for MBE is done in two ways. High-temperature baking between 1000 and 1250°C for up to 30 minutes decomposes the native oxide and removes other adsorbed species (notably carbon) by evaporation or diffusion into the wafer.[53] A better approach is to use a low-energy beam of an inert gas to sputter clean the surface. A short anneal at 800 to 900°C is sufficient to reorder the surface.

MBE doping has several distinguishing features. A wider choice of dopants can be used, compared to CVD epitaxy; more control of the doping profile is possible; and two doping processes are available. In principle the doping process is similar to the growth process. A flux of evaporated dopant atoms arrives at the growing interface, finds a favorable lattice site, and is incorporated. The doping

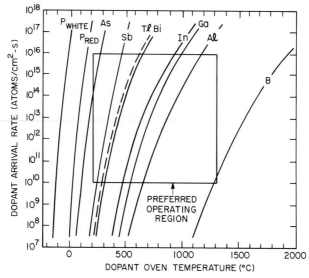

FIGURE 25
Flux of various dopant species versus oven temperature. (*After Bean, Ref. 50.*)

level is controlled by adjusting the dopant flux to the flux of silicon atoms. In practice a Knudsen effusion cell is used to evaporate dopants.[52] Unfortunately, desirable dopants such as As, P, and B evaporate too rapidly or too slowly for controlled use. As a result most workers use Sb, Ga, or Al, which compare favorably to other dopants as shown in Fig. 25.

Another complication is the temperature-dependent sticking coefficient shown in Fig. 26. A low value means re-evaporation occurs readily and incorporation of the dopant is more difficult. This temperature dependence requires precise control of substrate heating. A wide latitude in doping by evaporation has been demonstrated.[54,55] Values in the range 10^{13} to 10^{19} atoms/cm^3 have been reported with 1% radial uniformities.

Another doping technique uses ion implantation[56] (see Chapter 8). This technique uses a low-current (1μA), low-energy (0.1 to 3 keV) ion beam to implant dopant as the layer is growing. The low-energy beam places the dopant just below the growing interface, ensuring incorporation. Doping profiles not obtainable with CVD processes can be produced with ion implantation, as depicted in Fig. 27. This low-energy beam technique also allows the use of dopants such as B, P, and As. Because MBE is a vacuum process, it is very adaptable to ion-implant doping, and in situ monitoring of the beam is feasible.[52]

2.3.1 Equipment

An elementary MBE system is depicted in Fig. 23. It is, in essence, a UHV chamber where furnaces holding electronic-grade silicon and dopant direct a flux

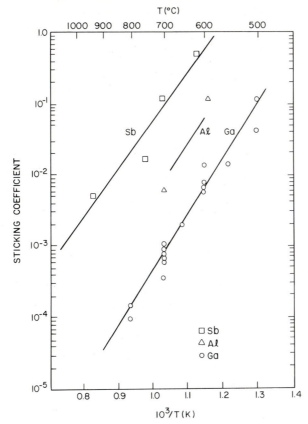

FIGURE 26
Sticking coefficient for Sb, Al, and Ga versus temperature. (*After Bean, Ref. 50.*)

of material to a heated substrate. Figure 28 illustrates the many components of a comprehensive system. A distinguishing feature of MBE is the ability to use sophisticated analytical techniques in situ to monitor the process. In contrast to the CVD process, MBE does not require extensive safety precautions, although solid arsenic dopant must be handled carefully.

The vacuum system is the heart of the apparatus. To consistently attain a vacuum level in the 10^{-10} Torr range, materials and construction choices must be carefully considered. Materials should have low vapor pressure and low sticking coefficients. Repeated exposure to air is detrimental to a UHV system because of the long bakes needed to desorb atmospheric species from the system walls. A load lock system minimizes this problem. Consistently low base pressure is needed to ensure overall film perfection and purity. These needs are best met with an oil-free pump design, such as a cryogenic pump.

Because of its high melting point, silicon is volatilized not by heating in the furnace, but by electron-beam heating. Dopants are heated in a furnace. A

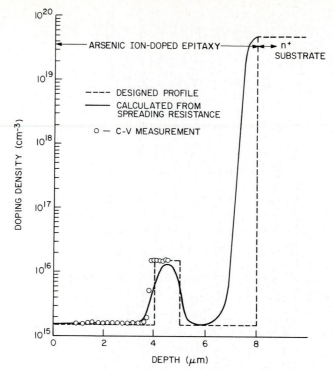

FIGURE 27
Doping profile obtained by ion implantation during MBE growth. (*After Ota, Ref. 56.*)

SILICON MBE SYSTEM

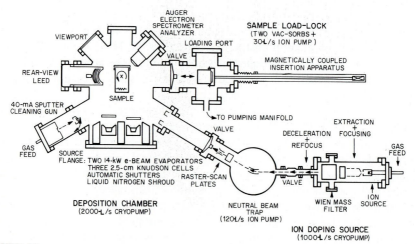

FIGURE 28
Schematic of practical MBE system. (*After Bean, Ref. 50.*)

constant flux is assured by the use of closed-loop temperature control. Baffles and shutters shape and control the flux, so uniformity of doping and deposition can be attained without boundary layer effects.

Substrates are best heated when they are placed in proximity to a resistance heater with closed-loop temperature control. Resistance heating generates temperatures over the range of 400 to 1100°C. A wide choice of temperature-sensing methods is available, including thermocouples, optical pyrometry, and infrared detection.

2.3.2 Film Characteristics

The preparation of high-quality films by MBE requires an in-situ cleaning process to remove adsorped contaminants and oxide films. Low base pressure is also a requirement to keep the surface clean. Lowering the pressure lessens the concentration of contaminants adsorbed on the substrate. These species would obstruct the single-crystal growth and nucleate a fault, as discussed in Section 2.2.7. The effects on dislocation density can also be seen in the pre-heat time for substrate bakeout prior to growth and in the temperature of growth (see Figs. 29 and 30).

2.4 SILICON ON INSULATORS

An all-silicon device structure has inherent problems that are associated with parasitic circuit elements arising from junction capacitance. These effects become a more severe problem as devices are made smaller (see Chapter 11). A way to

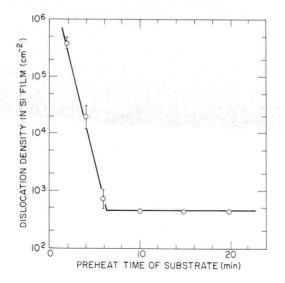

FIGURE 29
The dependence of film quality (dislocation density) on predeposition heating time. (*After Sugiura and Yamaguchi, Ref. 53.*)

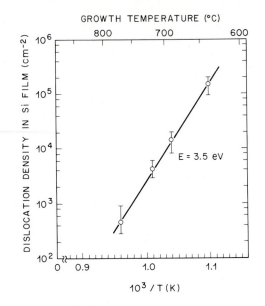

FIGURE 30
The dependence of film quality (dislocation density) on growth temperature. (*After Suguira and Yamaguchi, Ref. 53.*)

circumvent the problem is to fabricate devices in small islands of silicon on an insulating substrate as shown in Fig. 31. The initial approach to fabricating such a structure was to grow silicon epitaxially on a substrate of sapphire (Al_2O_3) or spinel ($MgAl_2O_4$). Since the substrate material differs from the layer, the process is termed heteroepitaxy. A more recent approach, yet to be perfected, is silicon on amorphous substrates.

2.4.1 Silicon on Sapphire

The processes and equipment used for silicon on sapphire (SOS) epitaxy are essentially identical to those employed for homoepitaxial growth. Silane is the

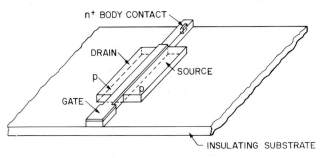

FIGURE 31
MOSFET device fabricated in silicon island on sapphire substrate. (*After Schlotter, Ref. 58.*)

favorite choice for the silicon source according to the pyrolysis reaction

$$SiH_4 \rightarrow Si + 2H_2 \tag{19}$$

in a carrier gas of hydrogen. Silane is chosen mainly for its low-temperature deposition capability, which is used in SOS to control autodoping of aluminum from the substrate. Common process parameters are deposition temperatures between 1000 and 1050°C and growth rates of 0.5 μm/min. Film thicknesses are on the order of 1 μm or less with film doping in the range of 10^{14} to 10^{16} atoms/cm^3. Various substrate (sapphire) orientations such as $<01\bar{1}2>$, $<10\bar{1}2>$, and $<1\bar{1}02>$, have been used to grow $<100>$ oriented silicon layers.[57,58] However, significant problems exist with the technology. Aluminum autodoping from the substrate restricts the choice of doping level, and the films are usually characterized by a high defect density. The latter characteristic results in very low minority-carrier lifetimes (1 to 10 ns).[58] The low minority-carrier lifetime also means that junction leakage currents could be higher than in comparable circuits in bulk wafers. As a result, only majority-carrier devices are practical. Both CMOS and NMOS circuits have been fabricated by using SOS epitaxy.

The defect structure of SOS devices has been studied by a number of workers.[57] The films are generally characterized by high densities of various defects such as stacking faults, misfit dislocations, and dislocations. A key finding was that defect density varies inversely with distance from the substrate (Fig. 32). This effect is related to the lattice mismatch between the layer and substrate. The strain caused by lattice mismatch is somewhat relieved by the formation

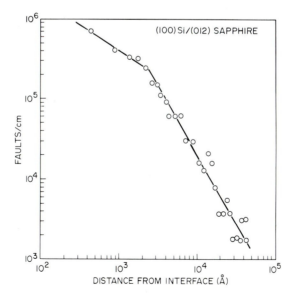

FIGURE 32
Stacking fault density as a function of distance above the substrate for an SOS structure. (*After Abrahams et. al., Ref. 57.*)

of misfit dislocations near the original layer-substrate interface. The transition layer between the epitaxial layer and the substrate is complicated, involving the formation of aluminum silicate from the outdiffusion of aluminum from the substrate. Such a layer is unavoidable in heteroepitaxy.[58] Another fundamental problem in SOS epitaxy is the thermal mismatch between the layer and the substrate. The thermal expansion of sapphire is approximately twice that of silicon. On cooling, this difference in thermal expansion causes a strain-induced change in the band structure that limits the carrier mobility to 80% of the bulk value. The carrier mobility is also reduced by the high defect densities.

These deficiencies have resulted in various attempts to improve SOS film quality. MBE is one solution, because of its lower process temperature, reduced autodoping, and stress. Some workers[59] have used laser annealing to improve the quality by melting and recrystallizing the layer. For example, a Q-switched ruby laser with energy densities greater than 1 J/cm^2 is used to reduce defect density and improve mobility.

2.4.2   Silicon on Amorphous Substrates

Silicon on insulator (SOI) is a recent nonepitaxial approach to providing single-crystal silicon. With this technology, amorphous or polycrystalline silicon is recrystallized on an amorphous substrate. Figure 33 shows a setup for recrystallization using a strip heater. The process is considered nonepitaxial because this silicon film is not single crystal as deposited. Energy for the process can also be supplied by electron beam[60] or laser.[61] The resultant structure is functionally similar to the heteroepitaxial SOS configuration, but without the attendant disadvantages. The recrystallized layers are potentially the equal of homoepitaxial silicon. SOI is not used commercially at present, but possible device applications include VLSI circuits, photovoltaic solar-energy conversion, and even three-dimensional ICs.

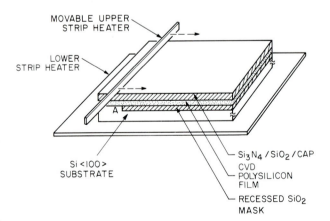

FIGURE 33
Schematic of one technique used to recrystallize polysilicon on SiO_2. Region *A* acts as a seed for the lateral recrystallization when the heater moves to the right. (*After Tsuar et. al., Ref. 63.*)

Several methods of preparing SOI have been investigated. Substrates can be conventional silicon wafers covered with silicon nitride, silicon dioxide, or even fused quartz substrates.[62] If the conventional silicon wafer is used, it is processed in a manner that yields a pattern of exposed silicon areas whose surface is coplanar with the surrounding oxide. This substrate is then coated with polysilicon in a low-pressure CVD process to a thickness of $0.5\mu m$. A movable strip heater (Fig. 33), positioned above one of the openings to the substrate, melts the polysilicon through to the substrate.[63] The heater is then moved laterally and, with the substrate acting as a seed, single-crystal silicon is grown laterally over the oxide-covered regions. The thermal stability of the molten zone is improved if it is capped with oxide and nitride layers. Capping also prevents contamination of the film. This technique is suitable for recrystallizing large areas, such as an entire wafer.

Another approach is to pattern a polysilicon layer on an amorphous substrate.[63] A laser is then rastered across the wafer to recrystallize the individual islands of silicon. This method does not need seeding from the substrate. Adjusting the energy parameters of the laser and its scan rate induces the islands to crystallize in a <100> orientation. High-quality n-channel depletion mode MOSFETs have been fabricated in recrystallized silicon.[64] The device structure (Fig. 34) is similar to that of SOS devices. In particular, the surface electron mobility was reported at 600 to 700 $cm^2/V\text{-}s$, a value near that of devices fabricated in single-crystal silicon. These results are better than those obtained from SOS devices.

Additional work is needed before this technology is the equal of homoepitaxial silicon technology, but it has the potential to revolutionize device design and fabrication.

2.5 EPITAXIAL EVALUATION

To evaluate epitaxial slices, layer doping and thickness which are easily quantified, are measured. Additionally, a cosmetic inspection is usually performed even though this evaluation is somewhat subjective. The prime requisites for routine measurements are high speed and repeatability. In an industrial environment, information is needed at relatively short intervals (< 1 h) to maintain pro-

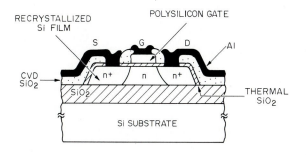

FIGURE 34
Cross section of a MOSFET formed in recrystallized polysilicon. (*After Tsuar et. al., Ref. 64.*)

cess control. Absolute accuracy is of lesser concern because material requirements are usually adjusted empirically to satisfy device needs. Only a few evaluation methods are commonly used.[65]

2.5.1 Epitaxial Thickness

Lightly doped silicon is transparent in the near infrared region, and heavily doped silicon ($> 1 \times 10^{18}$ atoms/cm^3) is an absorber. However, increased doping reduces the index of refraction (Fig. 35) below that of lightly doped silicon ($\bar{n} = 3.42$). As a result, interference fringes in the 5 to 50 μm wavelength range can be observed in the reflection spectra on a dispersive infrared spectrophotometer. The epitaxial layer thickness can be computed using the formula[66]

$$t = \frac{(P_n - 1/2 + P_i)W_n}{2(\bar{n}^2 - \sin^2 A)^{1/2}} \tag{20}$$

where W_n is the position of a maximum or minimum in the spectra in micrometers, $\bar{n}$ is the index of refraction, A is the angle of the incident light, P_n is the order of the maximum or minimum, and P_i is a correction factor that depends on the substrate used.

An automated approach to the measurement employs a Michaelson interferometer. This instrument samples all wavelengths simultaneously. Its output is called an interferogram, which is the Fourier transform of the reflectance spectra obtained on a spectrophotometer. A computer controls the interferometer and collects the data. The thickness can be computed from the interferogram,

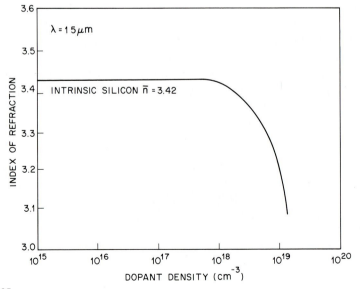

FIGURE 35
Typical index of refraction versus doping level for silicon at one wavelength.

or the computer can calculate the Fourier transform and then calculate the thickness using Eq. 20. Equipment for this second method is commercially available. Such equipment can measure thickness from less than 1 μm to more than several hundred micrometers. Measurement time is about 5 s with a measurement repeatability of $\pm 0.05 \mu$m.

The equivalent point of reflection of infrared measurements is at a heavily doped point on the outdiffusion tail. For common deposition conditions, this point is usually near the epi-substrate interface. Thus, an infrared thickness measurement is a reasonable monitor of the thickness added to the substrate, but is relatively insensitive to the shape or extent of the outdiffusion autodoping tail.[67]

For structures that are not amenable to infrared measurements, there are several alternatives. The length of the side of an epitaxial stacking fault, nucleated at the substrate, is linearly related to the layer thickness:[68]

$$t = C_1 l \tag{21}$$

where t is the layer thickness, l is the size of the fault, and C_1 is an orientation-dependent constant which is 0.707 for <100> and 0.816 for <111>.

Wafers can also be sectioned and stained with a number of chemical solutions to delineate the layer.[69] Spreading resistance[70] profiling (see Chapter 7) is particularly useful for structures that have multiple layers or structures where the total impurity profile is important.

2.5.2 Epitaxial Doping

The uncertainties of doping kinetics, background effects, and autodoping effects do not allow the doping in the layer to be established simply on the basis of the flows into the reactor. Three types of electrical measurements—sheet resistance, capacitance voltage, and spreading resistance—are used to measure doping levels.[71]

The control wafer technique is a widely used method that requires simple equipment. It involves placing in the reactor a lightly doped slice of a conductivity type opposite to the layer to be grown. After deposition a four-point probe measures the sheet resistance of the layer (see Chapter 1). The sheet resistance is converted to resistivity using the infrared thickness of an adjacent product slice.[72] This method is highly inaccurate in some cases.[71] Its suitability must be determined by correlating its measurements to measurements made on product slices by another method. In other cases, no correlation is possible due to a strong predeposition of substrate dopant onto the control wafer.

The second method, the preferred approach, is the use of diode C-V measurements (see Chapter 7). Implicit in the capacitance versus voltage characteristic of a reverse-biased step junction diode is the doping profile of the material according to the relationships

$$N(x) = C^3 \frac{dC}{dV}(CF_1)(CF_2)2qA^2\epsilon_s \tag{22}$$

$$x = \epsilon_s A / C \qquad (23)$$

where C is capacitance, V is voltage, q is charge, A is the diode area, ϵ_s is the dielectric permittivity of silicon, N is the doping density, and x is depth. CF_1 and CF_2 are correction factors for diffused-junction and depletion-layer widening effects.[73,74]

C-V measurements of a Schottky barrier diode, formed by using a mercury contact, are a rapid, nondestructive way to determine slice doping.[71,75] If the depletion layer can be spread to the substrate, some information on the autodoping tail can be obtained. The measurement can also be performed on mesa or planar junction diodes as a means of calibrating the area of the Schottky contact. The principle drawbacks of a C-V measurement are its high sensitivity to small errors in area and capacitance.[76]

The third method, spreading resistance measurements, was previously mentioned as a profiling technique. This method can determine a wafer's resistivity by surface measurements. The major difficulties are in maintaining accurate calibration because the probes wear with repeated usage, and in overcoming the influence of surface effects that affect the measured resistance.

Problem
A straight line is seen to result from a plot of $1/C^3$ versus V. What does this imply about the doping profile shape?

Solution
1 Since $(1/C^3) = mV + $ b, $dC/dV \propto \dfrac{1}{C^4}$

2 Substituting $1/C^4$ into Eq. 22 yields $N(x) \propto 1/C$, which by virtue of equation 23, means $N(x) \propto x$.

3 The doping profile is linearly graded.

2.5.3 Cosmetic Inspection and Perfection Evaluation

The wafer is usually examined with the unaided eye under high-intensity illumination to judge the quality of the deposit. Wafers may be rejected for any departure from a specular, smooth surface, including projections seen as bright spots of light, stains, haze, or scratches. The acceptance criteria are usually set empirically based on the type of device being fabricated. This inspection is often done using scanned laser or collimated light to detect light scattering centers. Additional inspection may be made at magnifications of from 50 to 200 to evaluate microdefect densities such as those of stacking faults and tripyramids using Nomarski phase-contrast microscopy. Another useful technique is to etch wafers in solutions such as Schimmel's, Secco's, or Sirtl's etch to determine dislocation and saucer pit densities.[77] The latter indicate that contamination is present in the process.

2.5.4 Lifetime

The lifetime of minority carriers is generally not a consideration in structures intended for IC fabrication, but could be of interest in some devices such as dynamic RAMs. Several measurements involving the transient response of diodes or MOS capacitors are applicable to epitaxial layers.[78,79] However, the diffusion length of carriers is often many times that of the layer thickness. This complicates the interpretation of the measurement results.

2.6 SUMMARY AND FUTURE TRENDS

Epitaxy as a process will remain integral to circuit manufacture. It offers doping profiles and material properties not obtainable otherwise. Homoepitaxial silicon structures will remain popular design choices. The advantages of SOI technologies are compelling for high-density and high-speed circuits. In particular, if silicon-on-SiO_2 can be perfected, it will offer the advantages of SOS without its problems. Lateral-seeded SOI will undoubtedly receive considerable research attention. MBE would be advantageous in fully ion-implanted VLSI circuits in which the total thermal cycle is minimized so that the doping capabilities of MBE can be exploited.

Although currently available equipment is adequate for most needs, several aspects of the epitaxial process are being improved. In keeping with automation elsewhere in the fabrication process, an autoloading epitaxial reactor is now available. This equipment exists in both multiwafer and single-wafer reactors that are fed from cassettes. Conceptually, a single-wafer reaction chamber could easily be optimized for temperature and gas flows to produce wafers having exceptional uniformity. The throughput of epitaxial reactors is less than that of LPCVD processes (Chapter 6) by a factor of 5 to 10. One difficulty is the low growth rates in the usual LPCVD temperature ranges (Fig. 8). An alternative reactor design, termed the rotary disc, is similar in load configuration to LPCVD equipment, and offers high capacity and efficiency.[80] Large-scale use of MBE will require equipment with throughputs comparable to present-day epitaxial reactors.

Although epitaxial processes are well characterized and understood, the trend to thinner layers for bipolar and unipolar ICs will result in incremental process improvements and the continued study of flatness changes, defect generation, and autodoping effects. Additionally, contamination, responsible for precipitates in epitaxial layers, needs to be reduced commensurate with the requirements of VLSI devices. Contamination-free epitaxy will be a worthwhile process improvement.

PROBLEMS

1 In a 1 h process at 1100°C using dichlorosilane, a 10 μm layer is grown on 20 substrates of 100 mm diameter in a horizontal reactor. Estimate the energy in kWh for the process. Assume a growth rate of 1 μm/min.

2 Determine the amount of mask compensation needed for an epitaxial wafer of <100> orientation containing an antimony buried layer with an epitaxial thickness of 7 μm.

3 Using the figures in the chapter, estimate the temperature of zero growth rate for each silicon source. Compare these temperatures to the nucleation temperature curve. What do you conclude?

4 Calculate activation energies from the Arrhenius plots for growth rate versus temperature and nucleation versus temperature. What do you conclude about the process?

5 A reverse-biased diode has a voltage capacitance characteristic defined by the relation $VC^2 = N$. What would you conclude about the shape of the doping profile? Suggest a graphical way to determine doping density from the C-V curve.

6 Using the diffusivity of boron at 1100°C (Chapter 7), calculate the minimum growth rate such that the condition of Eq. 14 is satisfied given a deposition time of 10 min.

7 Calculate the number of liters of hydrogen at STP that would have to be supplied *into* the reactor for the process of Problem 1. What do you conclude?

8 Does the thickness of the epitaxial wafer pose a problem in epitaxial processing from a stress viewpoint? Discuss your answer.

REFERENCES

1 H. C. Theuerer, J. J. Kleimack, H. H. Loar, and M. Christensen, "Epitaxial Diffused Transistors," *Proc. IRE*, **48**, 1642 (1960).

2 F. E. Holmes and C. A. T. Salama, "VMOS—A New MOS Integrated Circuit Technology," *Solid State Electron.*, **17**, 791 (1974).

3 D. S. Yaney and C. W. Pearce, "The Use of Thin Epitaxial Layers for MOS VLSI," *Proceedings of the 1981 International Electron Device Meeting*, IEEE, 1981, p. 236.

4 V. S. Ban, "Mass Spectrometric Studies of Chemical Reactions and Transport Phenomena in Silicon Epitaxy," *Proceedings of the Sixth International Conference on Chemical Vapor Deposition 1977*, Electrochem. Soc., Pennington, New Jersey, 1977, p. 66.

5 K. F. Jensen, "Modelling of Chemical Vapor Deposition Reactors," *Proceedings of the Ninth International Conference on Chemical Vapor Deposition 1984*, Electrochem. Soc., Pennington, New Jersey, 1984, p. 3.

6 R. M. Olsen, *Essentials of Engineering Fluid Flow*, International Textbook, Scranton, Pennsylvania, 1966.

7 C. W. Manke and L. F. Donaghey, "Numerical Simulation of Transport Processes in Vertical Cylinder Epitaxy Reactors," *Proceedings of the Sixth International Conference on Chemical Vapor Deposition 1977*, Electrochem. Soc., Pennington, New Jersey 1977, p. 151.

8 J. Bloem, "Silicon Epitaxy from Mixtures of SiH_4 and HCl," *J. Electrochem. Soc.*, **117**, 1397 (1970).

9 E. Sirtl, L. P. Hunt, and D. H. Sawyer, "High Temperature Reactions in the Silicon-Hydrogen-Chlorine System," *J. Electrochem. Soc.*, **121**, 919 (1974).

10 J. Nishizawa and M. Saito, "Growth Mechanism of Chemical Vapor Deposition of Silicon," *Proceedings of the Eighth International Conference on Chemical Vapor Deposition 1981*, Electrochem. Soc., Pennington, New Jersey, p. 317.

11 F. C. Eversteyn, "Chemical-Reaction Engineering in the Semiconductor Industry," *Philips Res. Rep.*, **29**, 45 (1974).

12 McD. Robinson, "CVD Doping of Silicon," in F. F. Y. Wang, Ed., *Impurity Doping Processes in Silicon*, North-Holland, Amsterdam, 1981.

13 J. Bloem, "The Effect of Trace Amounts of Water Vapor on Boron Doping in Epitaxially Grown Silicon," *J. Electrochem. Soc.*, **118**, 1837 (1971).

14 R. Reif, T. I. Kamins, and K. C. Saraswat, "A Model for Dopant Incorporation into Growing Silicon Epitaxial Films," *J. Electrochem. Soc.*, **126**, 644 and 653 (1979).

15 H. Basseches, R. C. Manz, C. O. Thomas, and S. K. Tung, *AIME Semiconductor Metallurgy Conference,* Interscience, New York, 1961, p. 69.

16 A. S. Grove, A. Roder, and C. T. Sah, "Impurity Distribution in Epitaxial Growth," *J. Appl. Phys.*, **36**, 802 (1965).

17 G. R. Srinivasan, "Autodoping Effects in Silicon Epitaxy," *J. Electrochem. Soc.*, **127**, 1334 (1980).

18 P. H. Langer and J. I. Goldstein, "Boron Autodoping during Silane Epitaxy," *J. Electrochem. Soc.*, **124**, 592 (1977).

19 G. R. Srinivasan, "Kinetics of Lateral Autodoping in Silicon Epitaxy," *J. Electrochem. Soc.*, **125**, 146 (1978).

20 W. H. Shepard, "Autodoping of Epitaxial Silicon," *J. Electrochem. Soc.*, **115**, 652 (1968).

21 W. C. Metz, "Autodoping of Boron Doped Epitaxial Layers for CMOS Devices," *Proceedings of the Ninth International Conference on Chemical Vapor Deposition 1984*, Electrochem. Soc., Pennington, New Jersey, 1984, p. 420.

22 R. Reif and R. W. Dutton, "Computer Simulation in Silicon Epitaxy," *J. Electrochem. Soc.*, **128**, 909 (1981).

23 M. L. Hammond, "Silicon Epitaxy," *Solid State Technol.*, **21**, 68 (1978). R. C. Rossi and K. K. Scheregraf, "Glassy Carbon-Coated Susceptors for Semiconductor CVD Processes," *Semicond. Int.*, **4**, 99 (1981).

24 W. Benzing, "A Near Future Perspective on Silicon Epitaxy," *Proceedings of the Ninth International Conference on Chemical Vapor Deposition 1984*, Electrochem. Soc., Pennington, New Jersey, 1984, p. 373.

25 B. T. Murphy, V. J. Glinski, P. A. Gary, and R. A. Pedersen, "Collector-Diffusion Isolated Integrated Circuits," *Proc. IEEE*, **57**, 1523 (1969).

26 J. Bloem, L. J. Giling, and M. W. M. Graef, "The Incorporation of Phosphorous in Silicon Epitaxial Layer Growth," *J. Electrochem. Soc.*, **121**, 1354 (1974).

27 P. Rai-Choudhury and E. I. Salkovitz, "Doping of Epitaxial Silicon," *J. Cryst. Growth*, **7**, 361 (1970).

28 J. Borkowicz, J. Korec, and E. Nossarzewska-Orlowska, "Optimum Growth Conditions in Silicon Vapour Epitaxy," *Phys. Status Solidi A*, **48**, 225 (1978).

29 A. Lekholm, "Epitaxial Growth of Silicon from Dichorosilane," *J. Electrochem. Soc.*, **120**, 1122 (1973).

30 G. Kosza, F. A. Kuznetsove, T. Kormany, and L. Nagy, "Optimization of Si Epitaxial Growth," *J. Cryst. Growth*, **52**, 207 (1981).

31 M. Ogirima, H. Saida, M. Suzuki, and M. Maki, "Low Pressure Silicon Epitaxy," *J. Electrochem. Soc.*, **124**, 903 (1977).

32 R. B. Herring, "Advances in Reduced Pressure Silicon Epitaxy," *Solid State Technol.*, **22**, 75 (1979).

33 S. P. Weeks, "Pattern Shift and Pattern Distortion during DVD Epitaxy on <111> and <100> Silicon," *Solid State Technol.*, **24**, 111 (1981).

34 C. M. Drum and C. A. Clark, "Anisotropy of Macrostep Motion and Pattern Edge Displacement on Silicon Near ‹100›," *J. Electrochem. Soc.*, **115**, 664 (1968). See also **117**, 1401 (1970).

35 P. H. Lee, M. T. Wauk, R. S. Rosler, and W. C. Benzing, "Epitaxial Pattern Shift Comparison in Vertical, Horizontal, and Cylindrical Reactor Geometries," *J. Electrochem. Soc.*, **124**, 1824 (1977).

36 K. V. Ravi, *Imperfection and Impurities in Semiconductor Silicon*, Wiley, New York, 1981.

37 C. M. Melliar-Smith, "Crystal Defects in Silicon Integrated Circuits—Their Cause and Effect," in *Treatise on Materials Science and Technology*, Academic, New York, 1977, Vol. 11, p. 47.

38 G. A. Rozgonyi, R. P. Deysher, and C. W. Pearce, "The Identification, Annihilation and Suppression of Nucleation Sites Responsible for Silicon Epitaxial Stacking Faults," *J. Electrochem. Soc.*, **123**, 1910 (1976).

39 L. E. Katz and D. W. Hill, "High Oxygen Czochralski Silicon Crystal Growth to Epitaxial Stacking Faults," *J. Electrochem. Soc.*, **125**, 1151 (1978).

40 C. W. Pearce and R. G. MacMahon, "Role of Metallic Contamination in the Formation of Saucer

Pit Defects in Epitaxial Silicon," *J. Vac. Sci. Technol.*, **14**, 40 (1977).

41 C. W. Pearce, "The Effect of Intrinsic Gettering on Dielectric Integrity," *Proceedings of the Conference on Low Temperature Processing for VLSI,* Electrochem. Soc., Pennington, New Jersey, 1986, p. 400.

42 S. K. Tung, "The Effects of Substrate Orientation on Epitaxial Growth," *J. Electrochem. Soc.*, **112**, 436 (1965).

43 B. J. Baliga, "Defect Control During Silicon Epitaxial Growth Using Dichlorosilane," *J. Electrochem. Soc.*, **129**, 1078 (1982).

44 J. Bloem and A. H. Goemans, "Slip in Silicon Epitaxy," *J. Appl. Phys.*, **43**, 1281 (1972).

45 Y. Sugita, M. Tamura, and K. Sugawara, "Misfit Dislocations in Bicrystals of Epitaxially Grown Silicon on Boron-Doped Silicon Substrates," *J. Appl. Phys.*, **40**, 3089 (1969).

46 J. Bloem, "Nucleation and Growth of Silicon by CVD," *J. Cryst. Growth*, **50**, 581 (1980).

47 H. M. Liaw et al., "Surface Morphology of Selective Epitaxial Silicon," *Semiconductor Silicon 1986*, Electrochem. Soc., Pennington, New Jersey, 1986, p. 260.

48 J. C. Sturm, C. M. Gronet, and J. F. Gibbons, "Minority Carrier Properties of Thin Epitaxial Silicon Films Fabricated by Limited Reaction Processing," *J. Appl. Phys.*, **59**, 4180 (1986).

49 T. J. Donahue and R. Reif, "Low Temperature Silicon Epitaxy Deposited by Very Low Pressure Chemical Vapor Deposition," *J. Electrochem. Soc.*, **133**, 1691, 1697, and 1701, (1986).

50 J. C. Bean, "Growth of Doped Silicon Layers by Molecular Beam Epitaxy," in F. F. Y. Wang, Ed., *Impurity Doping Processes*, North-Holland, Amsterdam, 1981.

51 H. Sugiura and M. Yamaguchi, "Growth of Dislocation-Free Silicon Films by Molecular Beam Epitaxy," *J. Vac. Sci. Technol.*, **19**, 157 (1981).

52 P. E. Luscher and D. M. Collins, "Design Considerations for Molecular Beam Epitaxy Systems," in B. R. Pamplin, Ed., *Molecular Beam Epitaxy*, Pergamon, London, 1981.

53 H. Sugiura and M. Yamaguchi, "Crystal Defects of Silicon Films Formed by Molecular Beam Epitaxy," *Jpn. J. Appl. Phys.*, **19**, 583 (1980).

54 U. Konig, H. Kibbel, and E. Kasper, "MBE: Growth and Sb Doping," *J. Vac. Sci. Technol.*, **16**, 985 (1979).

55 Y. Ota, "Si Molecular Beam Epitaxy (n on n^+) with Wide Range Doping Control," *J. Electrochem. Soc.*, **124**, 1795 (1977).

56 Y. Ota, "N-type Doping Techniques in Silicon Molecular Beam Epitaxy by Simultaneous Arsenic Ion Implantation and by Antimony Evaporation," *J. Electrochem. Soc.*, **126**, 1761 (1979).

57 M. S. Abrahams, C. J. Buiocchi, J. F. Corby, Jr., and G. W. Cullen, "Misfit Dislocation in Heteroepitaxial Si on Sapphire," *Appl. Phys. Lett.*, **28**, 275 (1976).

58 H. Schlotter, "Interface Properties of Sapphire and Spinel," *J. Vac. Sci. Technol.*, **13**, 29 (1976).

59 G. A. Sai-Halary, F. F. Fang, T. O. Sedgwick, and A. Segmuller, "Stress-Relieved Regrowth of Silicon on Sapphire by Laser Annealing," *Appl. Phys. Lett.*, **36**, 419 (1980).

60 K. Shibata, T. Inoue, and T. Takigawa, "Grain Growth of Polycrystalline Silicon Films on SiO_2 by CW Scanning Electron Beam Annealing," *Appl. Phys. Lett.*, **39**, 645 (1981).

61 D. K. Biegelsen, N. M. Johnson, D. J. Bartelink, and M. D. Moyer, "Laser-Induced Crystallization of Silicon Islands on Amorphous Substrates: Multilayer Structures," *Appl. Phys. Lett.*, **38**, 150 (1981).

62 R. A. Lemons and M. A. Bosch, "Periodic Motion of the Crystallization Front during Beam Annealing of Si Films," *Appl. Phys. Lett.*, **39**, 343 (1981).

63 B-Y. Tsuar, J. C. C. Fan, M. W. Geis, D. J. Silversmith, and R. W. Mountain, "Improved Techniques for Growth of Large Area Single Crystal Si Sheets over SiO_2 Using Lateral Epitaxy by Seeded Solidification," *Appl. Phys. Lett.*, **39**, 561 (1981).

64 B-Y. Tsuar, M. W. Geis, J. C. C. Fan, D. J. Silversmith, and R. W. Mountain, "N-Channel Deep-Depletion Metal-Oxide Semiconductor Transistors Fabricated in Zone-Melting-Recrystallized Polycrystalline Si Films in SiO_2," *Appl. Phys. Lett.*, **39**, 909 (1981).

65 P. H. Langer and C. W. Pearce, "Epitaxial Resistivity," *J. Test. Eval.*, **1**, 305 (1973).

66 W. G. Spitzer and M. Tannenbaum, "Interference Method for Measuring the Thickness of Epitaxial Grown Films," *J. Appl. Phys.*, **32**, 744, (1961).

67 K. Sato, Y. Ishikawa, and K. Sugaware, "Infrared Interference Spectra Observed in Silicon

Epitaxial Wafers," *Solid State Electron.*, **9**, 771 (1966).

68 Am. Soc. Test. Mater., ASTM Standard, F143, Part 43.

69 Am. Soc. Test. Mater., ASTM Standard, F110, Part 43.

70 I. Isda, H. Abe, and M. Kondo, "Impurity Profile Measurements of Thin Epitaxial Wafers by Multilayer Spreading Resistance Analysis," *J. Electrochem. Soc.*, **124**, 1118 (1977).

71 D. L. Rehrig and C. W. Pearce, "Production Mercury Probe Capacitance-Voltage Testing," *Semicond. Int.*, **3**, 151 (1980).

72 Am. Soc. Test. Mater., ASTM Standard, F374, Part 43.

73 M. G. Buehler, "Peripheral and Diffused Layer Effects on Doping Profiles," *IEEE Trans. Electron. Devices*, **ED-19**, 1171 (1972).

74 J. A. Copeland, "Diode Edge Effect on Doping-Profile Measurements," *IEEE Trans. Electron. Devices*, **ED-17**, 404 (1970).

75 P. J. Severin and G. J. Poodt, "Capacitance-Voltage Measurements with a Mercury-Silicon Diode," *J. Electrochem. Soc.*, **119**, 1384 (1972).

76 I. Amron, "Errors in Dopant Concentration Profiles Determined by Differential Capacitance Measurements," *Electrochem. Technol.*, **5**, 94 (1967).

77 D. G. Schimmel, "A Comparison of Chemical Etches for Revealing <100> Silicon Crystal Defects," *J. Electrochem. Soc.*, **123**, 734 (1976).

78 P. G. Wilson, "Recombination in P-I-N Diodes," *Solid State Electron.*, **10**, 145 (1967).

79 K. H. Zaininger and F. P. Herman, "The C-V Technique as an Analytical Tool," *Solid State Technol.*, **13**, 46 (1970).

80 V. S. Ban and E. P. Miller, "A New Reactor for Silicon Epitaxy," *Proceedings of the 7th International Conference on Chemical Vapor Deposition 1979*, Electrochem. Soc., Pennington, New Jersey, 1979, p. 102.

CHAPTER
3

OXIDATION

L. E. KATZ

3.1 INTRODUCTION

The oxidation of silicon is necessary throughout the modern integrated circuit fabrication process. Producing high-quality ICs requires not only an understanding of the basic oxidation mechanism, but the ability to form a high-quality oxide in a controlled and repeatable manner. In addition, to ensure the reliability of the ICs, the electrical properties of the oxide must be understood.

Silicon dioxide has several uses: to serve as a mask against implant or diffusion of dopant into silicon, to provide surface passivation, to isolate one device from another (dielectric isolation as opposed to junction isolation), to act as a component in MOS structures, and to provide electrical isolation of multi-level metallization systems. Several techniques for forming the oxide layers have been developed, such as thermal oxidation (including rapid thermal techniques), wet anodization, vapor-phase technique [chemical vapor deposition (CVD)], and plasma anodization or oxidation. When the interface between the oxide and the silicon requires a low-charge density level, thermal oxidation has been the preferred technique. However, since the masking oxide is generally removed, this consideration is not as important in the case of masking against diffusion of dopant into silicon. Obviously, when the oxide layer is required on top of a metal layer, as in the case of a multilevel metallization structure, the vapor-phase technique is uniquely suited. This chapter concentrates on thermal silicon oxidation, because it is the principal technique used in IC processing.

In this chapter we describe the oxidation process in order to provide a foundation for the understanding of the kinetics of growth and interface properties. Section 3.2 examines the oxidation model and its fit to experimental data; it also describes the effects of orientation, dopant concentration, addition of HCl to the ambient, and surface damage on the kinetics of oxidation. Section 3.3 concentrates on thin oxide growth techniques, kinetics, and electrical properties. Section 3.4 describes standard thermal oxidation techniques, such as dry, wet, and HCl dry, as well as the less familiar high-pressure and plasma oxidation techniques. Section 3.4 also describes the cleaning processes needed to remove surface contamination prior to oxidation. Section 3.5 covers the characteristics and properties of oxides, with emphasis on oxide masking, oxide charges, and stresses in thermal oxides. Sections 3.6 and 3.7 examine the redistribution of dopants at the Si–SiO₂ interface during thermal oxidation of single-crystal silicon and oxidation of polysilicon, respectively. Section 3.8 considers oxidation-induced stacking faults and oxide isolation defects. A summary and a discussion of future trends are presented in the last section, Section 3.9.

3.2 GROWTH MECHANISM AND KINETICS

Because a silicon surface has a high affinity for oxygen, an oxide layer rapidly forms when silicon is exposed to an oxidizing ambient. The chemical reactions describing the thermal oxidation of silicon[1] in oxygen or water vapor are given in Eqs. 1 and 2, respectively.

$$Si(solid) + O_2 \rightarrow SiO_2(solid) \tag{1}$$

$$Si(solid) + 2H_2O \rightarrow SiO_2(solid) + 2H_2 \tag{2}$$

The basic process involves shared valence electrons between silicon and oxygen; the silicon–oxygen bond structure is covalent. During the course of the oxidation process, the Si–SiO₂ interface moves into the silicon. Its volume expands, however, so that the external SiO₂ surface is not coplanar with the original silicon surface. Based on the densities and molecular weights of Si and SiO₂, we can show that for growth of an oxide of thickness d, a layer of silicon with a thickness of $0.44d$ is consumed (Fig. 1).

The framework of a model to describe silicon oxidation has been created. Radioactive tracer,[1] marker,[2] and infrared isotope shift[3] experiments have established that oxidation proceeds by the diffusion of the oxidizing species through the

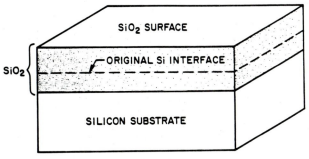

FIGURE 1
Growth of SiO₂.

oxide to the Si–SiO$_2$ interface, where the oxidation reaction occurs. Uncertainties exist, however, as evidenced by controversies in the literature as to whether charged or neutral species are transported through the oxide and as to the details of the reaction at the Si–SiO$_2$ interface.

3.2.1 Silicon Oxidation Model

Deal and Grove's model describes the kinetics of silicon oxidation.[4] The model is generally valid for temperatures between 700 and 1300°C, partial pressures between 0.2 and 1.0 atm (perhaps higher), and oxide thicknesses between 300 and 20,000 Å for oxygen and water ambients. Figure 2 shows the silicon substrate covered by an oxide layer that is in contact with the gas phase. The oxidizing species (1) are transported from the bulk of the gas phase to the gas–oxide interface with flux F_1 (the flux is the number of atoms or molecules crossing a unit area in a unit time), (2) are transported across the existing oxide toward the silicon with flux F_2, and (3) react at the Si–SiO$_2$ interface with the silicon with flux F_3.

For steady state, $F_1 = F_2 = F_3$. The gas-phase flux F_1 can be linearly approximated by assuming that the flux of oxidant from the bulk of the gas phase to the gas–oxide interface is proportional to the difference between the oxidant concentration in the bulk of the gas C_G and the oxidant concentration adjacent to the oxide surface C_S:

$$F_1 = h_G(C_G - C_S) \tag{3}$$

where h_G is the gas-phase mass-transfer coefficient.

To relate the equilibrium oxidizing species concentration in the oxide to that

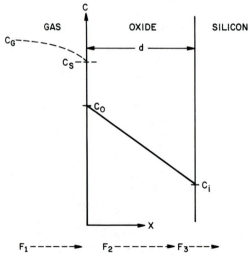

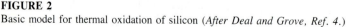

FIGURE 2
Basic model for thermal oxidation of silicon (*After Deal and Grove, Ref. 4.*)

in the gas phase, we invoke Henry's law,

$$C_o = Hp_S \tag{4}$$

and

$$C^* = Hp_G \tag{5}$$

where C_o is the equilibrium concentration in the oxide at the outer surface, C^* is the equilibrium bulk concentration in the oxide, p_S is the partial pressure in the gas adjacent to the oxide surface, p_G is the partial pressure in the bulk of the gas, and H is Henry's law constant. Using Henry's law along with the ideal gas law[5]

$$F_1 = h(C^* - C_o) \tag{6}$$

where h is the gas-phase mass-transfer coefficient in terms of concentration in the solid, given by $h = h_G/HkT$. Oxidation is a nonequilibrium process with the driving force being the deviation of concentration from equilibrium.[6] Henry's law is valid only in the absence of dissociation effects at the gas–oxide interface. This implies that the species moving through the oxide is molecular.

The flux of this oxidizing species across the oxide is taken to follow Fick's law at any point d in the oxide layer. Following the steady-state assumption, F_2 must be the same at any point within the oxide, resulting in

$$F_2 = \frac{D(C_o - C_i)}{d_o} \tag{7}$$

where D is the diffusion coefficient, C_i is the oxidizing species concentration in the oxide adjacent to the oxide–silicon interface, and d_o is the oxide thickness.

Assuming that the flux corresponding to the Si–SiO$_2$ interface reaction is proportional to C_i,

$$F_3 = k_s C_i \tag{8}$$

where k_s is the rate constant of chemical surface reaction for silicon oxidation.

After setting $F_1 = F_2 = F_3$, as dictated by steady-state conditions, and solving simultaneous equations, expressions for C_i and C_o can be obtained. Limiting cases arise when the diffusivity is either very small or very large. When the diffusivity is very small, $C_i \rightarrow 0$ and $C_o \rightarrow C^*$. This case is called the diffusion-controlled case. It results from the flux of oxidant through the oxide being small (due to D being small) compared to the flux corresponding to the Si–SiO$_2$ interface reaction. Hence, the oxidation rate depends on the supply of oxidant to the interface, as opposed to the reaction at the interface. In the second limiting case, where D is large, $C_i = C_o$. This is called the reaction-controlled case, because an abundant supply of oxidant is provided at the Si–SiO$_2$ interface, and the oxidation rate is controlled by the reaction rate constant k_s and by C_i (which equals C_o).

To calculate the rate of oxide growth, we define N_1 as the number of oxidant molecules incorporated into a unit volume of the oxide layer. Since the oxide has 2.2×10^{22} SiO$_2$ molecules/cm^3 and one O$_2$ molecule is incorporated into each

SiO_2 molecule, N_1 equals 2.2×10^{22} cm^{-3} for dry oxygen. The number for water-vapor oxidation is twice as big because two H_2O molecules are incorporated into each SiO_2 molecule. Combining various equations and assuming that an oxide may be present initially from a previous processing step or may grow before the assumptions in the model are valid, that is, $d_o = d_i$ at $t = 0$, the following equation can be generated:

$$d_o^2 + Ad_o = B(t + \tau) \tag{9}$$

where

$$A = 2D\left[\frac{1}{k_s} + \frac{1}{h}\right] \tag{9a}$$

$$B = \frac{2DC^*}{N_1} \tag{9b}$$

$$\tau = \frac{d_i^2 + Ad_i}{B} \tag{9c}$$

The quantity τ represents a shift in the time coordinate to account for the presence of the initial oxide layer d_i. Equation 9 is the well-known, mixed linear-parabolic relationship.

Solving Eq. 9 for d_o as a function of time gives

$$\frac{d_o}{A/2} = \left[1 + \frac{t + \tau}{A^2/4B}\right]^{1/2} - 1 \tag{10}$$

One limiting case occurs for long oxidation times when $t \gg \tau$ and $t \gg A^2/4B$.

$$d_o^2 = Bt \tag{11}$$

Equation 11 is the parabolic law, where B is the parabolic rate constant. The other limiting case occurs for short oxidation times when $(t + \tau) \ll A^2/4B$.

$$d_o = \frac{B}{A}(t + \tau) \tag{12}$$

Equation 12 is the linear law, where B/A is the linear rate constant. Equations 11 and 12 are the diffusion-controlled and reaction-controlled cases, respectively.

Problem

Show that results obtained for wet oxidation at 1000°C using Eq. 10 are similar to results obtained using Eq. 12 for short times and Eq. 11 for long times. Use values from Table 1 and let $\tau = 0$.

Solution

For 1000°C, $A = 0.226$ μm and $B = 0.287$ μm^2/h.

Equation 10 gives $d_o = 0.0121$ μm for $t = 0.01$ h and $d_o = 5.24$ μm for $t = 100$ h.

Equation 12 gives $d_o = 0.0127$ μm for $t = 0.01$ h.
Equation 11 gives $d_o = 5.35$ μm for $t = 100$ h.

Problem
Show that the equation $d_o^2 + Ad_o = B(t + \tau)$ can be used graphically to obtain the rate constants. Let $\tau =$ o.

Solution

$$d_o^2 + Ad_o = Bt$$

$$d_o = \frac{Bt}{d_o} - A$$

A chart is made of experimentally measured values of d_o versus t. Plotting d_o versus t/d_o gives a slope $= B$ and an intercept $= -A$.

3.2.2 Experimental Fit

This section compares Deal and Grove's model to their experimental measurements. Deal and Grove used <111> oriented, lightly boron-doped silicon wafers that were chemically cleaned prior to oxidation in purified dry oxygen (less than 5 ppm water content) or in wet oxygen (the partial pressure of water was 640 Torr). For wet oxygen oxidation they found that $d_i = 0$ at $t = 0$ by plotting oxide thickness versus oxidation time. Algebraically manipulating Eq. 9 and using the plot of wet oxygen data, they obtained the rate constants graphically. Table 1 lists the values of these rate constants for wet oxidation of silicon.[4] The absolute value of A increases with decreasing temperature, while the parabolic rate constant B decreases with decreasing temperature (Figs. 3a and b).

For dry O_2 a plot of oxide thickness versus oxidation time does not extrapolate to zero initial thickness, but instead to a value that equals about 250 Å for data spanning a range of 700 to 1200°C. The faster initial oxidation rate during the initial phase of oxidation implies a different mechanism in this region. This is discussed further in Sections 3.2.3 and 3.3.2. Thus, use of Eq. 9 for dry oxidation requires a value for τ that can be generated graphically by extrapolating the

TABLE 1
Rate constants for wet oxidation of silicon

Oxidation temperature (°C)	A (μm)	Parabolic rate constant B (μm^2/h)	Linear rate constant B/A (μm/h)	τ(h)
1200	0.05	0.720	14.40	0
1100	0.11	0.510	4.64	0
1000	0.226	0.287	1.27	0
920	0.50	0.203	0.406	0

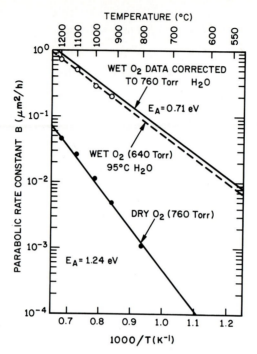

FIGURE 3
(a) The effect of temperature on the parabolic rate constant for dry and wet oxygen.

linear region back to the time axis. Problems arise at higher temperatures where the linear-parabolic or parabolic ranges are encountered, in which case the value of τ as defined in Eq. 9c must be used. Table 2 lists the values of rate constants for dry oxidation of silicon.[4]

Examination of Eq. 9b reveals that B is expected to be proportional to C^*, which, according to Henry's law, is proportional to the partial pressure of the oxidization species. However, A should be independent of the partial pressure.

TABLE 2
Rate constants for dry oxidation of silicon

Oxidation temperature (°C)	A (μm)	Parabolic rate constant B (μm^2/h)	Linear rate constant B/A (μm/h)	τ(h)
1200	0.040	0.045	1.12	0.027
1100	0.090	0.027	0.30	0.076
1000	0.165	0.0117	0.071	0.37
920	0.235	0.0049	0.0208	1.40
800	0.370	0.0011	0.0030	9.0
700	—	—	0.00026	81.0

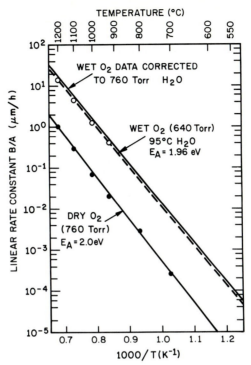

FIGURE 3
(b) The effect of temperature on the linear rate constant. (*After Deal and Grove, Ref. 4.*)

This has been confirmed experimentally for both wet and dry oxidations[4,7] in the temperature range between 1000 and 1200°C and between 0.1 and 1 atm. The pressure independence of A means that the linear rate constant B/A has the same linear pressure dependence as that of B.

Figure 3a shows the effect of temperature on the parabolic rate constant B for dry oxygen at 760 Torr, wet oxygen at 640 Torr, and for wet oxygen corrected to 760 Torr using the linear pressure dependence.[4] As might be expected from Eq. 9, the temperature dependence of B is similar to that of the diffusion coefficient D; that is, B increases exponentially with temperature. For dry oxygen the activation energy for B is 1.24 eV, which is comparable to the value of 1.17 eV for the diffusivity of oxygen through fused silica (similar in structure to thermal SiO_2). The wet oxygen activation energy (0.71 eV) also compares favorably with the activation energy for the diffusivity of water in fused silicon (0.80 eV).

Figure 3b shows the temperature dependence of the linear rate constant B/A for dry oxygen, wet oxygen at 640 Torr, and wet oxygen normalized to 760 Torr. Once again an exponential dependence is observed with activation energies 1.96 and 2.0 eV for wet and dry oxidation, respectively, although recent data show a different activation energy for oxidation below 900°C. Deal and Grove[4] show that these values reflect the temperature dependence of the interface reaction- rate constant, k_s. As stated previously, in the linear range the oxidation is reaction controlled. Similar values were obtained for the linear rate constants for both dry and wet oxidations, indicating a similar reaction or surface control mechanism. Interestingly, the above values are comparable to the 1.83 eV required to break a Si—Si bond.

The equilibrium concentration C^* of the oxidizing species in SiO_2 can be calculated from Eq. 9b by using appropriate values for B, D, and N_1. Table 3 gives two examples.[4]

Even though the diffusivity of water in SiO_2 is lower than that of oxygen, the parabolic rate constant B is substantially larger for wet oxidation than for dry.[4] This is the major reason that the parabolic oxidation rate in steam is faster than in dry oxygen; the flux of oxidant, and hence B, is proportional to C^*, which is approximately three orders of magnitude greater for water than for oxygen (see Table 3). Furthermore, since the linear rate constant B/A also is related to B and hence C^*, we can also attribute the faster linear oxidation rate for wet oxidation to this mechanism.

Deal and Grove's simple model (Eqs. 9 and 10) for thermal silicon oxidation provides excellent agreement with various normalized experimental data for both wet and dry oxidations.[4] The only exception is for SiO_2 films less than about 300 Å thick that are grown in dry oxygen. In this case, an anomalously high oxidation rate is observed with respect to the model.

3.2.3   Diffusing Species and Interface Considerations

The excellent agreement between the model and experimental observations supports the use of Henry's law.[4] This implies the lack of dissociative effects at the gas–SiO_2 interface, indicating the species diffusing in the oxide is molecular for both oxygen (dry) and steam (water-vapor) oxidations. Other results indicate that the oxidant is molecular for both water and oxygen oxidations, since good agreement is obtained between the calculated (for fused silica) and measured oxidation rates (for oxidation of silicon) with respect to absolute rate and pressure, and with respect to temperature dependence.

A proposed modification to the Deal–Grove model provides an excellent fit to the experimental data, including the thin, dry oxidation regime in which the Deal–Grove model breaks down.[8] The physical basis of the proposed model is that while diffusion through the oxide is still by molecular oxygen, the oxidation

TABLE 3
C^* values in SiO_2 at 1000°C

Species	C^* (cm^{-3})
O_2	5.2×10^{16}
H_2O	3.0×10^{19}

of silicon occurs by the reaction of a small concentration of atomic oxygen dissociated near the Si–SiO$_2$ interface.

As stated earlier, the question of whether the oxidizing species is charged or neutral is still a subject of controversy. While the above discussion favors diffusion of a molecular species, supportive evidence for a charged species arises from experiments showing that an applied electric field can influence the oxidation rate, either accelerating it or retarding it, depending on whether the silicon is positive or negative with respect to the oxide–gas interface.[2] Recent work, using an in situ corona discharge ion current source directed to the front of an oxidizing silicon wafer, studied the effect of the electric field induced in the oxide on the wafer back.[9] The results point to neutral O$_2$ as the dominant oxidant. These results are opposite to those favoring the charged species, in which results may have been influenced by the Pt electrodes forming PtSi, thereby disturbing the kinetics.[2] Other work based on studies of electrical conduction at elevated temperatures concludes that the species responsible for ionic conduction is doubly negative interstitial oxygen ions (O$_I^{2-}$).

We now shift our discussion from the unresolved question of the nature of the diffusant to the Si–SiO$_2$ interface. The structure and oxidation mechanism at the interface is particularly important since what occurs here from an atomistic point of view can influence not only the oxidation kinetics but also allied areas of interest, such as diffusion. Both the interface structure and its oxidation mechanisms are complicated, and are a continuing source of discussion in the literature.

The controversies as to whether charged or neutral species are transported through the oxide have been discussed using a different model,[10] which is based on a large molecular volume difference between Si and SiO$_2$. This difference must be accommodated to allow a newly formed SiO$_2$ molecule to fit into the normal SiO$_2$ structure. This leads to the proposal of an interface transition region, which consists of a network of extra half-planes terminating at the Si side of the Si–SiO$_2$ interface. Movement of this interface requires a supply of vacancies from the silicon to the interface, the movement of Si interstitials from the interface into the bulk Si, or free volume influx from the SiO$_2$ (i.e., viscous flow).[10] An additional proposal relating to the interface suggests that silicon is transformed to α-cristobalite plus interstitial Si ions.[10] Subsequent oxidation of the interstitials produces lattice distortion and transformation to vitreous silica. Hence, the crystalline SiO$_2$ phase exists only as a buffer between the Si and vitreous SiO$_2$. The proposed interface mechanisms are consistent with qualitative explanations related to oxidation-enhanced diffusion, stacking fault formation, interface charge, and oxidation velocity.

Additional mechanisms have been proposed to explain point-defect-related interface phenomena. The presence of doubly negative interstitial oxygen ions (O$_I^{2-}$) was discussed previously. Such ions at the Si–SiO$_2$ interface may react with silicon and displace it to an interstitial position in the lattice to form Si$_I$—O, which can combine to form SiO$_2$. A silicon interstitial flux can occur if the Si$_I$—O pair dissociates before forming SiO$_2$.[11] Such an incomplete oxidation occurs for one out of every thousand silicon atoms.[12] Although the interface reaction generally goes to completion, even a small flux of silicon interstitials into the silicon can have a large effect on defect formation or diffusion. The case of O_I^{2-} reacting with vacancies, as supplied from the silicon substrate, could lead to a vacancy flux, which can affect dopant diffusion.

Much of the recent work on interface mechanisms has been directed toward thin oxide growth. Proposed mechanisms include (1) control of oxide growth by the area density of silicon atoms initially, but by mechanical effects (stress) of the SiO_2 on Si when the oxide grows to become three-dimensional; (2) enhancement of the oxidation rate for dry O_2 by micropores in the SiO_2 with diameters of ~ 10 Å through which oxygen molecules can rapidly diffuse[13] (micropores presumably filled with OH radicals for wet oxidation, precluding an initial rapid growth); and (3) accumulation of fixed positive charge during oxidation, which reduces the concentration of holes at the Si–SiO_2 interface, thereby reducing the density of broken Si—Si bonds.[14] These broken bonds are believed to be a controlling factor in the initial stages of oxide growth.[15]

3.2.4   Orientation Dependence of Oxidation Rates

Experiments have indicated that the oxidation kinetics are a function of the crystallographic orientation of the silicon surface.[16] This relationship is attributed to the orientation dependence of k_s (Eqs. 8 and 9a) and manifests itself in an orientation-dependent linear rate constant. The linear rate constant is related to the interface reaction kinetics and depends on the rate at which silicon atoms are incorporated into the oxide. This depends on the silicon surface atom concentration, which is orientation dependent. As might be expected, the parabolic rate constant B is independent of silicon surface orientation, since B is diffusion limited. Figure 4 shows oxide thickness as a function of oxidation time in water at 640 Torr for both <111> and <100> oriented silicon.[17] Table 4 gives rate constants obtained from this data.[17] These data, along with those for dry oxygen, yield linear rate constants for <111> silicon that are an average of 1.68 times those for <100> silicon at corresponding temperatures.

A model has been presented[16] to explain how the linear oxidation rate of the silicon in steam depends on the orientation of the silicon surface. According to this model, a direct reaction occurs between a water molecule in the silica and a silicon–silicon bond at the Si–SiO_2 interface. At this interface all the silicon atoms are partially bonded to silicon atoms below and to oxygen atoms above. The orientation dependence of the oxidation rate comes from terms representing the activation energy for oxidation and the concentration of reaction sites. This concentration depends on the concentration of silicon–silicon bonds available for reaction at a given time. The bond is directional, so its availability depends on its angle relative to the surface plane, its position with respect to adjacent atoms, and the water molecule size, which must be such that it can screen adjacent bonds from other water molecules when reacting with some angled silicon–silicon bonds.[6] These and other geometric effects are called steric hindrances, and they result in

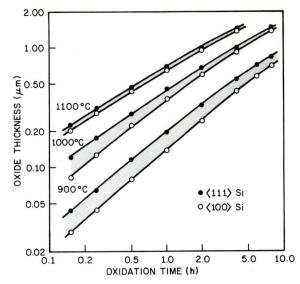

FIGURE 4
Oxide thickness versus oxidation time for silicon in H_2O at 640 Torr. (*After Deal, Ref. 17.*)

the linear oxidation rate's being orientation dependent. Table 5 lists calculated properties of three silicon planes.[16] The orientation dependence is related to the available bond density N and the orientation dependence of the activation energy.

As might be expected, steric hindrance results in higher activation energy. Experimental data have been analyzed to determine the apparent activation energy, which is the sum of two components: a term related to the enthalpy of solu-

TABLE 4
Rate constants for silicon oxidation in H_2O (640 Torr)

Oxidation temperature (°C)	Orien- tation	A (μm)	Parabolic rate constant B ($\mu m^2/h$)	Linear rate constant B/A (μm/h)	B/A ratio <111>/<100>
900	<100>	0.95	0.143	0.150	
	<111>	0.60	0.151	0.252	1.68
950	<100>	0.74	0.231	0.311	
	<111>	0.44	0.231	0.524	1.68
1000	<100>	0.48	0.314	0.664	
	<111>	0.27	0.314	1.163	1.75
1050	<100>	0.295	0.413	1.400	
	<111>	0.18	0.415	2.307	1.65
1100	<100>	0.175	0.521	2.977	
	<111>	0.105	0.517	4.926	1.65

Average 1.68

TABLE 5
Calculated properties of silicon crystal planes

Orientation	Area of unit cell (cm^2)	Si atoms in area	Si bonds in area	Bonds available	Bonds $\times$ 10^{14} cm^{-2}	Available bonds, N ($\times$ 10^{14} cm^{-2})	N relative to <110>
<110>	$\sqrt{2}\,a^2$	4	8	4	19.18	9.59	1.000
<111>	$1/2\,\sqrt{3}\,a^2$	2	4	3	15.68	11.76	1.227
<100>	a^2	2	4	2	13.55	6.77	0.707

tion of water in the silica films, and the orientation-dependent term related to the activation energy of oxidation. Table 6 lists the values of some apparent activation energies.[16] The interaction between the available bond density and the activation energy determines the orientation dependence of the linear oxidation rates. Experiments show that the oxidation rate v in steam is ordered in the following manner:[16]

$$v_{110} > v_{111} > v_{100}$$

Additional measurements in steam show the following oxidation rate sequence:

$$v_{111} > v_{110} > v_{100}$$

However, this set of measurements was made at a higher temperature than the former set.

For dry oxidation,

$$v_{110} > v_{111} > v_{100}$$

based on data at 800°C and thicknesses up to 150 Å. Above this thickness the order changes to

$$v_{111} > v_{110} > v_{100}$$

The model previously discussed, in which the growth rate is initially determined by the area density of atoms and subsequently by mechanical effects, can predict the relationship between <111> and <110> orientations, but is not valid for <100> silicon.[13] Steric effects may play a role.

TABLE 6
Apparent activation energies

Orientation	Activation energy (eV)
<110>	1.23 ± 0.02
<311>	1.30 ± 0.03
<111>	1.29 ± 0.03

3.2.5 Effect of Impurities and Damage on the Oxidation Rate

Because wet oxidation occurs at a substantially greater rate than dry oxidation, any unintentional moisture accelerates the dry oxidation. In fact, both the linear and parabolic oxidation rates are sensitive to the presence of water and other impurities. The effects of some of these impurities are discussed in this section.

Water. Experiments were done to study the effect of intentionally adding 15 ppm water vapor to a process that normally used less than 1 ppm water.[17] A significant acceleration in the oxidation rate was observed. For example, an 800°C oxidation of <100> silicon for 700 minutes grew an oxide approximately 300 Å thick with less than 1 ppm moisture and an oxide approximately 370 Å thick with 25 ppm moisture (23% thicker). In these experiments the oxygen was from a liquid source and the oxidation chamber was a double-wall, fused-silica tube with N_2 flowing between the walls. A precombustor and cold trap were used to achieve the less than 1 ppm moisture level.

Sodium. High concentrations of sodium influence the oxidation rate by changing the bond structure in the oxide, thereby enhancing the diffusion and concentration of the oxygen molecules in the oxide.[6]

Group III and V elements. The common dopant elements in this group, when present in silicon at high concentration levels, can enhance the oxidation behavior. The dopant impurities are redistributed at the growing Si–SiO_2 interface.[18] This effect is discussed in greater detail in Section 3.6, but we consider it from a mechanism standpoint here. The effect results in a discontinuous concentration profile at the interface; that is, the dopant segregates either into the silicon or into the oxide. The redistribution of the impurity at the interface influences the oxidation behavior. If the dopant segregates into the oxide and remains there (which is the case for boron in an oxidizing ambient), the bond structure in the silica weakens. This weakened structure permits an increased incorporation and diffusivity of the oxidizing species through the oxide, thus enhancing the oxidation rate. Impurities that segregate into the oxide but then diffuse rapidly through it (such as aluminum, gallium, and indium) have no effect on the oxidation kinetics. Figure 5 shows oxidation rate curves for various concentrations of boron in silicon for wet oxygen.[19] From the above discussion it is not surprising that an enhancement in the oxidation kinetics is observed where diffusion control predominates. For oxidation of phosphorus-doped silicon in wet oxygen,[19] a concentration dependence is observed only at the lower temperatures, where the surface reaction becomes important (Fig. 6). This dependence may be the result of phosphorus being segregated into the silicon. Figure 7 shows the oxidation rate constants for dry oxygen as a function of phosphorus-doping level.[20] Here B/A increases substantially at high concentration, thus reflecting the reaction-rate control, whereas B is relatively independent of concentration, thus reflecting the diffusion-limited control.

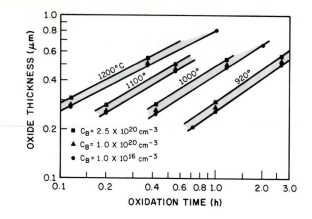

FIGURE 5
Oxidation of boron-doped silicon in wet oxygen (95°C H$_2$O) as a function of temperature and concentration. (*After Deal and Sklar, Ref. 19.*)

Problem

An oxide thickness difference of 500 Å is observed between boron- and phosphorus-doped regions in a silicon sample wet oxidized at 920°C for 60 minutes, the phosphorus region being thicker. The boron concentration is 1.0×10^{16} cm^{-3}. What is the phosphorus concentration?

Solution

From Fig. 5, the oxide grown for a boron concentration of 1.0×10^{16} cm^{-3} at 920°C for 60 minutes (wet) is ~2700 Å. Examination of Fig. 6 for an oxide thickness of 3200 Å, also for 920°C and 60 minutes, gives a phosphorus concentration ~3.7×10^{19} cm^{-3}.

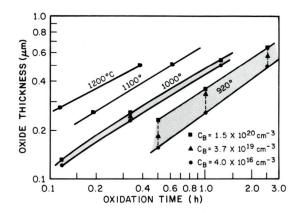

FIGURE 6
Oxidation of phosphorus-doped silicon in wet oxygen (95°C H$_2$O) as a function of temperature and concentration. (*After Deal and Sklar, Ref. 19.*)

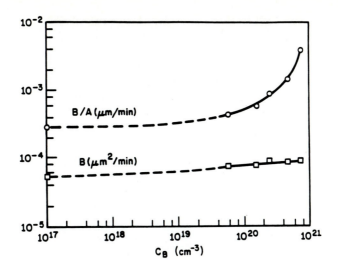

FIGURE 7
Oxidation rate constants for dry oxygen as a function of phosphorus-doping level at 900°C. (*After Ho et al., Ref. 20.*)

The oxidizing interface is a complicated and not fully understood region. Its high concentration of dopant provides further complications. A theoretical model has been developed to explain the concentration-enhanced oxidation.[21] According to the model, the high doping levels shift the position of the Fermi level, which results in enhanced vacancy concentrations. These point defects may provide reaction sites for the chemical reaction converting Si to SiO_2, thereby increasing the reaction rate.

Figure 6 shows the large increase in oxide thickness obtained for oxidation of heavily doped phosphorus in wet oxygen at lower temperatures. A dramatic example of this effect is seen in Fig. 8, which shows a bulk phosphorous-doped silicon wafer ($\sim 7 \times 10^{19}$ / cm^3) after oxidation at 750°C in steam at 20 atm pressure to accelerate the kinetics. The wafer was not preferentially etched. Phosphorus dopant variations (striations), incorporated into the Czochralski crystal during solidification (see Chapter 1), appear as color variations representing oxide thickness variations. These striations clearly correspond to the concentration-enhanced oxidation of the more heavily phosphorous-doped regions in the crystal.

Halogen. Certain halogen species are intentionally introduced into the oxidation ambient to improve both the oxide and the underlying silicon properties. Oxide improvements include a reduction in sodium ion contamination, an increased dielectric breakdown strength, and a reduced interface trap density. At or near the Si–SiO_2 interface, chlorine is instrumental in converting certain impurities in the silicon to volatile chlorides, resulting in a gettering effect. A reduction in oxidation-induced stacking faults is also observed. Chlorine is typically intro-

FIGURE 8
Concentration-enhanced oxidation, showing dopant variations in a heavily doped phosphorus substrate.

duced into dry oxygen ambients in the form of chlorine gas, anhydrous HCl, trichloroethylene, or trichloroethane.

Experimental results for dry O_2–HCl mixtures show that HCl additions increase the oxidation rate.[22] Typical HCl additions range from 1 to 5%. The parabolic rate constant B increases linearly with HCl additions above 1%. At 1000 and 1100°C large increases in B are initially observed. The linear rate constant B/A shows an initial increase when 1% HCl is added, but no further increase with subsequent HCl additions. The mechanisms associated with this enhanced growth rate are not fully understood. However, the generation of water upon adding HCl to dry oxygen does not fully account for the increased oxidation rate, since a similar increase occurs when chlorine is added (even though no water is generated in that case).[22]

For thermal oxidation of silicon in H_2O, adding 5 volume % HCl decreases the silicon oxidation rate by about 5%, apparently because of the reduced H_2O vapor pressure.[17] Although it is not common practice to add HCl to H_2O ambients, such addition appears to reduce impurity contamination from the oxidation system.

Thermal oxidation of silicon at 1100°C with additions of up to 1% trichloroethylene (TCE) yields oxidation rates comparable to similar concentrations of chlorine. At lower temperatures the values for O_2/TCE are larger. The mechanisms involved are complicated and not fully understood. Very low concentration additions of vapor from a compound with both chlorine and fluorine have resulted in substantial increase in the oxidation rate. For example, 0.055 volume % $C_2H_3Cl_2F$ in dry O_2 enhances the oxide thickness by a factor of ~5 over dry O_2 and by a factor of ~2–3 over 3 volume % Cl_2 or 10 volume % HCl for comparable time at 1000°C.[23]

Finally, a word of caution: Care must be taken in handling and using the halogens mentioned since the system's metallic parts and exhaust can corrode. Additionally, high concentrations of halogens at high temperatures can pit the silicon surface.

Effect of damage on oxidation rate. Determining how damage to the silicon affects the oxidation rate is not easy. To study these effects, the silicon is usually intentionally damaged by ion implantation of a nonelectrically active species (Si or Ar), or of a group III or V dopant. Separating damage effects from dopant effects is also difficult.

Enhanced thermal oxidation of implanted silicon as a function of ion species and concentration has been studied.[24] Implanted into <100> silicon were 80 keV arsenic, 60 keV boron, 106 keV antimony, and 48 keV argon with ion doses ranging from 4×10^{14} to 1×10^{16} cm^{-2}. For wet oxidation at 900°C the maximum enhancement of the oxidation rate was a factor of 1.1 for boron, 1.3 for argon, 3.5 for antimony, and 7.5 for arsenic. The higher enhancements occurred for the higher doses. The enhancement for argon is attributed to the damage effect; for the other cases the presence of the impurity atoms certainly contributes to the enhancement. Another study found a retardation effect for oxidation, following implantation of Ge, Si, and Ga into silicon. It also found an enhancement for B, Al, P, As, and Sb. Additionally, when thin dry oxides are grown after reactive ion etching at high power levels, thinner oxides result when compared with the non-plasma-etched case.[25]

3.3 THIN OXIDES

As design rules shrink to the submicron range for MOS VLSI, it becomes increasingly vital to grow oxides on crystalline silicon in the 50 to 200 Å range in a repeatable manner. These oxides must exhibit good electrical characteristics and provide long-term reliability. As an example, the dielectric material for MOS devices can be thin thermal oxide or a composite structure of thermal oxide (<100 Å) and silicon nitride. This dielectric is an active component of the storage capacitor in dynamic RAMs, and its thickness determines the amount of charge that can be stored.

In this section we will discuss growth techniques, kinetics, and thin oxide properties.

3.3.1 Growth Techniques

From a practical point of view, thin oxide growth must be slow enough to obtain uniformity and reproducibility. Various growth techniques include dry oxidation, dry oxidation with HCl, sequential oxidations using different temperatures and ambients, wet oxidation, reduced pressure techniques, and high pressure/low temperature oxidations. The oxidation rate will, of course, be lower at lower temperatures and at reduced pressures. Ultra-thin oxides (<50 Å) have been produced using hot nitric acid, boiling water, and air at room temperatures. New thin oxide growth techniques have recently been developed. One example is rapid thermal oxidation performed in a controlled oxygen ambient with heating provided by tungsten–halogen lamps.[26] Oxides of 40–130 Å (1150°C, 5 to 30 seconds) have been grown with good uniformity and breakdown fields of 13.8 MV/cm for 100 Å oxides. Another more complicated technique utilizes ultraviolet pulsed laser excitation in an oxygen environment.

3.3.2   Kinetics of Thin Oxide Growth

As noted earlier, the structure of the oxide very close to the silicon–oxide interface and the oxidation process itself both involve uncertainties. Our understanding is further complicated by the observation of an initial rapid oxidation for the case of dry oxide growth, which causes the linear portion of the oxide growth versus time curve to extrapolate to an initial thickness of about 200 Å.

In discussing the kinetics and properties obtained, it should be emphasized that thin oxide growth is influenced by the cleaning techniques used and the purity of the gases used (especially moisture content). Figure 9 shows an example of thin oxide growth on lightly doped substrates versus oxidation time in dry oxygen.[27] The wafers were chemically cleaned, dried in nitrogen, and loaded into the furnace in an argon ambient. Native oxide thickness and subsequent oxide growth when the ambient was switched to dry O_2 were measured using in situ ellipsometry. The enhancement in oxidation rate was found to decay exponentially with thickness. Additionally, the extent of thickness enhancement

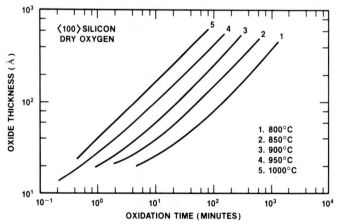

FIGURE 9
Oxidation of lightly doped <100> silicon in dry oxygen in the 800–1000°C range. (*After Massoud et al., Ref. 27.*)

is not a strong function of surface orientation, doping level, or oxygen partial pressure. The data in Figure 9 demonstrate that a set of time and temperature conditions can be chosen to grow thin oxides compatible with reasonable control and throughput.

Reduced pressure oxidation offers an attractive way of growing thin oxides in a controlled manner.[27, 28] Oxides between 30 and 140 Å thick have been grown at 900 to 1000°C using oxygen at a pressure of 0.25 to 2.0 Torr.[28] The observed kinetics are parabolic, and the rate constants agree with values extrapolated from atmospheric pressure. Oxides obtained by this technique etch at the same rate as dry oxides obtained at 950°C and 1 atm. The equal etch rate indicates a similar composition and structure between the two oxides. The intrinsic breakdown fields are high (10 to 13 MV/cm) and distributed over a narrow range. All indications are that the reduced pressure oxides are very uniform, homogeneous, and similar to thicker oxides prepared at atmospheric pressure.

A two-step process sequence has been devised in which a uniform, reproducible oxide of small defect density is formed at a moderate temperature (1000°C or less) using a dry O_2–HCl ambient.[29] The second step consists of a heat treatment in N_2, O_2, and HCl at 1150°C to provide passivation and to bring the oxide thickness to the desired level. Such a processing scheme takes advantage of beneficial effects occurring in both the lower and higher temperature ranges.

As a final example of thin oxide growth, we consider the use of high-pressure, low-temperature steam oxidation of silicon. At 10 atm pressure and 750°C, a 300 Å thick oxide can be grown in 30 min. Obviously the time, temperature, and pressure can be changed to vary the thickness. Such a technique has been applied to the growth of a thin gate oxide in the process to fabricate MOS dynamic RAMs.[30] At the same time the thin oxide layer was grown, a thick oxide layer was grown over a doped polysilicon layer as a result of concentration-enhanced oxidation. The properties of the oxides depended on the oxidation temperature rather than pressure. For example, oxide density and refractive index decreased, whereas chemical etch rate and residual stress increased with increasing temperature. The temperature and pressure ranges were 700 to 1000°C and approximately 5 to 10 atm, respectively. Results of 150 Å thick, high-pressure dry oxides grown at 800°C and 25 atm showed a breakdown field of 13.6 MV/cm, which was about 10% higher than the 1 atm control group. A growth rate of 10 Å /min was obtained for the same temperature and pressure.

3.3.3 Thin Oxide Properties

Processing conditions have an important impact on oxide properties. For example, oxide density increases as the oxidation temperature is reduced.[31] Additionally, HCl ambients have typically been used to passivate ionic sodium, improve the breakdown voltage, and getter impurities and defects in the silicon. This passivation effect begins to occur only in the higher temperature range. With a variety of thin oxide growth techniques emerging, it is vital that the resulting oxide properties be understood, so that reliable devices result. Aside from the kinetics of oxide growth, studied properties typically include refractive index, oxide composition, oxide charges, etch rate, density, susceptibility to pinholes, stress, leakage, and dielectric breakdown.

As exhibited in Figure 10, leakage of thin oxides is a concern where reduced oxide thickness results in increased leakage for a given voltage.[32] For 100 Å oxides there is a large contribution to leakage as a result of Fowler–Nordheim tunneling above five volts.[32]

The larger the capacitor area, the greater the chance of encountering a defect. It is convenient to change the area of a capacitor by taking a large number of capacitors in parallel. The weakest capacitors in this configuration determine the capacitor strength.

In evaluating dielectric breakdown, the typical measurements are of temperature and either field-dependent breakdown (which is breakdown in a ramping field) or time-dependent breakdown (which is breakdown at a constant field). This breakdown of the dielectric is a failure mode for MOS integrated circuits. Figure 11 gives examples of time-dependent breakdown as a function of oxide thickness for large area (0.3 cm^2). The purpose of this figure is to show that thinner oxides present greater failure risk.

Dielectric breakdown has been the subject of various breakdown modeling studies. One such study uses a hole-trapping model that allows a prediction of oxide reliability.[33] Based on the model and data from thin-oxide EERPOMs,

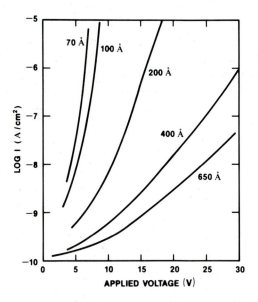

FIGURE 10
I-V characteristics for oxides of various thicknesses ranging from 70 Å to 650 Å. (*After Baglee, Ref. 32.*)

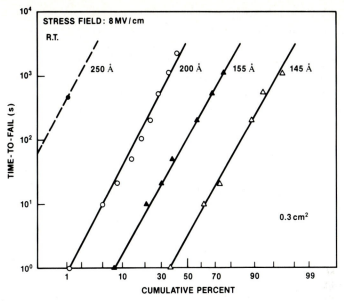

FIGURE 11
Time-Dependent Dielectric Breakdown. (*After A. M-R. Lin, private communication.*)

a 0.01% failure per thousand hours is estimated to be achievable with 95 Å oxides at 5 V. If a lower voltage is used, a thinner oxide would be acceptable.

3.4 OXIDATION TECHNIQUES AND SYSTEMS

The oxidation technique chosen depends upon the thickness and oxide properties required. Oxides that are relatively thin and those that require low charge at the interface are typically grown in dry oxygen. When sodium ion contamination is of concern, $HCl-O_2$ is the preferred technique. Where thick oxides (i.e., $>0.5\ \mu m$) are desired, steam is used ($\sim$1 atm or an elevated pressure of up to 25 atm). Higher pressure allows thick oxide growth to be achieved at moderate temperatures in reasonable amounts of time.

One atmosphere oxide growth, the most commonly used technique, is typically carried out in a horizontal diffusion tube, although vertical diffusion furnaces are being used more frequently. In the case of the horizontal furnace, the wafers are held vertically in a slotted paddle (boat), which is normally loaded using cassette-to-cassette equipment. Typical oxidation temperatures range from 700 to 1200°C and should be held to within ± 1°C to ensure uniformity. In a standard procedure the wafers are chemically cleaned, dried, loaded onto the paddle, and automatically inserted into the 700 to 900°C furnace, which is then ramped up to oxidation temperature. Ramping is used to prevent wafer warpage. Following oxidation, the furnace is ramped down and the wafers are removed.

Eliminating particles during oxidation is necessary to grow high-quality, reproducible oxides. In earlier procedures the paddle rested directly on the tube during insertion and withdrawal, or an integrated roller paddle design was used. In either case particulates were generated. Innovative designs now use a cantilevered arrangement or a soft contact system in which the paddle is inserted into the oxidation tube in a contactless manner and then lowered onto the tube. It is removed by reversing the steps.

The major new development in oxidation equipment is vertical diffusion furnaces, which have the processing tube in the vertical position. The wafers can be loaded from either the top or bottom, depending on the system, and are horizontal during oxidation, facing either up or down. In certain systems the wafers can be rotated to provide better uniformity. The claim of these systems is that they provide excellent thickness uniformity and low particle density. The quoted temperature uniformities range from $\pm 0.25°C$ to $\pm 1.0°C$ along the flat zone with oxide uniformity in the $\pm 1\%$ range.[34] In some systems the heating element can be moved relative to the wafers, allowing rapid cooldown. Wafers in the vertical systems are usually supported at three or four points or fully supported around most of the periphery. A controversy currently exists between vertical furnace and horizontal furnace vendors as to whether these arrangements result in wafer warpage.

3.4.1 Preoxidation Cleaning

Before being placed in a high-temperature furnace, wafers must be cleaned to eliminate both organic and inorganic contamination arising from previous processing steps and handling. Such contamination, if not removed, can degrade the electrical characteristics of the devices and can contribute to reliability problems.

Particulate matter is removed by either mechanical or ultrasonic scrubbing. Immersion processing techniques were the preferred chemical cleaning methods until the development of centrifugal spray methods, which eliminate the build-up of contaminants as cleaning progresses. The chemical cleaning procedure usually involves removing the organic contamination, followed by inorganic ion and atom removal.

A common cleaning procedure uses a H_2O–H_2O_2–NH_4OH mixture to remove organic contamination by the solvating action of the ammonium hydroxide and the oxidizing effect of the peroxide.[35] This process can also complex some group I and II metals. To remove heavy metals a H_2O–H_2O_2–HCl solution is commonly used. This solution prevents replating by forming soluble complexes with the removed ions, and the process is performed between 75 and 85°C for 10 to 20 minutes, followed by a quench, rinse, and spin dry.[35]

Many "optimum" cleaning procedures have evolved over the years. Reference 36 reviews the necessary consideration for optimizing the cleaning procedure for silicon wafers prior to high-temperature operations.

Modern diffusion (oxidation) furnaces are microprocessor controlled to provide repeatable sequencing, temperature control, and gas flow (mass flow control). The entire procedure previously described, from boat loading to boat withdrawal, is programmed. The microprocessor control provides a feedback loop for comparing the various parameters to the desired ones, and for making the appropriate changes. For example, the actual temperature of operation may change when the gas flow is changed. Direct control compares this temperature to the programmed temperature and automatically makes any necessary power changes.[37]

3.4.2 Dry, HCl Dry, Wet Oxidation

Dry oxidation or HCl dry oxidation is straightforward, using microprocessor-controlled equipment. The desired insertion and withdrawal rates, ramp rates, gas flows, and temperatures are all programmable. Care must be taken in handling HCl, especially with the exhaust, because HCl corrodes metal parts. Also remember that trace amounts of water vapor can drastically affect the oxidation rate.

Wet oxidation can be carried out conveniently by the pyrogenic technique, which reacts H_2 and O_2 to form water vapor. The microprocessor controls the H_2/O_2 mixture. The pyrogenic technique assures high-purity steam, provided high-purity gases are used. If wet oxidation by the bubbler technique is used, a carrier gas is typically flowed through a water bubbler maintained at 95°C. This temperature corresponds to a vapor pressure of approximately 640 Torr.

3.4.3 High-Pressure Oxidation

As we saw in Eq. 9b, the parabolic rate constant B is directly proportional to C^*, the equilibrium bulk concentration in the oxide, which in turn is proportional to the partial pressure of the oxidizing species in the gas phase. Oxidation in high-pressure steam produces a substantial acceleration in the growth rate.

High-pressure oxidation of silicon is particularly attractive, because thermal oxide layers can grow at relatively low temperatures in run times comparable to typical high-temperature, 1 atm conditions. The movement of previously diffused dopants can be minimized as can interface segregation of dopants. Low-temperature operating conditions also minimize lateral diffusion, which is of great importance as device dimensions get smaller. Another advantage is that oxidation-induced defects are suppressed (see Section 3.8). For higher-temperature, high-pressure oxidations, the oxidation time is reduced significantly.

High-pressure oxidation has been under investigation since the early 1960s.[16,38] Production equipment is now available, and the process has found device applications. The high-pressure technique has been used mostly in bipolar applications, although some companies have applied it to MOS products. For example, a high-speed, high-density, oxide-isolated bipolar process has been described.[39] In the MOS arena, application has been successfully made to the growth of a thick-field oxide layer in a dynamic RAM.[40] Many advanced MOS applications of the high-pressure technique are under development.

Figure 12 shows oxide thickness versus time data[41] for steam oxidation at various pressures and 900°C. The substantial acceleration in the oxidation rate caused by the increased pressure is apparent. In analyzing the kinetics of oxidation

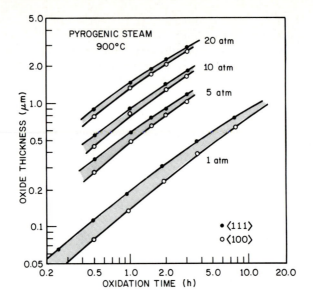

FIGURE 12

Oxidation thickness versus oxidation time for pyrogenic steam at 900°C for <100> and <111> silicon and pressures up to 20 atm. (*After Razouk, Lie, and Deal, Ref. 41.*)

at elevated pressure, several complications arise: continuous variations in pressure during pressurization, slightly variable pressurization times, small temperature variations that occur during pressurization and during the early part of the oxidation at full pressure, varying partial pressure of steam during depressurization, and thickness variations from run to run and across a wafer. A linear-parabolic model was used to analyze the data shown in Figure 12. A linear pressure dependence was observed for both the linear and parabolic rate constants.[41] Figure 13 shows the results for the parabolic rate constant,[41] where the dotted lines represent 5, 10, 15, and 20 times the parabolic rate constant at 1 atm. The figure shows that the rate constant is proportional to pressure, and also indicates the presence of a second activation energy below 900°C. This may be related to structural changes in the oxides.[41] A typical 10 atm oxidation cycle is shown in Figure 14.[42]

A similar kinetic study for dry high-pressure oxides has been reported.[43] Oxide thickness versus oxidation time for various pressures and 900°C is shown in Figure 15. Analysis of the results shows the parabolic and linear rate constants to be proportional to oxygen pressure to the power of 1 and 0.7, respectively.[43] As in the case of high-pressure steam oxidation, a change in activation energy at 900°C was observed. For thin high-pressure oxides, temperatures of 700-800°C, pressures of 10–20 atm, and short oxidation times are necessary.

Both pyrogenic and water-pumped equipment can provide steam oxidation to 25 atm pressure and 1100°C.[42] The water-pumped system alleviates the concern associated with using hydrogen at high pressure and temperature, but requires extra attention to purity, since the water quality and pumping apparatus determine

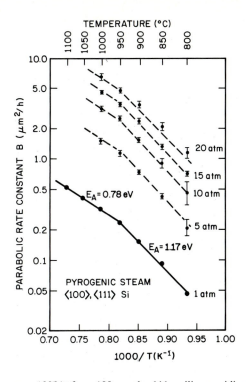

FIGURE 13
Parabolic rate constant versus $1000/T$ for $<100>$ and $<111>$ silicon oxidized at pressures of 1, 5, 10, 15, and 20 atm in pyrogenic steam. (*After Razouk, Lie, and Deal, Ref. 41.*)

the steam quality. Equipment for growing dry oxides at pressures up to 700 atm is in the developmental stages. High-pressure processing equipment and applications have recently been reviewed and are described in Reference 44.

Other uses of the high pressure equipment include flow and/or reflow of phosphosilicate glass. An improved flow results in smoother topography. Reflow after contact windows are open allows for better metal coverage on the sidewall of the window. Although little or no flow occurs at high pressure in an N_2 or dry O_2 ambient, an excellent flow of phosphosilicate glass can be obtained at temperatures of 800°C and lower in a high-pressure steam ambient.

3.4.4 Plasma Oxidation

The anodic plasma-oxidation process offers the possibility of growing high-quality oxides at temperatures even lower than those achieved with the high-pressure technique.[45] This process has all the advantages associated with low-temperature processing, such as minimized movement of previous diffusions and suppression of defect formation. Anodic plasma oxidation can grow reasonably thick oxides (on the order of $1 \mu m$) at low temperatures ($<600°C$) at growth rates up to about $1 \mu m/h$. Little production equipment currently exists for this process, which will become feasible only when such equipment is developed.

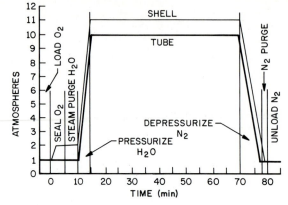

FIGURE 14
Typical 10 atm steam oxidation cycle. (*After Katz et al., Ref. 42.*)

Plasma oxidation is a low-temperature vacuum process,[45] usually carried out in a pure oxygen discharge. The plasma is produced either by a high-frequency discharge or a DC electron source. Placing the wafer in a uniform density region of the plasma and biasing it positively below the plasma potential allows it to collect active charged oxygen species. The growth rate of the oxide typically increases with increasing substate temperature, plasma density, and substrate dopant concentration.

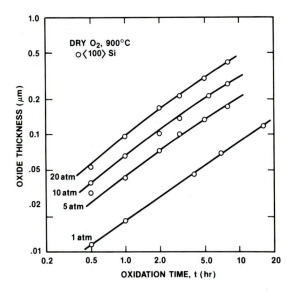

FIGURE 15
Oxide thickness versus oxidation time for <100> silicon oxidized in dry O_2 at 900°C and 1, 5, 10, and 20 atm. (*After Lie et al., Ref. 43.*)

The mechanisms involved with plasma oxidation are not fully understood. Uncertainty exists as to whether the oxide grows by the inward migration of oxygen species or by other, more complicated mechanisms. The quality of oxide grown by this process can be as good as that of thermally grown silicon dioxide.[45]

3.5 OXIDE PROPERTIES

Although the literature quotes specific values for various oxide properties, it is becoming apparent that these values are affected by the experimental conditions of oxide growth. For example, the index of refraction of dry oxides decreases with increasing temperature,[31] saturating above 1190°C at an index of 1.4620. Additionally, the apparent density of oxides grown at 800°C is 3% greater than of those grown above 1190°C.[31] The etch rate of thermal oxides at room temperature in buffered HF (49%) is generally quoted at about 1000 Å/min but varies with temperature and etch solution. The etch rate also varies with oxide density and thus with oxidation temperature. Measurements show that high-pressure oxides grown at 725°C and 20 atm exhibit a higher index of refraction, higher density, and slower etch rate in buffered HF than steam oxide grown at 900°C and 1 atm.[46] This difference is partially caused by the oxidation temperature effect.

For thin oxides, the role of the interface in determining oxide properties is important. Unanswered questions involve the effect of lattice mismatch on oxide structure, optical properties, oxidation kinetics, and oxide defects such as pinholes.

3.5.1 Masking Properties of SiO_2

A silicon dioxide layer can provide a selective mask against the diffusion of dopant atoms at elevated temperatures, a very useful property in IC processing. A predeposition of dopant by ion implantation, chemical diffusion, or spin-on techniques typically results in a dopant source at or near the surface of the oxide. During the high-temperature drive-in step, diffusion in the oxide must be slow enough with respect to diffusion in the silicon that the dopants do not diffuse through the oxide in the masked region and reach the silicon surface. The required thickness may be determined by experimentally measuring, at a particular temperature and time, the oxide thickness necessary to prevent the inversion of a lightly doped silicon substrate of opposite conductivity. A safety factor is added to this value. Typically, oxides used for masking common impurities in conventional device processing are 0.5 to 0.7 μm thick. The impurity masking properties result when the oxide is partially converted into a silica impurity oxide "glass" phase, and prevents the impurities from reaching the SiO_2–Si interface.

The values of diffusion constants for various dopants in SiO_2 depend on the concentration, properties, and structure of the SiO_2. Not surprisingly, quoted values may vary significantly. Table 7 lists diffusion constants for various common dopants.[47]

The commonly used n-type impurities P, Sb, and As, as well as the most

TABLE 7
Diffusion constants in SiO$_2$ (*Ref. 47*)

Dopants	Diffusion constants at 1100°C (cm^2/s)
B	3.4×10^{-17} to 2.0×10^{-14}
Ga	5.3×10^{-11}
P	2.9×10^{-16} to 2.0×10^{-13}
As	1.2×10^{-16} to 3.5×10^{-15}
Sb	9.9×10^{-17}

frequently used p-type impurity B, all have very small diffusion coefficients in oxides and are compatible with oxide masking. This is not true for gallium or aluminum (Al data not shown).

3.5.2   Oxide Charges

The Si–SiO$_2$ interface contains a transition region, both in terms of atom position and stoichiometry, between the crystalline silicon and amorphous silica. Various charges and traps are associated with the thermally oxidized silicon, some of which are related to the transition region. A charge at the interface can induce a charge of the opposite polarity in the underlying silicon, thereby affecting the ideal characteristics of the MOS device. This results in both yield and reliability problems.

Figure 16 shows general types of charges.[48] These charges are described by $N = Q/q$, where Q is the net effective charge per unit area (Coulombs/cm^2) at

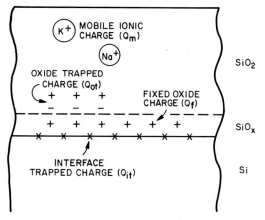

FIGURE 16
Charges in thermally oxidized silicon. (*After Deal, Ref. 48.*)

the Si–SiO$_2$ interface, N is the net number of charges per unit area (number/cm^2) at the Si–SiO$_2$ interface, and q is the electric charge. A brief description of the various charges follows.

Located at the Si–SiO$_2$ interface, interface-trapped charges Q_{it} have energy states in the silicon-forbidden bandgap and can interact electrically with the underlying silicon. These charges are thought to result from several sources, including structural defects related to the oxidation process, metallic impurities, or bond-breaking processes. A low-temperature hydrogen anneal (450°C) effectively neutralizes interface-trapped charge.[48] The density of these charges is usually expressed in terms of unit area and energy in the silicon bandgap (number/cm^2-eV). Capacitance-voltage (high-frequency, low-frequency, or quasistatic) and conductance-voltage techniques are typically used to determine the interface-trapped charges.[6] Values of 10^{10}/cm^2-eV and lower have been observed.

The fixed oxide charge Q_f (usually positive) is located in the oxide within approximately 30 Å of the Si–SiO$_2$ interface. Q_f cannot be charged or discharged. Its density ranges from 10^{10}/cm^2 to 10^{12}/cm^2, depending on oxidation and annealing conditions as well as orientation. Q_f is related to the oxidation process itself. For electrical measurements, Q_f can be considered as a charge sheet at the Si–SiO$_2$ interface. The value of this charge can be determined using the capacitance-voltage (C-V) analysis technique and the following equation:

$$\frac{Q_f}{q} = (-V_{FB} + \phi_{MS})\frac{C_o}{q} = (-V_{FB} + \phi_{MS})\frac{\epsilon_s}{qd_o} \tag{13}$$

where V_{FB} is the flatband voltage, ϕ_{MS} is the metal silicon work-function difference, ϵ_s is the dielectric permittivity of the semiconductor, d_o is the oxide thickness, and C_o is the oxide capacitance per unit area. Q_f values for <100> oriented silicon are less than those for <111> silicon. This difference is apparently related to the number of available bonds per unit area of silicon surface.

From a processing standpoint, both temperature and ambient determine Q_f.[49] In an oxygen ambient, the last high-temperature treatment determines Q_f; rapid cooling from high temperatures results in low values. Inert ambient annealing also results in low Q_f; however, at low temperatures enough time must be allowed to reach equilibrium.

The mobile ionic charge Q_m is attributed to alkali ions such as sodium, potassium, and lithium in the oxide as well as to negative ions and heavy metals. The alkali ions are mobile even at room temperature, when electric fields are present. Densities range from 10^{10}/cm^2 to 10^{12}/cm^2 or higher and are related to processing materials, chemicals, ambient, or handling. Because of larger ionic radii and lower mobility, the heavier elements contributing to this charge drift at a slower rate than the lighter elements. Measurements can be made by using the C-V technique, which involves a change in the silicon surface potential or current flow in the oxide as a result of ionic motion.

Since alkali ions can be present at various places in the oxide, the MOS capacitor is subjected to a temperature-bias stress test, which is compared to the standard C-V plot. The shift in flat-band voltage between the two curves allows the mobile ionic charge to be calculated. Common techniques used to minimize this charge include cleaning the furnace tube in a chlorine ambient, gettering with phosphosilicate glass, and using masking layers such as silicon nitride. Although chlorine in the oxidation ambient and hence in the oxide can complex sodium, the temperatures at which this is effective are higher than the normal processing temperatures.

The oxide-trapped charge Q_{ot} may be positive or negative, due to holes or electrons trapped in the bulk of the oxide. This charge, associated with defects in the SiO_2, may result from ionizing radiation, avalanche injection, or high currents in the oxide. It can be annealed out by low-temperature treatment (although neutral traps may remain).[48] Densities range from less than $10^9/cm^2$ to $10^{13}/cm^2$. Again the C-V technique can be used to measure the charge.

In addition to the earlier concerns, such as exposure of devices to ionizing radiation encountered in space flights, additional concerns arise from newer device-processing techniques such as ion implantation, e-beam metallization, plasma or reactive-sputter etching, and e-beam or x-ray lithography.

3.5.3 Oxide Stress

Understanding the stress associated with a film is important because high stress levels can contribute to wafer warpage, film cracking, and defect formation in the underlying Si. Room temperature measurements following thermal oxidation of silicon show SiO_2 to be in a state of compression on the surface. Stress values of 3×10^9 dyn/cm^2 are reported, with the stress attributed to the differences in thermal expansion for Si and SiO_2.[50] Viscous (shear) flow of thermally grown SiO_2 occurs at temperatures as low as 960°C, evidenced by the inability of the oxide–silicon structure (oxide on one side only) to remain thermally warped above that temperature.[51] In one experiment the stress present in thermal (wet) SiO_2 during growth was measured as a function of growth temperature in the range of 850 to 1030°C.[52] Growth at 950°C and below resulted in a compressive stress of approximately 7×10^9 dyn/cm^2 in the SiO_2. This at-temperature stress value is somewhat higher than the room temperature value of 3×10^9 dyn/cm^2 given above, indicating the possibility of some stress relief during cooldown. Stress-free growth at 975 and 1000°C was achieved.

During device processing, windows are cut into the oxide, resulting in a complex stress distribution. Exceedingly high stress levels can occur at these discontinuities. Such stress would typically be relieved by plastic flow or other stress-relief mechanisms. The stress reduction is further accomplished by shear components which average the load over adjacent areas.[50]

The possibility of structural damage in the silicon is very real. Shear stresses at the interface are comparable to the values of compressive stress given above.[51] These shear stresses are substantially higher than the values of 3.2×10^7 to 4.3×10^7 dyn/cm^2 given for the critical stress of shear flow for silicon at 800°C.[53] This leads to the possibility of plastic deformation in the silicon. The deleterious effect of structural damage in the silicon (particularly when decorated with impurities) on junction leakage and on other device properties is well documented. Additionally, viscous shear flow has been related to hole traps at the interface.

3.6 REDISTRIBUTION OF DOPANTS AT INTERFACE

When silicon is thermally oxidized, an interface is formed separating the silicon from the SiO_2. As oxidation proceeds, this interface advances into the silicon. A doping impurity (initially present in the silicon) will redistribute at the interface until its chemical potential is the same on each side of the interface. This redistribution may result in an abrupt change in impurity concentration across the interface. The ratio of the equilibrium concentration of the impurity (dopant) in silicon to that in SiO_2 at the interface is called the equilibrium segregation coefficient. (Note: In some literature an inverse definition is used, so care must be taken in using published values.) The experimentally determined segregation coefficient may differ from the equilibrium segregation coefficient. This will be determined primarily by the chemical potential differences and the kinetics of redistribution at the interface.

Two additional factors that influence the redistribution process are the diffusivity of the impurity in the oxide (if large, the dopant can diffuse through the oxide rapidly, affecting the profile near the Si–SiO_2 interface) and the rate at which the interface moves with respect to the diffusion rate. Figure 17 shows four different possibilities of impurity segregation;[54] segregation into the oxide with slow and fast diffusion, and segregation into the silicon with slow and fast diffusion.

The segregation coefficient determined experimentally is called the effective or interface segregation coefficient. It is particularly important to understand the concentration profile at the interface, since electrical characteristics are affected. In extreme cases inversion can occur.

Typically, to determine the segregation coefficient experimentally, a model for diffusion has been formulated, diffusion profiles have been experimentally determined in the silicon, and a segregation coefficient has been chosen to force the data to fit the model. Direct determination of the segregation coefficient is possible using the secondary-ion mass spectrometry (SIMS) technique to obtain concentration values in the oxide and in the silicon.

Most of the effort in segregation coefficient determination has been related to boron. The segregation coefficient, as defined above, increases with increasing temperature and is orientation dependent, with values for $<100>$ orientation being greater than for $<111>$ orientation. Reported coefficients[55-57] are generally 0.1 to approximately 1.0 over the temperature range 850 to 1200°C, although values greater than 1 have been obtained in special cases.[56] Figure 18 shows the results of some boron segregation determinations. Because very small amounts of moisture can greatly affect the segregation coefficient, a distinction must be made between dry, "near dry," and wet oxidations. A "dry oxidation" containing even 20 ppm moisture exhibits a segregation coefficient similar to that of wet

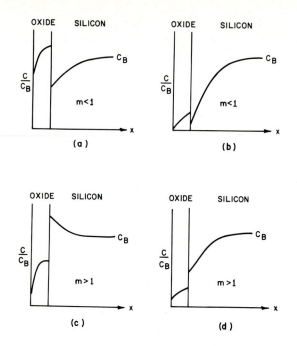

FIGURE 17
Impurity segregation at the Si–SiO$_2$ interface resulting from thermal oxidation. (a) Diffusion in oxide slow (boron), (b) diffusion in oxide fast (boron–H$_2$ ambient), (c) diffusion in oxide slow (phosphorus), and (d) diffusion in oxide fast (gallium). (*After Grove, Leistiko, and Sah, Ref. 54.*)

oxidation. The data in Fig. 18 show that "near dry" oxidation (obtained in a furnace without special drying precautions) and wet oxidation give virtually identical segregation coefficients. Larger segregation coefficient values are obtained when special drying precautions are taken.[57] Additionally, boron implanted through oxide, even when subsequent oxidations are performed in ambients with trace amounts of H$_2$O, has segregation coefficients equal to those for dry O$_2$. These effects are particularly important at lower temperatures. For example, at 900°C the surface concentration following a "near dry" oxidation is approximately one-half that of pure dry oxidation.[56] Quoted effective segregation coefficients (m_{eff}) for boron in silicon are[56]

1. Pure dry O$_2$, orientation independent:

$$m_{eff} = 13.4 \exp\frac{-0.33 \text{ eV}}{kT} \tag{14}$$

2. "Near dry" or wet O$_2$:

$$m_{111(eff)} = 65.2 \exp\frac{-0.66 \text{ eV}}{kT} \tag{15}$$

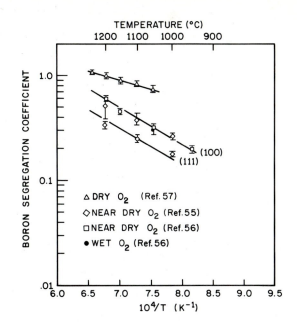

FIGURE 18
Boron segregation coefficient as a function of temperature for dry, near dry, and wet oxidations. (*After Fair and Tsai, Ref. 56.*)

$$m_{100(eff)} = 104.0 \exp \frac{-0.66 \, eV}{kT} \qquad (16)$$

For phosphorous, arsenic, and antimony, where the dopant segregates into the silicon (pile-up), segregation coefficient values of approximately 10 are usually quoted, although higher values (up to 800 at 1050°C) have been determined for arsenic.[54] With gallium, which diffuses rapidly in the oxide, a value of approximately 20 is given.[54]

3.7 OXIDATION OF POLYSILICON

Polycrystalline silicon has been used in IC technology to provide conducting lines between devices and gates. Thermal oxidation of polycrystalline silicon provides electrical isolation, which can be employed as an interlevel dielectric for multilayer structures. An understanding of the oxidation mechanisms is necessary because device reliability depends on the quality of the oxide. Various parameters of polycrystalline silicon including growth temperature, doping level, grain size, and morphology have been studied with respect to oxidation rate and oxide properties such as electrical conductivity, leakage current, and breakdown voltage. The goal is to obtain an interlevel dielectric whose properties are very close to those of oxides grown from single-crystal silicon.

In one study,[58] using CVD doped and oxidized polycrystalline films, the atmospheric-pressure polysilicon (deposited at 960°C) oxidized at the same rate as low-pressure polysilicon (deposited at 625°C). However, a substantial difference with respect to single-crystal silicon was observed. At moderate doping levels, the electrically active carrier concentration at the surface controlled the oxidation rate. While the total amount of dopant introduced into polysilicon and single-crystal samples was the same, the dopant diffused more deeply into the polysilicon, reducing the oxidation rate with respect to single-crystal silicon. This result should not be too surprising in light of our previous discussion of concentration-enhanced oxidation. Following a phosphorus predeposition having 70 $\Omega/\square$ sheet resistance and 850°C steam oxidation, oxide thickness values of approximately 3000 to 3200 Å on polysilicon, approximately 3850 Å on <100> single crystal, and approximately 4250 Å on <111> single crystal were obtained.[58] The ratio of polysilicon-consumed oxidation to oxide grown was about the same as for single-crystal silicon (0.44).

In another study,[59] using CVD (at 625°C) undoped polysilicon and lightly doped single-crystal silicon, the oxidation rate increased in the following order: <100>, <111>, polysilicon, and <110>. These observations are consistent with the transmission electron microscope determination that the polysilicon was oriented between <111> and <110>. For heavily phosphorous-doped polysilicon, the parabolic rate constant is saturated at concentrations greater than 2×10^{20} cm^{-3} while the linear rate constant continues to increase.

If the oxidation rate of polysilicon depended on the random orientation of the grains, which is true in the surface or reaction-controlled region, then a surface roughening would be expected. Surface roughening, however, is not as pronounced for oxidations at higher temperatures, where diffusion control is predominant. Transmission electron microscope results[60] show that the oxide exhibits thickness undulations coincident with previous grain boundaries. The oxide is thinner over grain boundaries by approximately 25% and forms intergranularly, in addition to forming on the surface. For higher-temperature oxidations, the thickness undulations are less severe because the oxide and silicon can flow and the reaction can enter the diffusion-controlled region.

Electrical parameters seem to degrade when the oxide–polysilicon interface shows a degree of roughness. Efforts directed at obtaining a smooth interface have shown LPCVD silicon deposited in the amphorous phase minimizes interface roughness. Properties of silicon dioxide grown from such an amphorous film were compared with similar properties of oxides grown from polycrystalline deposited films.[61] Results for dry-oxidized diffusion-doped films showed an improved oxide breakdown field, 6.2 MV/cm versus 3.0 MV/cm, and reduced leakage by two orders of magnitude for the amorphous case.

Additionally, the smoother interface improves the reliability of polysilicon–oxide–polysilicon structures with respect to electrical wear-out and time to failure.[61] Device reliability may be affected when the oxide is removed to open the contacts to the polysilicon; the oxide in the intergranular regions may also be removed unintentionally. Subsequent metallization can form a conducting path along the exposed regions between the grains in the polysilicon, and therefore causes electrical shorts.[60]

3.8 OXIDATION-INDUCED DEFECTS

3.8.1 Oxidation-Induced Stacking Faults

Thermal oxidation of silicon can produce stacking faults lying on <111> planes. These planar faults are structural defects in the silicon lattice that are extrinsic in nature and are bounded by partial dislocations. The growth mechanism may involve the coalescence of excess silicon atoms in the silicon lattice on nucleation sites, such as defects grown in during crystal growth, surface mechanical damage present prior to oxidation, chemical contamination, or defects referred to as saucer pits or hillocks. As a result of the oxidation process, excess interstitial silicon is present near the Si–SiO$_2$ interface. A small fraction of these silicon atoms flow into the bulk silicon. The silicon interstitial supersaturation in the silicon determines the stacking fault growth rate.[62] An alternative explanation involves a decreased vacancy concentration in the silicon near the Si–SiO$_2$ interface.

The deleterious nature of oxidation-induced stacking faults is well known. Examples include degraded junction characteristics in the form of increased reverse leakage current and storage time degradation in MOS structures. These problems occur when the stacking faults are electrically active as the result of being decorated with impurities, typically heavy metals. The decoration occurs both in the stacking fault itself and on the bounding dislocations. The dislocations in particular are favorable clustering sites because they represent a disarrayed high-energy region in the lattice. Diffusing impurity atoms prefer to reside in such a region because they distort the lattice less here than in the perfect lattice.

The growth of oxidation-induced stacking faults is a strong function of substrate orientation, conductivity type, and defect nuclei present. Observations show that the growth rate is greater for <100> than <111> substrates. Additionally, the density is greater for n-type conductivity than for p-type conductivity. Figure 19 shows that stacking fault length is a strong function of oxidation temperature.[63] The curves in Fig. 19 clearly show two regions: a growth region and a retrogrowth region. In the retrogrowth region, stacking fault formation is suppressed while preexisting stacking faults shrink. (The addition of HCl to the ambient can also suppress stacking fault formation.) The activation energy in the growth region is 2.3 eV independent of surface orientation and ambient (dry or wet). In the retrogrowth temperature range, stacking faults initially grow and then begin to shrink as oxidation proceeds.

Typically, the distribution of surface stacking fault lengths is very tight, except for an anomalous few percent that exhibit substantially greater lengths. Shorter-length stacking faults are usually bulk-nucleated stacking faults intersecting the surface. The length-to-depth ratio of the surface-oxidation stacking fault is approximately 3 to 10.

Additional observations show that for comparable oxide thickness shorter stacking faults are grown (in the growth region) when the oxidation temperature is lower. Indeed, even for oxides as thick as 1 μm, stacking fault formation is completely suppressed when the temperature is reduced below 950°C.[64] Shrinkage of preexisting stacking faults can also be accomplished by high-temperature inert

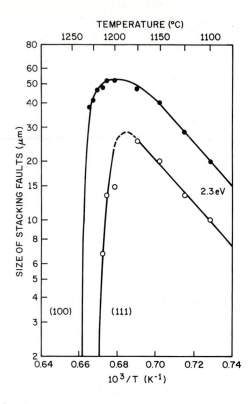

FIGURE 19
Growth of oxidation-induced stacking faults versus temperature for 3 h of dry oxidation. (*After Hu, Ref. 63.*)

ambient heat treatment, N_2 for example, with an activation energy of 2.3 eV initially and 4.3 eV upon further annealing.[65] The fast shrinkage during the initial stage is attributed to self-interstitial capture at the $Si–SiO_2$ interface,[65] while the subsequent slower shrinkage rate, with its activation energy being close to that of silicon self-diffusion, suggests that the later shrinkage is related to the diffusion of silicon atoms.

Experimental observations show that at comparable temperature and time, the oxidation stacking fault length is greater for steam ambients than for dry ambients. This suggests that the oxidation rate strongly influences the point-defect mechanism responsible for stacking fault growth. Equation 17 is a proposed model in which the oxidation rate is the controlling parameter in oxidation stacking fault length:[62]

$$\frac{dl}{dt} = K_1 \left[\frac{dT_{ox}}{dt} \right]^n - K_2 \tag{17}$$

where l is the stacking fault length, T_{ox} is the oxide thickness, t is the time, n is the power dependence, K_1 is related to the growth mechanism and defect generation at the Si–SiO$_2$ interface, and K_2 is related to the retrogrowth mechanism. Applying this equation to experimental data gives values for n, K_1 and K_2. A 0.4 power (less than linear) dependence of oxidation stacking fault growth rate on the oxidation rate is observed.[62] Therefore, smaller stacking faults will result for a higher oxidation rate at the same temperature for the same oxide thickness. This, of course, is the case with high-pressure oxidation, where the oxidation rate is increased. Figure 20 shows an experimental result for a 950 to 1100°C temperature range at both 1 and 6.4 atm pressure.[64] These results confirm the proposed model. Additional results[66] at 700°C and 20 atm steam pressure show complete stacking fault suppression for all oxide thicknesses studied (up to 5 μm).

3.8.2   Oxide Isolation Defects

Selective oxidation of silicon represents an important part of IC processing. For VLSI, oxide isolation is preferred to junction isolation. Stress along the edge of an oxidized area, especially in recessed oxides (that is, where the silicon has been trenched prior to oxidation to produce a reasonably planar surface), may produce

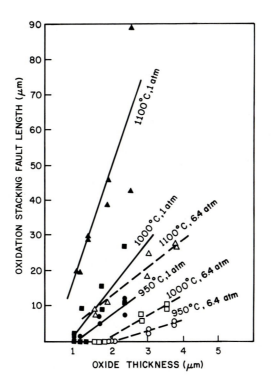

FIGURE 20
Length of oxidation-induced stacking faults versus oxide thickness for 1 atm and 6.4 atm steam oxidations. (*After Tsubouchi, Miyoshi, and Abe, Ref. 64.*)

severe damage in the silicon. Such defects result in increased leakage in nearby devices. The stress generated by the growing oxide, whose volume is over twice that of the consumed silicon, must be relieved without damaging the silicon. Various parameters have been examined for recessed isolation processes, with the conclusion that the oxidation temperature must be sufficiently high to allow the stress in the oxide to be relieved by viscous flow. High temperatures (around 950°C) will prevent stress-induced defect formation in a recessed structure (recess approximately 1.2 μm and oxide growth approximately 2.2 μm). This critical temperature correlates well with that for stress-free growth in oxides at 1 atm.

Problem
Devise a processing scheme to generate selectively a planar recessed oxide in silicon.

Solution
Use silicon nitride as a mask for oxidation. Use photolithographic techniques to pattern the nitride. After removing the photoresist, either wet, chemically etch, or plasma etch the exposed silicon to a depth such that after oxidation the surface will be planar. Remember that a layer of silicon 0.44 d_o is consumed when an SiO_2 layer of thickness d_o is formed.

In many applications a selective oxide is grown without first trenching the silicon. A two-dimensional oxidation model that simulates both diffusion of oxidant into the oxide and viscoelastic deformation has been used to analyze stress in the oxide during thermal oxidation of silicon for a LOCOS (local oxidation) structure. The results indicate the stressed region, that covered by the nitride mask, suffered a retardation of oxidation.[67] While the mechanisms are not yet fully understood, it is clear that oxide growth is affected by stress.

3.9   SUMMARY AND FUTURE TRENDS

The ability to mathematically describe the oxidation process reasonably well in the simplest form has been demonstrated. Our understanding of the oxidizing species and the point-defect mechanisms in the vicinity of the oxidizing interface is still evolving. We can determine experimentally the effect of impurity species, dopant concentration, and orientation on the oxidation kinetics, but we are somewhat less able to explain some of the mechanisms involved.

An understanding of oxide charge is necessary in order to fabricate highly reliable devices. This is particularly important with the new processing techniques used for VLSI fabrication. An understanding of how to form oxides without damaging the underlying silicon is necessary when fabricating advanced structures, such as dielectrically isolated devices that may require thick recessed oxides. Oxide viscosity is a first-order effect, and oxidation temperatures above 950°C minimize stress-related defect formation.

Polycrystalline silicon usage has become increasingly important, and has attracted more study recently both in its formation and oxidation. The polysilicon deposition technique, polysilicon grain size, orientation, and doping level all affect oxidation. Formation of oxide in intergranular regions and its removal when contacts to the polysilicon are opened lead to the possibility of electrical shorts during metallization.

The impact of continually shrunken vertical and lateral dimensions, tighter design rules, and lower-temperature processing cannot be overlooked in future development. A very promising technique emerging for growing thin oxides is rapid thermal processing, in which the surface of the silicon is subjected to a high temperature, 1150°C, for a very short period of time in an oxidizing ambient.

High-pressure oxidation allows thick oxides to be grown at low to moderate temperatures while suppressing oxidation-induced stacking faults. Additionally, it is being examined for growing thin oxides. High-pressure oxidation has found use in bipolar devices and is emerging as a useful technique in MOS processing.

The prevention of lateral oxidation (bird's beak) during selective oxidation of silicon has received much attention. Schemes using a protective sidewall trench to prevent this effect have been encouraging. Another approach to avoid this problem is to pattern a uniform oxide layer resulting in vertical oxide steps, the so-called nonselective oxide scheme.

Oxide requirements for advanced structures are changing. As discussed earlier, these requirements range from highly reliable thin oxides to thick isolation oxides that can be grown at moderate temperatures. Emphasis on oxidation techniques such as high-pressure and rapid thermal oxidation has occurred. It is inevitable that further advances will be made in growth techniques, processing schemes, and understanding of oxidation mechanisms.

PROBLEMS

1 Show from the densities and molecular weights of Si and SiO_2 that a layer of silicon of thickness approximately equal to $0.44\,d_o$ is consumed when a SiO_2 layer of thickness d_o is formed. Use density values of 2.27 grams/cm^3 for SiO_2 and 2.33 grams/cm^3 for Si.

2 Show that Eq. 10, $\dfrac{d_o}{A/2} = \left[1 + \dfrac{t + \tau}{A^2/4B}\right]^{1/2} - 1$, reduces to $d_o^2 = Bt$ for long times and to $d_o = \dfrac{B}{A}(t + \tau)$ for short times.

3 Use $d_o^2 + Ad_o = B(t + \tau)$ graphically to generate a plot for the 1100°C oxidation data of Fig. 4. Use $\tau = 0$ and <100> orientation to obtain rate constants. Compare your results to those of Fig. 3.

4 Using Eq. 9 and Table 1, determine how long it will take to grow 2.0 μm of SiO_2 at 920°C and 25 atm steam pressure.

5 Define a set of conditions to minimize the chance of inverting the surface of an n-type substrate (containing a boron diffusion) when oxidizing the wafer.

6 List possible ways of growing an oxide on a substrate without forming oxidation-induced stacking faults.

7 Solve Eq. 9 for oxide thickness as $f(t, \tau, A, B)$.

8 Make use of the equation derived in Problem 7, and the data in Tables 1 and 2, to generate oxide thickness versus time curves for wet and dry oxidations at 1100°C. Assume $\tau = 0$.

9 For selective oxidation, show how you might prevent lateral oxidation during the oxide growth.

REFERENCES

1 M. M. Atalla, "Semiconductor Surfaces and Films; the Si–SiO$_2$ System," in H. Gatos, Ed., *Properties of Elemental and Compound Semiconductors*, Interscience, New York, 1960, Vol. 5, pp. 163–181.

2 P. J. Jorgensen, "Effect of an Electric Field on Silicon Oxidation," *J. Chem. Phys.*, **37**, 874 (1962).

3 J. R. Ligenza and W. G. Spitzer, "The Mechanisms for Silicon Oxidation in Steam and Oxygen," *J. Phys. Chem. Solids*, **14**, 131 (1960).

4 B. E. Deal and A. S. Grove, "General Relationship for the Thermal Oxidation of Silicon," *J. Appl. Phys.*, **36**, 3770 (1965).

5 A. S. Grove, *Physics and Technology of Semiconductor Devices*, Wiley, New York, 1967, Chapter 2.

6 E. H. Nicollian and J. R. Breuws, *MOS Physics and Technology*, Wiley, New York, 1982, Chapter 9.

7 P. S. Flint, "The Rates of Oxidation of Silicon," Abstract 94, *The Electrochem. Soc. Extended Abs., Spring Meeting*, Los Angeles, May 6–10, 1962.

8 J. Blanc, "A Revised Model for the Oxidation of Si by Oxygen," *Appl. Phys. Lett.*, **33**, 424 (1978).

9 D. N. Modlin and W. A. Tiller, "Effects of Corona-Discharge Induced Oxygen Ion Beams and Electric Fields on Silicon Oxidation Kinetics," *J. Electrochem. Soc.*, **132**, 1659 (1985).

10 W. A. Tiller, "On the Kinetics of the Thermal Oxidation of Silicon," I. *J. Electrochem. Soc.*, **127**, 619 (1980); II. **127**, 625 (1980); III. **128**, 689 (1981).

11 R. B. Fair, "Oxidation, Impurity Diffusion, and Defect Growth in Silicon—An Overview," *J. Electrochem. Soc.*, **128**, 1361 (1981).

12 S. M. Hu, "Formation of Stacking Faults and Enhanced Diffusion in the Oxidation of Silicon," *J. Appl. Phys.*, **45**, 1567 (1974).

13 E. A. Irene, H. Z. Massoud, and E. Tierney, "Silicon Oxidation Studies: Silicon Orientation Effects on Thermal Oxidation," *J. Electrochem. Soc.*, **133**, 1253 (1986).

14 V. K. Samalam, "Theoretical Model for the Oxidation of Silicon," *Appl. Phys. Lett.*, **47**, 736 (1985).

15 S. A. Schafer and S. A. Lyon, "New Model of the Rapid Initial Oxidation of Silicon," *Appl. Phys. Lett.*, **47**, 154 (1985).

16 J. R. Ligenza, "Effect of Crystal Orientation on Oxidation Rates in High Pressure Steam," *Phys. Chem.*, **65**, 2011 (1961).

17 B. E. Deal, "Thermal Oxidation Kinetics of Silicon in Pyrogenic H$_2$O and 5% HCl/H$_2$O Mixtures," *J. Electrochem. Soc.*, **125**, 576 (1978).

18 M. M. Atalla and E. Tannenbaum, "Impurity Redistribution and Junction Formation in Silicon by Thermal Oxidation," *Bell Syst. Tech. J.*, **39**, 933 (1960).

19 B. E. Deal and M. Sklar, "Thermal Oxidation of Heavily Doped Silicon," *J. Electrochem. Soc.*, **112**, 430 (1965).

20 C. P. Ho, J. D. Plummer, J. D. Meindl, and B. E. Deal, "Thermal Oxidation of Heavily Phosphorus Doped Silicon," *J. Electrochem. Soc.*, **125**, 665 (1978).

21 C. P. Ho and J. D. Plummer, "Si–SiO$_2$ Interface Oxidation Kinetics: A Physical Model for the Influence of High Substrate Doping Levels. I. Theory," *J. Electrochem. Soc.*, **126**, 1516 (1979); "II. Comparison With Experiment and Discussion," *J. Electrochem. Soc.*, **126**, 1523 (1979).

22 D. W. Hess and B. E. Deal, "Kinetics of the Thermal Oxidation of Silicon in O$_2$/HCl Mixtures," *J. Electrochem. Soc.*, **124**, 735 (1977).

23 P. F. Schmidt, R. J. Jaccodine, C. H. Wolowodiuk, and T. Kook, "Chemically Enhanced Thermal Oxidation of Silicon," *Materials Lett.*, **3**, 235 (1985).

24 J. F. Gotzlich, K. Haberger, H. Ryssel, H. Kranz, and E. Traumuller, "Dopant Dependence of the Oxidation Rate of Ion Implanted Silicon," *Radiat. Eff.*, **47**, 203 (1980).

25 A. Bhattacharyya, T. Bril, C. Vorst, B. Westlund, and F. Van Roosmalen, "Effect of Dry Etching of a Thermal Oxide on Subsequent Growth and Properties of Thin Oxides," *J. Electrochem. Soc.*, **133**, 1670 (1986).

26 J. Nulman, J. P. Krusius, and A. Gat, "Rapid Thermal Processing of Thin Gate Dielectrics. Oxidation of Silicon," *IEEE Electron Devices Letters*, **EDL-6**, 205 (1985).

27 H. Z. Massoud, J. D. Plummer, and E. A. Irene, "Thermal Oxidation of Silicon in Dry Oxygen Growth-Rate Enhancement in the Thin Regime," *J. Electrochem. Soc.*, **132**, 2685 (1985).

28 A. C. Adams, T. E. Smith, and C. C. Chang, "The Growth and Characterization of Very Thin Silicon Dioxide Films," *J. Electrochem. Soc.*, **127**, 1787 (1980).

29 C. Hashimoto, S. Muramoto, N. Shiomo, and O. Nakajima, "A Method of Forming Thin and Highly Reliable Gate Oxides," *J. Electrochem. Soc.*, **127**, 129 (1980).

30 M. Hirayama, H. Miyoshi, N. Tsubouchi, and H. Abe, "High Pressure Oxidation for Thin Gate Insulator Process," *IEEE Trans. Electron. Devices*, **ED-29**, 503 (1982).

31 E. A. Taft, "The Optical Constants of Silicon and Dry Oxygen Oxides," *J. Electrochem. Soc.*, **125**, 968 (1978).

32 D. A. Baglee, "Characteristics and Reliability of 100 Å Oxides," *Proc. 22nd IEEE Rel. Phys. Symp.*, 152 (1984).

33 C. Hu, "Thin Oxide Reliability," *Proc. Int. Electron. Devices Meeting*, 368 (1985).

34 P. H. Singer, "Trends in Vertical Diffusion Furnaces," *Semicond. Int.*, **9**, 56 (1986).

35 W. Kern and D. A. Puotinen, "Cleaning Solutions Based on Hydrogen Peroxide for Use in Silicon Semiconductor Technology," *RCA Rev.*, **31**, 187 (1970).

36 D. Burkman, "Optimizing the Cleaning Procedure For Silicon Wafers Prior to High Temperature Operations," *Semicond. Int.*, **4**, 103 (1981).

37 P. S. Burggraff, "The Case for Computerized Diffusion Control," *Semicond. Int.*, **4**, 37 (1981).

38 J. R. Ligenza, "Oxidation of Silicon by High Pressure Steam," *J. Electrochem. Soc.*, **109**, 73 (1962).

39 J. Agraz-Guerena, P. T. Panousis, and B. L. Morris, "OXIL, A Versatile Bipolar VLSI Technology," *IEEE Trans. Electron Devices*, **ED-27**, 1397 (1980).

40 N. Tsubouchi, H. Miyoshi, H. Abe, and T. Enomoto, "The Applications of High Pressure Oxidation Process to the Fabrication of MOS LSI," *IEEE Trans. Electron Devices*, **ED-26**, 618 (1979).

41 R. R. Razouk, L. N. Lie, and B. E. Deal, "Kinetics of High Pressure Oxidation of Silicon in Pyrogenic Steam," *J. Electrochem. Soc.*, **128**, 2214 (1981).

42 L. E. Katz, B. F. Howells, L. P. Adda, T. Thompson, and D. Carlson, "High Pressure Oxidation of Silicon by the Pyrogenic or Pumped Water Technique," *Solid State Technol.*, **24**, 87 (1981).

43 L. N. Lie, R. R. Razouk, and B. E. Deal, "High Pressure Oxidation of Silicon in Dry Oxygen," *J. Electrochem. Soc.*, **129**, 2828 (1982).

44 S. P. Tay and J. P. Ellul, "High Pressure Technology for Silicon I. C. Fabrication," *Semicond. Int.*, **9**, 122 (1986).

45 V. Q. Ho and T. Sugano, "Selective Anodic Oxidation of Silicon in Oxygen Plasma," *IEEE Trans. Electron Devices*, **ED-27**, 1436 (1980).

46 L. E. Katz and B. F. Howells, "Low Temperature, High Pressure Steam Oxidation of Silicon," *J. Electrochem. Soc.*, **126**, 1822 (1979).

47 M. Ghezzo and D. M. Brown, "Diffusivity Summary of B, Ga, P, As, and Sb in SiO_2," *J. Electrochem. Soc.*, **120**, 146 (1973).

48 B. E. Deal, "Standardized Terminology for Oxide Charges Associated with Thermally Oxidized Silicon," *IEEE Trans. Electron Devices*, **ED-27**, 606 (1980).

49 B. E. Deal, "The Current Understanding of Charges in the Thermally Oxidized Silicon Structure," *J. Electrochem. Soc.*, **121**, 198C (1974).

50 R.J. Jaccodine and W. A. Schlegel, "Measurement of Strains at Si–SiO_2 Interface," *J. Appl.*

Phys., **37**, 2429 (1966).

51 E. P. EerNisse, "Viscous Flow of Thermal SiO_2," *Appl. Phys. Lett.*, **30**, 290 (1977).

52 E. P. EerNisse, "Stress in Thermal SiO_2 During Growth," *Appl. Phys. Lett.*, **35**, 8 (1979).

53 S. M. Hu, "Temperature Dependence of Critical Stress in Oxygen Free Silicon," *J. Appl. Phys.*, **49**, 5678 (1978).

54 A. S. Grove, O. Leistiko, and C. T. Sah, "Redistribution of Acceptor and Donor Impurities During Thermal Oxidation of Silicon," *J. Appl. Phys.*, **35**, 2695 (1964).

55 J. W. Colby and L. E. Katz, "Boron Segregation at $Si–SiO_2$ Interface as a Function of Temperature and Orientation," *J. Electrochem. Soc.*, **123**, 409 (1976).

56 R. B. Fair and J. C. C. Tsai, "Theory and Direct Measurement of Boron Segregation in SiO_2 during Dry, Near Dry and Wet O_2 Oxidation," *J. Electrochem. Soc.*, **125**, 2050 (1978).

57 S. P. Murarka, "Diffusion and Segregation of Ion-Implanted Boron in Silicon in Dry Oxygen Ambients," *Phys. Rev. B*, **12**, 2502 (1975).

58 T. I. Kamins, "Oxidation of Phosphorous-Doped Low Pressure and Atmospheric Pressure CVD Polycrystalline-Silicon Films," *J. Electrochem. Soc.*, **126**, 838 (1979).

59 H. Sunami, "Thermal Oxidation of Phosphorus-Doped Polycrystalline Silicon in Wet Oxygen," *J. Electrochem. Soc.*, **125**, 892 (1978).

60 E. A. Irene, E. Tierney, and D. W. Dong, "Silicon Oxidation Studies: Morphological Aspects of the Oxidation of Polycrystalline Silicon," *J. Electrochem. Soc.*, **127**, 705 (1980).

61 L. Farone, R. D. Vibronek, J. McGinn, "Characterization of Thermally Oxidized n$^+$ Polycrystalline Silicon," *IEEE Trans. Electron Devices*, **ED-32**, 577 (1985).

62 A. M-R. Lin, R. W. Dutton, D. A. Antoniadis, and W. A. Tiller, "The Growth of Oxidation Stacking Faults and the Point Defect Generation at $Si–SiO_2$ Interface during Thermal Oxidation of Silicon," *J. Electrochem. Soc.*, **128**, 1121 (1981).

63 S.M. Hu, "Anomalous Temperature Effect of Oxidation Stacking Faults in Silicon," *Appl. Phys. Lett.*, **27**, 165 (1975).

64 N. Tsubouchi, H. Miyoshi, and H. Abe, "Suppression of Oxidation-Induced Stacking Fault Formation in Silicon by High Pressure Steam Oxidation," *J. Appl. Phys.*, **17**, 223 (1978).

65 K. Nishi and D. A. Antoniadis, "Fast Shrinkage of Oxidation Induced Stacking Faults in Silicon at the Initial Stage of Annealing in Nitrogen," *Appl. Phys. Lett.*, **46**, 516 (1985).

66 L. E. Katz and L. C. Kimerling, "Defect Formation During High Pressure, Low Temperature Steam Oxidation of Silicon," *J. Electrochem. Soc.*, **125**, 1680 (1978).

67 S. Isomae, S. Yamamoto, S. Aoki, and A. Yajima, "Oxidation Induced Stress in Locos Structure," *IEEE Electron Device Letters*, **EDL-7**, 368 (1986).

CHAPTER
4

LITHOGRAPHY

R. K. WATTS

4.1 INTRODUCTION

A lithograph, for people interested in art, is a less expensive picture made by impressing in turn several flat, embossed slabs, each covered with greasy ink of a particular color, onto a piece of stout paper. The various colors or levels must be accurately aligned, or registered, with respect to one another within some registration tolerance to produce the effect intended by the artist who carved or etched the slabs. Many "originals" can be made from the same slabs as long as the yield (ratio of good pictures to the total) remains adequately high. The technique was first used in the late eighteenth century, but was perhaps suggested by scenes carved in bas relief on more or less flat stones by many ancient peoples. For centuries it has been common practice for scholars and art lovers to copy these scenes by pressing a paper onto the relief to transfer the image to the paper.

Figure 1 illustrates schematically the lithographic process used to fabricate circuit chips. The exposing radiation is transmitted through the "clear" parts of a mask. The circuit pattern of opaque chromium blocks some of the radiation. This type of chromium/glass mask is used with ultraviolet (UV) light. Other types of exposing radiation are electrons, x-rays, or ions. Shadow (proximity) printing may be employed where the gap between mask and wafer is small. In the case of a nonexistent gap, the method is called contact printing. Or some sort of image-forming system (a lens, for example) may be interposed between mask and wafer.

Therefore, lithography for integrated circuit manufacturing is analogous to the lithography of the art world. The artist corresponds to the circuit designer. The slabs are masks for the various circuit levels. The press corresponds to the exposure system, which not only exposes each level but also aligns it to

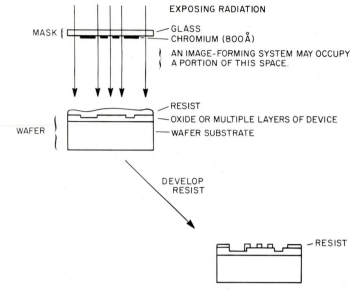

FIGURE 1
Device lithography generalization.

a completed level. The ink may be compared either with the exposing radiation or with the radiation-sensitive resist. And the paper can represent the wafer into which the patterns will be etched, using the resist as a stencil.

Optical lithography has penetrated the "1 μm barrier" of resolution. As other lithographic techniques such as those using electron, x-ray, or ion radiation have improved, so has optical lithography. For these competing methods large teams of capable people have been working steadily for ten years or more. Several methods have only one or two key problems remaining to be solved. Solution generally requires application of standard, but difficult and meticulous, engineering.

4.2   OPTICAL LITHOGRAPHY

Optical lithography comprises the formation of images with visible or ultraviolet radiation in a photoresist using contact, proximity, or projection printing (discussed in Section 4.2.3). For integrated circuit production the linewidth limit of optical lithography lies near 0.4 μm, although 0.2 μm features may eventually be printed under carefully controlled conditions. Optical lithography will continue to occupy the primary position for the foreseeable future.

4.2.1   Optical Resists

Photoresists are of two types. A negative resist on exposure to light becomes less soluble in a developer solution, while a positive resist becomes more soluble. Commercial negative resists, such as Kodak Microneg 747, consist of two parts: a chemically inert polyisoprene rubber, which is the film-forming component, and a photoactive agent. The photoactive agent on exposure to light reacts with the rubber to form crosslinks between rubber molecules, making the rubber less soluble in an organic developer solvent. The reactive species formed during the exposure can react with oxygen and be rendered ineffective for crosslinking. Therefore the resist is usually exposed in a nitrogen atmosphere.

The developer solvent dissolves the unexposed resist. The exposed resist swells as the uncrosslinked molecules are dissolved. The swelling distorts the pattern features and limits resolution to 2 to 3 times the initial film thickness.

Positive resists have two components: a resin and a photoactive compound dissolved in a solvent. The photoactive compound is a dissolution inhibitor. When it is destroyed by exposure to light, the resin becomes more soluble in an aqueous developer solution. The unexposed regions do not swell much in the developer solution, so higher resolution is possible with positive resists. The development process of projection printed images in positive resists has been modeled theoretically.[1] It is an isotropic etching process. The sensitivity of most standard resists peaks in the 300 to 400 nm spectral range. Two examples of commercially available positive resists are MP-2400 and HPR-206.

The light intensity $I(\lambda, z)$ which is effective in exposing a volume element of resist at height z above the substrate depends on the reflectivity of the substrate, the thickness T of the resist $(T > z)$, and the convolution of the absorption spectrum of the resist with the spectrum of the incident light. The refractive index of most resists is about 1.6, leading to an optical mismatch at the air/resist interface. If the substrate is highly reflective, the interference pattern of standing waves that are set up has a node at the substrate and a peak amplitude that depends on resist thickness T. Figure 2 shows the standing waves. This variation in peak intensity with resist thickness becomes less with decreasing substrate reflectivity and increasing absorption by the resist. Since resist thickness varies at a step in substrate topology, the resulting difference in effective exposure leads to size variations in the resist image. Because of the high resolution of positive resists, the standing wave patterns are often well resolved, and the resulting corrugations are seen in the resist edge profile.

Photoresists are being developed for exposure at shorter wavelengths where higher resolution is possible. A few such deep UV resists are poly-methyl methacrylate (PMMA), sensitive for $\lambda < 250$ nm; polybutene sulfone, sensitive for $\lambda \leq 200$ nm; and MP-2400, sensitive for $\lambda = 250$ nm. At these shorter wavelengths, the radiation quantum is large enough to produce scission (breakage) of the molecular chain.

Other properties of resists that are quite important are good adhesion to the substrate and resistance to wet and dry etch processes. In general, the commercially available optical resists are compatible with such processes.

The unwanted variation of feature size in the resist image is due to many effects, some related to resist properties and resist processing and others to the exposure tool. Variation in feature size can be minimized by proper control of exposure and development.

Computer modeling of resist images is an important field of study. For an introduction to this subject, see Chapter 10 of this book and Chapters 2 and 4 of Reference 2.

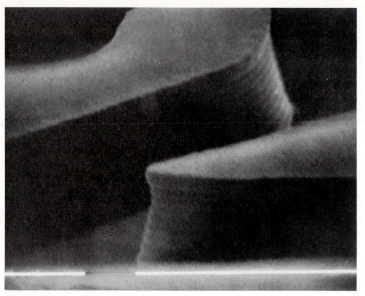

FIGURE 2
Standing waves in resist profile.

4.2.2   Contact and Proximity Printing

In contact printing a photomask is pressed against the resist-covered wafer with pressures typically in the range of 0.05 atm to 0.3 atm and exposed by light of wavelength near 400 nm. Very high resolution (≤ 1 μm linewidth) is possible, but because of spatial nonuniformity of the contact, resolution may vary considerably across the wafer. To provide better contact over the whole wafer, a thin (0.2 mm) flexible mask has been used; 0.4 μm lines have been formed in a 0.98 μm resist.[3] The contact produces defects in both the mask and the wafer so that the mask, whether thick or thin, may have to be discarded after a short period of use. For example, one worker reports mask defect densities increasing to 37 defects/cm^2 after 15 exposures from 13 defects/cm^2 after 5 exposures.[4] Defects include pinholes in the chromium film, scratches, intrusions, and star fractures.

Contact printing nevertheless continues to be widely used. Features as small as 0.25 μm have been produced in 1.8 μm thick PMMA resist using 200 to 260 nm radiation.[5] Quartz or Al$_2$O$_3$ mask substrates must be used to pass these shorter wavelengths, since the usual borosilicate glass strongly absorbs wavelengths less than 300 nm.

Proximity printing has the advantage of longer mask life because there is no contact between the mask and the wafer. Typical separations between mask and wafer are in the range of 20 to 50 μm. Resolution is not as good as in contact printing or projection printing.

Figure 3 shows proximity printing in the schematic form of a mask with a long slit of width W separated from a parallel image plane (wafer) by a gap g. We assume that g and W are larger than the wavelength λ of the imaging light and that $\lambda \ll g < W^2/\lambda$—the region of Fresnel diffraction. Then the diffraction that forms the image of the slit is a function only of the particular combination of $\lambda, W,$ and g which we shall call the parameter Q where

$$Q = W\sqrt{2(\lambda g)} \qquad (1)$$

The diffraction patterns are well known. Some are shown in Fig. 4 for $Q = 2, 3,$ and 7 (solid curves).[6] The limit $g < W^2/\lambda$ corresponds to $Q > \sqrt{2}$. The dashed rectangles in Fig. 4 show the light intensity at the mask. The larger the value of Q, the more faithful the image. Thus the resolution becomes better at smaller gaps and shorter wavelengths.

Two other noteworthy features of the diffraction patterns are the ragged peaks and the slope near $x = W/2$. The peaks can be smoothed by use of diverging rather than collimated light. The dotted curves are for light with a divergence half-angle $\alpha = 1.5\lambda/W$ radians. If α becomes large, the edge slope is reduced further, increasing linewidth control problems. Some smoothing also occurs because of the spread of wavelengths in the source, but this is a smaller effect. The illumination system must provide an apparent source size large enough to give a value of α— typically a few degrees—optimized for the smallest features to be printed. This is illustrated in Fig. 5. The mercury arc used as source is too small to yield the required α, so a scheme like that illustrated in Fig. 5c is used. The optical system must also minimize nonuniformity of intensity across the field; typically this may be 3 to 5%. Figure 6 shows the illumination system of a proximity printer. The illumination is telecentric, or normally incident, at the mask to prevent runout (magnification) errors. With a Hg arc source the strong lines at 436 nm, 405 nm,

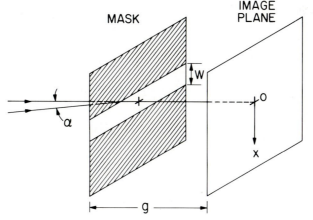

FIGURE 3
Proximity printing idealization.

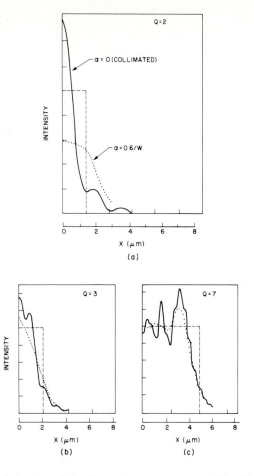

FIGURE 4
Image of slit with $Q = 2, Q = 3$, and $Q = 7. g = 10$ μm; $\lambda = 0.4$ μm. (*After Watts, Ref. 47.*)

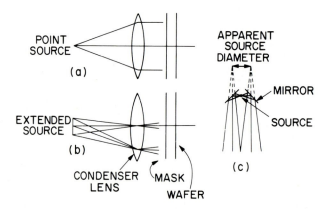

FIGURE 5
Proximity printing with collimated and uncollimated light. (*After Watts, Ref. 47.*) (a) shows a point source. (c) shows a way of obtaining an extended source like that depicted schematically in (b).

and 365 nm provide exposure flux. The same printer is available with Xe–Hg source for enhanced output in the 200–300 nm spectral region. Exposure times of 1 min (a very long exposure compared with standard resists) are required with PMMA resists, and 2 μm resolution is obtained with $g = 10$ μm to 20 μm.[7]

The edge of a feature in developed resist occurs at the position where the product of light intensity and exposure time equals the resist threshold dose. Linewidth control is in general more difficult in proximity printing than in contact printing. Linewidth is influenced by variations in light intensity, gap g, and resist properties. Linewidth has been measured as a function of light intensity I and mask-wafer gap.[8] The linewidth variation, defined as the printed width minus the width of the line on the mask, is proportional to $\ln I$ and to $\sqrt{g}$, and for each value of g there is an optimum exposure intensity which minimizes the effect of gap variation on linewidth. The mask-wafer gap can vary across a wafer and from one wafer to another because of wafer and mask bowing and dirt particles between the wafer and wafer chuck. This last problem is solved with a pin chuck, in which only a number of points contact the wafer. For exposure intensity near the optimum value, control of exposure to within 13% should provide linewidth control of ± 0.5 μm, even for large variations of the gap g. Linewidth control to within ± 0.25 μm with a 50 μm gap has been reported.[4]

Because of the sloping edges of the diffraction pattern of the slit, the image of a mask object consisting of equally spaced parallel lines will lose definition as the spacing between lines decreases and the edge tails begin to overlap. If I_M is the peak intensity in the image (on the lines) and I_m is the minimum intensity

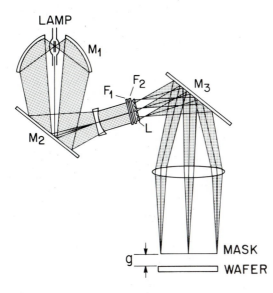

FIGURE 6
Illumination system of Canon PLA 500F proximity printer. F = filter. M = mirror. L = fly's eye lens. (*After Watts, Ref. 47.*)

in the image (between lines, where this quantity ideally should be zero), the modulation of the image M_i is given by

$$M_i \equiv \frac{I_M - I_m}{I_M + I_m} \qquad (2)$$

M_i is a function of linewidth, line spacing, gap, and wavelength.[9] For proper exposure of resist, a ratio I_M/I_m of at least 4 is desirable. This corresponds to $M_i = 0.6$, a value which can be exceeded only with very small gaps. Further loss of modulation can be reduced by use of Fe_2O_3 or Cr_2O_3–Cr masks, which have much lower reflectivity than Cr and reduce scattered light under the opaque parts of the mask. (Cr_2O_3–Cr means that the chromium is covered with a thin chromium oxide layer to reduce reflectivity.)

4.2.3   Projection Printing

Projection printing offers higher resolution than proximity printing together with large separation between mask and wafer. Four important performance parameters of a printer are resolution, level-to-level alignment accuracy, throughput, and depth of focus. The resolution of an optical imaging system of numerical aperture $NA = \sin \alpha$ with light of wavelength λ is, according to Rayleigh's criterion, $0.61\lambda/\sin \alpha$—the separation of two barely resolved point sources.[10] The Rayleigh depth of focus[10] is given by $\pm \lambda/(2 \sin^2 \alpha)$. However, near the limiting resolution of the system, the contrast in the image is uselessly small. We shall see how optical performance can be specified in a more meaningful way by means of the system transfer function.

We consider a general optical system (Fig. 7) with no aberrations. An aberration is a departure of the imaging wavefront from the spherical form; that

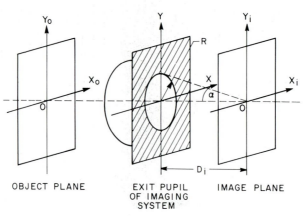

FIGURE 7
Projection printing idealization. (*After Watts, Ref. 47.*)

is, a spherical wave diverging from a point in the object plane is converted to a spherical wave converging to a point in the image plane. The f number of the system is $F \equiv D_i/2R$, and the numerical aperture is $NA = \sin \alpha = R/\sqrt{(D_i^2 + R^2)} \approx R/D_i$. R is the radius of the exit pupil, and D_i is the distance from this pupil to the image plane.

We want to recall briefly some results from the theory of image formation and to show how the transfer functions differ for coherent and incoherent imaging systems.[10] For coherent object illumination all points in the object have wave amplitudes with fixed phase relationships, and all phases have the same time dependence. For example, in Fig. 5a the phase varies across the mask plane at right in a way simply determined by the path lengths from the point source. On the other hand, if there is no phase correlation and the phases vary independently from point to point across the object, the illumination is spatially incoherent. This is the situation in Fig. 5b. We consider first the coherent case. It is not necessary that the radiation be monochromatic, only that the wavelength spread be small compared with the average wavelength. The wavelength λ can stand for the average wavelength.

The field amplitude at point x_i, y_i in the image plane (the wafer) will be called $U_i(x_i, y_i)$. It is caused by a field distribution $U_o(x_o, y_o)$ in the object plane (the mask). We define a pupil function $P(x, y)$ to represent the transmission of the round exit pupil, or hole, shown in Fig. 7. For points (x, y) within the pupil, $P = 1$; for points outside, $P = 0$. The simplest object is a point source at $x_o, y_o = 0$. The image of this source has the form

$$I_i(x_i, y_i) = \frac{\pi}{(2\lambda F)^2}[2J_1(s)/s]^2, s = \left(\frac{\pi}{\lambda F}\right)\sqrt{x_i^2 + y_i^2} \qquad (3)$$

This is the well-known Airy pattern. J_1 is the first-order Bessel function. It is also called the point-spread function because the projection system spreads the point into the circular image represented by Eq. 3. The diameter (twice the distance from the origin to the first zero of J_1) of the pattern is 2.4 λF. Thus, the smaller the wavelength and the f number (or the larger the NA) of the projection system, the better the resolution.

High fidelity audio components are characterized in large part by the frequency response function. For a good loudspeaker the frequency response is flat over the whole audio range, dropping to zero near 20 Hz and 20 kHz. The frequency response function indicates how the speaker degrades the incoming signal, whose Fourier transform is a frequency spectrum. The output spectrum is obtained by multiplying each component of the input by the value of the frequency response function at that frequency. Similarly, optical systems are characterized by a frequency response function H, which is called in this case the optical transfer function. In optics the spatial frequency, or number of lines per mm in a grating pattern of equal lines and spaces, is the analog of temporal frequency, and a spatial coordinate, say x, is the analog of the time coordinate t. If $G_i(u, v)$ and $G_o(u, v)$ are the Fourier transforms of $U_i(x_i, y_i)$ and $U_o(x_o, y_o)$ respectively, then $H(u, v)$ is given by

$$G_i(u, v) = H(u, v)G_o(u, v) \qquad (4)$$

Just as in the electrical case, the output (image) spectrum is obtained from H and the input (object) spectrum. It can be shown that

$$H(u, v) = P(u\lambda D_i, v\lambda D_i) \qquad (5)$$

$$= 1, \quad \text{if } \sqrt{u^2 + v^2} \le R/\lambda D_i$$

$$= 0, \quad \text{if } \sqrt{u^2 + v^2} > R/\lambda D_i$$

The transfer function is equal to the pupil function. It is plotted in Fig. 8, where the cutoff (spatial) frequency is written $u_m/2 = R/\lambda D_i = 1/2\lambda F$.

Now let us consider spatially incoherent illumination, in which the phases vary randomly across the mask. The frequency response function H can be defined by an equation like Eq. 4 relating Fourier transforms of image and object *intensities*. (The intensity I is the time average of the squared amplitude, $I = <UU^*>$.) In the incoherent case H is given by the overlap of two displaced pupil functions, as shown in Fig. 9. It can be calculated simply to be

$$|H(u)| = \frac{2}{\pi}[\cos^{-1}(u/u_m) - (u/u_m)\sqrt{1 - (u/u_m)^2}], \qquad (6)$$

for $u \le u_m$. The limiting spatial frequency is twice as great as for coherent illumination, but image contrast or modulation is less at the lower frequencies. The transfer function describes how the system degrades the image of a sinusoidal object grating. In general there is a reduction in modulation and a phase shift. The modulation transfer function (MTF) at frequency u is the ratio of image modulation (see Eq. 2.) to object modulation for an object grating of periodicity u^{-1}.

$$|H(u)| = M_i(u)/M_o(u) \qquad (7)$$

The authors of Reference 11 describe MTF measurements for projection, prox-

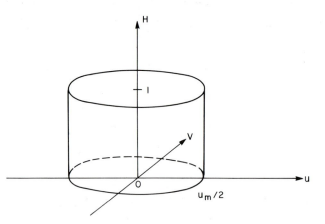

FIGURE 8
Coherent or optical transfer function for a round pupil.

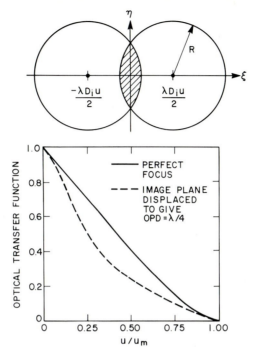

FIGURE 9
Incoherent transfer function for a round pupil, both in focus and out of focus.

imity, and contact printers using a square-wave bar target. The sine wave MTF can be calculated from the measured square-wave response.

In modern printers the illumination is intermediate between the coherent and the incoherent limits. The separation between points in the image which have correlated phases is neither infinite nor near zero. It can sometimes be varied by means of a diaphragm in the condenser illuminating the object. This is a familiar phenomenon to anyone who has used a microscope. As the aperture stop diaphragm is closed, the image looks "sharper," but interference rings appear around edges, and the light level drops. This is the near-coherent case. Opening the diaphragm causes the rings to disappear, and the image becomes brighter. The ratio S of the numerical aperture of the condenser to that of the projection optics is often used as a measure of coherence. (A slightly different definition is needed for the Kohler illumination often employed in printers.) $S = 0$ implies near coherence; if $S = 1$ the apertures are matched and the entrance pupil of the projection optics is filled to give nearly incoherent illumination. Further increase of S just causes more light scattering. Partial coherence has some advantages over incoherent illumination. The useful range, MTF > 0.6, is extended to higher spatial frequencies; edge gradients in the image become steeper; and the image is a little less sensitive to defocusing. Figure 10 shows a transfer function for a lens with $NA = 0.28$ and partial coherence $S = 0.75$ for the wavelength 436 nm. This is compared with a curve for a quartz lens with $NA = 0.38$ and incoherent

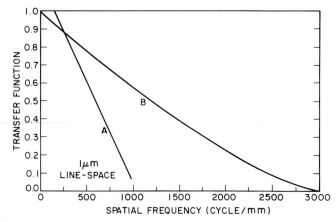

FIGURE 10
Comparison of two transfer functions. A. $NA = 0.28$; $S = 0.75$; $\lambda = 0.436$ μm. B. $NA = 0.38$; $S = 1$; $\lambda = 0.248$ μm.

illumination ($S = 1$). The wavelength is 248 nm. Obviously the second lens is capable of much higher resolution.

Since the transfer functions with round pupil are symmetrical under rotation about the vertical axis, they can be written and displayed as functions of a single variable u as in Eqs. 6 and 7 and Figs. 8 and 9. But it is understood that u stands for $\sqrt{u^2 + v^2}$. This two-dimensional nature explains why a small contact hole with predominant spatial frequency components $u = v = u_o$ requires a different exposure from a long line of the same width with $u = u_o$, $v \approx 0$, for $|H(\sqrt{2}u_o)|$ is less than $|H(u_o)|$. If both these types of features occur on the same mask, both types of resist image will not have correct dimensions. In general, for very small features representing high spatial frequencies, the required exposure depends on the shape of the feature and proximity to other features, setting a practical resolution limit much less than u_m.

Aberrations can be treated by including a phase error $\phi(x, y)$ in the pupil function: $P(x, y) \rightarrow P(x, y)\exp[i\,\phi(x, y)]$. This reduces the magnitude of the transfer function, except in the coherent case. Most printers have very nearly diffraction limited optics.

The focus error is a simple but important aberration. The error is a displacement of best focus away from its intended position. In Fig. 9 the dashed curve shows the effect of a displacement of the image plane or wafer from the focal plane by one Rayleigh unit $w = 2\lambda F^2$, corresponding to a phase error of $\pi/2$ at the edge of the pupil. Defocus affects different linewidths differently. [12,13]

Some information on several projection printers is collected in Table 1. They are of two types. The Perkin-Elmer 600HT uses reflective optics, as illustrated in Fig. 11. A curved lamp source (not shown) illuminates an arc on the mask, and this arc is imaged onto the wafer with unity magnification ($m = 1$). In this arrangement only a small zone of the spherical primary mirror is used, providing nearly diffraction-limited imaging. Mask and wafer are swept through the arc to

TABLE 1
Some projection printers

	m	NA	Field (mm^2)	$l_{0.6}$ (μm)	w (μm)	Align (3σ) (μm)	Throughput (wafers/hr)
Perkin Elmer 600HT	1	0.16	150ϕ	1.9	± 7.8	± 0.35	100
GCA 8000/1635	0.2	0.35	11 × 11	0.8	± 1.5	± 0.2	13(150mm)
GCA 8000/52529	0.2	0.29	18 × 18	1.1	± 2.6	± 0.2	37(150mm)
Nikon NSR 1505G4C	0.2	0.42	15 × 15	0.8	± 1.2	± 0.15	40(125mm)
Perkin Elmer SRA9535	0.2	0.35	17 × 17	0.9	± 1.7	± 0.15	50(125mm)
AT&T DUV	0.2	0.38	10 × 10	0.5	± 0.9	~± 0.3*	~30(100mm)

*2σ

form an image of the whole mask. The other printers use refractive optics: high quality lenses with many elements. They project a 5X reduced image onto the wafer.

As rough measures of resolution and depth of focus, $l_{0.6}$—the linewidth of the equal line/space pattern for which the MTF of an incoherent system would have a value 0.6—and w, the Rayleigh unit of defocus for an incoherent system, are listed in Table 1. Alignment tolerances and throughputs are those given by the manufacturer. If the whole image field cannot be filled with a large chip or several smaller ones and must be reduced in size, then more steps are needed to cover the wafer, and throughput decreases. The very small depth of focus of the high-resolution optics requires close automatic control of lens-to-wafer separation.

Because the size of the smallest feature resolvable with an optical system is proportional to λF, $(u_m = 1/\lambda F)$, higher resolution is obtained by working at shorter wavelengths. There is more room for improvement here in projection printing than in proximity printing, with which resolution is proportional to $\lambda^{1/2}$. (See the discussion of proximity printing.) F can also be reduced, but useful field size decreases with decreasing F. In addition, depth of focus is proportional to λF^2. Thus it is better to reduce λ rather than F. The reduced depth of focus and degree of linewidth control accompanying higher resolution seem to represent an important practical limit. Some other problems encountered in the 200 to 300 nm

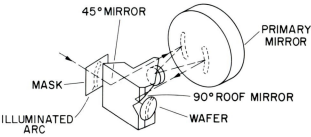

FIGURE 11
Projection optics, Perkin-Elmer printer.

wavelength range are increased Rayleigh scattering ($\sim \lambda^{-4}$), greater difficulty in reducing aberrations, and a need for new resists. Reflective optics are more suitable;[14] there are few optical glasses available for use in this wavelength range. Step-repeat exposure offers level-to-level registration precision that is independent of wafer size by separate alignment of each exposure field. The mask pattern dimensions are larger and therefore more convenient in the reduction projection systems than in those imaging with unity magnification.

The last entry in the table is an experimental stepper.[15] The source is an excimer laser operating at 248 nm. The lens is all silica. A next generation lens will provide the same resolution over a much larger field. Clearly this machine is a portent of things to come. The measure of resolution $l_{0.6}$ tabulated in Table 1 is meant to be an indication of the resolution obtained under production conditions. Ultimate resolution obtainable under carefully controlled conditions is higher. For example, the AT&T DUV stepper can image 0.35 μm line/space patterns in thin resist over a central part of the exposure field. More advanced systems will extend the ultimate resolution to about 0.2 μm.

Problem

Using the data of Table 1, divide field area by $l_{0.6}^2$ for each optical system shown. This ratio is the number of resolution elements. Explain why it is an order of magnitude larger for the first entry in the table than for the others despite the fact that *NA* is smaller for the first entry.

Solution

The whole field listed in the table for the first entry is not really comparable with the other smaller fields. The field considered for this purpose should be the (smaller) area of the arc-shaped image, which is scanned to fill a larger circular field.

Problem

1 A proximity printer operates with a 10 μm mask-wafer gap, and a wavelength of 430 nm. Another printer uses a 40 μm gap with wavelength 250 nm. Which offers higher resolution?

2 Why must the reticle (mask) used in a wafer stepper be completely free from defects? Why can some defects be tolerated in systems exposing the entire wafer at once?

Solution

1 $Q = W\sqrt{\dfrac{2}{\lambda g}}$.
   A. $\lambda = 0.43$ μm, $g = 10$ $\mu m \rightarrow Q = 0.68$ W
   B. $\lambda = 0.25$ μm, $g = 40$ $\mu m \rightarrow Q = 0.45$ W
   Therefore **A** gives higher resolution.

2 If a reticle contains a defect, that defect will appear in every exposed field on the wafer. If the field contains a single chip and the defect is fatal, then every chip will be inoperative. A mask projected onto the whole wafer at once need not be entirely defect free; a single defect will affect only one chip in this case.

4.3   ELECTRON LITHOGRAPHY

Electron lithography offers higher resolution than optical lithography because of the small wavelength of the 10–50 keV electrons. The resolution of electron lithography systems is not limited by diffraction, but by electron scattering in the resist and by the various aberrations of the electron optics. Scanning electron-beam systems have been under development for two decades, and commercial systems are available. The EBES (electron beam exposure system) machine has proved to be the best photomask pattern generator. It is widely used in mask shops. Because of the serial nature of the pattern writing, throughput is much less than for optical systems. However, some special products such as microwave transistors have for many years been manufactured by direct wafer patterning. In the first application to low-volume integrated circuits, some levels were patterned optically and some by electron beam.[16]

4.3.1   Resists

Electron exposure of resists occurs through bond breaking (positive resist) or the formation of bonds or crosslinks between polymer chains (negative resist). The incident electrons have energies far greater than bond energies in the resist molecules, and so all these energies are effective. Both bond scission and bond formation occur simultaneously. Which predominates determines whether the resist is positive or negative.

In a negative resist, electron-beam-induced crosslinks between molecules make the polymer less soluble in the developer solution. One crosslink per molecule is sufficient to make the polymer insoluble. Resist sensitivity increases with increasing molecular weight. If the molecules are larger, then fewer crosslinks are required per unit volume for insolubility. The polymer molecules in the unexposed resist will have a distribution of lengths or molecular weights and thus a distribution of sensitivities to radiation. The narrower the distribution, the higher will be the contrast γ. The exposure dose q has units of charge deposited by the beam per unit area—C/cm^2. Figure 12 shows thickness remaining after development for COP negative resist. γ is the slope and is called the contrast.

In a positive resist the scission process predominates, the exposure leading to lower molecular weights and greater solubility. Again, high molecular weight and narrow distribution are advantageous.

Two factors are of major importance in limiting resist resolution: swelling of the resist in the developer (more important for negative resists) and electron scattering. Swelling of negative resists, whether optical, electron, or X-ray, has two deleterious effects. Two adjacent lines of resist may swell enough that they touch. On contracting in the rinse they may not completely separate, leaving a "bridge" here and there. Secondly, this expansion and contraction weakens adhesion of very small resist features to the substrate and can cause small undulations in narrow (0.5 μm) lines. Both problems become less severe as resist thickness is reduced.

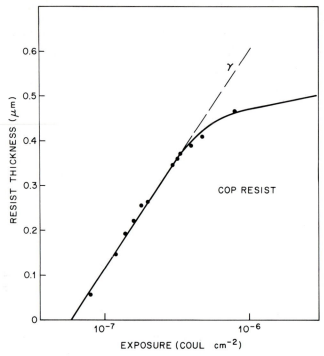

FIGURE 12
Exposure curve for COP negative resist.

When electrons are incident on a resist or other material, they enter the material and lose energy by scattering, thus producing secondary electrons and X-rays.[17] This fundamental process limits resolution of electron resist to an extent that depends on resist thickness, beam energy, and substrate composition. More electrons are scattered back into the resist from a GaAs substrate than from a silicon substrate, for example. The envelope of the electron cloud in the material can be pictured as an onion bulb pulled closer to the surface as beam voltage decreases. At higher beam voltage the electrons penetrate farther before being scattered over large lateral distances.

For an electron beam of zero width incident at position r on the resist-covered substrate, the distribution of energy deposited in the resist at depth z is closely approximated by

$$f(r, z) = a_1 \exp[-r^2/\beta_f^2(z)] + a_2 \exp[-r^2/\beta_b^2(z)] \qquad (8)$$

where $\beta_f(z)$ is the width of the forward-scattered beam in the resist, and $\beta_b(z)$ represents the back scattering from the substrate, with $\beta_b(z) >> \beta_f(z)$. $f(r, z)$ is the point-spread function, like that of Eq. 3 for optics. Generally the value of z of interest is that corresponding to the resist/substrate interface z_i. For a 25 keV electron beam penetrating 0.5 μm thick resist on a silicon substrate, for example,

$\beta_f(z_i) = 0.06 \ \mu m$, $\beta_b(z_i) = 2.6 \ \mu m$, and $a_2/a_1 = 2.7 \times 10^{-4}$. As separation between lines decreases, the back-scattered electrons contribute a greater dose between the lines where the dose should be zero. This is somewhat similar to the reduced modulation in an optical image at higher spatial frequencies. Indeed, electron exposure of resist may be treated by introducing a modulation transfer function, as in light optics. This formalism can also be applied to other types of lithography.

An exposed pattern element adjacent to another element receives exposure not only from the incident electron beam but also from scattered electrons from the adjacent element. This is called the proximity effect and is, of course, more pronounced the smaller the space between pattern elements. For example, an isolated 0.5 μm line requires 20–30% more exposure than 0.5 μm lines separated from each other by 0.5 μm. Thus, as pattern density increases, it becomes necessary to adjust exposure for various classes of elements, or in the extreme case, for different parts of elements.

Resist resolution is better in thinner resist layers. Minimum thickness is set by the need to keep defect density sufficiently low and by resistance to etching in device processing. For photomasks where the surface is flat and only a thin layer of chrome must be etched with a liquid etchant, resist thicknesses in the range 0.2–0.4 μm are used. For device processing in which topographic steps must be covered and more severe dry gas plasma etching is used, thicknesses of 0.5 μm to 2 μm are required. Most electron resists are not as resistant to dry etching as optical resists. One way of alleviating the problems of proximity effect, step coverage, and process worthiness is through use of a multilayer resist structure in which the thick bottom layer consists of a process-resistant polymer. In one realization of a three-layer structure, the uppermost layer of electron resist is used to pattern a thin intermediate layer, such as 1200 Å of SiO_2, which serves as a mask for etching the thick polymer below.[18] For electron lithography a conducting layer can be substituted for the SiO_2 to prevent charge buildup that can lead to beam placement errors. In another two-layer resist structure, both the thin upper layer and the thick lower layer are positive electron resists, but they are developed in different solvents. The thick layer can be overdeveloped to provide the undercut profile that is ideal for lift-off processing.

Table 2 lists a few readily available electron resists.[17] Many other resists are under development. PMMA stands for polymethyl methacrylate, the highest resolution resist known. MP-2400 is an example of an optical resist that is also electron sensitive. Values for sensitivity and resolution are approximate. Because faster electrons penetrate more deeply, more current is required at higher voltages. A resist is about one-half as sensitive for 20 keV electrons as it is for 10 keV electrons.

4.3.2 Mask Generation

The first widespread use of electron-beam pattern generators has been in photo-mask making. Let us see what advantages they have in this application. Figure

TABLE 2
Some electron resists

Resist	Polarity	Sensitivity (C/cm^2) @ 20 kV	Resolution (μm)	γ
PBS (Mead Tech.)	+	1.8×10^{-6}	0.5	1.7
PMMA (KTI Chem.)	+	1×10^{-4}	<0.1	2
EBR-9 (Toray Ind.)	+	1.2×10^{-6}	0.5	3
FBM-110 (Daikin Ind.)	+	1.5×10^{-6}	1.5	5
AZ 2400 (Shipley Co.)	+	2×10^{-4}	0.5	2
COP (Mead Tech.)	−	5×10^{-7}	1.5	0.8
OEBR-100 (Tokyo Okha)	−	5×10^{-7}	1.5	0.8
SEL-N (Somar Ind.)	−	1×10^{-6}	1	0.6
GMCIA (AT & T)	−	7×10^{-6}	0.5	1.7
CMS (Toyo Soda)	−	2×10^{-6}	0.7	1.5
RE-4000 N (Hitachi Chem.)	−	3.5×10^{-6}	1	1.3

13 shows two methods of making a photomask. On the left a reticle is patterned with an optical pattern generator—a machine which under computer control exposes and places pattern elements to form the chip image at 10× scale and which can make 60 to 100 exposures per minute. The pattern element is the image of an illuminated aperture of variable size. The reticle is then used as the object in a step-repeat camera, which steps the reduced image to fill the desired mask area. Stepping accuracy is interferometrically controlled. The master mask produced may be used in a projection printer or copied and the copies used in a contact printer. If an electron-beam pattern generator is used to make the reticle, the main advantage is speed in the case of complex chips. A large, dense chip can require 20 hours or more of optical pattern generator time, but only two hours or less of electron-beam pattern generator time.

On the right, Fig. 13 shows the more efficient method of electron-beam patterning. The mask may be either a 1× mask or a 5× or 10× reticle for use in

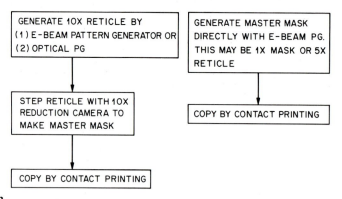

FIGURE 13
Two paths for generation of a photomask.

a wafer stepper. Table 3 shows some specifications of two types. CD stands for critical dimension.

Pattern placement errors (displacements of patterns from the desired locations on a mask) are measured with scanning, computer-controlled x–y optical measuring machines or with an electron-beam pattern generator. Figure 14 shows measurements at 100 points of a 100 mm × 100 mm array. For this mask maximum placement error was 0.099 μm. Stacking of two masks is often measured with an overlay comparator which superimposes the images of two masks. Fused silica mask substrates, with their low coefficient of thermal expansion, can reduce thermal contributions to stacking errors—the relative placement errors between two masks.

Misregistration of one circuit level to another on the wafer can be an important source of yield reduction. There are many contributions Δ_i to the total registration error Δ_t. If the contributions are mutually independent, then

$$\Delta_t = (\sum_i \Delta_i^2)^{1/2} \tag{9}$$

If the two masks used to image the level patterns do not overlay perfectly, there will be a registration error on the wafer. Thus the mask stacking errors contribute to Δ_t. The alignment tolerance of the printer that forms the image of the mask also makes a contribution. Another contribution to misregistration comes from wafer processing (etch tolerances, process-induced wafer distortion). If each of these three errors—mask stacking, alignment, and processing—has the small value 0.2 μm, then the total registration error would be $\Delta_t \geq 0.3$ μm. Since the ratio of minimum feature size to registration tolerance is usually in the range 3 to 5, registration of this precision would be barely adequate for circuits with 1 μm minimum features. There may be other contributions to Δ_t in addition to the three mentioned. For example, a change in the adjustment of the printer that transfers the mask image can lead to a change in image distortion with resulting placement errors.

Mask defects are either opaque spots in areas that should be transparent or pinholes where the chromium layer should be continuous. Sometimes critical masks are repaired to reduce the number of defects. Most vendors offer "zero

TABLE 3
Specifications of quartz master masks (Ultratech, Inc.), 125 mm plate, 100 mm array

	E-beam generated		Optically generated
	1X mask	5X reticle	1X mask
defects:	0.08/cm^2	0	0.03/cm^2
CD uniformity:	0.25μm*	0.25μm*	0.125μm*
CD tolerance:	±0.15μm*	±0.15μm*	±0.1μm*
stacking error:	±0.2μm*	±0.2μm*	±0.2μm*

*3σ

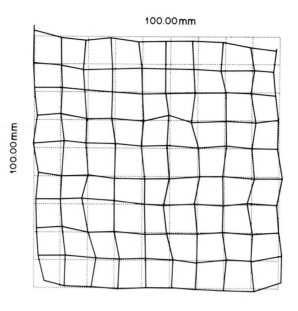

100.00mm

100.00mm

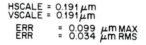

HSCALE = 0.191 μm
VSCALE = 0.191 μm
ERR = 0.099 μm MAX
ERR = 0.034 μm RMS

FIGURE 14
Photomask placement errors measured by MEBES at 100 points. The distortion scale is 0.191 μm per square.

defect" masks. If a master mask is copied by contact printing, the defect density of the copy will exceed that of the master. Defect density is generally measured by an optical instrument that compares one chip with a similar one on the mask and records differences or defects greater than a minimum size (~ 0.6 μm). Not every mask defect results in an inoperable chip. A reticle, however, must have no defects at all.

4.3.3 Electron Optics

Scanning electron-beam pattern generators are similar to scanning electron microscopes, from which they are derived. Figure 15 shows the basic probe-forming electron optical system. Two or more magnetic lenses form a demagnified image of the source on the wafer image plane. Provisions for scanning the image and blanking the beam are not included in the figure. The cathode is generally a thermionic emitter—either a tungsten hairpin or a pointed LaB_6 rod. Field emitters are also being used. In a field emitter, a strong electric field "pulls" the electrons out. The rod may be sintered material or a crystal. Emission current density from the cathode J_c is given by

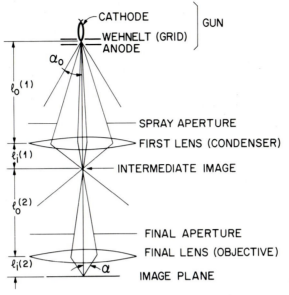

FIGURE 15
A simple two-lens probe-forming electron optical system.

$$J_c = AT^2\exp(-E_w/kT) \tag{10}$$

LaB$_6$ has a lower work function E_w than tungsten and about the same value of A (Table 4). A is called the Richardson constant and k is Boltzmann's constant. The electrons are accelerated through the voltage $V = 10$ to 50 kV and focused by the gun to a spot called the "crossover" of diameter $d_o = 10$ to 100 μm. The crossover is seen near the anode in Fig. 15. If I_b is the beam current at the crossover, the brightness—current density per unit solid angle—is

$$\beta = 4I_b/(\pi d_o \alpha_o)^2 \tag{11}$$

where d_o is as defined in Fig. 15. The maximum value of the brightness is

$$\beta = J_c e V/ \pi kT \tag{12}$$

TABLE 4
Thermionic emitters

	E_w (eV)	A (A/cm^2-K^2)	T (K)	J_c (A/cm^2)
Sintered LaB$_6$	2.4	40	1923	65
LaB$_6$ xtl.<100>*	2.47	14	1923	17
W	4.5	60	2700	1.75

* Single crystal with <100> orientation.

The thermal field emission (TFE) source consists of a tungsten tip of radius 0.5 to 1 μm heated sufficiently (1100°C–1400°C) to provide annealing of sputtering damage. The TFE source permits relaxed vacuum requirements and better stability compared to a cold field emitter. Equation 11 does not apply and there is no gun crossover. The source, in this case, is a small (100–1000 Å) virtual one located inside the tip. Thus, the gun and the imaging optics are very different from those of the thermionic source we consider. Thermal field emitters are of interest because for small image spots they give higher currents than thermionic sources.

The lenses in the electron optical column (Fig. 15) are magnetic. The field of such a lens, $\vec{B} = \hat{e}_r B_r(r, z) + \hat{e}_z B_z(r, z)$, has cylindrical symmetry about the optical axis z. If a parallel beam of radius r_o enters the field of the lens, the electrons experience a force that causes those not on axis to rotate about the axis and turn toward it. For a thin lens the electron path beyond the lens is given by

$$dr/dz \approx -(r_o e/8mV) \int_{-\infty}^{\infty} B_z^2 dz \qquad (13)$$

The constant $dr/dz = -r_o/f$, where f is the focal length of the lens and e is the electronic charge. Many rules of light optics apply here also, such as the thin-lens law relating object distance, image distance, and focal length:

$$1/l_o + 1/l_i = 1/f \qquad (14)$$

The magnification of a lens is $M = l_i/l_o$. An intermediate image is formed by the first lens with magnification M_1, and this is the object which is further demagnified by the second lens to form a spot in the image plane of diameter

$$d_i = Md_o, \quad M = M_1 M_2 \qquad (15)$$

where d_o is the object diameter and M_2 is the magnification of the second lens. Typically $M \approx 10^{-3}$–10^{-1}. Although the lengths $l_o^{(1)}$ and $l_i^{(2)}$ are fixed, $l_i^{(1)}$ and $l_o^{(2)}$ are variable. If current through the windings of the first lens of Fig. 15 is increased, $l_i^{(1)}$ decreases, M_1 is reduced, and the beam current passing through the final aperture is reduced because of the increased divergence of the beam at the intermediate image. The current density and current in the image plane are

$$J \leq \pi \beta \alpha^2 \quad \text{and} \quad I = J(\pi d_i^2/4) \qquad (16)$$

Spot sizes of interest are in the range from 0.1 to 2 μm. This is far from the diffraction limit. From Section 4.2.3 the diameter of the central spot of the Airy pattern is $1.2\lambda/\alpha$. $NA (\simeq \sin \alpha)$ is the numerical aperture. For 15 keV electrons the wavelength $\lambda = 0.1$ Å. Taking $\alpha = 10^{-2}$ radians, we have a diffraction spot width

$$d_{diff} = 1.2\lambda/\alpha = 10^{-3} \; \mu m \qquad (17)$$

Thus diffraction can be ignored. However, aberrations of the final lens and of the deflection system will increase the size of the spot and can change its shape as well. Aberrations of the other lenses are not so important since the intermediate images are larger. Figure 16 shows a typical double-deflection system above the final lens, arranged so that the deflected beam always passes through the center of the lens. In some designs the deflected beam is incident normal to the wafer surface (telecentric) to minimize runout error for warped wafers. The deflection coils provide a magnetic field perpendicular to the beam axis. There will be another set of deflection coils to provide deflection perpendicular to the plane of the figure as well.

The aberrations are treated as independent contributions to spot broadening, each contributing a circle of confusion or a blur to the spot. The actual spot size in the presence of aberrations is then the square root of the sum of the squares of the independent contributions. The aberrations are of two types: aberrations of the undeflected beam and aberrations that are functions of the distance r in the image plane from the axis to the position of the deflected beam. The aberrations of the first type are

$$\text{Spherical aberration: } d_s = (1/2\,C_s)\alpha^3$$

$$\text{Chromatic aberration: } d_c = C_c(\Delta V/V)\alpha$$

Spherical aberration is the focusing of rays passing through different parts of the lens at different distances; a ray near the axis has a longer focal length than one farther from the axis. $e\Delta V$ is the spread of electron energies, usually a few eV. C_s and C_c are constants that characterize the aberrations. Astigmatism, a third type of aberration, results from the breaking of cylindrical symmetry in the column and can be removed by introducing compensating fields with stigmator coils (Fig. 16). Deflection aberrations, such as coma and field curvature, are proportional to $r^2\alpha$ or $r\alpha^2$. We lump them all together and call their contribution to the spot

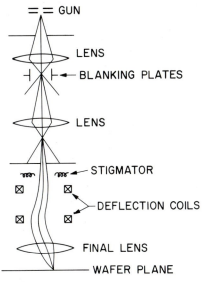

FIGURE 16
Electron optical system with double deflection.

spread d_{df}. In addition to the deflection aberrations, there is field distortion, but distortion is a shift of the image point, not a broadening of the spot. Table 5 shows these aberrations for the IBM VS1 machine.[19] We note that deflection aberrations dominate. Distortion and some of the deflection aberrations can be reduced by addition of correction signals to the deflection fields. Distortion with this machine is ± 2.5 μm at the edges of a 5 mm $\times$ 5 mm field, or $\pm 5 \times 10^{-4}$ expressed as a fraction of the field. For optical printers this ratio is typically much smaller.

Another source of spot broadening is the mutual Coulomb repulsion of the electrons as they traverse the column. For total column length L this contribution is approximately given by

$$d_{ee} \approx [LI/(\alpha V^{3/2})] \times 10^8 \mu \text{m} \qquad (18)$$

This spreading becomes important only at large currents when small spot sizes or sharp edges of large spots are sought. For example, if $I = 500$ nA, $L = 70$ cm, $V = 20$ kV, and $\alpha = 10^{-2}$ rad, then Eq. 18 gives $d_{ee} = 0.12$ μm. Another effect of the interelectronic interactions is an increase of energy spread ΔV of the electrons. Through chromatic aberration of the optics, this leads also to broadening of the spot. In the presence of aberrations Eq. 16 becomes

$$I = (\pi^2/4)\beta \alpha^2 [d^2 - d_s^2 - d_c^2 - d_{df}^2 - \ldots] \qquad (19)$$

d is the beam diameter in the presence of aberrations,

$$d^2 = d_i^2 + d_s^2 + d_c^2 + d_{df}^2 + \ldots \qquad (20)$$

Problem
An electron beam exposure system operates at 25 kV accelerating voltage. Column length is 1 m. Spot current is 300 nA, and numerical aperture of the final lens is 10^{-2} rad. The energy spread at the cathode is 0.2 V. If the coefficients of spherical and chromatic aberration are 10 cm and 62.5 cm respectively, what is the resolution limit at the center of the exposure field?

Solution

$$d_{ee} = (LI/\alpha V^{3/2}) \times 10^8$$

$$= 0.08 \ \mu m$$

$$d_s = \frac{1}{2} C_s \alpha^3$$

$$= 0.05 \ \mu m$$

$$d_c = C_c \left(\frac{\Delta V}{V} \right) \alpha$$

$$= 0.05 \ \mu m$$

$$d = (d_{ee}^2 + d_s^2 + d_c^2)^{1/2} = 0.11 \ \mu m$$

TABLE 5
**Aberrations at the corners of 2 mm × 2 mm and 4 mm × 4 mm exposure
fields for VS1.**

	2 mm field	4 mm field
Spherical, d_s	0.01 μm	0.01 μm
Chromatic, d_c	0.04	0.04
Deflection, d_{df}	0.087	0.35
Total, no dynamic correction	0.095	0.35
Total, dynamic correction	0.043	0.050

4.3.4 Raster Scan and Vector Scan

In raster scan the beam is deflected repetitively over the exposure field, as in a
television raster. The beam is turned on at various points in the scan to expose
the desired pattern. The EBES machine, developed by Bell Laboratories and
available commercially in the form of the MEBES system from Perkin Elmer,
uses beam deflection in one dimension (mainly).[20] The writing scheme is shown
in Fig. 17. The stage moves continuously in a direction perpendicular to the
writing direction. The pattern data are decomposed into a number of stripes, and
one stripe is written over all chips of the same type before the next stripe is begun.
    The stripe is 2048 addresses wide, an address corresponding to a width from
0.1 to 1.0 μm. Beam diameter can be varied from 0.1 to 1.0 μm. The pattern

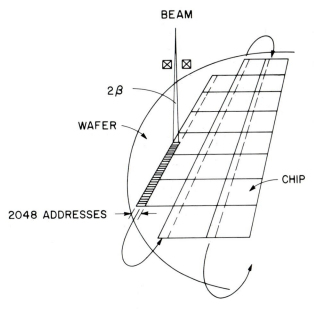

FIGURE 17
EBES / MEBES writing scheme. Curved arrows indicate the serpentine stage motion.

information comes to the blanking plates from a shift register at a 40 or 80 MHz rate. Since total time to write the 2048 scan is 31.6 μs $-$ 12.5 ns per spot plus 6 μs for flyback, the writing rate is approximately 8 cm^2/min with a 0.5 μm address and 2.4 cm^2/min with 0.25 μm address.

In a vector-scan pattern generator the beam is directed sequentially to the parts of the chip pattern to be exposed. The pattern is decomposed into a number of elements (rectangles, triangles, etc.) and each one is filled in by the writing beam. Many vector-scan machines expose in step-repeat fashion. Figure 18 shows an exposure field of dimensions F $\times$ F. The alignment marks in the field corners are first scanned to set deflection amplitude, offset, rotation, and x–y scan axis orthogonality. Marks may be etched trenches in silicon, or they may be metal layers or other features that give sufficient contrast. A mark position is sensed by the change produced in the number of back-scattered and secondary electrons on scanning over the mark edges. Marks may be etched into a wafer and several circuit levels overlayed by reference to the marks. The accuracy of the overlay depends on many variables such as field size, alignment mark condition, resist thickness, and stability of field distortion from one level to the next. Reported values of overlay error lie in the range ± 0.02 μm to ± 0.6 μm. The beam is scanned over each pattern element in turn. Dotted lines indicate where the beam is turned off, that is, deflected by voltages applied to the blanking plates (Fig. 16) and prevented from passing through an aperture. For example, the rectangle shown might be specified in the program by an instruction such as

$$(x_1, y_1)\text{RECT}(x_2, y_2)\text{R}(\theta) \tag{21}$$

Software, as in optical pattern generators, consists of an operating system and a high-level system in which pattern elements and commands are simply expressed. In one such machine the point (x_1, y_1) is determined by 16-bit digital-to-analog

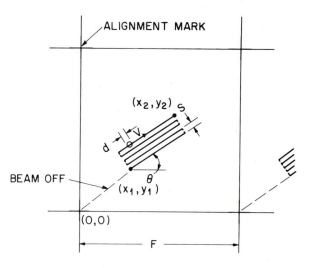

FIGURE 18
Vector scan writing of a rectangular pattern element. The scan starts at (x_1, y_1) and ends at (x_2, y_2).

converters, or DACs (65536 point resolution in x and y), and scan line lengths are determined by other DACs.

After all elements in the field are exposed, the stage is stepped to bring the next field under the beam and the exposure process is repeated. The stage need not provide highly precise motion. Stage position is generally monitored by a laser interferometer, and small differences from desired stage location are compensated by small offsets of the beam. If a chip is larger than an exposure field, then several fields can be used to expose the chip, but scan distortion at the field edges must be small enough so that patterns in adjacent fields are properly joined. If the machine is used to pattern masks rather than wafers, no alignment marks are used on the mask. In this case the deflection parameters are set by periodic reference to a mark fixed on the stage.

Once the beam current is set, exposure is controlled by varying the scan speed v and scan spacing s. These are set to the desired values for the whole pattern, or v might be set to different values for some pattern elements to compensate for proximity effects. From Eq. 21 it is obvious that few pieces of data are needed to trace out a pattern element that may contain very many beam diameters or "pixels." The proper exposure is found by exposing a test pattern at a series of scan speeds.

The scan field must be as large as allowed by deflection aberration and distortion because stage-stepping time affects throughput. Reported exposure fields lie in the range $F = 0.25$ mm to 6 mm — much smaller than for optical step-repeat machines.

Commercially available vector-scan machines are JEOL JBX6A3, EBMF of Cambridge Scientific Instruments, and Beamwriter of Philips. Vector-scan systems have also been developed by Hughes Research, IBM, AT&T, and Texas Instruments.

The EBES4 machine developed by AT&T Bell Laboratories has a continuously moving stage, unlike the step-repeat systems. Software is compatible with the other EBES/MEBES systems.[21] The coordinate system can be distorted digitally to correct for errors or to improve registration to existing levels. Column length was kept to 0.6 m to reduce Coulomb distortion of the spot (see Eq. 18). Three hierarchical deflection systems are used. Magnetic deflection provides coverage of a 0.28 mm square field with a 5 μs settling time. Deflection over a 32 μm square field is electrostatic with a 100 ns settling time. Microfigures are generated over a 2 μm square field by another set of electrostatic deflectors with settling time less than 0.5 ns. The combination of the three deflection systems is designed to be accurate to within $\pm$ 0.016 μm.

4.3.5 Variable Beam Shape

In the JEOL JBX6A3 machine shown schematically in Fig. 19, the spot is rectangular with variable size and shape. The image of the first square aperture may be shifted in two dimensions to cover various portions of the second aperture, which is the object imaged on the mask or wafer. Minimum rectangle width is 0.3 μm; maximum width is 12.5 μm. Maximum current density is 2A/cm^2

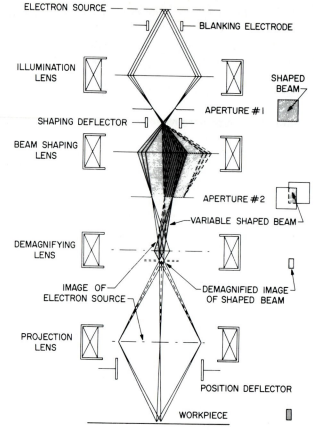

FIGURE 19
JEOL JBX6A3 electron optical system.

with the LaB$_6$ cathode. Electrostatic beam deflection is employed. Proximity effect corrections are easily made by variation of the exposure time of individual rectangles. The maximum exposure field is 2.5 mm × 2.5 mm. Thus, many fields must be stepped to expose a mask or wafer. For a chip containing 9×10^4 rectangles the patterning time, assuming a time of 1.6 μs per shot, is 0.14 seconds. The time to step and align each exposure field, including a 0.25 s alignment time, is 0.39 seconds. Thus, for each chip containing four fields we have

$$t_e = 0.14 \text{ s}$$
$$t_{sr} = 1.56 \text{ s}$$
$$\overline{ 1.7 \text{ s}}$$

Total time to pattern a 100 mm wafer containing 282 chips would be 8.0 min. Thus, 7.5 wafers/hour could be patterned. Many wafers can be held at one time in the vacuum system so that the load/unload time for each wafer is small.

The shaped-beam methods offer a way to project many image points in parallel and to achieve faster exposure without the need for the fast deflection system required with a small scanned beam.[22] Design of any electron-beam pattern generator necessarily is based on many compromises. The designers of EBES4 first considered the variable-shaped-beam approach. However, the machine is intended for sub-0.75 μm patterns.

Problem

Some high-resolution photoresists are also good electron resists with $q \approx 20\mu$ C/cm^2. Suppose 20% of the chip area is scanned by the beam, that the field size is $F = 6$ mm, that the step-repeat and overhead times are $t_{sr} = 0.4$ s/field and $t_{oh} = 0.3$ s/field, and that there are $N = 190$ chips/wafer. Calculate throughput for $I = 10^2$ nA.

Solution

$$t = N(t_e + t_{sr} + t_{oh}). \quad t_e = \eta F^2 q / I.$$

$$\eta = 0.2, F = 0.6 \text{ cm}, N = 190, t_{sr} = 0.4 \text{ s}, t_{oh} = 0.3 \text{ s}.$$

$$t_e = (0.2)(0.6)^2(2 \times 10^{-5})(10^7)$$

$$= 0.144 \times 10^2.$$

$$t = (190)(14.4 + 0.4 + 0.3)$$

$$= 0.797 \text{ hr.}$$

$$t^{-1} = 1.26 \text{ wafers / hr.}$$

4.3.6 Electron Projection

Electron projection systems are another method of achieving high resolution over a large field with high throughput. Rather than a small beam writing the pattern in serial fashion, a large beam provides parallel exposure of a large area pattern.

In a 1:1 projection system parallel electric and magnetic fields image electrons onto the wafer. The "mask" is of quartz patterned with chrome and covered with CsI on the side facing the wafer. Photoelectrons are generated on the mask/-cathode by backside UV illumination. Westinghouse, Thomson CSF, Philips,[23] Toshiba, and Radiant Energy Systems have developed systems. The advantages of the system include stable mask, good resolution, fast step-repeat exposure with low sensitivity electron resists, large field, and fast alignment. Proximity effects can be compensated for by undersizing or oversizing the features on the mask. Or the electron energy can be increased to 50 keV or more. Apparently neither method is entirely satisfactory.

Another problem remains to be solved before this system can become a strong contender for the high-volume production of advanced chips. The cathode has an unacceptably short life, only 50 exposures, before the CsI must be replenished by evaporation of fresh material.

4.3.7   Electron Proximity Printing

An electron proximity printing system has been under development for a number of years by IBM Deutschland.[24] This is a step-repeat system in which a silicon membrane stencil mask containing one chip pattern is shadow printed onto the wafer. Since the mask cannot accommodate re-entrant geometries (doughnut, for example), these are printed with two masks. Registration is accomplished by reference to alignment marks on each chip. Overlay error is less than 0.05 μm (3σ). An advantage of this system is its ability to measure and compensate for mask distortions. Proximity effects must be treated by changing the size of pattern elements. The major disadvantage of the system is the need for two masks for each pattern.

4.4   X-RAY LITHOGRAPHY

Since the proposal for X-ray lithography in 1972, the technique has been under development in many laboratories.[25] We saw how diffraction effects are reduced and resolution improved by reducing the wavelength in optical lithography. If the wavelength is reduced further, all optical materials become opaque because of the fundamental absorption, but transmission increases again in the X-ray region. In X-ray lithography an X-ray source illuminates a mask, which casts shadows into a resist-covered wafer. Materials useful for the absorptive and transmissive parts of the mask, the atmosphere in the exposure chamber, and the resist are in large part determined by the absorption spectra of these materials in the X-ray region.

The X-ray absorption of several materials is shown in Fig. 20. Over wide ranges of wavelength the absorption coefficient of an elemental material of density ρ and atomic number Z is proportional to $\rho Z^4 \lambda^3$. As λ increases, the proportionality constant decreases in step-function fashion at the "absorption-edge" wavelengths corresponding to the ionization energies of the inner electrons of the K, L, and other shells. Note the large differences in the absorption coefficient observed for different materials at the same wavelength.

4.4.1   Resists

An electron resist is also an X-ray resist, since an X-ray resist is exposed largely by the photoelectrons produced during X-ray absorption. The energies of these photoelectrons are much smaller (0.3 keV to 3 keV) than the 10 keV to 50 keV energies used in electron lithography, making proximity effects negligible in the X-ray case and promising higher ultimate resolution.

On traversing a path of length z in resist or any other material an X-ray flux is attenuated by the factor $\exp(-\alpha z)$. For most polymer resists containing only H, C, and O and with density $\rho \approx 1$ gm/cm^3, a 1 μm thick resist film absorbs about 10% of the incident X-ray flux at the Al$_{K\alpha}$ X-ray wavelength $\lambda = 8.3$Å. This small absorption has the advantage of providing uniform exposure throughout the resist thickness z and the disadvantage of reduced sensitivity. As in the optical case, X-ray resist sensitivity is generally quoted in terms of incident dose q(J/cm^2) required for exposure; sometimes absorbed dose αq(J /cm^3) is used.

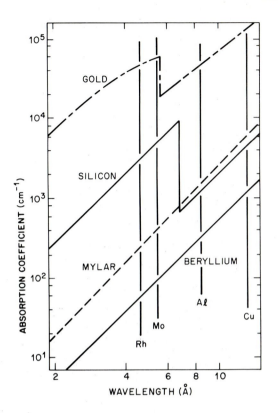

FIGURE 20
Absorption coefficients of several materials in the soft X-ray region. Sharp lines in the spectra of four sources (Rh, Mo, Al, Cu) are indicated.

For the shorter X-ray wavelengths the λ^3 dependence of the absorption coefficient leads to low sensitivity. This can be offset by incorporating heavier elements to increase absorption. For Pd$_{L\alpha}$ radiation, $\lambda = 4.37$ Å, the DCOPA resist incorporates chlorine. The Cl$_K$ absorption edge at 4.40 Å provides higher absorption, and chlorine is a chemically reactive species. Thus, two effects contribute to higher sensitivity.

The negative resists are limited in resolution by swelling during development, as in optical and electron lithography. Thus, minimum features of only 0.75 μm can be resolved in DCOPA of final thickness 0.5 μm. Improved resolution can be obtained using thinner resist in a multilayer structure as in optical and electron lithography. Conventional X-ray sources produce flux at the wafer that is typically less than that available in optical lithography by a factor of $\sim 10^3$. Resist sensitivity must be $\lesssim 10$ mJ/cm^2 to prevent long exposure times.

4.4.2 Proximity Printing

Because the X-ray wavelength is small, diffraction effects can be largely ignored and simple geometrical considerations can be used in relating the image to the pattern on the mask. The opaque parts of the mask cast shadows onto the wafer below. The edge of the shadow is not absolutely sharp because of the finite extent of the X-ray source (diameter of the focal spot of the electrons on the anode) at distance D from the mask. If the gap between mask and wafer, as shown in Fig. 21, is called g, this blur δ is given by

$$\delta = Sg/D. \tag{22}$$

Typical values are $g = 20\ \mu m$, $S = 3$ mm, $D = 40$ cm, $\delta = 0.15\ \mu m$. Resolution is determined by δ, by the minimum linewidth achievable in mask fabrication, and by properties of the resist used.

The angle of incidence of the X-rays on the wafer varies from 90° at the center of the wafer to $\tan^{-1}(D/R)$ at the edge of the exposure field of radius R. The shadows are slightly longer at the edge by the amount

$$\Delta = g(R/D). \tag{23}$$

This small magnification is generally of no concern. In the special cases where it may be undesirable, it can be compensated for when the mask is patterned. For multilevel devices the magnification must have the same value for each level, or at least its variation must be within the registration tolerance. This implies stringent control of the gap g. Wafer warping in processing depends on wafer thickness and orientation and on the nature of the process; it can be nearly eliminated with a proper vacuum chuck. It is not necessary that the gap have the same value at

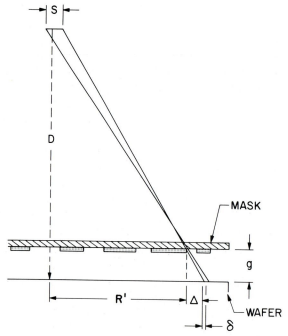

FIGURE 21
X-ray proximity printing.

all points on the wafer, only that the spatial variations be the same, within close tolerance, for all levels.

Automatic registration is desirable. Automatic systems with precision better than 0.1 μm have been reported. Such precision places stringent requirements on the mechanical design of the alignment stage; the mechanisms found in optical proximity printers are in general not adequate.

Machine imprecision is only one component of the total registration error. Such others as mask stacking errors due to pattern generator placement errors, mask distortion, and wafer process-related contributions also play a role.

4.4.3 X-ray Sources

X-rays are produced by the interaction of incident electrons and a target material. The maximum X-ray energy is the energy E of the incident electrons. If E is greater than the excitation energy E_C of the characteristic line radiation of the atoms of the material, the X-ray spectrum will contain these lines.

In early X-ray lithography experiments, the X-ray source was often an electron-beam evaporator with the chamber modified to accept a mask and wafer. The target metal could be changed easily to modify the X-ray spectrum. X-ray generation by electron bombardment is a very inefficient process; most of the input power is converted into heat in the target. The X-ray flux is generally limited by the heat dissipation in the target. With electrons focused to a spot of diameter 1 mm on an aluminum target on a water-cooled stem, 400–500 W is a typical upper limit for the input power. The X-ray power produced is only about 10 mW, and this power is distributed over a hemisphere. The X-ray power is proportional to the electron current. The power in the line radiation is also proportional to $(E - E_C)^{1.63}$ for a target thin enough that absorption of X-rays by the target itself can be neglected. However, as E increases, the electrons penetrate a thick target more deeply. The characteristic X-rays produced must pass, on the average, through more material on the way out, and are absorbed more strongly. This effect and the extension of the high energy cutoff of the X-ray spectrum set an upper limit for E.

Much higher X-ray fluxes are available from generators which have high-speed rotating targets.[26] The heat dissipation is spread over a larger area; the electron focal spot is on the rim of a spinning, water-cooled wheel. The wheel may be tilted or the rim beveled so that an elliptical focal spot appears round when viewed from the direction of interest. Because more power can be dissipated in the larger spot, source brightness can be increased in this way, at the expense of a relative enhancement of the higher energy portion of the spectrum and some increase in the flux spatial nonuniformity. Reference 26 reviews the status of high-power X-ray generator development.

Another type of source, which is capable of an order of magnitude greater flux, is the plasma discharge source. There are several versions, but all function by heating a plasma to a temperature high enough to produce X radiation. The radiation consists of strong lines superimposed on a weak continuum. The source is pulsed at a low repetition rate. In one embodiment the source size is 2 mm. Repetition rate is 3 Hz.[33] For FBM-G resist the total time to expose nine 2 cm × 2 cm fields is 120 seconds. If overhead is 30 seconds, then throughput is 24 wafers / hr. Special problems with such a source are reliability and contamination produced in the plasma chamber.

4.4.4 X-ray Masks

An X-ray mask consists of an absorber on a transmissive membrane substrate. The ratio of metal thickness to substrate thickness is greater than for a photomask. These thicknesses are determined by the transmission of the materials for the X-ray wavelength of interest.

Of the heavy metals with larger ρZ^4 absorbers, gold has been widely used because it is easily patterned. The thicknesses of gold necessary for absorption of 90% of the incident X-ray flux are 0.7 μm, 0.5 μm, 0.2 μm, or 0.08 μm for the X-ray wavelengths 4.4 Å(Pd_L), 8.3 Å(Al_K), 13.3 Å(Cu_L), or 44.8 Å(C_K), respectively. Thus, in general the metal is considerably thicker than the chromium layer on a photomask. Methods for patterning the gold with high resolution include electroplating and ion milling. Electroplating produces excellent definition with vertical walls, but requires a vertical-wall primary pattern in a resist that has a thickness equal to that of the metal to be plated. More often, a subtractive process has been employed in which a thinner resist layer is used to pattern a thin layer of a refractive metal; the refractive metal serves as a mask for ion milling the underlying gold. With this method it has been possible to form walls that depart from the vertical by 20° or less. The minimum linewidth attainable by ion milling in 0.5 μm thick gold is ~ 0.4 μm. For higher resolution, longer wavelengths such as the 13 Å Cu_L radiation may be used where gold thickness can be reduced. Lines as small as 0.16 μm have been replicated with this type of radiation.[34]

The membrane forming the mask substrate should be as transparent to the X-rays as possible, smooth, flat, dimensionally stable, reasonably rugged, and transparent to visible light if an optical registration scheme is used. Materials that have been used include polymers such as polymide and polyethylene terephthalate, silicon, SiC, Si_3N_4, Al_2O_3, and a Si_3N_4–SiO_2–Si_3N_4 sandwich structure. Although different mask substrates are appropriate for different portions of the soft X-ray spectrum, there is not yet general agreement on the best material for any particular wavelength.

The major questions remaining about X-ray masks concern their dimensional stability, minimum attainable defect densities, and ease of handling. Dimensional stability can be degraded by radiation damage produced by the X-ray flux. This also makes the mask substrate optically opaque. Polymer membrane substrates can be distorted locally by the absorber metallization. They are impractical. For greater dimensional integrity a stiffer substrate material is favored. If a plot like that of Fig. 14 is made for a BN X-ray mask, the pattern distortion is typically worse than for the photomask used to produce Fig. 14.

4.4.5   Synchrotron Radiation

Some experiments have been reported in which the synchrotron radiation from electron synchrotrons and storage rings was used for X-ray lithography. The small angular divergence of the radiation simplifies mask-wafer registration, and the high intensity of the radiation leads to short exposure times. A single storage ring could provide radiation to a large number of exposure stations.

In synchrotrons and storage rings, high-energy electrons are forced into closed curved paths by magnetic fields. An electron moving through a perpendicular magnetic field has an acceleration directed toward the center of the orbit and emits radiation. For the high-energy electrons of interest, which have velocities very nearly equal to that of light, the radiation is emitted in a narrow cone in the forward direction of motion of the electron. An observer looking along a tangent to the orbit sees a bright spot. The radiation is very different from that of a point source because of the narrow beam from each electron. The radiation from a circular machine would come from all tangents and have the shape of a disc. The angular divergence of the radiation in the vertical direction is $\psi \approx (1957E)^{-1}$, where E is the electron energy in 1 GeV and ψ is in radians. Thus, for a 1 GeV machine the vertical divergence is only 0.5 mradians.

High-energy electrons are provided to the storage ring by a microtron, a small synchrotron, or a small linear accelerator. The ring is briefly operated as a synchrotron to boost the electron energy to the final value. Then, the electrons may circulate for several hours in a stable orbit. The loss due to the power radiated as synchrotron radiation is compensated by one or more acceleration cavities around the ring. Nevertheless, the current slowly decays because electrons collide with the walls.

The peak of the power spectrum of the synchrotron radiation occurs at wavelength λ_p. This is related to the electron energy E (in GeV) and magnet-bending radius R (in m) by

$$\lambda_p = 2.35 \; R/E^3 \qquad (24)$$

A 0.83 GeV machine with $R = 2.1$ m would have a power spectrum peaked at $\lambda_p = 8.4$ Å.

During the last few years a Fraunhofer Institute in Berlin has been developing a small storage ring called COSY for X-ray lithography.[35] Figure 22 shows a sketch of this ring. Total floor space required is 30 m^2. Other components of the exposure system are under development elsewhere. For example, the stepper has been constructed by Suss. Alignment accuracy is ± 0.02 μm. COSY will provide a flux density of 250 mW/cm^2. There will of course be several beam lines, each with its own stepper. A similar effort is underway in Japan. The cost of such a facility for volume production of advanced chips would be high, but not prohibitive. The cost of a modern production line is as high as five times the cost of COSY. Special problems include reliability and safety. However, the goal for COSY is 95% up time. The resolution limit of COSY is determined by diffraction. Recall from section 4.2.2 the image of an edge of a feature on a mask used in proximity printing is spread over a distance $\approx 1.3 \sqrt{g\lambda/2}$. For COSY this

LARGER STORAGE RING

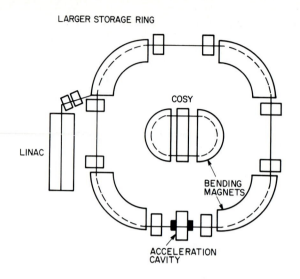

FIGURE 22
Sketch of COSY storage ring for lithography compared in size with a larger conventional ring.

diffraction blur is 0.2 μm. The parameters δ and Δ are much smaller because of the small source size and the small angular divergence of the radiation.

4.4.6 Commercial X-ray systems

Table 6 lists specifications for three commercially available X-ray step-repeat exposure systems.[36,37,38] The first two use conventional electron bombardment X-ray sources. These are automatic systems that can be compared with the optical exposure tools described in a previous section. Resolution of these systems is limited by penumbral blur and by the low-resolution resists that must be used. The third entry in the table has a laser-driven plasma source of very small diameter.[38] The larger X-ray flux of this machine allows the use of better resists. Ultimate

TABLE 6
Three commercial X-ray steppers

	XLS-1000 Perkin Elmer	MX-1600 Micornix	XRL5000 Hampshire
Source	Rotating, W	Pd	plasma
Wavelength (Å)	7.0	4.4	12-14
Gap (μm)	20	25	20
Blur (μm)	0.17	0.35	0.05
Field (mm^2)	≤30 × 30	≤ 50 × 50	14 × 14
Overlay error (μm)	±0.15(3σ)	±0.1(2σ)	±0.1(3σ)
Throughput (wafer/hr)	40, 100 mm*	10, 150 mm†	15–40, 150 mm

*10 mJ/cm^2 resist.

†20 mJ/cm^2 resist (DCOPA).

resolution is limited by diffraction. But practically usable resolution is limited by the accuracy of registration, which is in turn limited by control of the mask-wafer gap (± 0.5 μm).

4.5 ION LITHOGRAPHY

Ion lithography systems are of two types: a scanning focused-beam system and a masked-beam system. When an ion beam is used to expose resist, higher resolution is potentially possible than with an electron beam because of less scattering. In addition, resists are more sensitive to ions than to electrons. There is also the possibility of wafer processing without resists if the ion beam is used to implant or sputter selected areas of the wafer. The most important application is repair of photomasks, a task for which commercial systems are available.

The sensitivity for PMMA resist has been measured for 30 keV, 60 keV, and 200 keV He$^+$ ions and for 100 keV and 150 keV Ar$^+$ ions. The dose required for exposure is nearly two orders of magnitude less than with 20 keV electrons.[39] The perpendicular straggle of the path of an ion penetrating material and the range of low-energy secondary electrons produced are less than the range of back-scattered electrons produced in electron lithography. The ion energies for exposing resist depend on the ion. If the ion must penetrate 2500 Å of resist, then a proton would need 14 keV and a Au ion 600 keV for the projected range to be 3000 Å.

For a beam used in fabrication by sputtering, lower energies are generally of interest. When an ion beam is incident on a material, the sputtering yield increases with beam energy for beam energy larger than some small threshold value. There is an energy limit beyond which the yield decreases as the ions penetrate more deeply and fewer surface atoms receive enough energy to leave the surface. For example, the peak in the sputtering curve for Ar$^+$ ions incident on Cu occurs at 23 keV. For ion implantation, energies from 30 keV to 500 keV are used, and doses range up to 10^{15} ions / cm^2 (or 1.6×10^{-4} C/cm^2 for monovalent ions, representing a much larger dose than for resist exposure).

The problems of ion optics for scanning ion systems are more severe than for electron optics. The brightest sources are the two types of field ionization sources in which ions are produced in the strong field near a pointed tungsten tip. The source of ionized material is a gas surrounding the tip or a liquid metal that flows to the tip from a reservoir. The largest current densities obtained in the focused image of such a source are 1.5 A / cm^2 for Ga$^+$ in a 0.1 μm spot and 15 mA/cm^2 for H$^+$ in a 0.65 μm spot.[40,41] Total beam current is severely limited. There are no bright sources for such useful implant species as B and P. Electrostatic lenses rather than magnetic must be used for focusing ion beams. If a magnetic lens were used to focus an ion beam, the field would have to be much larger than in the electron optics case since from Eq. 13 the required field is proportional to $(mV/f)^{\frac{1}{2}}$, where m is the particle mass. Different isotopes would be focused to different points. Similarly, magnetic deflection is much less practical than electrostatic. Electrostatic optical systems generally have higher aberrations, necessitating small aperture α and small scan fields.

A prototype scanning system has been reported in which a beam of 57 keV Ga$^+$ ions is focused to a 0.1 μm diameter spot with current density 1.5 A/cm^2.[40] Spot size is apparently limited by chromatic aberration of the electrostatic lens and the large 14 eV energy spread of the source. There is also work on use of collimated ion beams with stencil masks. Some workers have employed a collimated 500 eV Hg$^+$ beam to transfer a pattern by sputter etching.[42]

Mask-based systems are of two types. One type is a 10X-reduction projection step-repeat system which projects 60–100 keV light ions through a stencil mask.[43] Overlay accuracy is $\pm 0.05\mu$m. The major disadvantage is the fragile foil mask which can treat re-entrant patterns only by double exposure with two masks.

The second type is a step-repeat proximity printer in which 300 keV protons are projected through the 0.5 μm thick portions of an all-silicon mask.[44] The mask is aligned so that the ions travel along the channels in the <100> direction. Projected throughput is sixty 100-mm wafers/hr, and the overlay error is $\sim 0.1\mu$m—barely adequate for 0.5 μm lithography. Resolution is set by scattering of the ions as they emerge from the channels. Edge resolution is 0.1 μm for the beam-scattering angle of 0.3°. The main disadvantage is the fragile mask with the 0.5 μm membrane. However, if a large fraction of the mask consists of the thicker silicon absorber, then the mask may be sufficiently rugged. The highest resolution in resists is obtained with a stencil mask and a beam of protons of energy 40 to 80 keV. Lines as fine as 400 Å have been printed in this way.[45]

4.6   SUMMARY AND FUTURE TRENDS

Optical lithography will continue to improve with wavelengths approaching 190 nm, the limit for silica. Even if reflective optics are employed, the radiation must still penetrate the mask. The optical wafer stepper will be the lithography system of choice for many years because of its relative simplicity, convenience, and reasonably high throughput. The practical resolution limit in production applications will be 0.5 μm or slightly lower. This resolution will be needed for MOS production by 1992. It will be available. As we have seen, the main barriers to higher optical resolution are (1) optical materials, (2) small depth of focus, and (3) the difficulty of obtaining diffraction-limited imaging over a large field.

Custom circuits and experimental devices for which high throughput is not needed will continue to be patterned by scanning electron-beam systems. In these applications fine definition, good overlay, flexibility, and quick turnaround are of primary importance. The main disadvantages of scanning systems are complexity and low throughput. Although impressive increases in throughput and reliability have been made by Japanese manufacturers of these systems, there is still the formidable problem that throughput is roughly inversely proportional to the square of the linewidth.

Finally, there is the application of a new lithographic method to high-volume production of advanced circuits with dimensions beyond the optical limit. There

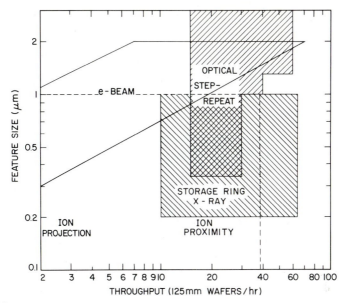

FIGURE 23
Resolution and throughput in the submicron regime. The boundaries of each region should be considered fuzzy.

are several candidates for this task: X-ray lithography with storage ring source, masked ion-beam lithography, and possibly electron proximity printing. The advantages and disadvantages of each have been discussed in previous sections. It is not yet possible to predict which will meet with success, just as it is not yet possible to predict with certainty the limit of miniaturization for the much more familiar MOS circuit technology. Some projections of resolution and throughput are indicated in Fig. 23. The rectangle bordered by the dashed lines indicates the regime of ion lithography. No boundary is shown between the ion projection and ion proximity printing regions. In general, the present projection machines have lower throughput than the proximity printers.

At present, resists are needed for all forms of lithography. Multilayer resist schemes are capable of high resolution. They will not find wide acceptance unless a rapid, convenient, production-worthy method of processing them is developed. Single-layer resist is also capable of high resolution with useful thickness.[46] As higher resolution is sought, resists may disappear. Indeed, lithography itself would then disappear and be combined with processing. New schemes combining patterning and formerly separate processing techniques would be both additive and subtractive in contrast with the present, nearly exclusive, use of subtractive etching. This development will be slow because the disdavantage of resist processing (application, developing) will be replaced by the necessity of working out a new technique for each material (polysilicon, SiO_2, $Ga_xAl_{1-x}As$, etc.). The new methods required for the sub-0.1 μm region will of course depend on the nature of the new circuit technologies that will replace MOS, bipolar, and GaAs MESFET designs.

PROBLEMS

1 Derive Eq. 6, $H(u) = \frac{2}{\pi}[\cos^{-1}(u/u_m) - (u/u_m)\sqrt{1 - (u/u_m)^2}]$, using the figure and simple geometrical considerations.

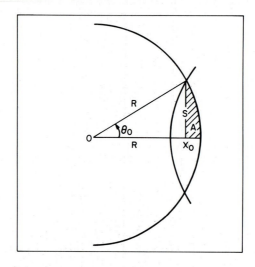

2 Suppose that you are required to specify the resist thickness that will be used in a production lithographic process. The following data are available:
- 1.5 μm minimum features must be printed. Resolution is adequate when the resist thickness T is in the range 0.5 to 2.0 μm, but feature size control is better for thinner resists.
- Each wafer has 150 chip sites; each chip has a 0.2 cm^2 active area.
- Five mask levels are required to complete the device.
- 2000 finished wafers must be produced each day (20 hr per day = 3 shifts).
- The resist defect density D_0 increases as the resist is made thinner, where D_0 is the number of defects per square centimeter, and is approximated by $D_0 = 1.4\ T^{-3}$. T is in microns.
- The chip yield (percentage good) can be approximated at each mask level by $y = (1 + qD_0a)^{-1}$, where q is the fraction of defects that render a chip inoperable (fatality rate) and a is the active area of the chip.
- On average, 50% of the defects are fatal defects.
- More time is needed to expose thick resist than to expose thin resist. The exposure tool throughput in wafers/hr is approximated by $125 - 50T$ for $(0.5 \leq T \leq 2.0\ \mu\text{m})$.
- **(a)** Specify the resist thickness to be used and justify your recommendation with tabular and graphical data.
- **(b)** If exposure tools cost $350,000 each, what is the difference in equipment cost for a process using 1 μm and 1.5 μm of resist?

3 In electron beam lithography the term *Gaussian beam diameter* (d_G) describes the diameter of an electron beam in the absence of system aberrations, that is, a beam distorted only by the thermal velocities of the electrons. The current density in a Gaussian beam is given by $J = J_p \exp\left[-(r/\sigma)^2\right]$, where J_p is the peak current density, r is the radius from the center of the beam, and σ is the standard deviation of electron distribution in the beam. Defining $d_G = 2\sigma$, derive an expression relating d_G to the peak current density J_p and the total current in the electron beam I.

Answer: $I = (\pi/4)J_p d_G^2$

4 The maximum current density J_m that can be focused towards a spot with a convergence half-angle α is limited by the transverse thermal emission velocities of the electrons in a Gaussian electron beam. J_m is given by the *Langmuir limit equation*

$$J_m = J_c\left(1 + \frac{eV_0}{kT_c}\right) \sin^2\alpha$$

where J_c is the cathode (source) current density, T_c is the temperature corresponding to the electron energy, k is Boltzmann's constant (1.38×10^{-23}J/K), and e is the electronic charge (1.6×10^{-19}C). For small convergence angles α, derive an expression that relates the Gaussian beam diameter d_G to the electron source parameters J_c, T_c, and V_0.

Answer: $d_G^2 \geq IkT_c/[(\pi/4)J_c eV_0\alpha^2]$

5 (a) The brightness β of a source of electrons is defined as the current density J emitted per unit solid angle Ω, that is, $\beta = J/\Omega$. The units of β are amperes per square centimeter per steradian. Assume that the current is emitted from (or coverages toward) a small area through a cone of included half-angle α and that α is small. Derive an expression relating the maximum source brightness to the source parameters J_c, T_c, and V_0.

Answer: $\beta \approx J_c eV_0/\pi kT_c$

(b) Assuming that brightness is conserved in the electron beam column, show that the Gaussian beam diameter d_G is related to the source brightness.

Answer: $d_G \approx (2/\pi)(1/\alpha)(I/\beta)^{1/2}$

6 Suppose that an x-ray resist must see a mask modulation greater than or equal to 0.6 in order to form useful resist images. What is the minimum gold thickness required on an x-ray mask to satisfy this requirement if the exposure wavelength is 4 Å?

Answer: $T \leq 0.31 \ \mu m$

REFERENCES

1 F. H. Dill, W. P. Hornberger, P. S. Hauge, and J. M. Shaw, "Characterization of Positive Photoresist," *IEEE Trans. Electron Devices*, **ED-22**, 445 (1975).

2 R. K. Watts and N. G. Einspruch, Eds., *Lithography for VLSI, VLSI Electronics: Microstructure Science*, Vol. 16, Academic Press, New York, 1987.

3 H. I. Smith, N. Enfremow, and P. L. Kelley, "Photolithographic Contact Printing of 4000 Å Linewidth Patterns," *J. Electrochem. Soc.*, **121**, 1503 (1974).

4 W. N. Jones, "A Far Proximity Photolithographic Process for Semiconductor Manufacture," *Proc. Microelectronics Sem. Kodak Interface,* **75**, 49 (1976).

5 B. J. Lin, "Deep UV Lithography," *J. Vac. Sci. Technol.*, **12**, 1317 (1975).

6 J. G. Skinner, "Some Relative Merits of Contact, Near-Contact, and Projection Printing," *Proc. Microelectronics Sem. Kodak Interface*, **73**, 53 (1974).

7 Y. Nakane, T. Tsumori, T. Mifune, "Deep UV Photolithography," *Proc. Microelectronics Sem. Kodak Interface*, **78**, 32 (1978).

8 D. A. McGillis and D. L. Fehrs, "Photolithographic Linewidth Control," *IEEE Trans. Electron Devices*, **ED-22**, 471 (1975).

9 R. C. Heim, "Practical Aspects of Contact / Proximity Photomask / Wafer Exposure," *Proc. SPIE*, **100**, 104 (1977).

10 M. Born and E. Wolf, *Principles of Optics*, 5th Ed., Pergamon Press, New York, 1975.

11 M. C. King and M. R. Goldrick, "Optical MTF Evaluation Techniques for Microelectronic Printers," *Solid State Technol.*, **20**, 37 (1977).

12 M. A. Narasimham and J. H. Carter, Jr., "Effects of Defocus on Photolithographic Images Obtained with Projection Printing Systems," *Proc. SPIE*, **135**, 2 (1978).

13 H. Moritz, "High Resolution Lithography with Projection Printing," *IEEE Trans. Electron Devices*, **ED-26**, 705, (1979).

14 J. H. Bruning, "Performance Limits in 1:1 UV Projection Lithography," *J. Vac. Sci. Technol.*, **16**, 1925 (1979).

15 V. Pol, J. H. Bennewitz, G. C. Escher, M. Feldman, V. A. Firtion, T. E. Jewell, B. E. Wilcomb, and J. T. Clemens, "Excimer Laser-Based Lithography: A Deep Ultraviolet Wafer Stepper," *Proc. SPIE*, **633**, 6 (1986).

16 R. D. Moore, "E-Beam Direct Writing in Manufacturing," *Proc. Symp. Electron and Ion Beam Technol.*, St. Louis, May 1980, p. 126.

17 L. F. Thompson, C. G. Wilson, and M. J. Bowden, Eds., *Introduction to Microlithography*, ACS, Washington, D.C., 1983.

18 J. M. Moran and D. Maydan, "High Resolution, Steep Profile Resist Patterns," *J. Vac. Sci. Technol.*, **16**, 1620 (1979).

19 T. H. P. Chang, A. D. Wilson, C. H. Ting, R. Viswanathan, M. Parikh, and E. Munro, "The Probe Forming and Deflection System for Vector Scan 1 E-B Lithographic System," *Proc. Symp. Electron Ion Beam Sci. Technol.*, Princeton, May 1976, p. 377.

20 D. R. Herriott, R. J. Collier, D. S. Alles, and J. W. Stafford, "EBES: A Practical Electron Lithographic System," *IEEE Trans. Electron Devices*, **ED-22**, 385 (1975).

21 D. S. Alles, C. J. Biddick, J. H. Bruning, J. T. Clemens, R. J. Collier, E. A. Gere, L. R. Harriott, F. Leone, R. Liu, T. J. Mulrooney, R. J. Nielsen, N. Paras, R. M. Richman, C. M. Rose, D. P. Rosenfeld, D. E. A. Smith, and M. G. R. Thomson, "EBES4-A New Electron Beam Exposure System," *J. Vac. Sci. Technol.*, **B5**, 47 (1987).

22 H. C. Pfeiffer, "Recent Advances in Electron-Beam Lithography for the High-Volume Production of VLSI Devices," *IEEE Trans. Electron Devices*, **ED-26**, 663 (1979).

23 R. Ward, A. R. Franklin, I. H. Lewin, P. A. Gould, and M. J. Plummer, "A 1:1 Electron Stepper," *J. Vac. Sci. Technol.*, **B4**, 89 (1986).

24 P. Nehmiz, W. Zapka, U. Behringer, M. Kallmeyer, and H. Bohlen, "Electron Beam Proximity Printing: Complementary Mask and Level-to-Level Overlay with High Accuracy," *Proc. Symp. Electron, Ion, Photon Beams*, Tarrytown, NY, May 1980, p. 136.

25 D. L. Spears and H. I. Smith, "High Resolution Pattern Replication Using Soft X-rays," *Electronics Lett.*, **8**, 102 (1972).

26 M. Yoshimatsu and S. Kozaki, "High Brilliance X-ray Sources," in *X-ray Optics*, H. J. Queisser, Ed., Springer, Berlin, 1977.

27 S. Yamazaki, S. Nakayama, T. Hayasaka, and S. Ishihara, "X-ray Exposure System Using Finely Position Adjusting Apparatus," *J. Vac. Sci. Technol.*, **15**, 987 (1978).

28 T. Hayashi, "Electron Beam and X-ray Lithography for VLSI Devices," *Proc. Symp. Electron Ion Beam Technol.*, Seattle, May 1978, p. 85.

29 R. K. Watts, K. E. Bean, and T. L. Brewer, "X-ray Lithography with Aluminum Radiation and SiC Mask," *Proc. Symp. Electron Ion Beam Technol.*, Seattle, May 1978, p. 453.

30 J. H. McCoy, "X-ray Lithography for Integrated Circuits—A Review," *Proc. SPIE*, **100**, 162 (1977).

31 J. R. Maldonado, M. E. Poulsen, T. E. Saunders, F. Vratny, and A. Zacharias, "X-ray Lithography Source Using A Stationary Solid Pd Target," *J. Vac. Sci. Technol.*, **16**, 1942 (1979).

32 G. A. Wardley, R. Feder, D. Hofer, E. E. Castelanni, R. Scott, and J. Topalian, "X-ray Lithography," *Circuits Mfg.*, **15**, 30 (1978).

33 I. Okada, Y. Saitoh, S. Itabashi, and H. Yoshihara, "A Plasma X-ray Source for X-ray Lithography," *J. Vac. Sci. Technol.*, **4**, 243 (1986).

34 D. C. Flanders and H. I. Smith, "Surface Relief Gratings of 3200 A Period, Fabrication Techniques and Influence on Thin Film Growth," *J. Vac. Sci. Technol.*, **15**, 1001 (1978).

35 A. Heuberger, "X-ray Lithography," in *Microcircuit Engineering 85*, K. B. van der Mast and S. Radelaar, Eds., Elsevier, Amsterdam, 1985.

36 B. S. Fay and W. T. Novak, "Advanced X-ray Alignment System," *Proc. SPIE*, **632**, 146 (1986).

37 R. B. McIntosh, Jr., G. P. Hughes, J. L. Kreuzer, and G. R. Conti, "X-ray Step-and-Repeat Lithography System for Submicron VLSI," *Proc. SPIE*, **632**, 156 (1986).

38 R. D. Frankel, J. P. Drumheller, A. S. Kaplan, and M. J. Lubin, "X-Ray Lithography Process Optimization Using a Laser-Based X-Ray Source," *Proc. Microelectronics Sem. Kodak Interface*, **86**, 82 (1987).

39 M. Komuro, N. Atoda, and H. Kawakatsu, "Ion Beam Exposure of Resist Materials," *J. Electrochem. Soc.*, **126**, 483 (1979).

40 R. L. Seliger, J. W. Ward, V. Wang, and R. L. Kubena, "A High Intensity Scanning Ion Probe with Submicrometer Spot Size," *Appl. Phys. Lett.*, **34**, 310 (1979).

41 J. Orloff and L. W. Swanson, "A Scanning Ion Microscope with a Field Ionization Source," in *Scanning Electron Microscopy*, ITT Res. Inst., Chicago, 1977, p. 57.

42 B. A. Free and G. A. Meadows, "Projection Ion Lithography with Aperture Lenses," *J. Vac. Sci. Technol.*, **15**, 1028 (1978).

43 G. Stengl, H. Löschner, W. Maurer, and P. Wolf, "Ion Projection Lithography Machine IPLM-01: A New Tool for Sub 0.5 μm Modification of Materials," *J. Vac. Sci. Technol.*, **B4**, 194 (1986).

44 J. L. Bartelt, "Masked Ion Beam Lithography: An Emerging Technology," *Solid State Technol.*, **28**, 215 (1986).

45 J. N. Randall, D. C. Flanders, N. P. Economou, J. P. Donnelly, and E. I. Bromley, "Masked Ion Beam Resist Exposure Using Grid Support Stencil Masks," *J. Vac. Sci. Technol.* **B3**, 58 (1985).

46 M. Sasago, M. Endo, Y. Hirai, K. Ogawa, and T. Ishihara, "Half-Micron Photolithography Using A KrF Excimer Laser Stepper," *Proc. IEDM*, 316 (1986).

47 R. K. Watts, "Advanced Lithography" *Very Large Scale Integration, Fundamentals and Application*, Ed. D F. Barbe, Springer, New York, 1982, Chapter 2.

CHAPTER
5

REACTIVE
PLASMA
ETCHING

R. J. SCHUTZ

5.1 INTRODUCTION

Integrated circuit fabrication processes that use reactive plasmas are commonplace in today's semiconductor production lines. The term reactive plasma is meant to describe a discharge in which ionization and fragmentation of gases take place and produce chemically active species, frequently oxidizers and/or reducing agents. Such plasmas are reactive both in the gas phase and with solid surfaces exposed to them. When these interactions are used to form volatile products so that material is removed or etched from surfaces that are not masked by lithographic patterns, the technique is known as reactive plasma etching.

When reactive plasmas were first used in semiconductor factories, in the early 1970s, little was known about the chemistry and physics of these discharges. Today, there is a broad base of empirical knowledge, some qualitative understanding of active mechanisms, and even some detailed understanding of specific isolated phenomena. This insight now allows for the tailoring of plasma processes to meet the more stringent requirements of today's submicron device features.

This chapter starts with a discussion of plasma discharge properties and the diagnostic techniques used to analyze them. The replication of lithographic patterns by etching is then described along with mechanisms that lead to patterns etched straight into the surface with no lateral material removal under the mask. This property is perhaps the one most important attribute of reactive plasma etching. Other process parameters are then discussed, followed by descriptions of etching equipment currently used in factories and research and development

184

laboratories. Finally, specific plasma processes used to etch material important to the semiconductor industry are reviewed.

5.2 PLASMA PROPERTIES

This section deals with fundamental aspects of chemically reactive plasmas. The definition of a plasma will be discussed along with basic parameters used to describe plasma properties. Techniques to measure these parameters will be reviewed, and finally some properties of plasmas used in typical etching processes will be discussed. Processing plasmas are extremely complex, and many of the details of physical and chemical interactions both within the plasma and with surfaces exposed to the plasma are not yet understood. Therefore, this section will not attempt to give a comprehensive review of plasma physics but will try to give the reader some understanding of the concepts involved. Finally, the interaction of plasmas with surfaces will be discussed with the goal of describing some of the mechanisms thought to be active in etching. For other comprehensive reviews of this subject see, for example, References 1–3.

5.2.1 DC Plasma Excitation

A plasma can be thought of as a collection of electrons, singly and multiply charged positive and negative ions along with neutral atoms, and molecules and molecular fragments. For most plasmas discussed in this chapter, these particles are confined in a reaction chamber and held at a pressure of 0.5 to 25.0 Pascals (1.0 Pa = 7.5 mTorr). The charged particles result from the interaction of the initially introduced gas with an applied electric field. A simple method of "exciting" a plasma is to place a dc potential across two conducting electrodes in the chamber as shown in Fig. 1. At a high local field, which might occur any place in the chamber when the potential is first applied, ionization of a neutral gas molecule, perhaps by photoionization or field emission, will occur. The released electron is then accelerated toward the positive electrode, or anode, and along the way undergoes a series of collisions. Elastic collisions simply deflect the electron, but many types of inelastic collisions can occur that serve to further ionize or excite neutral species in the plasma. For example, electron collisions such as

$$\text{Dissociative attachment} \qquad e + AB \rightarrow A^- + B^+ + e \qquad (1)$$

$$\text{Dissociation} \qquad e + AB \rightarrow e + A + B \qquad (2)$$

$$\text{or Ionization} \qquad e + A \rightarrow A^+ + 2e \qquad (3)$$

occur, resulting in atoms and molecules or molecular fragments in various states of ionization. Some of these collisions yield more electrons, which raise the state of ionization or the density of the plasma. Additional electrons are generated by secondary emission from energetic positive ions colliding with the negative electrode or cathode. Inelastic electron collisions can also cause neutrals and ions to be raised to excited electronic states that later decay by photoemission. These interactions cause the characteristic plasma glow.

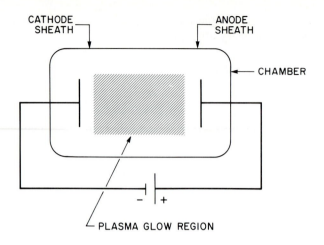

CATHODE SHEATH

ANODE SHEATH

CHAMBER

PLASMA GLOW REGION

FIGURE 1
Schematic representation of a dc glow discharge showing the glow and sheath regions when the cathode and anode are closely spaced. As the spacing is increased, the plasma structure become more complex.

Other types of collisions lower the ionization level in the plasma. The major loss of charged particles is by neutralization through collisions with the chamber wall. Rarer events, such as

$$\text{ion–ion recombination} \qquad A^- + B^+ \rightarrow A + B \qquad (4)$$

$$\text{and electron–ion recombination} \qquad e^- + A^+ \rightarrow A \qquad (5)$$

also decrease the plasma density. If a dynamic equilibrium is reached in which the number of charged particles remains constant on the average, then the plasma is self-sustaining. For typical plasmas used in semiconductor processes, this state is reached when less than 0.1% of the gas is ionized.

Neglecting the typically small concentration of negative ions, the electron and positive ion concentrations may be considered equal. Many of the electrical characteristics of plasmas are caused by the great disparity between the mass of the electrons and ions. For example, Fig. 1 shows a region adjacent to the cathode called the cathode sheath. The high mobility of light electrons and the repulsive field cause the electron population in this region to be so depleted that few electronic excitations occur and a glow is therefore not observed. For this reason a sheath is also frequently referred to as the *dark space*. Sheaths form not only at the cathode surface but also at the anode or at any conducting or insulating surface exposed to the plasma. The formation of these sheaths is caused by the much higher speeds of the electrons caused by their low mass. The higher electron speeds lead to a higher negative than positive charge flow through any given surface in any one direction. This higher rate of impingement of electrons onto any surface causes the surface to become negatively charged with respect to the plasma, thus repelling electrons and attracting positive ions that form a space-charged region by the surface. This region extends from the surface to where the field from the negative surface is totally shielded and the electron and positive ion concentrations are again equal. While the surfaces just discussed could be either insulating or conducting, it is important to note that both electrodes in a dc plasma discharge must be conductors. Covering either electrode with an insulating layer allows the surface of the insulator to charge negatively, independent of the electrode potential, to the isolated surface potential just discussed. The net electrode current through the plasma will fall to zero, the ionization rate in the plasma will decrease, and the plasma will extinguish.

5.2.2 AC Plasma Excitation

If the polarity on conducting electrodes is alternated at a relatively low frequency ($<10^4$ Hz), the effects just discussed will remain unchanged except for the periodic change in position of the cathode and anode dark spaces. If the rate of reversal, or frequency, is increased, the plasma characteristics will change when the period of oscillation is shorter than the time required for the plasma to reach equilibrium condition. This process is dominated by the heavy, slow ions. For example, if a radio frequency (rf) signal in the megahertz range is applied to an electrode through an impedance matching network, as shown in Fig. 2a, the responses of the plasma to the positive and negative cycles are different. When the electrode is positive, many highly mobile electrons are accelerated toward the electrode, causing a significant accumulation of negative charge. When the electrode is negative, heavy, immobile ions are accelerated toward it; however, significantly fewer of these ions strike the electrode than did electrons on the

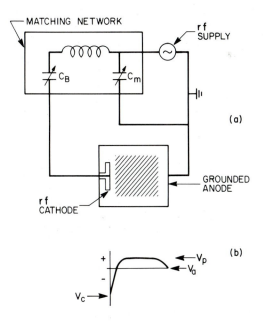

FIGURE 2

(a) A schematic representation of an rf plasma discharge where the power is supplied to the rf cathode through a matching network. (b) A plot of the average potential between the anode (V_a), the cathode (V_c), and the plasma (V_p). The horizontal position axis is meant to coincide with (a).

previous cycle. The plasma therefore acts electrically as a diode with a net negative charge building on the electrode. Note that Fig. 2a shows the electrode dc isolated from the power supply by a capacitor C_B. This "blocking" capacitor does not allow the electrode to discharge through the power supply.

Figure 2b shows the average dc negative potential of the two electrodes and the plasma. It should be kept in mind that there is actually an rf signal super-imposed on this dc potential. The negative dc potential on the "driven" smaller electrode, or cathode, is larger than that on the counter electrode, or anode. If the two electrodes were equal in area and the reaction chamber were made of a dielectric material or an isolated conductor, then from symmetry arguments V_a should be equal to V_c. An early classic theory of plasma characteristics predicts[4]

$$\frac{V_c}{V_a} = \left(\frac{A_a}{A_c}\right)^4 \tag{6}$$

where A_a and A_c are the anode and cathode areas, respectively. The plasma model used in this analysis treats the sheaths as pure capacitors (see Section 5.2.3 for a more detailed equivalent circuit model) so that larger areas have larger capacitances. The electrode/plasma/electrode circuit then acts as a capacitance voltage divider with the larger voltage across the larger impedance (smaller area). There is little conclusive evidence to support this relationship; in fact, most workers in the field have pieces of evidence which suggest the voltage ratio varies as the first, not the fourth, power of the area ratio.

Figure 2b indicates that there is no dc electric field in the plasma glow region. It can be considered a conductor which is at a dc voltage called the plasma potential, V_p. The voltage drops between this potential and the electrode potentials V_c and V_a are sustained across the sheaths. If no collisions occur in the sheath regions, then the average kinetic energy of singly ionized positive ions striking the cathode is $(V_p - V_c)$eV, where the units of V_p and V_c are volts.

An important difference between this high-frequency discharge and a dc plasma is that if either electrode is totally covered by an insulator, then the discharge will not be extinguished if the capacitor formed by the insulating surface is large enough to couple energy into the plasma. In fact, the effect of a cathode insulating surface is as if a capacitor were added in series to C_B in Fig. 2. This is the principal reason that high-frequency plasmas can be used so effectively in semiconductor processing. Replacing the cathode in Fig. 2 with an oxidized, and therefore insulating, wafer results in the surface of the wafer being raised to approximately the same average potential, as measured by a dc voltmeter, as when the cathode was a conductor. However, with the insulated cathode the voltmeter can no longer accurately measure this potential because the small current required to drive the meter cannot be supplied from the plasma through the insulating cathode. If the oxidized wafer covers only part of the conducting cathode, then once again the meter can accurately reflect the dc bias voltage at the cathode.

High-frequency electric fields cause more efficient ionization in a discharge than dc fields, as demonstrated by the decrease in minimum operating pressure as the excitation frequency increases. For example, a 13.56 MHz discharge can typically be sustained at 0.5 Pa, while 9.0 Pa might be necessary with a dc field. The mechanism of the enhanced ionization rate is thought to be caused by timely electron-atom collisions[1] when the velocities of the oscillating electron and the colliding electron add constructively.

5.2.3 Equivalent Circuits

An equivalent circuit for an ac plasma, neglecting its diode characteristic, is shown in Fig. 3. One may actually estimate the values of these resistive and capacitive components by noting the values of C_m and C_B in the matching network (Fig. 2a) at the condition of perfect tune with minimum reflected rf power back into the power supply.[5] Assuming the conductivity in the glow region of the plasma is dominated by fast electrons, then[6]

$$\sigma_e = ne^2/m_e \nu_e \qquad (7)$$

where σ_e is the electronic conductivity in the plasma, n is the charge density in the plasma, e is the electronic charge, m_e is the mass of the electron, and ν_e is the electron momentum transfer collision frequency. Then the resistance of this region is

$$R = \frac{\ell}{\sigma_e A} \qquad (8)$$

where ℓ and A are the length and cross section area respectively of the glow region. The sheath conductivity can be estimated in the same manner, replacing m_e and ν_e with m_i and ν_i, the ionic counterparts, since ion conductivity dominates in the sheath region. The sheath capacitances can be simply estimated from

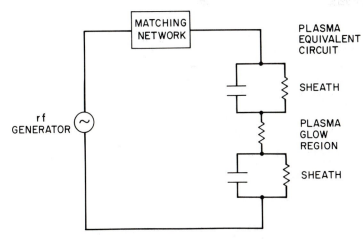

FIGURE 3
An equivalent circuit for an rf plasma discharge.

$$C_s = \frac{\epsilon_o A_s}{\ell_s}$$

(9)

where ℓ_s and A_s are the sheath length and cross section area, respectively. Figure 4 shows how these impedances change with frequency. The plasma body is assumed to be only resistive at these frequencies and therefore frequency independent. However, the sheaths change character; they are mostly resistive at low frequency and capacitive at high frequency. Recent experimental work indicates when the equivalent circuit model of Fig. 3 is valid.[7] How the transition with frequency affects the energy distribution of ions accelerated across the sheath can be seen[8] in Fig. 5. Here the time-averaged relative flux of ions is plotted versus energy for Cl_2^+ and Cl^+ in a Cl_2 ac discharge for two frequencies. At the low frequency (100 KHz), the maximum ion energy corresponds approximately to that furnished by the maximum sheath field because there is sufficient time for the ions to

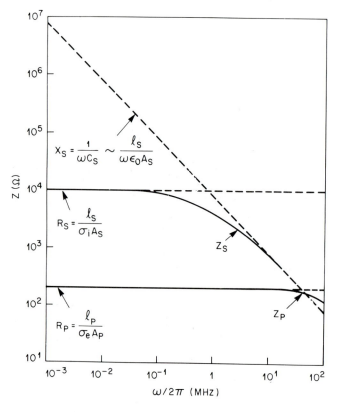

FIGURE 4

Calculated sheath and plasma impedances which show how the sheath impedance changes from resistive to capacitive with increasing frequency. The plasma remains resistive over the frequency range of interest. R and X are the resistive and reactive components respectively of the total impedance, Z. The subscripts s and p denote the sheath and plasma respectively. (*After Dautremont-Smith, Gottscho, and Schutz, Ref. 6.*)

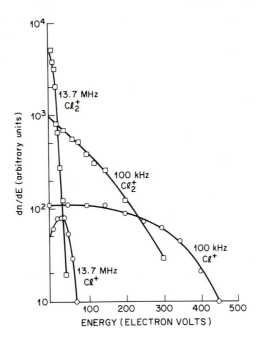

FIGURE 5
Ion bombardment energy distribution in a 40 Pa Cl_2 plasma with 0.6 W/cm^2 and a 1.0 cm electrode spacing. (*After Bruce, Ref. 8.*)

respond to the instantaneous field. The higher sheath impedance at lower frequencies causes this field to be higher. Collisions and response to the instantaneous field in the sheath cause the observed distribution to lower energies. At the higher frequency (13.7 MHz), the energy distribution is much narrower and at significantly lower energy than that predicted by the maximum field because the immobile ions cannot respond to the instantaneous field.

5.2.4 Plasma Analytical Techniques

This section will briefly outline some experimental techniques that are commonly used to measure plasma parameters.

Emission Spectroscopy. Most processing plasmas emit optically from the infrared to the ultraviolet. A simple analytical technique is to measure the intensity of these emissions versus wavelength with the aid of a spectrometer.[7] Using observed spectral peaks, it is usually possible to determine the presence of neutral and ionic species by correlating these emissions with previously determined spectral series. Relative concentrations of species can be obtained by correlating changes in intensity with plasma parameter changes; however, absolute concentrations cannot be obtained.

Actinometry. Quantitative concentrations can be obtained if the observed spectral intensities are compared to emissions from a small amount of inert gas (usually Ar), which is added to the plasma for this purpose. This technique, known as actinometry, has recently been used extensively to study plasma chemistry.[7]

Laser Induced Fluorescence. This technique again identifies and measures relative concentrations of ground state species by their emission spectra; however, here the plasma constituents are pumped up to excited states not by the plasma itself, but by a tunable dye laser.[7]

Mass Spectrometry. A self-contained vacuum mass spectrometer, which ionizes gas constituents with an electron beam and then mass separates them, can be used to study plasmas. Because this technique has its own ionization source, information on neutral species can be obtained, but little can be learned about actual ion concentrations in the plasma. To eliminate this problem some investigators sample impinging ions through a hole in one of the electrodes and then mass separate these ions which are produced in the plasma. In this way they can identify and estimate concentrations of ions striking the electrode.[9]

Langmuir Probe. A conducting probe smaller than the particle mean free paths and placed directly in the plasma is called a Langmuir probe.[10] Plasma electron density, electron temperature, and plasma potential can be measured from the current versus voltage characteristics of this probe. In this way the energy of ions striking the cathode can be inferred without the more difficult direct measurement discussed in the next section.

Retarding Grid. Another method of directly measuring impinging ion energy is to place a retarding grid and collector behind a sampling hole in an rf electrode.[11] The grid voltage versus collector current can yield information about the distribution of ion energy in addition to the average ion energy.

5.2.5   Plasma Chemistry

As suggested previously, the plasma discharge contains many neutral and ionic species that are continuously created and annihilated by collisions in the gas phase and with the chamber walls. The concentration of any species in the plasma must therefore be thought of as a dynamic equilibrium value. The modeling and calculating of these concentrations require the knowledge of many parameters— such as collision cross sections, dissociation rates, and recombination rates— which are often not known. Though the task is formidable, some recent work suggests that a continuum approach to the problem can yield reasonable results.[12] This model assumes that the properties of charged particles, such as diffusivity and mobility, can be described by continuum variables. Then continuity equations describe the creation, loss, and transport of all charge carriers.

An often quoted example of plasma chemistry that has been only qualitatively explained but has a significant impact on semiconductor processing is that of CF_4–O_2 plasmas. CF_4 is an example of a Freon®(see Table 1) which is normally extremely stable but which dissociates into F atoms and fluorinated fragments (CF_x) in a plasma. When CF_4 is used as the feed gas for an rf plasma etch, no appreciable etching of either Si or SiO_2 on the wafer occurs.[13] The etch rate of Si increases dramatically, as seen in Fig. 6, as O_2 is added to the feed gas. The etch rate stops increasing at about 12% O_2, at which point it starts to decrease with increasing O_2 concentration. A similar effect is observed for SiO_2 except that the peak occurs at 20% O_2. These effects can be explained by considering the dissociation and recombination of CF_4. Atomic F is the active etchant for Si and SiO_2 by the formation of the volatile SiF_4 and O_2.

$$Si + 4F \rightarrow SiF_4 \qquad\qquad (10)$$

$$SiO_2 + 4F \rightarrow SiF_4 + O_2 \qquad\qquad (11)$$

CF_4 continuously undergoes dissociative collisions in the plasma to form $CF_x(x \leq 3)$ radicals and F. At the same time, recombination of these two species is also occurring. The F concentration is therefore directly related to the difference in the rates of the two types of reactions. Adding O_2 to the plasma leads to the formation of COF_2, CO, and CO_2, which decreases the concentration of CF_x thereby decreasing the CF_x/F recombination rate. This results in a higher F concentration.[13] The etch rate increases with F concentration until the effect of O_2 diluting the CF_4 becomes significant. This point is reached sooner for Si etching because of surface effects. The O_2 molecules are thought to chemisorb on the surface of Si more readily than on SiO_2, thus blocking Si–F interaction.

If H_2 is added to the plasma instead of O_2, quite different results are observed.[14] Figure 7 shows the etch rates of Si and SiO_2 in a CF_4 + H_2 plasma as a function of H_2 concentration. The SiO_2 etch rate is relatively insensitive to H_2 concentration over the range investigated; however, the Si etch rate decreases dramatically with increasing H_2 concentration. This effect is explained by two mechanisms. First, the H_2 can react with F to form HF. This reaction competes with the Si–F interaction (Eq. 10) thus reducing the Si etch rate. In addition, H_2 affects the carbon-related chemistry. Equations 10 and 11 neglect the presence of carbon. The plasma etching of Si with fluorinated Freons is also thought to involve the reaction of CF_x radicals with the surface.[15] In an oxidizing ambient

TABLE 1
Chemical formulas of
Freons® commonly used as
reactive plasma feed gases.

Freon®	Formula
11	$CFCl_3$
12	CF_2Cl_2
13	CF_3Cl
14	CF_4
23	CHF_3
115	C_2ClF_5

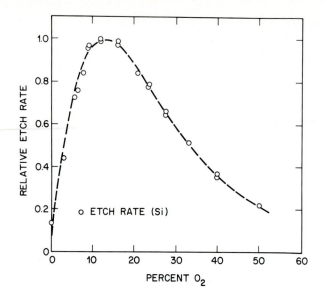

FIGURE 6
Relative etch rate of Si in a $CF_4 + O_2$ plasma vs. concentration of O_2. (*After Mogab, Adams, and Flamm, Ref. 13.*)

or in the case of SiO_2 these reactions result in etching

$$CF_x + Si + O_2 \rightarrow SiF_y + (CO, CO_2) \tag{12}$$

$$CF_x + SiO_2 \rightarrow SiF_y + (CO, CO_2) \tag{13}$$

in contrast to a reducing ambient (with H_2 present), which results for Si in the

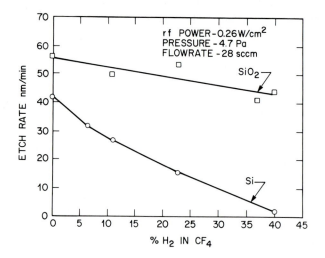

FIGURE 7
Etch rates of Si and SiO_2 in a $CF_4 + H_2$ plasma versus concentration of H_2. (*Ater Ephrath, Ref. 14.*)

production of CF_x or $(CHF_x)_y$ residues. For example:

$$CF_x + Si \rightarrow C + SiF_y \tag{14}$$

The reaction of the C on the etching surface with F and/or H results in fluorocarbons and hydrocarbons, which polymerize on the surface of the Si. The formation of this polymer on Si reduces the etch rate. In the case of SiO_2 in a reducing ambient, etching proceeds as in Eq. 13 with no polymer formation.

5.2.6   Plasma Surface Interactions

The interaction of reactive plasmas with surfaces is often divided into two components: physical and chemical. A physical interaction refers to the surface bombardment by energetic ions accelerated across the sheath. Here the ions' loss of kinetic energy in the surface dominates the interactions. Chemical interactions are standard electronic bonding processes that result in the formation or dissociation of chemical species on the surface. These interactions do not rely on the kinetic energy of impinging ions. This section will show that while this distinction is often convenient when discussing reactive plasma processes it is too simplistic a model of plasma surface interactions.

Sputtering. Sputtering is the ejection of material from a surface caused by bombardment by energetic inert ions such as Ar^+ or Xe^+. (Sputtering as a source for thin film deposition in an Ar plasma is discussed in Section 9.4.1.) Since these billiard-ball-like collisions never result in ambient gases bonding with ejected surface atoms, the interaction is purely physical. An important sputtering parameter is the yield, which is defined as the number of ejected surface atoms per incoming ion at a given ion energy. Yield plots for many ion–surface combinations have been determined.[16] One important aspect of sputtering is that the yield changes with the angle of the ion flux, as shown in Fig. 8. The maximum sputtering yield occurs because fewer collisions close to the surface are required to eject an atom as the flux angle is moved away from the surface normal. This preferential material removal often results in resist faceting, which is discussed in Section 5.3.5.

Plasma Surface Chemistry. A reactive plasma discharge generates many chemically active neutral species in addition to ions. Because these neutrals are frequently more active etchants than the original molecular source gas, it is impossible to attribute a plasma-enhanced etch rate simply to ion bombardment. For example, chlorinated and fluorinated plasmas contain atomic chlorine and fluorine, respectively, which are far more chemically active than the naturally occuring molecular species. Experiments were performed that attempted to isolate the contributions of these various etch components. For example, atomic F and molecular F_2 etch Si spontaneously at room temperature without a plasma. This apparently occurs through an initial addition of fluorine to the surface with a later desorption of the volatile SiF_4.[17] When ions strike a surface undergoing such a F–Si interaction, the increased etch rate is larger than can be accounted for by simply adding a sputtering component to the chemical

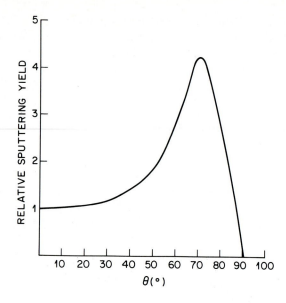

FIGURE 8

A typical curve of relative sputtering yield versus incidence angle, θ, of ion beam. (*After Townsend, Kelly, and Hartley, Ref. 16.*)

rate.[18] This synergistic interaction could be caused by the ions accelerating the desorption[17] of SiF_4 or by the ions damaging the near surface, thus increasing its reactivity with F.[19] This type of reaction is usually referred to as ion-enhanced etching. Cl_2, on the other hand, does not etch silicon spontaneously, but a similar chlorination of the surface occurs. In the presence of ion bombardment, such as in a plasma, this surface desorbs products, some of which are volatile ($SiCl_4$) and some of which are precursors ($SiCl_2$) to volatile products.[20] Ion bombardment contributes to this etch mechanism again not by simply adding a sputtering component but by synergistically interacting with surface chemistry to produce etching. This interaction was shown by studying reactive etch rate dependence on the angle of ion bombardment.[21] The angular dependence observed was that expected if the rate were dependent only on ion flux and not on the sputtering yield of Fig. 7. This type of reaction is often referred to as ion-induced etching because it occurs only in the presence of ion bombardment.

5.3 FEATURE-SIZE CONTROL AND ANISOTROPIC ETCH MECHANISMS

5.3.1 Pattern Transfer

Many semiconductor IC fabrication processes involve the sequential formation of patterned layers of materials on the wafer surface. Usually these layers alternate between insulators and conductors. Insulating layer patterns normally cover most of the surface with windows opened in the layer through which contacts can be made or in which devices are formed on the Si surface. Conducting layers normally appear as series of lines which can contact active device terminals through windows in the insulator and serve as interconnects between devices (see Section 9.1). As described in the previous chapter, the pattern that will eventually be in these layers is first defined in a polymeric resist film and then transferred to the layer. As shown schematically in the cross section views of Fig. 9, this pattern transfer process is accomplished with one of two techniques. For the lift-off technique (Fig. 9a) the patterned resist film is formed first followed by a blanket deposition of the layer. Dissolving away the resist then "lifts off" the unwanted material. Two disadvantages of this technique are a rounded feature profile and temperature limitations. Because of shadowing the deposited feature profile made with the lift-off technique has a rounded top. A more desirable profile has a rectangular cross section that maximizes cross sectional area and minimizes electrical resistance for a conducting line. With lift-off, the deposition technique is also limited to temperatures below 200–300°C, at which point resist begins to degrade.

For these reasons most patterns are transferred using the etching technique.

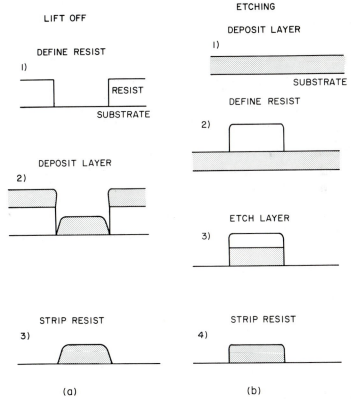

FIGURE 9
A schematic representation of two techniques for transferring resist features into a layer. (a) Shows the resist/deposition strip sequence of lift off, and (b) shows the deposit/resist/etch/strip sequence of etching.

Here the layer is first blanket deposited; then the patterned resist is formed and the layer is etched using the resist as a mask. If the etching process attacks the layer surface equally in all directions, the etch is said to be isotropic. Most liquid etches are isotropic and result in the undercut of the mask and the narrowing of the feature (Fig. 10a). For feature linewidths and line spaces significantly larger than the feature height, isotropic etching presents only a minor problem. Once the amount of the feature narrowing for an isotropic etch is known, then the designed feature size can be achieved by making the mask larger, as shown in Fig. 10b. Mask compensation combined with isotropic etching is a common practice in LSI technology. If the etching process attacks the layer significantly faster in the vertical direction than in the horizontal, the etch is said to be anisotropic (Fig. 10c). For VLSI devices, in which feature heights have comparable dimensions to feature linewidths and spaces, mask compensation is impossible. For this reason plasma etch techniques, which are capable of anisotropic etching, are so important to VLSI.

The degree of anisotropy of an etch process can be expressed as

$$A = 1 - \frac{V_h}{V_v} \tag{15}$$

where A is the degree of anisotropy, V_h is the horizontal etch rate and V_v is the vertical etch rate. Isotropic etching is represented by $A = 0$ and anisotropic etching by $A = 1$. One must be careful when trying to extract this parameter from the profile of an etched feature. The profile in Fig. 10d represents a feature etched to completion and then left in the plasma for an additional "overetch" time. Here V_h / V_v is not simply the feature height divided by the undercut. For this to be true

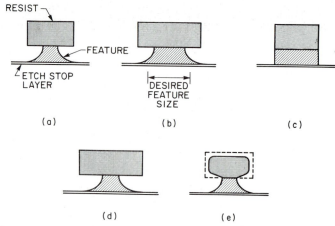

(a)   (b)   (c)

(d)   (e)

FIGURE 10

A schematic representation of some commonly observed etched profiles: (a) purely isotropic etch, (b) isotropic etch with a compensated mask, (c) anisotropic etch with no horizontal component, (d) isotropic etch with overetch, and (e) isotropic etch with isotropic etching of the mask.

the feature must be etched with no overetch. Figure 10e shows a feature etched with no overetch; however, the mask etched horizontally and again the undercut observed after etching are not indications of V_h. Only from Fig. 10a, for which the etch was accomplished just to completion and no horizontal resist erosion occurred, can A truly be determined from a final feature profile.

Figure 11 is a scanning electron micrograph (SEM) of an anisotropically etched pattern of polycrystalline silicon. The patterned conducting layer is used here as the gates for MOS transistors.

5.3.2 Ion Enhanced and Ion Induced Etching

As discussed in section 5.2.6, ions accelerated across the sheath in an rf plasma cause either ion-enhanced or ion-induced etching. The paths of these ions as they are accelerated across the sheath potential are essentially vertical, perpendicular to a horizontal wafer. There are two sources of horizontal velocity components of these ions. Thermal velocities are random and are therefore a source of horizontal motion. These velocities are small at room temperature compared to vertical velocities resulting from sheath voltages, which normally range from 50–500 volts. Collisions within the sheath can also cause off-vertical motions. Neglecting

FIGURE 11
An SEM of an anisotropically etched MOS gate-level pattern of polycrystalline silicon on a patterned field oxide used to isolate different transistor regions. (*SEM courtesy of A. Kornblitt, AT&T Bell Laboratories.*)

these two minor effects, ion paths can be assumed to be vertical. Only horizontal surfaces, then, experience ion bombardment. Ion-assisted processes are expected to produce anisotropic etching with some degree of mask undercut while ion-induced processes should produce perfectly anisotropic etching and result in the vertical profile of Fig. 10c.

5.3.3 Recombinant Species Mechanism

Another important mechanism for anisotropic etching was first recognized when phosphorus doped Si, which etched isotropically in a Cl_2 plasma, would etch anisotropically if C_2F_6 were added to the source gas.[22] This effect was qualitatively accounted for by assuming that the two gases dissociated in the plasma by

$$C_2F_6 + e \rightarrow 2CF_3 + e \tag{16}$$

$$Cl_2 + e \rightarrow 2Cl + e \tag{17}$$

Two possible surface interactions are therefore

$$xCl + Si \rightarrow SiCl_x \text{ (etching)} \tag{18}$$

and

$$CF_3 + Cl \rightarrow CF_3Cl \text{ (recombination)} \tag{19}$$

Equation 18 results in etching of the Si surface while Eq. 19 results in recombination of gaseous species with no material removal. It was speculated that ion bombardment enhanced the reaction in Eq. 18 and promoted etching while Eq. 19 was the dominant reaction on the feature sidewalls. The recombinant therefore acts as a sidewall passivant in the presence of an etch which might otherwise be isotropic.

5.3.4 Sidewall Mechanism

This idea of sidewall passivation can be extended to account for effects that were recently observed. For example, some plasma etch processes result in the formation of stable films on the sidewalls of etched features. These films are also capable of passivating or protecting features from etching species.[23] There are at least two mechanisms that can account for this sidewall buildup. The first is the deposition of a polymeric material that is known to occur in plasma discharges with carbon-containing source gases.[15] In the case of fluorine-containing Freons as source gases, this polymer deposition is linked to the formation of unsaturated CF_2 radicals generated by the plasma.

   The second source of material on feature sidewalls is the etch product species generated at horizontal surfaces exposed to ion bombardment. As mentioned in

Section 5.2.6, these products are frequently nonvolatile and can stick to and react with vertical surfaces not exposed to ion bombardment. This source of sidewall buildup is termed redeposition and is'commonly observed in sputter etching where all etch products are nonvolatile.

What feature profile might then be expected for pure anisotropic etching in the presence of plasma deposition and/or redeposition from etching surfaces? A simplified model of such a profile can be obtained if one assumes that (1) the deposition rate is constant onto all vertical surfaces, (2) the deposit also etches purely anisotropically at the same rate as the etching layer, (3) the simultaneous etching and deposition can be approximated by sequential etching and deposition processes, and (4) a finite element two-dimensional model of a feature cross section can be used.

Problem
Using the above assumptions, generate on a piece of graph paper a series of sequential cross sections depicting the anisotropic etching of a feature with sidewall deposition. Also assume the initial resist profile is vertical and the resist etches at half the rate of the etching layer.

Solution
The six sequential profiles shown in Fig. 12 result from five etch-deposition-etch steps.

This simplified model predicts an outwardly tapered feature with a trapezoidally shaped deposited layer on the feature sidewalls as in Fig. 13a. Figure 14 shows that such profiles have, in fact, been observed. As the assumptions change about the deposited layer, so does the final profile. For example, if the redeposition etches slower than the resist, it will appear as "horns" sticking up on the sides of the resist as in Fig. 13b. Assuming that the deposition rate is not constant on all vertical surfaces, other effects can occur. For example, if deposition from the plasma is the dominant mechanism, then the rate will be higher toward the tops of closely spaced features. This leads to profiles not unlike those observed when physical vapor-deposited layers are formed on closely spaced patterned structures (see Fig. 13c). This inwardly tapered profile might be mistaken for isotropic etching if the original resist feature is not carefully measured. If redeposition

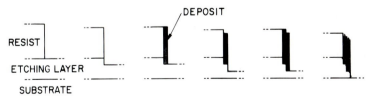

FIGURE 12
The sequential formation, from left to right, of an etching feature profile in the presence of deposition. The etching of horizontal surfaces and deposition onto vertical surfaces is assumed to occur sequentially.

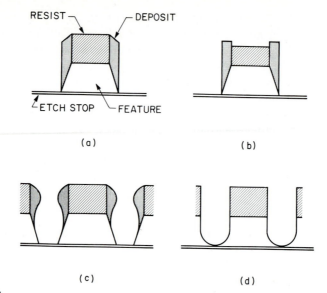

FIGURE 13
Some feature profiles that can be caused by deposition during anisotropic etching. (a) A trapezoidally shaped deposit caused by a uniform deposition rate which can result from widely spaced features. (b) This same type of deposit can result in "horns" if the resist etches faster than the deposit. (c) Inwardly bowed walls can be caused by a plasma source of deposit on closely spaced features. (d) Rounded bottoms can be caused by redeposition, which has a much higher rate toward the bottom of a small etching area. Such rounding is often observed for submicron window etches and is removed during an overetch, leaving vertical walls.

from etching areas dominates, then the deposition rate is higher at the bottom of closely spaced features. This might lead to a rounding at the bottom of the feature (see Fig. 13d) and the eventual etching away of the redeposit because the outward taper leads to a net removal rather than a net deposition of material. This last effect can be understood with the help of the following example.

Problem
If a given process etches all material purely vertically with a given rate, E, and deposits material uniformly onto vertical surfaces with a deposition rate D and a flat surface is exposed to this process at an angle θ from horizontal, then at what angle θ will the deposition rate equal the etch rate and the surface remain unchanged?

Solution
If the etch rate is E in a vertical direction, then on an angled surface it is decreased to E_θ where

$$E_\theta = E\cos\theta$$

This result makes the assumption that the etch rate is dependent on a vertical flux of etchant. With the same reasoning, the deposition rate onto this surface will be

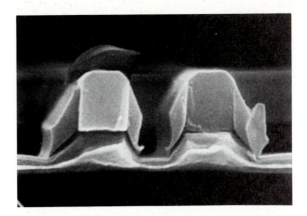

FIGURE 14
An SEM of the cross section of a feature etched in the presence of an extremely high deposition rate. The trapezoidally shaped deposits are as predicted by a uniform deposition rate. Between the closely spaced features the deposition is thinner and less uniform. (*SEM courtesy of C. W. Van Hise, AT&T Bell Laboratories.*)

$$D_\theta = D\sin\theta$$

Therefore at $E_\theta = D_\theta$

$$\theta_{Equal} = \tan^{-1}\left(\frac{E}{D}\right) \tag{20}$$

For $\theta < \theta_{Equal}$ the surface will be etched, and for $\theta > \theta_{Equal}$ a film on the surface will grow.

As feature profiles evolve under conditions of etch with varying deposition rates, this effect could generate unexpected shapes if one more mechanism is assumed. Suppose, as postulated in the recombinant model of anisotropy, that the chemistry of the etch environment is active enough to cause isotropic etching but that this is prevented by deposit (previously considered to be recombinant species) onto the sidewalls. Figure 15 is a computer simulation that shows that redeposition-type deposits with higher rates toward the bottom of the feature can lead to situations in which deposition causes an outward tapering which is subsequently etched away because $\theta < \theta_{Equal}$ of Eq. 20. Isotropic etching of the layer might then proceed, causing an undercut that is more severe toward the bottom of the feature rather than at the top as in simple isotropic etching with no deposition. This type of feature, undercut more toward the bottom than the top, has been reported and is often observed when the mechanism of anisotropy is related to material on the feature sidewalls and not simply ion-assisted or ion-induced etching.[24]

Equation 20 can also be used to interpret results obtained with different initial resist profiles. It was shown that if an initial resist edge is inclined at some angle less than a critical angle then it will etch away at the corner and

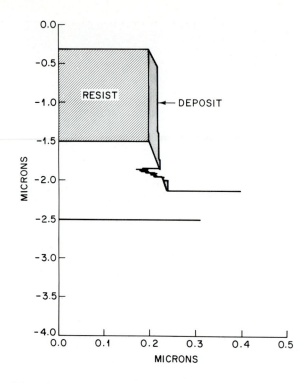

FIGURE 15
A computer simulation of isotropic etching with deposition. This deposition occurs at a higher rate toward the bottom of the feature which results in the deposition etching away. The subsequently formed deep notch is caused by the high isotropic etch rate in the model.

linewidth loss occurs.[23] If, however, a more vertical resist profile exists, then deposition onto the resist will occur and anisotropy might be preserved. This is illustrated by the data presented in Fig. 16.

One last effect that can be attributed to deposition is an etch proximity effect. For both plasma deposition and redeposition types, it is expected that higher rates of sidewall deposition should occur on features adjacent to large, open etching areas. Such areas provide a larger source of redeposition and an unimpeded view of plasma sources. This effect leads to more feature-size increase for isolated features than for closely packed features. Figure 14 illustrates this etch proximity effect. It is even possible to observe closely spaced features that are undercut and isolated features that have outward tapers and significant growth. When feature size must be held constant with close tolerances, such as for MOS gate etching, this type of etch process is to be avoided.

5.3.5 Other Effects Influencing Edge Profile

Faceting is the result of physical sputtering the original resist feature.[16] This effect is therefore most noticeable at high bias-voltage conditions. As mentioned

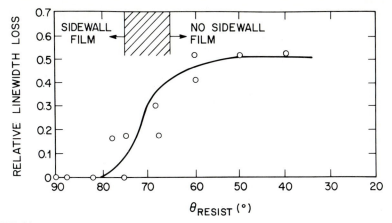

FIGURE 16
A plot of relative linewidth loss vs. initial resist angle for poly-Si etched in a CCl_4 plasma. The angle θ is measured with respect to a lateral line. The data indicates a critical angle below which linewidth loss is observed.

in section 5.2.6, sputtering rates are dependent on the ion angle of incidence and for most materials peak at an angle between 0 and 90°. This effect causes the propagation of a surface that is inclined to the angle of incidence with the highest sputtering rate. This surface or facet is shown in Fig. 17. If the edge of the facet propagates from the resist corner down to the etching layer, then the profile of the feature will also exhibit a taper at an angle modified by the ratio of resist to layer-etch rates.

Another phenomenon related to ion energy is *trenching*,[16] also depicted in Fig. 17. Here the etch rate is enhanced in the immediate vicinity of a masked feature. There are two related mechanisms that lead to trenching. The first involves the low-angle reflection of ions from the wall of the resist feature, leading to a higher ion flux at the feature wall. For this mechanism to be active there must be a sizable fraction of ions with at least slightly off-vertical trajectories or the sidewall must have a slight taper, as shown in the figure. Such a taper could result from deposition during anisotropic etching.

The second mechanism should be active only for etching conductors because

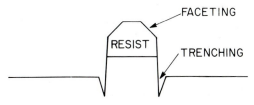

FIGURE 17
A schematic illustration of resist faceting and trenching.

it involves the bending of electric field lines due to surface topography. This off-vertical field also has the effect of enhancing ion flux at feature edges and leads to trenching. Since trenching is a small effect ($\approx$ 5%), it is often not noticed unless thick layers are etched. Deep etching also enhances the mechanisms of ion reflection and field bending.

5.4   OTHER PROPERTIES OF ETCH PROCESSES

The size and profile of the final etched feature are the most important properties of an etching process. Other process properties, however, such as those discussed in the following sections, also affect the value of a process and can determine if it is suitable for manufacturing integrated circuits.

5.4.1   Selectivity

The selectivity of a process is a measure of the etch rate of the layer to be etched relative to other material on the wafer surface. The selectivity is normally expressed simply as the ratio of these two rates:

$$S_{AB} = \frac{E_A}{E_B} \qquad (21)$$

where E_A is the etch rate of the layer to be etched and E_B is the etch rate of a second material. The selectivity to the masking material and to the layer immediately under the layer to be etched are two important process parameters. The former usually determines the initial height of resist necessary to accurately transfer a pattern into a layer. For example, with a resist selectivity of 0.8, at least a 3750 Å thick resist is necessary to mask a 3000 Å layer. In practice, resist faceting and overetching demand an even thicker resist layer. The selectivity to the resist can also affect the final etched feature size if the mask does not have a perfectly vertical profile and if the resist etches isotropically. Problem 6 at the end of the chapter deals with calculating this effect. Multilayer resist schemes that result in thick (1.0–2.0 μm), vertical masks minimize the need for high resist selectivity and allow longer overetches.[25]

Overetching is a term used to describe the common practice of submitting a layer to the etch process for a time longer than is necessary to just etch through the layer at one part of a wafer. It is because of this overetch that the selectivity to the then exposed underlying layer is also important. Overetching is necessary because all etch processes have some degree of etch rate nonuniformity and many etching layers start with nonuniform thicknesses because of nonuniform deposition conditions and underlying topography.

Figure 18 shows how topography can affect the required selectivity of a process. This figure illustrates the fabrication of an MOS transistor. Figure 18a shows the structure after the isolating field oxide is patterned, in this case by anisotropically etching down to the substrate in transistor areas. The gate oxide is

GATE ETCH

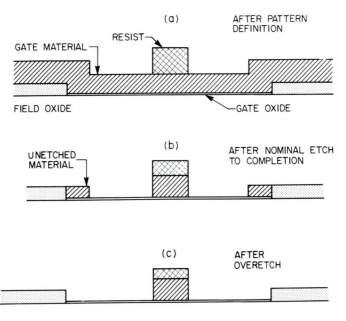

FIGURE 18
A schematic illustration of how underlying topography can increase the selectivity demanded of an etch process. (a) Before etching. (b) After etching to nominal completion. (c) After overetching to remove unetched stringers.

then grown, the gate material blanket deposited, and gate lithography completed. Figure 18b shows the structure after etching to nominal completion with a totally anisotropic etch. The unwanted gate material is etched from all horizontal surfaces, but it remains along the vertical surfaces of the field oxide. Removing the material, frequently called a stringer, requires an overetch determined by the height of the field oxide. The underlying layer exposed to this overetch is the thin oxide, which cannot be totally removed during the overetch or the source and drain region will be rapidly etched away. It is apparent that this problem is severe because the walls of the field oxide are vertical. If these walls were tapered outward, then the necessary overetch would be diminished and the selectivity requirement relaxed. Tapered field oxide walls can be formed by selected area oxide, discussed in Chapter 11. Unfortunately, such a taper decreases the allowable density of transistors because it increases the minimum spacing between transistors.

Problem
Plot required gate etch selectivity and minimum MOS transistor spacing versus angle of field oxide wall on the same graph. Assume length of full thickness field oxide (s) is 0.5 μm, gate thickness (g) is 0.4 μm, field oxide thickness (f) is 0.3

μm, gate oxide thickness is 200 Å, maximum gate oxide etched during overetch is 100 Å, and the gate etch is perfectly uniform.

Solution

Figure 19 below illustrates a cross sectional view of the problem. Real conformal deposition of gate material would have rounded corners as shown by the dotted lines, but for simplicity the solid line profile is assumed. The transistor spacing, x, can be expressed as

$$x = s + 2(f \text{ctn} \theta)$$

where $s = 0.5 \mu$m and $f = 0.3 \mu$m. The thickness of gate material removed during the overetch is

$$(w - g) = g(\sec \theta - 1) \le f$$

where $g = 0.4 \mu$m. The condition that this quantity never exceed the field oxide thickness must be imposed for θ approaching 90°. The etch selectivity of gate material to gate oxide is S and 0.01 μm of gate oxide may be removed during overetch so $(w - g)$ can also be expressed as .01 S. Therefore

$$S = 100g(\sec \theta - 1)$$

where $g(\sec \theta - 1) < f$. The *plot* of S and x versus θ shows close transistor spacing can only be achieved with an extremely selective gate etch (see Fig. 20). It also shows that above 60° the gate etch selectivity requirements are determined by the field oxide and gate oxide thicknesses alone.

In practice high selectivity along with a vertical etch profile is not easy to attain. High selectivities are typically achieved in systems where chemical etch rates dominate and ion-activated or -enhanced mechanisms are minimized. Such mechanisms, which are usually more active at high sheath potential, yield anisotropic etching but also result in significant physical sputtering that is not material selective.

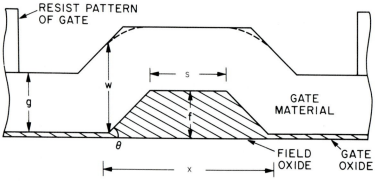

FIGURE 19

A cross-sectional view of the field oxide region between two MOS transistors. The gate material has been deposited but not yet etched.

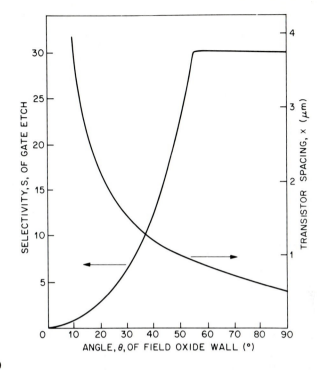

FIGURE 20
A plot of gate etch selectivity and transistor spacing versus angle of field oxide wall for the conditions given in the problem.

5.4.2   Uniformity

The etch rate uniformity refers to the constancy of an etching process at every site on every wafer in the process. Intrawafer uniformity refers to rates measured within a single wafer while interwafer uniformity refers to rates measured from wafer to wafer. This parameter is usually determined by measuring etch rates at random sites, assuming a Gaussian distribution of rates and calculating the standard deviation, σ, of the distribution. Overetch times used to compensate for the nonuniformity are typically $\approx 3\sigma$. For single wafer etchers, interwafer nonuniformity is minimized and only the intrawafer uniformity need be discussed; however, for batch etchers, both must be considered. The two major causes of nonuniform etch rates are electric field gradients along the plasma sheath and differences in the concentrations of active chemical species in the plasma. Field gradients lead to nonuniform etching for surface-reaction-controlled processes while concentration gradients in the plasma lead to nonuniform etching for transport-controlled processes. Etchant concentration variations can be caused by local depletion of active species and nonuniform introduction or extraction of gas. Most commercial etchers attempt to achieve more uniform etchant distribution by distributing a series of gas inputs throughout the chamber. Distributing large diameter pumping ports is not easily accomplished, so pumping normally takes place through one large hole in the chamber. An etch nonuniformity affected by changes in flow rate, or residence time, usually indicates a nonuniform gas composition.

Electric field gradients are more difficult to reduce since they are the result of the interactions between all plasma parameters. Usually an etch nonuniformity, which degrades with increasing rf power, indicates a field-related mechanism. Improvements can then be made by modifying the geometry of the rf cathode or grounded chamber surfaces. These modifications are easier to make in a single-wafer etcher than in a batch etcher.

5.4.3   Throughput

The number of wafers etched per unit time is referred to as the throughput of a process. When the technology exists, only economic factors can determine the acceptable throughput. For batch etchers which process many wafers simultaneously, typically acceptable throughputs of about 60 wafers/hr require relatively low etch rates of 100–600 Å/min. Single-wafer etchers must achieve rates at least 10–20 times faster for equivalent throughput.

5.4.4   Defect and Impurity Introduction

An ideally clean etching system faithfully transfers the lithographic pattern into the etching layer, stops etching at the underlying layer, and leaves no residue on the wafer. If, however, a small particle falls onto the wafer, it can act as a mask during the etch and form an unwanted feature or defect. Acceptable defect densities for a total patterning process, lithography plus etching, are typically less than 1 defect/cm^2.

One source of particles is the flaking of material deposited onto etching chamber walls. Many plasma processes etch wafer surfaces, which have relatively high plasma sheath potentials, while depositing polymer-like coatings onto surfaces such as the chamber walls, which have lower sheath potentials. If the chamber must be vented to atmosphere and evacuated between runs, high-velocity gas passing over the chamber walls can cause flaking. Cleaning the chamber at regular intervals either with a high-power O_2 plasma or with a solvent can minimize the problem. Processing chambers where the wafers are introduced and removed through vacuum load locks often introduce fewer defects because they eliminate the vent and evacuation steps.

Reactive plasma processes establish a dynamic equilibrium of deposition and/or removal of materials at various chamber surfaces.[26] The concentration of active etchant in the system is one of the parameters established by this equilibrium. Changing the nature of the chamber surface, by cleaning, can change equilibrium values, thereby changing process properties such as etch rate and anisotropy. For this reason chambers are typically "conditioned" simply by prerunning the process immediately after cleaning to reestablish equilibrium surfaces. Minimizing the formation of the deposit, for example by adding small amounts of O_2 to the plasma, is another approach. This can, however, reduce the selectivity of the process to the photoresist mask.

Unwanted material of molecular dimensions can also deposit onto wafer surfaces. This material can be either polymer-like or redeposit of unetchable material from chamber walls or wafer holders. A well-known example of the latter occurs when silicon immediately adjacent to an aluminum holder is etched in a chlorinated or fluorinated plasma.[27] Nonvolatile aluminum or aluminum compounds are sputtered from the holder and redeposited onto the wafer. These redeposits mask any further etching. The result is a mottled surface, frequently called grass, that is shown in Fig. 21.

Even if redeposition does not occur at this magnitude and a "clean" etch is observed, careful surface analysis reveals that bombardment by reactive ions modifies the surface. Deep-level transient spectroscopy[28] (DLTS), reflected high-energy electron diffraction (RHEED),[29] electron spin resonance (ESR),[30] Auger electron spectroscopy (AES),[30] and MOS charge retention time[31] are among the techniques used to study surfaces exposed to reactive plasmas. These and other similar references show that during etching, impurities are implanted into the surface down to a depth of about 200 Å. The depth is dependent on the mass of the implant and the plasma sheath potential. These impurities sometimes originate in the chamber walls (e.g., Fe and Ni from stainless steel chambers) and sometimes originate in the plasma source gas (e.g., H, C, Cl, or F). When the etching stops at the single crystal substrate, these implanted impurities cause crystalline damage in a layer about as deep as the implants. While this damage can be annealed out by furnace annealing or rapid thermal annealing, the implanted impurities can be removed only by wet chemical etching. These results are especially important for dielectric etches that stop at transistor regions in the substrate.

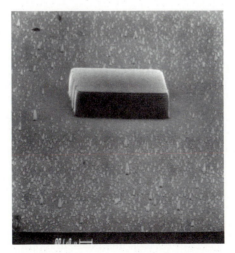

FIGURE 21
An SEM of unetched residue, or "grass," around a patterned feature.

5.4.5 Radiation Damage to Dielectrics

When IC wafers are etched in a reactive plasma, they are exposed to accelerated positive ions and electrons and to photons from the infrared to the soft X-ray regime. Insulating dielectrics exposed to this environment experience changes in their electrical properties. When the dielectric is directly exposed to the plasma, all of this bombardment including the implanted ions contributes to the damage. However, when the dielectric is beneath other layers, only the more penetrating radiation affects the dielectric. Studies of thermally grown SiO_2 directly exposed to reactive plasmas indicate the presence of fixed positive charge and neutral electron traps in the oxide. Interface charges can also be created at the $Si–SiO_2$ interface.[31]

5.5 REACTIVE PLASMA-ETCHING TECHNIQUES AND EQUIPMENT

When plasma-etching techniques were first investigated for applications in IC fabrication processes, the anisotropy of the technique was of prime importance. As the IC industry and fabrication equipment industry matures, more effort is being placed on optimizing the other process parameters discussed in the last sections. For example, today's etchers emphasize automation to decrease defect density and increase throughput. This section will discuss etching equipment and techniques from a historical perspective. Older equipment will be discussed first, then currently used batch and single-wafer etchers and finally newer techniques, some of which are now incorporated in commercially available equipment.

5.5.1 Plasma Etchers and Barrel Reactors

The first technologically significant, reactive plasma etcher design is typically referred to as a parallel-plate etcher. Its configuration is very similar to that shown in Fig. 2. A round, rf-driven electrode is positioned parallel to another round, grounded electrode on which many wafers are placed.[32] As with most other commercial etchers the excitation frequency used is 13.56 MHz, which has been allotted by the Federal Communication Commission for industrial and medical use.[33] The gas-pumping system consists only of an oil-immersed mechanical pump so that operating pressures are typically 0.1–2.0 Torr. Just as in more modern equipment, fluorine- and chlorine-containing Freons, among other gases, are used as sources. Under these conditions, the plasma sheath potential to the grounded electrode is low (<20 V) and the chemical reactivity of the plasma is high. As might be expected, the etch selectivity of this system could also be high, but because of low ion energies ion bombardment plays a minor role in the surface chemistry. Only under special conditions[22] could this arrangement etch anisotropically. For historical reasons this configuration is usually referred to as plasma etching.

A barrel reactor holds the wafers vertically in an electrically floating cassette in a quartz chamber held at 0.1–2.0 Torr. Here the rf power is capacitively coupled to the plasma through large plates held against the chamber wall. Plasma sheath potentials to the wafer are again small, as for plasma etching, and ion bombardment plays a small role in the surface chemistry. Barrel reactors are relatively cheap and can process many wafers simultaneously. They are used principally as isotropic resist removers (also called strippers or ashers) with O_2 as the source gas. Even though the average ion bombardment energy is low, barrel reactors can still cause significant damage to dielectrics (oxide). A technique that totally eliminates radiation damage is to strike the plasma in a chamber that is separated and shielded from the wafers and then flow the product gases over the wafers. This downstream etching is accomplished by chemically active atoms and radicals generated by the plasma and is totally isotropic. [34]

5.5.2 Reactive Ion Etching

Reactive ion etching (RIE), sometimes called reactive sputter etching (RSE), is the name used to describe an arrangement that differs from plasma etching in a few respects. Here the rf-driven electrode instead of the grounded electrode holds the wafer. This allows the grounded electrode to have a significantly larger area because it is in fact the chamber itself. This larger grounded area combined with the lower operating pressures (5.0–100.0 mTorr) leads to significantly higher plasma-sheath potentials (20–500 V) at the wafer surface, which result in higher energy ion bombardment.

Early reactive ion etchers were made simply by modifying existing plasma etcher designs. The first significant deviation from this design is the hexagonal cathode etcher[35] shown in Fig. 22. Here wafers are first mounted onto rectangular holders, or "trays," that are subsequently clamped onto the six vertical facets of a hexagonal prism. The whole metal "bell jar" which is the top of the process chamber acts as the grounded anode. This geometry has an even higher cathode/anode area ratio, which leads to even higher sheath potentials at the wafer surfaces. The facets can hold more wafers than the parallel plate design so throughput is increased. Etching nonuniformities caused by electric field gradients are reduced by the almost cylindrical geometry. A disadvantage of this arrangement is that it is not easily automated. Most systems are capable only of automatically loading wafers onto the trays. The etcher must then be vented to atmosphere to manually load the trays.

All etchers discussed to this point use "batch" processes that etch many wafers simultaneously to increase throughput. For modern circuits with submicron feature sizes, etching processes are more critical. More vertical profiles, better linewidth control, higher selectivity, and better uniformity are necessary. One approach to this problem is to use single-wafer etchers, which etch one wafer at a time. A typical design is shown in Fig. 23. These machines are easily automated to perform wafer cassette-to-cassette operations so that no operator handling

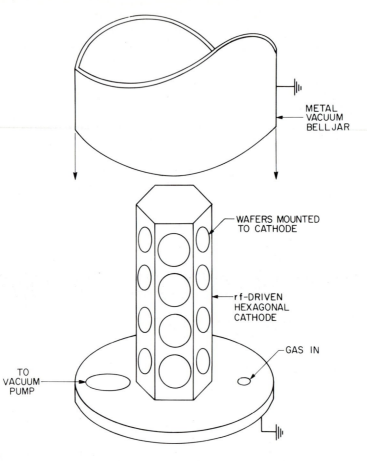

FIGURE 22
(a) A schematic representation of a hexagonal cathode etcher with the vacuum bell jar raised.

is required. They can incorporate a load-locked chamber so that the process chamber need not be vented under normal use. Single-wafer etchers can also tailor the electrode geometry and gas flow to maximize etch uniformity across the wafer. This enhanced uniformity combined with automatic endpoint detection and microprocessor control can provide good process control.

A drawback of single-wafer etchers is that they must etch at higher rates to compete with the throughput of batch etchers. This constraint forces commercial single-wafer etchers to operate at higher rf power density and sometimes higher pressure where process control and selectivity is more difficult to achieve. For this reason some manufacturers are offering hybrid reactors that combine a few single-wafer etchers in one machine. This approach has the added advantage of dedicating chambers to specific chemistries for multiple-etch processes such as the polycide etch discussed in Section 5.6.1.

FIGURE 22
(b) A commercial hexagonal cathode etcher with the bell jar lowered. To the right of the etching chamber is an automatic loading station.

5.5.3 Reactive Ion-Beam Etching (RIBE)

The complex interrelationships between parameters in a reactive plasma make it difficult to optimize an etch process. One approach to isolate these parameters is the RIBE arrangement shown in Fig. 24. Here the reactive ions are generated in one section of the chamber and then accelerated through one or more grids toward the wafer surface. In this way the energy and the density of ions can be separately controlled. Actual semiconductor etch processes have been developed using this technique, which is also used in the laboratory to study mechanisms active in reactive plasma processes.[36] Gas pressures are kept low (< 2 mTorr) so that scattering does not decollimate the beam; ion energies are high (600–1200 V), and the same types of reactive gases are used as in reactive plasma processes. Radiation damage because of the high-energy ions can present a problem in some applications.

Early RIBE machines used hot cathodes to emit electrons, which then generate reactive ions. The chlorinated and fluorinated hydrocarbons attack, or deposit onto, this cathode thus decreasing its useful life. A microwave discharge using electron cyclotron resonance (ECR) to produce a dense plasma (see Fig. 24) was demonstrated to eliminate this problem. ECR uses a magnetic field to change

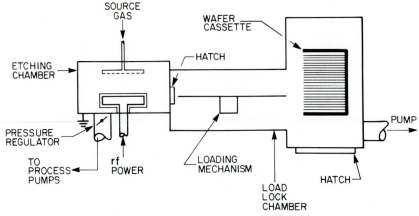

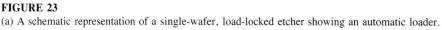

FIGURE 23
(a) A schematic representation of a single-wafer, load-locked etcher showing an automatic loader.

the normal straight line trajectories of electrons into circular or helical orbits. This increases the average lifetimes of electrons by increasing the average time of flight before a collision with the chamber wall, a major cause of recombination.

5.5.4 Microwave Plasma Etching

High-frequency microwave discharges with or without magnetic enhancement can produce dense plasmas with very low sheath potentials. Without an accelerating grid to increase the energy and directionality of the ions, this type of reactive discharge can etch isotropically, at high etch rate, and with a minimum of radiative-damage effects. Some degree of anisotropic etching is achieved if the system is operated at extremely low pressure ($< 10^{-3}$ Torr) and the plasma volume is contained in a chamber remote from the wafer[37] (see Fig. 25). This arrangement achieves some directionality of ion and reactive neutral bombardment by decreasing the number of gas-phase collisions between the microwave source and the wafer. Another method of achieving directional bombardment is to apply a low-power rf signal to the wafer. The rf plasma at the water surface develops a higher sheath potential, which results in anisotropic etching.

5.5.5 Magnetic Confinement

Just as a magnetic field is used in the ECR microwave discharge to increase the ion density, magnetic enhancement can also be used in rf discharges. When this technique is used to increase rates in thin-film sputter-deposition systems, it is called magnetron sputtering. In magnetron sputtering, a magnet is placed behind the source material or target. This same technique can be applied to reactive plasma etching by placing a magnet behind the etching wafer. This does indeed lead to higher plasma densities and higher etch rates, but because of

FIGURE 23
(b) A commercial single-wafer etcher having four separate etching chambers which can operate simultaneously.

the nonuniform magnetic fields the etch rates are extremely nonuniform across the wafer.[38] One approach to this problem is to position the magnet on the chamber wall across from the wafer.[39] Enhanced etch rates are still observed while uniformity is more acceptable. Uniformity can be further improved by rotating the magnets to provide a more uniform time-averaged plasma density.[38]

5.5.6 Effect of Plasma Parameters

Excitation Frequency. The frequency of the alternating current discharge affects the relative production rate of reactive radicals, the ionization level of the plasma,

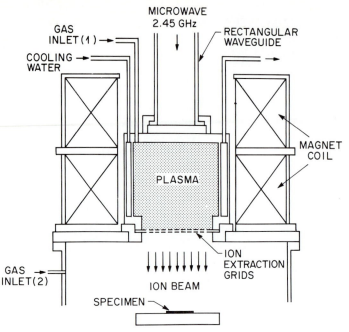

FIGURE 24
A schematic representation of a reactive ion-beam etcher using an electron cyclotron resonance chamber as an ion source. (*After Matsuo and Adachi, Ref. 36.*)

and the sheath potentials at the wafer surface. Ionization and dissociation that leads to chemical activity generally increase with increasing frequency. As discussed in Section 5.2.3, the sheath potential decreases with increasing frequency so that in general very high frequency microwave (10^9 Hz) processes etch isotropically at very high rates while lower frequency processes (10^5 Hz) etch anisotropically at lower rates. However, etch rate dependence on frequency cannot always be predicted from these trends. For example, the etch rate of Si in a CCl_4 plasma was found to decrease by more than an order of magnitude as the discharge frequency was raised to above 7 MHz, and the etch rate of aluminum increased by a factor of four over the same range of frequency increase.[40] Cl_2 etching of Si is ion-induced while Cl_2 etching of Al is ion-enhanced. In the former case the etch rate depends mostly on ion energy, and in the latter case it depends mostly on the density of reactive species.

Flow Rate. All etching environments are dynamic in that feed gas is continuously introduced into the chamber and gases are continuously pumped away. An important process parameter is the flow rate of feed gases into the system. Another more meaningful way of expressing the flow rate is the residence time or the average length of time a molecule of gas spends in the chamber, disregarding any chemical reactions that might occur:

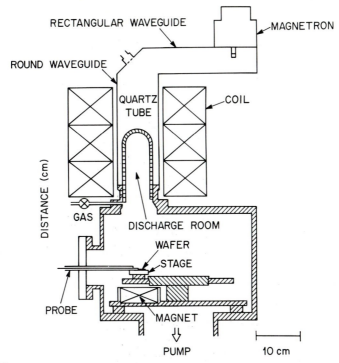

FIGURE 25
Diagram of microwave plasma-etching equipment. (*After Suzuki, Okudaira, and Kanomata, Ref. 37.*)

$$R = \frac{PV}{r} \tag{22}$$

where R = residence time in seconds, P = pressure in Torr, V = volume of the processing chamber in liters and r = flow rate of feed gas in Torr-liters per second. Most gas flow meters are calibrated in standard cubic centimeters per minute (sccm), and it should be kept in mind that 1.0 Torr-l/s = 79.0 sccm. If the active etchant species has an average lifetime, due to loss processes, significantly shorter than the average residence time, then changing the flow rate over a limited range should have a minimal effect on the process. If, however, the intrinsic average lifetime of an active etchant species is longer than the average residence time, then the actual lifetime value in the reaction chamber will be the residence time. In such a situation, changing flow rates can change the observed loading effect.

A loading effect exists if the etch rate for a material decreases as the amount of this material in the etching chamber increases. The theoretical dependence of etch rate on material surface area for a single etchant species is[41]

$$E = \frac{\beta \tau G}{1 + K\beta \tau \Phi} \tag{23}$$

where E = etch rate, β = reaction rate constant, G = generation rate of active species, K = a constant dependent on material and geometry, τ = actual lifetime of the active species, and Φ = the surface area of the etching material. This relationship indicates that the loading effect will be minimized if $K\beta\tau\Phi \ll 1$.

Pressure. It should be apparent by now that all plasma parameters are interrelated and that attributing a specific process characteristic to an isolated parameter change is too simplistic. Keeping these interdependences in mind, a few general effects of pressure can still be isolated. Holding all other parameters constant, etch rate generally increases with increased pressure in the 5–50 mTorr range. This can be attributed to an increased concentration of both neutral and ionized species. However, one study of downstream product concentration in a reactive plasma indicates that while the concentrations of active species do increase with increasing pressure, these concentrations are constant with residence time.[42] This suggests that an observed etch rate increase with pressure is actually an etch rate increase with residence time. Increasing pressure also changes the electrical properties of the plasma.

Sheath potentials decrease with increasing pressure. If an ion-induced process is raised to a higher pressure, it is possible to observe a decrease in etch rate. Also, if an ion-enhanced process is raised to a higher pressure, etch profiles become more isotropic. Pressure also affects the energy of ions accelerated across the sheath in a more subtle way. If the mean free path is smaller than the sheath, then a significant amount of collision and ionization will occur in the sheath region. Many ions will therefore not be accelerated across the full sheath potential. So at higher pressures not only is the maximum ion energy decreased but the distribution of ion energies is increased to include more at lower energies.

Power. Increasing the rf power dissipated in the plasma increases the sheath potential and increases the concentration of reactive neutrals and ions. Etch rates therefore normally increase almost linearly with increasing power. The thrust of many reactor modifications discussed later in this section is to isolate these two effects. For a given chemistry it would be an advantage to hold the reactive species concentration constant while adjusting the sheath potential just high enough to achieve anisotropic etching.

5.5.7 Process Monitoring and Endpoint Detection

The time to completion of any etch process can be calculated by determining the average etch rate of the process and calculating the time necessary to etch through the layer. Etching is then allowed to continue for an overetch period necessary to compensate any etch rate nonuniformity, layer thickness nonuniformity, or underlying topography. It is always desirable to minimize the overetch time so that the erosion of the underlying layer is minimized. A method of determining the nominal endpoint of the process allows such a reduction in overetch time.[43]

A simple, frequently applicable technique is to monitor the dc bias potential while holding the rf power constant. Since the voltage/power dependence is related to plasma chemistry, a change in this relationship frequently occurs when the plasma contribution from the etching layer is no longer present.

The intensities of spectral lines emitted from the plasma also change when the etching layer is totally removed. An example of this effect is the sharp decrease in the 261.4 nm line, a signature for AlCl, when an aluminum layer is totally removed.[44]

Laser interferometry of the wafer surface is used not only to determine endpoint but also to continuously monitor etch rates. During etching, the intensity of laser light reflected off a thin film surface oscillates. This oscillation occurs because of the phase interference between the light reflected from the outer and inner interfaces of the etching layer. This layer must therefore be optically transparent or semitransparent to observe the oscillation. Figure 26 shows a typical signal from a silicide/polycrystalline Si gate etch.[45] The period of the oscillation is related to the change in film thickness by

$$\Delta d = \lambda/2n \tag{24}$$

where Δd is the change in film thickness for one period of reflected light, λ is the wavelength of laser light, and n is the refractive index of the etching layer. Table 2 lists Δd for various materials under typical conditions.

5.6   SPECIFIC ETCH PROCESSES

5.6.1   Polycrystalline Silicon/Polycide

Low-pressure, chemical-vapor-deposited (LPCVD) polycrystalline silicon (polysilicon) is typically used as an MOS gate material because of the stability of the

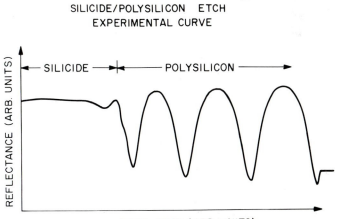

FIGURE 26
The relative reflectance of the etching surface of a composite silicide/poly-Si layer. Endpoint of the etch is indicated by the cessation of the reflectance oscillation. (*After Heimann and Schutz, Ref. 45.*)

TABLE 2
Change in film thickness for
one period of reflected light
using a $\lambda = 6328$ Å helium–
neon laser source.

SiO_2	2200 Å
Si	800 Å
Si_3N_4	1500 Å
Photoresist	~1900 Å

polysilicon–SiO_2 interface at high temperatures. Recently, composite layered structures of metal silicides over polysilicon (polycides) have also been used for gates because of the added electrical conductivity of the silicide layer. Important etch parameters for this gate etch are selectivity to SiO_2, anisotropy, and line profile control. Table 3 shows that most chlorinated or fluorinated gases can etch silicon in a reactive plasma, but the tendency of fluorinated gases to etch Si isotropically and to also etch SiO_2 limits the choice to pure chlorine or chlorine plus oxygen. This mixture enhances the poly-Si/SiO_2 selectivity. Unfortunately, most refractory metal silicides do not etch cleanly at high enough rates in Cl_2, so most reported polycide etches use two sequential processes. The first usually uses a Freon or $BCl_3 + Cl_2$ to etch the silicide.[46] The second process uses Cl_2 to selectively etch the poly-Si. Figure 27 shows an SEM cross section of polysilicon gate etch with features 0.5 μm wide.

An increase in the concentration of n type dopant in polysilicon is known to decrease the degree of anisotropy of most reactive plasma etches. The reason that heavily doped polysilicon etches at a higher rate with a tendency toward isotropic etching is thought to be related to the free electron concentration in the silicon. It was shown, however, that this correlation is not exact.[47]

TABLE 3
Some source gases used to reactive plasma
etch silicides, silicon, and silicon dioxide.

Silicides	Si	SiO_2
$CFCl_3$	Cl_2	CHF_3
CF_2Cl_2	F_2	$CF_4 + O_2$
CCl_4	HF	$CF_4 + H_2$
$BCl_3 + Cl_2$	$CFCl_3$	$SiCl_4$
$CF_4 + O_2$	CF_2Cl_2	
SF_6	CCl_4	
NF_3	$BCl_3 + Cl_2$	
	$CF_4 + O_2$	
	SF_6	
	NF_3	

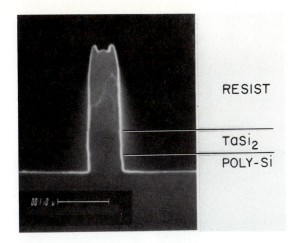

RESIST

TaSi$_2$

POLY-Si

FIGURE 27
An SEM of a 0.5 μm wide polysilicon gate etch. The etch stop was 100 Å gate oxide in the transistor source and drain regions. (*Courtesy of W. E. Willenbrock, Jr., AT&T Technologies.*)

5.6.2 Trench Etching

Another application of silicon etching has recently emerged with the need for deep (>3.0 μm) narrow (<1.25 μm) slots or trenches in the silicon substrate. These trenches can be used to isolate devices in CMOS and bipolar circuits.[48] Vertical capacitors or even transistors can also be fabricated inside the trenches.[49] All of these applications allow for denser packing of devices.

Figure 28a schematically shows the profile of an ideal trench. It has inwardly sloped walls and the bottom is rounded. The sloped walls, caused by redeposition during etching, are necessary to eliminate the formation of a void in the trench during any conformal deposition process used to refill the trench. Most trench processes require such a deposition to at least planarize the surface and possibly to form a capacitor plate if the trench is lined with a capacitor dielectric. Figure 28b shows such a void caused by an undercut trench etch. The rounded bottom of the ideal profile is desirable to eliminate an electric field concentration caused by sharp corners. Figure 28c shows a tapered trench with no void in the refill. It does, however, have the deeply cut corners with the unfortunate name of "trenching" discussed in Section 5.3.5.

5.6.3 Silicon Dioxide and Silicon Nitride

Two typical SiO$_2$ etch processes are the opening of transistor regions in field oxide, as discussed in the last problem (Section 5.4.1), and the etching of contact windows down to an underlying conductor. Since this conductor is often silicon, the selectivity of SiO$_2$ to Si must be high. An undercut contact window makes the subsequent conductor deposition into the window extremely difficult, so vertical etching is another important process parameter. Fortunately the etching of silicon oxides, either thermal or LPCVD, is ion-induced in most RIE chemistries so that undercut is rarely observed and vertical contact walls are common.

(a)

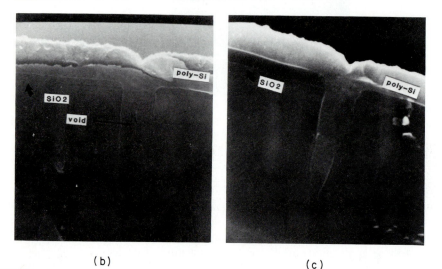

(b)   (c)

FIGURE 28

(a) A schematic representation of an ideal trench cross section. (b) An SEM cross-sectional view of a refilled trench which does not have tapered walls. A void in the LPCVD poly-Si is evident. (c) The same view of a tapered trench which shows no void.

Outwardly tapered contact walls allow for a more uniform physical deposition (e.g., evaporation or sputtering) of a metal into the contact windows. This is illustrated in Figs. 29a and 29b. Such a taper can be etched in silicon oxides if the initial resist profile is tapered either by special resist exposure and development[50] or by flowing the resist at high temperature. Figure 29c shows this tapered resist before oxide etch, and Fig. 29d shows the tapered window after oxide etch. The angle of the final taper and the size of the window depend on the angle of the initial resist, the oxide/resist etch selectivity and the amount of overetch. This process therefore requires tight control over these parameters.

Table 3 lists some source gases typically used for oxide etching. It was suggested very early that CF_3 radicals might interact preferentially with SiO_2 over Si to form volatile products.[51] There is still general agreement that selectivity is achieved by increasing the ratio of CF_x radicals to F atoms in the plasma. An example of this effect is given in Section 5.2.5: H_2 is added to CF_4 to achieve SiO_2/Si selectivities of 20.0. Another source gas that achieves a similar result is CHF_3. For both these chemistries the addition of O_2 or H_2 decreases or increases

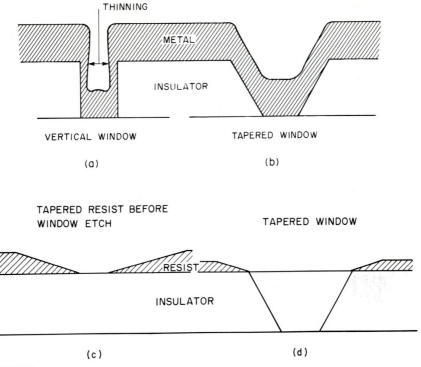

FIGURE 29
(a) A schematic cross-sectional view of a vertical contact window in which a metal conductor is vapor deposited. (b) This same deposition into a tapered contact window. (c) A tapered resist pattern before oxide etch. (d) The tapered profile transferred into the oxide.

respectively the SiO_2/Si selectivity. The addition of O_2 also significantly decreases the high polymer-formation rate on the reaction chamber walls.

Window etch profiles in oxide, before reaching endpoint, often have rounded bottoms as in Fig. 13d. These profiles might be related to the observation that etch rates in submicron windows are often lower than in large, open areas. It is possible that redeposition is at least partly responsible for both these effects.

Two types of deposited films are usually referred to as silicon nitride. The first is deposited using plasma-enhanced CVD at below 350°C. This material has a significant hydrogen concentration and has polymer-like properties. The second material, Si_3N_4, is deposited at much higher temperatures using CVD or LPCVD. Some applications of these two materials in IC fabrication do not require accurate profile control during etching. An example is the patterning of plasma-enhanced CVD films used as a final passivation layer over completed circuits. Here the etching of large area windows down to contact pads requires little feature-size control. A typical etch for this application uses $CF_4 + O_2$ as the source gas in the "plasma etch" mode.[52] This etch suffers poor selectivity to both Si and SiO_2. The selectivity to Si is improved in plasmas that generate CF_x radicals,[15] and generally parameters that lead to high SiO_2/Si selectivity also yield high Si_3N_4/Si selectivity.[53] A more recent report shows that Si_3N_4 can be etched anisotropically in the RIE mode with a selectivity[54] to Si and SiO_2 of 20. Either of two source gases, CH_2F_2 or CH_3F, can produce these results. This high selectivity is important when patterned Si_3N_4 is used as an oxidation mask to define transistor regions in thermally grown field oxide. In this process, a thin ($\approx$1000 Å) Si_3N_4 film is deposited over a thin ($\approx$200 Å) SiO_2 layer on the subtrate. The nitride is patterned and thick ($\approx$4000 Å). SiO_2 is grown in unmasked regions by high-temperature oxidation. The nitride etch must not totally erode the thin oxide used as an etch stop.

5.6.4   Aluminum

Aluminum metallization can take the form of a pure Al film, an Al/polysilicon composite structure, an Al–Cu alloy, an Al–Si alloy, or an Al–Si–Cu alloy. Silicon is added to the aluminum to minimize "spiking," which is the nonuniform interdiffusion of silicon and aluminum in contact regions. Copper is added to enhance the electromigration resistance of aluminum conductors. Though each of these metallizations have slightly different etching characteristics, they neverthe-less share the many problems[55] of reactive plasma etching Al.

One such problem is caused by the stable native oxide surface always found on Al films. This oxide frequently etches at a much slower rate than the Al film itself. At best this leads to an irregular incubation period that is difficult to take into account when timing is used to predict endpoint. At worst, the etching surface can be extremely rough because of slight nonuniformities in the oxide thickness. A reproducible etch process does not have a high Al/Al_2O_3 selectivity. Either CCl_4 or a mixture of BCl_3 and Cl_2 can meet this criterion and can anisotropically etch Al. Pure Cl_2 source gas, on the other hand, etches Al quickly and isotropically but etches Al_2O_3 at a very low rate.

The etched profile and etch rate are strong functions of the concentration of the active species, Cl, in the plasma. High Cl concentrations lead to high chemical-etch rates, which in turn lead to undercut Al profiles. A batch process that depends on a fully loaded system to deplete the Cl concentration enough to produce an anisotropically etched profile might undercut if less than a full load of wafers is etched.[56] The recombinant, or sidewall, passivation effect is the generally accepted mechanism of anisotropic etching of Al. Porous aluminum chlorides form readily on the reactor walls. This film absorbs variable amounts of water vapor on exposure to atmosphere leading to etch irreproducibilities. Heating the reactor walls to minimize the chloride growth or, better, using a load-locked reactor are two approaches to this problem.

Once etched, an Al pattern cannot simply be removed from the reactor because exposure to atmosphere causes rapid corrosion of the Al. A chlorinated residue on both the Al and the resist reacts with water vapor to form HCl, which attacks the Al. Two ways of avoiding this problem are to rinse the wafers immediately in water or to proceed, in situ, with an O_2 plasma etch, which removes the resist and passivates the Al surfaces.[57]

5.6.5 Planarization

The conformal deposition of a dielectric such as phosphorus-doped silicon dioxide over a patterned conducting layer results in a stepped profile similar to that in Fig. 18. In order to ensure uniform coverage of the next conducting layer, the dielectric surface must be smoothed. This can be accomplished by flowing the dielectric at high temperatures (>800°C). When the highest allowable substrate temperature is less than the dielectric flow temperature, two etching techniques can be used to smooth the surface. The first, called planarization,[58] is accomplished by spinning a resist or any other polymeric layer onto the irregular dielectric film, which is deposited thicker than necessary for the final structure. This results in a smoothing of the new top surface. This new surface is then transferred into the dielectric by etching in a reactive plasma that etches the resist and the dielectric at the same rate. Smoothing of a dielectric surface can also be accomplished by depositing more dielectric than necessary and then simply etching in an RIE mode. Faceting (Fig. 17) can then cause rounding of corners that allow better metal step coverage.

5.7 SUMMARY AND FUTURE TRENDS

VLSI process requirements can now be met by current low pressure, rf, RIE etch processes. Important process parameters are faithful reproduction of lithographic feature size, feature-profile control, selectivity, uniformity, and defect or impurity introduction. The limitations on current processes are the interdependence of plasma parameters. For example, in most rf systems, as the power is raised to increase the plasma density, the sheath bias also increases with a functional dependence determined by the chamber geometry, pressure, and residence time. This inflexibility limits process optimization.

Future device-fabrication demands will be different only in that the wafer size will be larger and these same process parameters will have even smaller tolerances. These two factors make single-wafer etch chambers with individual process endpoint control mandatory. The reproducibility gained from a load-locked system might also be necessary. An ideal etcher with universal application would be capable of controlling neutral species adsorbed onto wafer surfaces as well as the type, density, energy, and trajectories of incoming ions. Figure 30 shows that many workers in the field believe that a microwave discharge, perhaps enhanced with ECR, might be the ion source for such an etcher. Ion energy could then be separately controlled with a grid or an independent rf field on the wafer. This arrangement, as pointed out by the figure, provides better ion energy control, which should lead to lower radiation damage and high selectivity because of minimized physical sputter etching. The figure also points out that a single wafer etcher could, in principle, be built smaller, with a smaller "footprint" providing smaller equipment area per wafer than batch etchers. This would, however, depend on the details of the physical design. The low particles generation and uniformity dependencies proposed in the figure might not yet be generally accepted as intrinsic properties of the four process types represented.

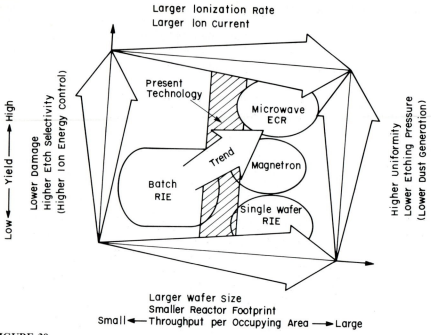

FIGURE 30

A map of plasma parameters which predicts microwave plasmas to dominate future processes. (*After Saikai, Shibata, Okudaira, and Hamazaki, Ref. 59.*)

To be technologically significant any new etcher design must also be cost-effective, so at this point no one design is obviously superior. One possibility is that different designs will be more suited to different processes so that a production line might have different types of dedicated etchers for each process.

While future etcher designs might still be uncertain, it is clear that increased understanding of plasma physics and chemistry are necessary to guide the way toward optimal plasma processes.

PROBLEMS

1 Assuming a mask that cannot erode, sketch the edge profile of an isotropically etched feature in a film of thickness h_f on an unetchable substrate for (a) etching just to completion, (b) 100% overetch, and (c) 200% overetch. What shape does the profile tend toward as overetching proceeds? Comment on the advisability of estimating the degree of anisotropy of etching from scanning electron micrographs of edge profiles taken after removal of the masking layer.

2 Is the plasma distribution shown in Fig. 2 realistic? How can plasma be eliminated in the reaction chamber where it is not desirable?

3 Calculate the etch rate selectivity required for a layer of average thickness h_f to be etched over the substrate resulting in a maximum penetration into the substrate of h_s. Assume the layer is deposited nonuniformly so that its thickness is $h_f(1 + \delta)$ at the thickest and $h_f(1 - \delta)$ at the thinnest. Analogously the etch rate of the film is nonuniform; its average value is v_f and it varies over the range $R_f(1 \pm \phi_f)$. Assume the worst case, in which the thinnest film region is in the fastest etching position. Also, the time to completion, at which point the thickest part of the film is etching at the slowest rate, is t_c, and the total etch time is $t_c (1 + \Delta)$ where Δ is a fractional overetch time.

4 Consider a discharge in CF_4. Assume that electrons and gas molecules can be treated as hard spheres with masses m and M, respectively. Calculate the maximum fractional loss of kinetic energy for an electron striking a CF_4 molecule. Consider the CF_4 molecule initially at rest, and the collision to be elastic. Repeat the calculation for an inelastic collision for which the potential energy of the CF_4 molecule increases by the maximum amount possible.

5 A CF_4–O_2 rf plasma is operated at 300 W and 0.5 Torr, with a feed-gas flow rate of 100 cm^3/min (STP). The plasma occupies a volume of 4000 cm^3. Under these conditions atomic fluorine is generated at a rate of 10^{16} cm^{-3} s^{-1} in the plasma. The combined effect of loss mechanisms for F atoms results in a rate of loss proportional to the F atom concentration. In steady state, this concentration is measured as 3×10^{15} cm^{-3}. What is the mean lifetime of F atoms for these conditions? How does it compare to the residence time of an average molecule in the plasma? How would the etch rate of Si be affected if the flow rate were increased tenfold while holding other parameters constant?

6 In a situation similar to the evolution between Figs. 29c and 29d, show that W, the difference between the window opening in the resist feature and the maximum window opening after completion of the etch, is

$$W = 2(V_v \mathrm{ctn}\theta + V_e)t_t$$

where V_v = vertical component of mask etch rate, V_e = lateral component of mask etch rate, t_t = total etch time, and θ = angle of resist taper to a horizontal line. Note in this case the resist is assumed to also have a lateral etch component. How does this change the shape of the etched feature from that of Fig. 29d?

7 What are the major distinctions between reactive ion etching and parallel-plate plasma etching? Compare the advantages and limitations of these techniques.

8 The enthalpy of the exothermic reaction

$$Si + 4F \rightarrow SiF_4$$

is 370 kcal / g-mole at 25°C. At what rate is heat generated when a 100 mm diameter, 0.5 mm thick Si wafer is etched on one face at a rate of 1.0 μm/min in an F-generating plasma? Suppose that the wafer is thermally isolated during etching. By how much will its temperature rise if 5.0 μm are etched away?

9 Explain why endpoint detection by monitoring of reactant species requires a loading effect.

REFERENCES

1 B. Chapman, *Glow Discharge Processes*, Wiley, New York, 1981.
2 J. W. Coburn, "Physical and Chemical Mechanisms in the Ion-Enhanced Etching of Silicon," *Vacuum*, **34**, 157 (1984).
3 J. A. Mucha and D. W. Hess, "Plasma Etching," in L. F. Thompson, C. G. Willson, and

M. J. S. Bowden, Eds., *Introduction to Microlithography: Theory, Materials, and Processing*, American Chemical Society, (1983), p. 215.

4 H. R. Koenig and L. I. Maissel, "Application of R. F. Discharges to Sputtering," *IBM J. Res. Develop.*, **14**, 168 (1970).

5 A. J. van Roosmalen, "Plasma Parameter Estimation from R. F. Impedance Measurements in a Dry Etching System," *Phys. Lett.*, **42**, 416 (1983).

6 W. C. Dautremont-Smith, R. A. Gottscho and R. J. Schutz, "Plasma Processing: Mechanisms and Applications," in G. E. McGuire, Ed., *Semiconductor Materials and Process Technologies*, Noyes Publications, Park Ridge, NJ, to be published (1988).

7 R. A. Gottscho and T. A. Miller, "Optical Techniques in Plasma Diagnostics," *Pure and Applied Chem.*, **56**, 189 (1984).

8 R. H. Bruce, "Anisotropy Control in Dry Etching," *Solid State Technol.*, **24**, 64 (1981).

9 J. W. Coburn and E. Kay, "Some Chemical Aspects of the Fluorocarbon Plasma Etching of Silicon and Its Compounds," *Solid State Technol.*, **22**, 117 (1979).

10 See for example C. H. Steinbruchel, "Langmuir Probe Measurements on CHF_3 and CF_4 Plasmas: The Role of Ions in the Reactive Sputter Etching of SiO_2 and Si," *S.S. Sci. and Tech.*, **130**, 648 (1983).

11 B. Drevillon, J. Perrin, J. M. Siefert, J. Huc, A. Lloret, G. de Rosny, and J. P. M. Schmitt, "Growth of Hydrogenated Amorphous Silicon Due to Controlled Ion Bombardment from a Pure Silane Plasma," *Appl. Phys. Lett.*, **42**, 801 (1983).

12 A. D. Richards, B. E. Thompson and H. H. Sawin, "Continuum Modeling of Argon Radio Frequency Glow Discharges," *Appl. Phys. Lett.*, **50**, 492 (1987).

13 C. J. Mogab, A. C. Adams, and D. L. Flamm, "Plasma Etching of Si and SiO_2, the Effect of Oxygen Additions to CF_4 Plasmas," *J. Appl. Phys.*, **49**, 3796 (1978).

14 L. M. Ephrath, "Selective Etching of Silicon Dioxide Using Reactive Ion Etching with CF_4-H_2," *J. Electrochem. Soc.*, **126**, 1419 (1979).

15 S. Matsuo, "Selective Etching of SiO_2 Relative to Si by Plasma Reactive Sputter Etching," *J. Vac. Sci. Technol.*, **17**, 587 (1980).

16 P. D. Townsend, J. C. Kelly, and N. E. W. Hartley, *Ion Implantation, Sputtering and Their Applications*, Academic Press, New York, 1976.

17 H. F. Winters, J. W. Coburn, and T. J. Chuang, "Surface Processes in Plasma-Assisted Etching Environments," *J. Vac. Sci. Technol.*, **B1**, 469 (1983).

18 J. W. Coburn and H. F. Winters, "Ion and Electron Assisted Gas-Surface Chemistry—an Important Effect in Plasma Etching," *J. Appl. Phys.*, **50**, 3189 (1979).

19 V. M. Donnelly and D. L. Flamm, "Anisotropic Etching in Chlorine-Containing Plasmas," *Solid State Technol.*, **24**, 161 (1981).

20 D. L. Smith and R. H. Bruce, "Si and Al Etching and Product Detection in a Plasma Beam Under Ultra High Vacuum," *J. Electrochem. Soc.*, **129**, 2045 (1982).

21 H. Okano and Y. Horiike, "Si Etch Rate and Etch Yield with Ar^+/Cl_2 System," *Jpn. J. Appl. Phys.*, **20**, 2429 (1981).

22 C. J. Mogab and H. J. Levinstein, "Anisotropic Plasma Etching of Polysilicon," *J. Vac. Sci. Technol.*, **17**, 721 (1980).

23 S. E. Bernacki and B. B. Kosicki, "Controlled Film Formation During CCl_4 Plasma Etching," *J. Electrochem. Soc.*, **131**, 1926 (1984).

24 A. C. Adams and C. D. Capio, "Edge Profiles in the Plasma Etching of Polycrystalline Silicon," *J. Electrochem. Soc.*, **128**, 366 (1981).

25 J. M. Moran and D. Maydan, "High Resolution, Steep Profile Resist Patterns," *Bell Sys. Tech. J.*, **58**, 1027 (1979).

26 C. B. Zarowin, "Relation Between the R.F. Discharge Parameters and Plasma Etch Rates Selectivity and Anisotropy," *J. Vac. Sci. Technol.*, **A2**, 1537 (1984).

27 L. M. Ephrath, "The Effect of Cathode Materials on Reactive Ion Etching of Silicon and Silicon Dioxide in a CF_4 Plasma," *J. Electron. Mater.*, **7**, 415 (1978).

28 H. Matsumoto and T. Sugano, "Characterization of Reactive Ion Etched Silicon Surface by Deep Level Transient Spectroscopy," *J. Electrochem. Soc.*, **129**, 2823 (1982).

29 A. Rohatgi, P. Rai-Choudhyry, S. J. Fonash, P. Lester, Ranbir Singh, P. J. Caplan, and E. H. Poindexter, "Characterization and Control of Silicon Surface Modification Produced CCl$_4$ Reactive Ion Etching," *J. Electrochem. Soc.*, **133**, 408 (1986).

30 R. G. Frieser, F. J. Montillo, N. B. Zingerman, W. K. Chu, and S. R. Mader, "Silicon Damage Caused by Hydrogen-Containing Plasmas," *J. Electrochem. Soc.*, **130**, 2237 (1983).

31 R. A. Gdula, "The Effects of Processing on Radiation Damage in SiO$_2$," *IEEE Trans. Electron. Devices*, **26**, 644 (1979).

32 A. R. Reinberg, "R.F. Plasma Deposition of Inorganic Films for Semiconductor Applications," *Extended Abstracts Electrochem. Soc.* (1974).

33 See the latest "Revision of the Rules Regarding Industrial, Scientific and Medical (ISM) Equipment," in GEN Docket No. 20718 before the Federal Communications Commission, Washington, D.C., adopted Aug. 5, 1985.

34 J. J. Hannon and J. M. Cook, "Oxidative Removal of Photoresist by Oxygen/Freon®116 Discharge Products," *J. Electrochem. Soc.*, **131**, 1164 (1984).

35 D. Maydan, U.S. patent 4,298, 443, Nov. 3, 1981.

36 S. Matsuo and Y. Adachi, "Reactive Ion Beam Etching Using a Broad Beam ECR Ion Source," *Jap. J. Appl. Phys.*, **21**, L4 (1982).

37 K. Suzuki, S. Okudaira, and I. Kanomata, "The Role of Ions and Neutral Active Species in Microwave Plasma Etching," *J. Electrochem. Soc.*, **126**, 1024 (1979).

38 Y. Horiike, H. Okano, T. Yamazaki, and H. Horie, "High-Rate Reactive Ion Etching of SiO$_2$ Using a Magnetron Discharge," *Jap. J. Appl. Phys.*, **20**, L817 (1981).

39 M. Sekine, T. Arikado, H. Okano, and Y. Horiike, "High-Rate Reactive Ion Etching Using Collimated Beam Produced by 10^{-3} Torr Magnetron Discharge," *VLSI Symposium Digest*, IEEE, San Diego, 1986, p. 5.

40 R. H. Bruce, "Frequency Dependence of CCl$_4$ Etching," in R. G. Frieser and C. J. Mogab, Eds., *Proceedings of Symposium on Plasma Etching and Deposition*, Electrochemical Society, 1981, p. 243.

41 C. J. Mogab, "The Loading Effect in Plasma Etching," *J. Electrochem. Soc.*, **124**, 1262 (1977).

42 E. A. Truesdale, G. Smolinsky, and T. M. Mayer, "The Effect of Added Acetylene on the R.F. Discharge Chemistry of C$_2$F$_6$. A Mechanistic Model for Fluorocarbon Plasmas," *J. Appl. Phys.*, **51**, 2909 (1980).

43 P. J. Marcoux and P. D. Foo, "Methods of End Point Detection for Plasma Etching," *Solid State Technol.*, **24**, 115 (1981).

44 B. J. Curtis, "Optical End-Point Detection for the Plasma Etching of Aluminum," *Solid State Technol.*, **23**, 129 (1980).

45 P. A. Heimann and R. J. Schutz, "Optical Etch-Rate Monitoring: Computer Simulation of Reflectance," *J. Electrochem. Soc.*, **131**, 881 (1984).

46 See for example R. W. Light and H. B. Bell, "Patterning of Tantalum Polycide Films," *J. Electrochem. Soc.*, **131**, 459 (1984).

47 G. C. Schwartz and P. M. Schiable, "The Effects of Arsenic Doping in Reactive Ion Etching of Silicon in Chlorinated Plasmas," *J. Electrochem. Soc.*, **130**, 1898 (1983).

48 R. D. Rung, "Trench Isolation Prospects for Application in CMOS VLSI," in *Proc. IEDM*, 1985, p. 574.

49 T. Furuyama, T. Ohsawa, Y. Watanabe, T. Tanaka, K. Natori, and O. Ozawa, "An Experimental 4-Mbit CMOS DRAM," *IEEE J. Solid State Circuits*, **SC-21**, 605 (1986).

50 J. A. Bondur and H. A. Clark, "Plasma Etching for SiO$_2$ Profile Control," *Solid State Technol.*, **23**, 122 (1980).

51 R. A. H. Heinecke, "Control of Relative Etch Rates of SiO$_2$ and Si in Plasma Etching," *Solid State Electron.*, **18**, 1146 (1975).

52 R. A. H. Heinecke, "Plasma Reactor Design for the Selective Etching of SiO$_2$ on Si," *Solid State Electron.*, **19**, 1039 (1976).

53 D. L. Flamm and V. M. Donnelly, "The Design of Plasma Etchants," *Plasma Chem. Plasma Proc.*, **1**, 317 (1981).

54 Y. Kawamoto, T. Kure, N. Hashimoto, and T. Takaichi, "Highly Selective Dry Etching of

Si$_3$N$_4$," *Electrochem. Soc. Meeting*, **84-2**, abs. 395 (1984).

55 D. W. Hess, "Plasma Etching of Aluminum," *Solid State Technol.*, **24**, 189 (1981).

56 R. W. Light, "Reactive Ion Etching of Aluminum/Silicon," *J. Electrochem. Soc.*, **130**, 2225 (1983).

57 See for example W. Y. Lee, J. M. Eldridge, and G. C. Schwartz, "Reactive Ion Etching Induced Corrosion of Al and Al-Cu Films," *J. Appl. Phys.*, **52**, 2994 (1981).

58 A. C. Adams and C. D. Capio, "Planarization of Phosphorus-Doped Silicon Dioxide," *J. Electrochem Soc.*, **128**, 423 (1981).

59 S. Saikai, F. Shibata, S. Okudaira, and R. Hamazaki, "Plasma Etching Device Using Microwaves," *Semiconductor World* (April 1986), p. 44 (in Japanese).

CHAPTER
6

DIELECTRIC
AND POLYSILICON
FILM DEPOSITION

A. C. ADAMS

6.1 INTRODUCTION

Deposited films are widely used in the fabrication of modern VLSI circuits. These films provide conducting regions within the device, electrical insulation between metals, and protection from the environment. Deposited films must meet many requirements. The film thickness must be uniform and reproducible over each device and over all the wafers processed at one time. The structure and composition of the film must be controlled and reproducible. Finally, the method for depositing the film must be safe, reproducible, easily automated, and inexpensive.

The most widely used materials for film deposition are polycrystalline silicon, silicon dioxide, stoichiometric silicon nitride, and plasma-deposited silicon nitride. The most common deposition methods are atmospheric-pressure chemical vapor deposition (APCVD), low-pressure chemical vapor deposition (LPCVD), and plasma-enhanced chemical vapor deposition (PECVD or plasma deposition). Several reviews of these materials and their preparation are available. [1−3]

Polycrystalline silicon, usually referred to as polysilicon, is prepared by pyrolyzing silane at 575 to 650°C. Polysilicon is used as the gate electrode material in MOS devices, as a conducting material for multilevel metallization, and as a contact material for devices with shallow junctions. Polysilicon is usually deposited without dopants. The doping elements—arsenic, phosphorus, or boron—reduce the resistivity of the polysilicon and are added later by diffusion

233

or ion implantation. The dopants can also be added during deposition, which is better for some device structures. Polysilicon containing several percent oxygen is a semi-insulating material that is used for circuit passivation.

Dielectric materials are used for insulation between conducting layers, for diffusion and ion implantation masks, for diffusion from doped oxides, for capping doped films to prevent the loss of dopants, for gettering impurities, and for passivation to protect devices from impurities, moisture, and scratches. Phosphorus-doped silicon dioxide (usually called P-glass, phosphosilicate glass, or PSG) is especially useful, because it inhibits the diffusion of sodium impurities and because it softens and flows at 950 to 1100°C, creating a smooth topography that is beneficial for depositing metals. Borophosphosilicate glass (BPSG), formed when boron is added in addition to phosphorus, flows at lower temperatures, 850 to 950°C.

Silicon nitride is a barrier to sodium diffusion, is nearly impervious to moisture, and has a low oxidation rate. Stoichiometric silicon nitride (Si_3N_4), deposited at 700 to 900°C, is used as an oxidation mask to create planar structures and as a gate dielectric with thermally grown silicon dioxide in dual dielectric devices. Plasma-deposited silicon nitride (plasma nitride or SiNH) forms at much lower temperatures, 200 to 350°C, and provides passivation and protection for the device. The low deposition temperature allows this material to be used over aluminum or gold metallization.

Of the many methods available for depositing thin films, various chemical vapor deposition (CVD) techniques are most often used for semiconductor processing.[4] These chemical depositions occur under a variety of conditions. Deposition temperatures vary from 100 to 1000°C, and pressures range from atmospheric down to about 7 Pa (0.05 Torr). The energy for the reaction can be supplied thermally, by photons (photochemically), or by a glow discharge.

Historically, dielectric and polysilicon films have been deposited at atmospheric pressure by using a variety of reactor geometries. These include horizontal reactors with the wafers lying on a hot susceptor and the reactant gases flowing over the surfaces, usually at high velocities. The susceptor is heated by radiation using high-intensity lamps, radio frequency induction, or electrical resistance. Various vertical reactors also exist, usually consisting of a bell jar reaction chamber with samples oriented in a vertical direction on a rotating assembly. As in the horizontal reactors, the susceptor is heated by radiation, induction, or resistance. All these atmospheric-pressure reactors tend to have low wafer throughput, require excessive wafer handling during loading and unloading, and provide thickness uniformities that are no better than ±10%. Consequently, they have been replaced by low-pressure, hot-wall reactors. Plasma-assisted depositions in hot-wall reactors or with parallel-plate geometries are also available for applications that require low sample temperatures, 100 to 350°C.

The potential advantages of the low-pressure deposition processes are (1) uniform step coverage, (2) precise control of composition and structure, (3) low-temperature processing, (4) fast deposition rates, (5) high throughput, and (6) low processing costs. Compromises and trade-offs are made among these properties. For instance, low deposition rates may be tolerated to achieve low deposition temperatures. A goal in developing a deposition process is to best use the advantages of CVD and to find the optimal compromise for specific device structures.

6.2 DEPOSITION PROCESSES

6.2.1 Reactions

Table 1 lists some typical reactions that may be used to deposit films on device wafers. The choice of a particular reaction is often determined by the deposition temperature (which must be compatible with the device materials), the film properties, and certain engineering aspects of the deposition (wafer throughput, safety, and reactor maintenance).

The most common reactions for depositing silicon dioxide for VLSI circuits are oxidizing silane, SiH_4, with oxygen at 400 to 450°C; decomposing tetraethoxysilane, $Si(OC_2H_5)_4$, at 650 to 750°C; and reacting dichlorosilane, $SiCl_2H_2$, with nitrous oxide at 850 to 900°C. Doped oxides are prepared by adding a dopant to the deposition reaction. The hydrides—arsine, phosphine, or diborane—are often used because they are readily available gases; however, halides or organic compounds such as phosphorus tribromide or trimethylphosphite can also be used. Silicon nitride is prepared by reacting silane and ammonia at atmospheric pressure at 700 to 900°C, or by reacting dichlorosilane and ammonia at reduced pressure at 700°C. Plasma-deposited silicon nitride is deposited by reacting silane with ammonia or nitrogen in a glow discharge between 200 and 350°C. This reaction is useful for depositing passivation coatings over finished devices where higher temperatures cause unwanted reactions between the silicon and the metal conductors. Similarly, plasma-deposited silicon dioxide is formed

TABLE 1
Typical reations for depositing dielectrics and polysilicon.

Product	Reactants	Deposition temperature, °C
Silicon dioxide	$SiH_4 + CO_2 + H_2$	850–950
	$SiCl_2H_2 + N_2O$	850–900
	$SiH_4 + N_2O$	750–850
	$SiH_4 + NO$	650–750
	$Si(OC_2H_5)_4$	650–750
	$SiH_4 + O_2$	400–450
Silicon nitride	$SiH_4 + NH_3$	700–900
	$SiCl_2H_2 + NH_3$	650–750
Plasma silicon nitride	$SiH_4 + NH_3$	200–350
	$SiH_4 + N_2$	200–350
Plasma silicon dioxide	$SiH_4 + N_2O$	200–350
Polysilicon	SiH_4	575–650

from silane and nitrous oxide in a glow discharge. Polysilicon is prepared by pyrolyzing silane at 575 to 650°C.

6.2.2   Equipment

Figures 1 and 2 give schematics of four reactors commonly used for depositions. Figure 1a shows a hot-wall, reduced-pressure reactor used to deposit polysilicon, silicon dioxide, and silicon nitride. The reactor consists of a quartz tube heated by a three-zone furnace, with gas introduced into one end and pumped out the other. Pressures in the reaction chamber are typically 30 to 250 Pa (0.25 to 2.0 Torr); temperatures range between 300 and 900°C; and gas flows are between 100 and 1000 std. cm^3/min. (Gas flows are always reported at standard conditions, 0°C and 1 atm pressure.) Wafers stand vertically, perpendicular to the gas flow, in a quartz holder. Special inserts that alter the gas flow dynamics are sometimes used. Each run processes 50 to 200 wafers with thickness uniformities of the

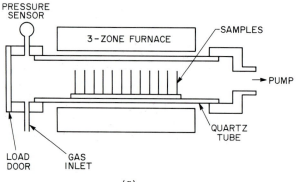

(a)

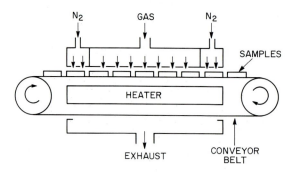

(b)

FIGURE 1
Schematic diagrams of CVD reactors: (a) Hot-wall, reduced-pressure reactor. (b) Continuous, atmospheric-pressure reactor.

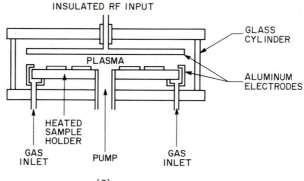

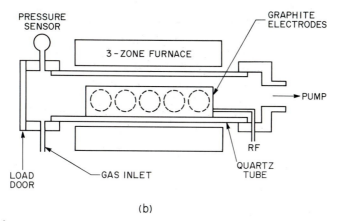

FIGURE 2
Schematic diagrams of plasma-deposition reactors: (a) Parallel-plate. (b) Hot-wall.

deposited films within ±5%. Hot-wall, reduced-pressure reactors are easily scaled to hold 150 mm diameter wafers. The major advantages of these reactors are excellent uniformity, large load size, and ability to accommodate large diameter wafers. The disadvantages are low deposition rates and the frequent use of toxic, corrosive, or flammable gases.

Figure 1b shows a continuous throughput, atmospheric-pressure reactor used to deposit silicon dioxide. The samples are carried through the reactor on a conveyor belt. Reactant gases flowing through the center of the reactor are contained by gas curtains formed by a fast flow of nitrogen. The samples are heated by convection. The advantages of these continuous reactors are high throughput, good uniformity, and ability to handle large diameter wafers. The major disadvantages are that fast gas flows are required and that these reactors must be cleaned frequently.

Figure 2a shows a radial-flow, parallel-plate, plasma-assisted CVD reactor. The reaction chamber is a cylinder, usually glass or aluminum, with aluminum plates on the top and bottom. Samples lie on the grounded bottom electrode. A radio frequency voltage, applied to the top electrode, creates a glow discharge between the two plates. Gases flow radially through the discharge. They are usually introduced at the outer edge and flow toward the center, although the opposite flow pattern can be used. Resistance heaters or high-intensity lamps heat the bottom, grounded electrode to a temperature between 100 and 400°C. This reactor is used for the plasma-assisted deposition of silicon dioxide and silicon nitride. Its main advantage is low deposition temperature. There are three disadvantages: capacity is limited, especially for large-diameter wafers; wafers must be loaded and unloaded individually; and wafers may be contaminated by loosely adhering deposits falling on them.

The hot-wall, plasma-deposition reactor shown in Fig. 2b solves many of the problems encountered in the radial-flow reactor. The reaction takes place in a quartz tube heated by a furnace. The samples are held vertically, parallel to the gas flow. The electrode assembly contains long graphite or aluminum slabs to support the samples. Alternating slabs are connected to the power supply to generate a discharge in the space between the electrodes. Advantages of this reactor are its high capacity and low deposition temperatures. Its drawbacks, however, are that particles can be formed while the electrode assembly is being inserted, and that wafers must be individually handled during loading and unloading.

6.2.3 Safety

Many of the gases used to deposit films are hazardous. The safety problems are more severe for low-pressure depositions because the processes often use concentrated gases. For instance, 100% silane is used for polysilicon depositions at reduced pressure, compared to only 3% silane in nitrogen for the same deposition at atmospheric pressure. Low-pressure depositions have additional safety problems because hazardous gases can dissolve or react in the vacuum pump oil.

The hazardous gases fall into four general classes: poisonous; pyrophoric, flammable, or explosive; corrosive; and dangerous combinations of gases. Table 2 lists hazardous properties of gases commonly used in CVD. Examples of dangerous gas combinations that may be encountered are silane with halogens and oxygen with hydrogen.

Many of the gases react with air to form solid products. Consequently, small leaks cause particles to form within the gas lines. These particles eventually plug the line or the gas metering equipment. Detailed safety precautions for CVD processes have been published and should be consulted.[5]

6.3 POLYSILICON

Polysilicon is used as the gate electrode in MOS devices. A metal or metal silicide, such as tungsten or tantalum silicide, may be deposited over the polysilicon gate to increase the electrical conductivity. Polysilicon is also used for resistors, conductors, and to ensure ohmic contact to shallow junctions. The polysilicon is

TABLE 2
Properties of common gases used in CVD

Gas	Formula	Hazard	Flammable limits in air (vol%)	Exposure limit (ppm)
Ammonia	NH_3	toxic, corrosive	16–25	25
Argon	Ar	inert	—	—
Arsine	AsH_3	toxic	—	0.05
Diborane	B_2H_6	toxic, flammable	1–98	0.1
Dichlorosilane	SiH_2Cl_2	flammable, toxic	4–99	5
Hydrogen	H_2	flammable	4–74	—
Hydrogen chloride	HCl	corrosive, toxic	—	5
Nitrogen	N_2	inert	—	—
Nitrogen oxide	N_2O	oxidizer	—	—
Oxygen	O_2	oxidizer	—	—
Phosphine	PH_3	toxic, flammable	pyrophoric	0.3
Silane	SiH_4	flammable, toxic	pyrophoric	0.5

deposited by pyrolyzing silane between 575 and 650°C in a low-pressure reactor (Fig. 1a). The chemical reaction is

$$SiH_4 \rightarrow Si + 2H_2 \tag{1}$$

Further processing for polysilicon gates involves doping, etching, and oxidation.

Two low-pressure processes are common for depositing polysilicon. One uses 100% silane at a pressure of 25 to 130 Pa (0.2 to 1.0 Torr). The other process is performed at the same total pressure but uses 20 to 30% silane diluted in nitrogen. Both processes deposit polysilicon on 10 to 200 wafers per run with thickness uniformities of ±5%. The deposition rates are 10 to 20 nm/min.

6.3.1 Deposition Variables

Temperature, pressure, silane concentration, and dopant concentration are important process variables in the deposition of polysilicon; wafer spacing and load size have only minor effects.[6,7] Figure 3 shows that the deposition rate increases rapidly as the temperature increases. The temperature dependence is usually exponential and follows the Arrhenius equation:

$$R = A \exp{(-qE_a/kT)} \tag{2}$$

where R is the deposition rate, E_a is the activation energy in eV, T is the absolute temperature in K, A is the frequency factor, k is Boltzmann's constant (1.381×10^{-23}J/K), and q is the electronic charge (1.602×10^{-19} coulomb). When the logarithm of the deposition rate is plotted against the reciprocal of the absolute temperature, a straight line results with a slope equal to $-qE_a/k$. The activation energies for polysilicon deposition, calculated from the slopes of the lines in Fig. 3, are about 1.7 eV (40 kcal/mole).

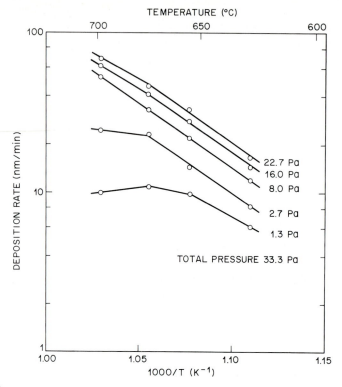

FIGURE 3
Arrhenius plot for polysilicon deposition for different silane partial pressures.

Equation 2 shows that the deposition rate increases with increasing temperature. However, at high temperature the reaction becomes faster than the rate at which unreacted silane arrives at the surface. When this occurs the deposition rate no longer increases with temperature and the reaction is said to be mass-transport limited. Some of the high temperature data in Fig. 3 show this condition. Depositions occurring in the mass-transport limited regime depend on the reactant concentration, reactor geometry, and gas flow. When the rate of reaction is slower than the rate of reactant arrival, the deposition is surface-reaction limited, and the critical variables are reactant concentration and temperature. The linear portion of the lines in Fig. 3 show the surface-reaction limited condition. Depositions that are surface-reaction limited have excellent thickness uniformity and step coverage.

The Arrhenius equation is useful for finding the deposition rate at different temperatures. For instance, if polysilicon deposits at 10 nm/min at 600°C (873 K) with an activation energy of 1.7 eV, the rate at 550°C is found by substituting in Eq. 3:

$$\log(R_1/R_2) = \frac{qE_a}{2.3k}\left(\frac{T_1 - T_2}{T_1 T_2}\right) \tag{3}$$

Solving for R_2 gives 2.5 nm/min.

Polysilicon depositions at reduced pressure for VLSI manufacturing are limited to temperatures between 575 and 650 °C. At higher temperatures, gas-phase reactions, which result in a rough, loosely adhering deposit, and silane depletion, which causes poor uniformity, become significant. At temperatures much lower than 575 °C, the deposition rate is too slow to be practical.

Polysilicon depositions often use a temperature ramp with the rear furnace zone 5 to 15 °C hotter than the front and center zones. The higher temperature increases the deposition rate in the rear zone and compensates for the silane depletion. Under optimal conditions the increased deposition rate results in a uniform thickness throughout the deposition zone. However, the structure of polysilicon is strongly influenced by temperature, so a temperature ramp may cause a variation in structure and film properties.

Pressure can be varied in a low-pressure reactor by changing the gas flow into the reactor while keeping the pumping speed constant, or by changing the pumping speed at a constant inlet gas flow. If the inlet gas is a mixture of silane and nitrogen, the nitrogen flow can be changed while keeping the silane flow constant, or the silane and nitrogen can both be changed while keeping the ratio constant. All three methods—changing pumping speed, changing nitrogen flow, or changing total gas flow with a constant ratio—are used to control the reactor pressure. If the total gas flow is varied (constant ratio and pump speed), the deposition rate is a linear function of pressure. But if the pumping speed or the nitrogen flow is changed, the rate depends only slightly on pressure (see Fig. 4). Deposition reproducibility is best when the inlet gas flows are kept constant and the pressure is controlled by the pumping speed.

The polysilicon deposition rate is not a linear function of the silane concentration. Figure 5 gives representative data for four deposition temperatures and for a total pressure of 33 Pa (0.25 Torr). The nonlinear behavior is caused by a sequence of surface processes. First silane adsorbs on the surface. The adsorbed silane then decomposes to form SiH_2 and adsorbed hydrogen. The SiH_2 decomposes further to form polysilicon and more adsorbed hydrogen. Finally, the adsorbed hydrogen leaves and more silane adsorbs to continue the reaction. The reaction sequence is

$$SiH_{4(g)} \rightleftharpoons SiH_{4(ad)} \tag{4}$$

$$SiH_{4(ad)} \rightleftharpoons SiH_{2(ad)} + H_{2(ad)} \tag{5}$$

$$SiH_{2(ad)} \rightleftharpoons Si + H_{2(ad)} \tag{6}$$

$$H_{2(ad)} \rightleftharpoons H_{2(g)} \tag{7}$$

If the last step, desorption of the hydrogen, is slow compared to the other steps, the overall reaction rate is given by

$$rate = K_1 p_s^{1/2} / (1 + K_2 p_s^{1/2}) \tag{8}$$

where p_s is the partial pressure of silane and K_1 and K_2 are derived from the equilibrium constants for the individual reaction steps. The data in Fig. 5 closely follow the solid lines obtained from Eq. 8. The agreement supports but does not prove the suggested reaction sequence.

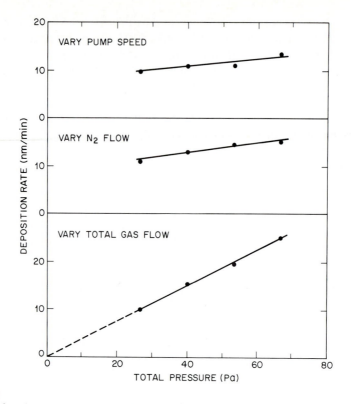

FIGURE 4
The effect of total pressure on the polysilicon deposition rate.

Polysilicon is doped during deposition by adding phosphine, arsine, or diborane to the reactants. Figure 6 shows how the dopant affects the deposition rate. Adding diborane causes a large increase in the deposition rate. In contrast, adding phosphine or arsine causes a rapid decrease in the deposition rate. The phosphine or arsine strongly adsorb on the surface, blocking the adsorption of silane and decreasing the deposition rate. Diborane forms borane radicals, BH_3, that catalyze gas-phase reactions and increase the deposition rate. The thickness uniformity across a single wafer degrades when dopants are added. Uniformity can be maintained by using an insert to control the flow of reactant gases around the samples.

6.3.2 Structure

The structure of polysilicon is strongly influenced by dopants, impurities, deposition temperature, and post-deposition heat cycles.[8-10] Polysilicon deposited below 575 °C is amorphous with no detectable structure. Polysilicon deposited above 625 °C is polycrystalline and has a columnar structure. Crystallization and grain growth occur when either amorphous or columnar polysilicon is heated. Figure 7 illustrates all three structures, showing transmission electron microscope

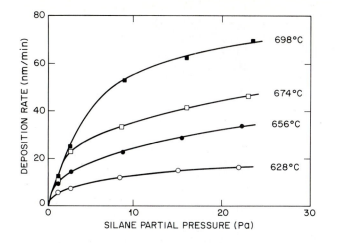

FIGURE 5
The effect of silane concentration on the polysilicon deposition rate.

(TEM) cross sections of polysilicon deposited at 605°C (amorphous) and 630°C (columnar), and annealed at 700°C (crystalline grains). After high-temperature heat cycles, there are no significant structural differences between polysilicon that is initially amorphous or columnar.

The deposition temperature at which the transition from amorphous to columnar structure occurs is well defined but depends on many variables, such as

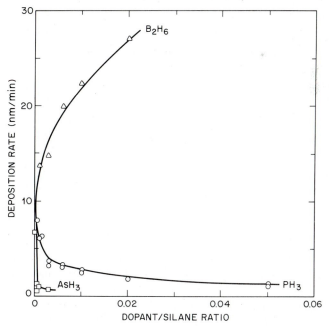

FIGURE 6
The effect of dopants on the polysilicon deposition rate at 610°C.

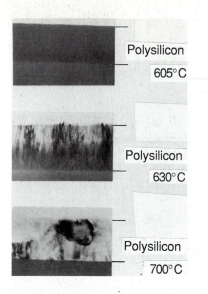

FIGURE 7
TEM cross sections (60,000X) of polysilicon. (a) Amorphous structure deposited at 605 °C. (b) Columnar structure deposited at 630 °C. (c) Crystalline grains formed by annealing an amorphous sample at 700 °C.

deposition rate, partial pressure of hydrogen, total pressure, presence of dopants, and presence of impurities (O, N, or C). The transition temperature is between 575 and 625 °C for depositions in an LPCVD reactor. After deposition, polysilicon recrystallizes when heated; however, the crystallization temperature is strongly influenced by dopants and impurities. Oxygen, nitrogen, and carbon impurities stabilize the amorphous structure to temperatures above 1000 °C.

The average diameter of the column, that is, the columnar grains, can be measured by TEM surface replication. The size depends on film thickness and is typically between 0.03 and 0.3 μm. The grain size after crystallization depends on heating time, temperature, and dopant concentration. Polysilicon doped with a high concentration of phosphorus and heated between 900 and 1000 °C for 20 min has an average grain size between 0.5 and 1 μm.

Polysilicon deposited at 600 to 650 °C has a $\{110\}$-preferred orientation. At higher deposition temperatures the $\{100\}$ orientation predominates, but the structure contains significant contributions from other orientations, such as $\{110\}$, $\{111\}$, $\{311\}$, and $\{331\}$. Dopants, impurities, and temperature influence the preferred orientation.

The structural changes in polysilicon during typical device processing can be summarized as follows. Polysilicon deposited between 600 and 650 °C has a columnar structure with grain sizes between 0.03 and 0.3 μm and a $\{110\}$-preferred orientation. During phosphorus diffusion at 950 °C, the polysilicon recrystallizes to form crystallites with an average size of 0.5 to 1.0 μm. The grains continue to grow during oxidation at 1050 °C to a final size of 1 to 3 μm.

Polysilicon deposited at temperatures below 600 °C behaves similarly, except the initial film is amorphous.

6.3.3 Doping Polysilicon

Polysilicon can be doped by diffusion, implantation, or the addition of dopant gases during deposition (in situ doping).[11-16] All three methods are used for device fabrication. Figure 8 shows the resistivity of polysilicon doped with phosphorus by these three methods. The diffusion data show the resistivity after a 1 h diffusion at various temperatures. The implantation data show the resistivity after a 1 h, 1100 °C activation. The resistivities for the in situ doped samples are measured after deposition at 600 °C and after a 30 min anneal at the indicated temperature. Diffusion is a high-temperature process that results in low resistivities. The dopant concentration in diffused polysilicon often exceeds the solid solubility limit, with the excess dopant segregated at the grain boundaries. Hall mobilities for heavily diffused polysilicon are usually 30 to 40 cm^2/V-s.

The resistivity of implanted polysilicon depends primarily on implant dose, annealing temperature, and annealing time. The high resistivity in lightly implanted polysilicon (Fig. 8) is caused by carrier traps at the grain boundaries.

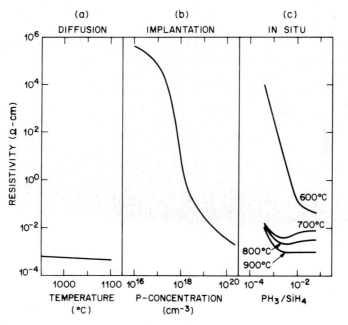

FIGURE 8

Resistivity of P-doped polysilicon. (a) Diffusion: 1 h at the indicated temperature. (*After Kamins, Ref. 11.*) (b) Implantation: 1 h anneal at 1100 °C. (*After Mandurah, Saraswat, and Kamins, Ref. 12.*) (c) In situ: as deposited at 600 °C and after a 30-min anneal at the indicated temperature. (*After A. C. Adams, unpublished data.*)

Once these traps are saturated with dopants, the resistivity decreases rapidly and approaches the resistivity for implanted single-crystal silicon. The mobility for heavily implanted polysilicon is about 30 to 40 cm^2/V-s, similar to the values for diffused polysilicon. Implanted polysilicon has about ten times higher resistivity than diffused polysilicon because of the differences in dopant concentrations: about 10^{20} cm^{-3} for a heavy implant and greater than 10^{21} cm^{-3} for a heavy diffusion.

Polysilicon films doped during deposition by adding phosphine, arsine, or diborane have resistivities that are strong functions of deposition temperature, dopant concentration, and annealing temperature. Figure 9 shows the resistivities for in situ doped polysilicon deposited at different temperatures. The transition from high resistivity at low deposition temperature to low resistivity at high temperature corresponds to the change from an amorphous to columnar structure. The phosphorus-doped films in the figure (denoted by the circles) change structure at 625°C; the boron-doped polysilicon (denoted by the triangles) changes at temperatures between 525 and 550°C. The resistivity of doped amorphous polysilicon decreases during annealing, mainly because of crystallization (Fig. 8). After annealing, the resistivity is not a strong function of the initial dopant concentration. Doped polysilicon that is crystalline when deposited shows almost no change in resistivity after annealing. The dopant concentration in in situ doped

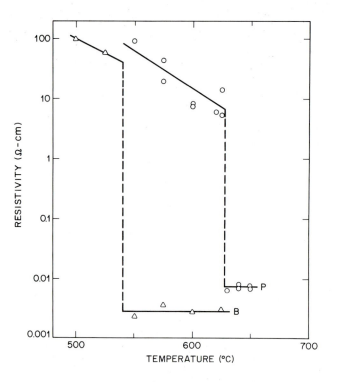

FIGURE 9
Resistivity of in situ doped polysilicon deposited at different temperatures. The triangles denote boron-doped polysilicon and the circles denote phosphorus-doped polysilicon.

polysilicon is high, 10^{20} to 10^{21} cm^{-3}, but the mobility is often low, 10 to 30 cm^2/V-s.[14]

A comparison of the three doping processes shows that the major differences are lower resistivity for diffusion, lower dopant concentration for implantation, and lower mobility for in situ doping. Implantation and in situ doping, however, offer the advantage of lower processing temperatures, often the dominant consideration in VLSI processing.

6.3.4 Oxidation of Polysilicon

The details of polysilicon oxidation are discussed in Chapter 3. Polysilicon is usually oxidized in dry oxygen at temperatures between 900 and 1000°C to form an insulator between the doped-polysilicon gate and other conducting layers.[17-19] Undoped or lightly doped silicon oxidizes at a rate between the rates for {111} and {100} crystalline silicon. Phosphorus-doped polysilicon oxidizes faster than undoped polysilicon, and the rate of oxidation is determined by the dopant concentration at the polysilicon surface.

The silicon dioxide grown on polysilicon has lower breakdown fields, higher leakage currents, and higher stress than oxides grown on single-crystal silicon.[20-22] Typical breakdown fields are 4 to 5 MV/cm. The degraded oxide properties are related to the rough polysilicon–oxide interface that is caused by the initial roughness of the polysilicon surface. Preferential oxidation at grain boundaries also contributes to the rough interface. Oxidation of amorphous polysilicon produces a smooth interface, and the oxide has high breakdown strength (7 to 9 MV/cm) and low leakage currents.

6.3.5 Properties of Polysilicon

The chemical and physical properties of polysilicon often depend on the film structure (amorphous or crystalline) or on the dopant concentration. The etch rate of polysilicon in a plasma (discussed in Chapter 5) depends on the dopant concentration. Polysilicon that is heavily phosphorus-doped etches at higher rates than undoped or lightly doped polysilicon. Polysilicon is usually etched in a $CF_4 + O_2$ or a $CF_3Cl + Cl_2$ plasma.

The optical properties of polysilicon depend on its structure. The imaginary part of the dielectric function is particularly structure sensitive.[23,24] Crystalline polysilicon has sharp maxima in the dielectric function near 295 and 365 nm (4.2 and 3.4 eV). Amorphous polysilicon has a broad maximum without sharp structure. In addition, amorphous polysilicon has a higher refractive index throughout the visible region than crystalline polysilicon. The difference in refractive index must be considered when the thickness of polysilicon is measured by optical techniques. Approximate refractive index values at a wavelength of 600 nm are 4.1 for crystalline polysilicon and 4.5 for amorphous material.

Other properties of polysilicon are its density, 2.3 g/cm^3; coefficient of thermal expansion, 2×10^{-6}/°C; and temperature coefficient of resistance, 1×10^{-3}/°C. These are important for modeling heat dissipation in devices.

6.3.6 Oxygen-doped Polysilicon

The addition of oxygen to polysilicon increases the film resistivity. The resulting material, semi-insulating polysilicon (SIPOS), is used as a passivating coating for high voltage devices.[25,26] SIPOS is deposited when silane reacts with small amounts of nitrous oxide at temperatures between 600 and 700°C. The two simultaneous reactions are

$$SiH_4 \rightarrow Si + 2H_2 \tag{9}$$

$$SiH_4 + xN_2O \rightarrow SiO_x + 2H_2 + xN_2 \tag{10}$$

The quantity of nitrous oxide in the reaction determines the film composition and resistivity (see Fig. 10). SIPOS has a multi-phase microstructure containing crystalline silicon, amorphous silicon, silicon dioxide, and silicon monoxide. The specific composition depends on the deposition temperature, the amount of nitrous oxide used in the reaction, and the time and temperature of post-deposition anneals. SIPOS used in VLSI circuits contains between 25 and 40 at. % oxygen.

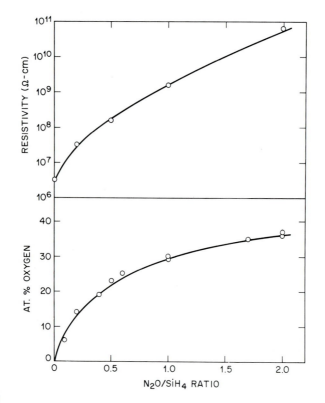

FIGURE 10
The effect of reactant concentrations on the resistivity and composition of SIPOS. (*After Mochizuki, et al., Ref. 25, and Knolle and Maxwell, Ref. 26.*)

6.4 SILICON DIOXIDE

Silicon dioxide films can be deposited with or without dopants. Undoped silicon dioxide is used as an insulating layer between multilevel metallizations, an ion-implantation or diffusion mask, a capping layer over doped regions to prevent outdiffusion during heat cycles, and to increase the thickness of thermally grown oxides. Phosphorus-doped silicon dioxide is used as an insulator between metal layers, as a final passivation over devices, and as a gettering source. Oxides doped with phosphorus, arsenic, or boron are used occasionally as diffusion sources.

The processing sequence for silicon dioxide depends on its specific use in the device. Oxides used as insulators between conducting layers are deposited, densified by annealing, and etched to open windows. A solution containing fluoride or a CHF_3 plasma etches the oxide. Phosphorus-doped oxides, used in the flowed glass process, are heated to a temperature between 950 and 1100°C so that the oxide softens and flows, providing a smooth topography which improves the step coverage of the next metallization. Phosphorus-doped oxides used for passivation are deposited at temperatures lower than 400°C, and areas for bonding are opened by etching.

6.4.1 Deposition Methods

Several deposition methods are used to produce silicon dioxide. They are characterized by different chemical reactions, reactors, and temperatures. Films deposit at low temperatures, lower than 500°C, by reacting silane, dopant, and oxygen.[27-29] The chemical reactions for phosphorus-doped oxides are

$$SiH_4 + O_2 \rightarrow SiO_2 + 2H_2 \tag{11}$$

$$4PH_3 + 5O_2 \rightarrow 2P_2O_5 + 6H_2 \tag{12}$$

Under normal deposition conditions, hydrogen is formed rather than water. The deposition can be carried out at atmospheric pressure in a continuous reactor (Fig. 1b) or at reduced pressure in an LPCVD reactor (Fig. 1a). The main advantage of silane–oxygen reactions is the low deposition temperature allowing films to deposit over aluminum metallization. Consequently, these films provide passivation coatings over the final device and insulation between aluminum levels. The main disadvantages of silane–oxygen reactions are poor step coverage (see Section 6.4.3) and particles caused by loosely adhering deposits on the reactor walls.

Silicon dioxide also deposits at 650 to 750°C in an LPCVD reactor by decomposing tetraethoxysilane, $Si(OC_2H_5)_4$.[30-32] This compound, abbreviated TEOS, is vaporized from a liquid source. The overall reaction is

$$Si(OC_2H_5)_4 \rightarrow SiO_2 + \text{by-products} \tag{13}$$

where the by-products are a complex mixture of organic and organosilicon compounds. The decomposition of TEOS is useful for depositing insulators over polysilicon gates, but the high deposition temperature precludes its use over aluminum. The advantages of TEOS deposition are excellent uniformity, conformal step coverage, and good film properties. The disadvantages are the high temperature and liquid source requirements.

Silicon dioxide deposits at temperatures near 900°C and at reduced pressure by reacting dichlorosilane with nitrous oxide:[33]

$$SiCl_2H_2 + 2N_2O \rightarrow SiO_2 + 2N_2 + 2HCl \tag{14}$$

This deposition gives excellent uniformity and is used to deposit insulating layers over polysilicon; however, this oxide frequently contains small amounts of chlorine that may react with the polysilicon and cause film cracking.

Doped oxides are formed by adding small amounts of the dopant hydrides (phosphine, arsine, or diborane) during the deposition. Because the dopant hydrides are very toxic, other dopant compounds are often used. Dopant halides or organic compounds are safer than hydrides but less convenient because they must usually be vaporized from solid or liquid sources.

Dopant concentrations are reported as weight percent (wt. %), atom percent (at. %), or mole percent (mol %). The relationships between these units for phosphorus-doped oxides are

$$\text{mole } \% \ P_2O_5 = \frac{6010W}{6200 - 81.9W} \tag{15}$$

$$\text{atom } \% \ P = \frac{12000W}{18600 - 81.9W} \tag{16}$$

where W is weight percent of phosphorus. The relationships for boron-doped oxides are

$$\text{mole } \% \ B_2O_3 = \frac{6010W}{2160 - 9.5W} \tag{17}$$

$$\text{atom } \% \ B = \frac{12020W}{6480 + 91.7W} \tag{18}$$

where W is now the weight percent of boron. Occasionally, weight percent, atom percent, and mole percent are written as w/o, a/o, and m/o.

The doped oxides used as diffusion sources contain 5 to 15 wt. % of the dopant. Doped oxides used for passivation or for interlevel insulation contain 2 to 8 wt. % phosphorus. The phosphorus prevents the diffusion of ionic impurities to the device. Doped oxides used for the P-glass flow process (described in Section 6.4.4) contain 6 to 8 wt. % phosphorus. Glass with lower phosphorus concentrations will not soften and flow; higher concentrations react slowly with atmospheric moisture to form acid products that corrode the aluminum metallization. Borophosphosilicate glass used for low temperature flow contains 4 to 6 wt. % phosphorus and 1 to 4 wt. % boron.

6.4.2   Deposition Variables

The deposition of silicon dioxide depends on the same variables that are important for polysilicon, that is, temperature, pressure, reactant concentration, and presence of dopants. In addition, other variables, such as wafer spacing and total gas flow, are important for some silicon dioxide depositions. The deposition rate for the reaction of silane with oxygen increases with temperature, but the apparent activation energy is low, less than 0.4 eV (10 kcal/mole). This activation energy is much less than the values usually observed for chemical reactions and is similar to the values for adsorption on a surface or for gas-phase diffusion. The deposition has a complicated dependence on oxygen concentration. Namely, if the oxygen concentration is increased at a constant temperature, the deposition rate increases rapidly, goes through a maximum, and then slowly decreases. Figure 11 shows representative data. Surface reactions cause this behavior; at high oxygen concentrations the surface is nearly saturated with adsorbed oxygen and further deposition reactions with silane are blocked. When phosphine is added to the reaction, the rate rapidly decreases and then slowly increases. This deposition behavior is also attributable to surface adsorption effects since phosphine also strongly adsorbs on the surface and blocks reactions with silane. At high partial pressures of silane, the deposited silicon dioxide is hazy, probably because of gas-phase reactions. The gas-phase transport of material to the wafer surface is important in the silane–oxygen reaction. A special sample holder that directs the gas to the wafers is required for uniform depositions.

The deposition of silicon dioxide by decomposing TEOS occurs at temperatures between 650 and 750 °C. Figure 12 shows deposition rates as a function of

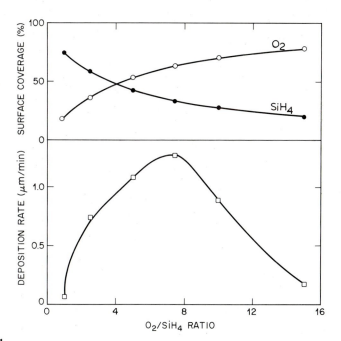

FIGURE 11
The deposition rate and surface coverage for the silane-oxygen reaction at atmospheric pressure and 350 °C. (*After Cobianu and Pavelescu, Ref. 28.*)

temperature for the TEOS decomposition[30] and for the silane–oxygen reaction.[29] The activation energy for the TEOS reaction is about 1.9 eV (45 kcal/mole) and decreases to 1.4 eV (32 kcal/mole) when phosphorus-doping compounds are present. Note the contrast between these activation energies and the low activation energies in silane–oxygen reactions. Figure 13 shows how the deposition rate depends on the TEOS partial pressure.[30,32] The nonlinear behavior is similar to effects observed in polysilicon deposition and is caused by surface-catalyzed reactions. At low TEOS partial pressures the deposition rate is determined by the rate of the surface reaction. At high partial pressures, the surface becomes nearly saturated with adsorbed TEOS, and the deposition rate starts to become independent of the TEOS pressure.

The deposition of silicon dioxide at 900°C using dichlorosilane and nitrous oxide has a strong nonlinear pressure dependence that is a function of wafer position in the reactor. Gas transport and depletion are significant in this deposition.

Phosphorus-doped oxides are deposited by adding phosphorus compounds, usually phosphine or trimethylphosphite, to the silane–oxygen or TEOS reaction. Doping is difficult with the dichlorosilane–nitrous oxide reaction because the phosphorus oxides are volatile at the high deposition temperature. Adding phosphorus to the low-pressure depositions causes the thickness uniformity to degrade. The deposition of phosphorus-doped silicon dioxide requires inserts to ensure uniform gas flow over the wafer surfaces. Boron-doped oxides are deposited by adding diborane, boron trichloride, or trimethylborate to the silane–oxygen or TEOS reactions. Addition of boron does not degrade thickness uniformity or deposition rate.

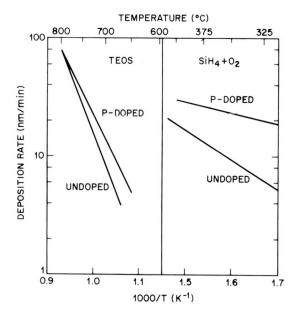

FIGURE 12

Arrhenius plots for the low-pressure deposition of SiO_2. (*After Adams and Capio, Ref. 30, for the TEOS data and after Learn, Ref. 29, for the silane–oxygen data.*)

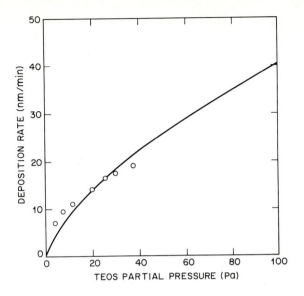

FIGURE 13
Deposition rate for different TEOS concentrations. (*After Adams and Capio, Ref. 30, for the data points and after Huppertz and Engl, Ref. 32, for the solid line.*)

6.4.3 Step Coverage

Three general types of step coverage are observed for deposited silicon dioxide. They are schematically diagrammed in Fig. 14. Figure 14a shows a completely uniform or conformal step coverage; the film thickness along the walls is the same as the film thickness at the bottom of the step. Uniform step coverage results when reactants or reactive intermediates adsorb on the surface and then rapidly migrate along the surface before reacting. The rapid migration results in a uniform surface concentration, regardless of the topography, and gives a completely uniform thickness.

When the reactants adsorb and react without significant surface migration, the deposition rate is proportional to the arrival angle of the gas molecules. Figure 14b gives an example in which the mean free path of the gas is much larger than the dimensions of the step. The arrival angle in two dimensions at the top horizontal surface is 180°. At the top of the vertical step, the arrival angle is only 90°, so the film thickness is reduced by half. Along the vertical walls the arrival angle, ϕ, is determined by the width of the opening and the distance from the top. The arrival angle determines the film thickness and is calculated from

$$\phi = \arctan \frac{w}{z} \tag{19}$$

where w is the width of the opening and z is the distance from the top surface. This type of step coverage is thin along the vertical walls and may have a crack at the bottom of the step caused by self-shadowing.

Figure 14c gives a diagram for no surface migration and for a short mean free

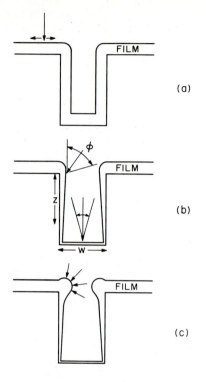

FIGURE 14
Step coverage of deposited films. (a) Uniform coverage resulting from rapid surface migration. (b) Nonconformal step coverage for long mean free path and no surface migration. (c) Nonconformal step coverage for short mean free path and no surface migration.

path. Here the arrival angle at the top of the step is 270°, giving a thicker deposit. The arrival angle at the bottom of the step is only 90°, so the film is thin. Gas depletion effects are also observed along the step walls. The thick cusp at the top of the step and the thin crevice at the bottom combine to give a re-entrant shape that is particularly difficult to cover with metal.

Figure 15 gives examples of the different types of step coverage. The samples are prepared by etching {110} single-crystal silicon in hot potassium hydroxide to form vertical grooves 5 μm wide and 50 μm deep. After about 1 μm of oxide has been deposited, the samples are cleaved and a cross section examined to determine the step coverage. A nearly uniform coverage is observed for the TEOS deposition at reduced pressure (Fig. 15a). The mean free path of the reactants in the gas at the deposition conditions (700°C and 30 Pa) is several hundred microns, much larger than the dimensions of the groove. Consequently, gas-phase diffusion (without surface interactions) into the groove is negligible. However, surface diffusion of the TEOS is rapid, resulting in the uniform coverage.

Figure 15b shows silicon dioxide deposited from silane and oxygen at reduced pressure. The mean free path is still large, several hundred microns,

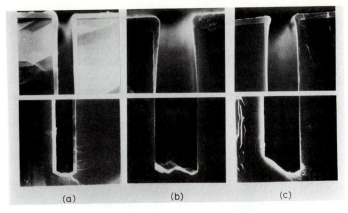

(a)   (b)   (c)

FIGURE 15
SEM cross sections (5000X) showing step coverage of deposited oxides. (a) TEOS deposition at 700°C. (b) Silane-oxygen reaction at 450°C and reduced pressure. (c) Silane–oxygen reaction at 480°C and atmospheric pressure.

but no surface migration takes place, and the step coverage is determined by the arrival angle. Figure 15c shows silicon dioxide deposited by reacting silane and oxygen at atmospheric pressure. The film builds up at the top of the step because of the short mean free path (less than $0.1\ \mu m$) at atmospheric pressure. The nonconformal step coverage with re-entrant angles causes metallization failures by preventing uniform deposition of evaporated or sputtered aluminum. In addition, the region at the bottom of the step often etches rapidly, causing serious problems in later processing.

Other materials besides deposited silicon dioxide have the types of step coverage shown in Figs. 14 and 15. Most evaporated or sputtered metals have step coverage similar to that shown in Fig. 15b. Chemically deposited polysilicon and silicon nitride have uniform coverage. Plasma-deposited silicon dioxide and silicon nitride are similar to Fig. 15b.

6.4.4 P-Glass Flow

Phosphorus-doped silicon dioxide is frequently used as an insulator between polysilicon gates and the top metallization. A re-entrant shape in the oxide covering the polysilicon gate makes uniform deposition of the metal film impossible. The poor step coverage of the phosphorus-doped silicon dioxide can be corrected by heating the samples until the oxide softens and flows. This process is called P-glass flow.[34,35]

P-glass flow is illustrated in the scanning electron microscope (SEM) photographs in Figs. 16 and 17. Figure 16 shows a polysilicon line crossing an oxide step with the entire surface covered with 4.6 wt. % P-glass. The samples have been heated in steam at 1100°C for four different lengths of time from 0 (Fig. 16a) to 60 min (Fig. 16d). Flow is shown by the progressive loss of detail. The

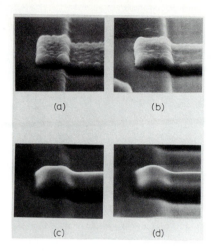

FIGURE 16
SEM photographs (3200X) showing surfaces of 4.6 wt. % P-glass annealed in steam at 1100 °C for the following times: (a) 0 min; (b) 20 min; (c) 40 min; (d) 60 min. (*After Adams and Capio, Ref. 34.*)

SEM cross sections in Fig. 17 show P-glass covering polysilicon. The samples contain from 0 to 7.2 wt. % phosphorus and have been heated in steam at 1100 °C for 20 min. Samples without phosphorus do not flow (Fig. 17a). The re-entrant shape, thick at the top and thin at the bottom, is easily seen. As the phosphorus concentration in the oxide increases, flow increases, thus decreasing the angle made by the P-glass going over the step. As these figures prove, P-glass flow is

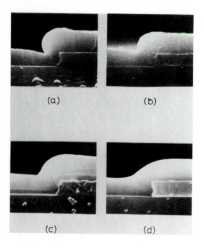

FIGURE 17
SEM cross sections (10,000X) of samples annealed in steam at 1100 °C for 20 min for the following weight percent of phosphorus: (a) 0.0 wt. % P; (b) 2.2 wt. % P; (c) 4.6 wt. % P; (d) 7.2 wt. % P. (*After Adams and Capio, Ref. 34.*)

a time-dependent phenomenon. Samples usually do not reach an equilibrium state during flow. Flow depends on several variables: annealing time, temperature, rate of heating, phosphorus concentration, and annealing ambient.

Figure 18 summarizes many of these effects by showing the angle made by the P-glass after different flow treatments and for different phosphorus concentrations. As deposited the steps are re-entrant with 120° angles. Flow is measured by the decrease in the angle. P-glass flow is greatest for high phosphorus concentrations, steam ambient, and high temperatures.

The P-glass flow process requires temperatures to be as high as 950 to 1100°C. It also requires phosphorus concentrations of 6 to 8 wt. %. P-glass with less phosphorus requires higher temperature for flow. Phosphorus concentrations greater than 8 wt. % may cause corrosion of the aluminum metallization by the acid products formed from the reaction between the phosphorus in the oxide and atmospheric moisture. Addition of boron to the phosphorus-doped silicon dioxide lowers the softening temperature.[36−38] Flow occurs at temperatures between 850 and 950°C even with phosphorus concentrations as low as 4 wt. %. Typical dopant concentrations are 1 to 4 wt. % boron and 4 to 6 wt. % phosphorus. The borophosphosilicate glass (BPSG) has the advantage of lower flow temperatures; however, bubbles of volatile phosphorus oxides and crystallites of boron-rich phases can occur unless the dopant concentrations are carefully controlled.

The step coverage of deposited oxides can be improved by planarization or

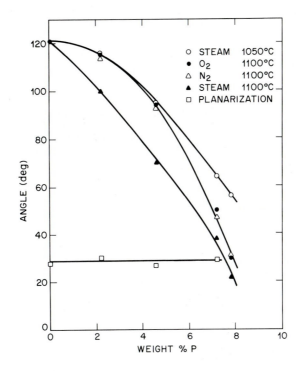

FIGURE 18
Step angles made by P-glass after different flow treatments. (*After Adams and Capio, Ref. 34.*)

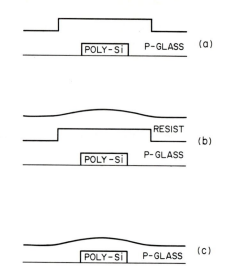

FIGURE 19
Schematic representation of the planarization process. (a) Polysilicon step covered with P-glass. (b) Coated with resist. (c) After etching the resist leaving a smooth P-glass surface. (*After Adams and Capio, Ref. 34.*)

etch-back techniques.[34] Figure 19 shows the planarization process. An abrupt step in phosphorus-doped silicon dioxide (P-glass) is covered with a sacrificial organic coating (resist). Since the organic material has a low viscosity, flow occurs during application or during a low temperature bake to give a smooth surface. The sample is plasma etched to remove all the organic coating but only part of the P-glass. The etch conditions are selected to remove the organic material and the P-glass at equal rates. Since plasma etching preserves the original smooth surface of the coating, the P-glass is left with the same smooth contour. The planarization process reduces step heights by at least 50%, gives surface angles of 5 to 30°, is independent of phosphorus concentration, and requires temperatures less than 200°C. The disadvantages are the increased thickness for the original oxide film and the need for additional processing steps.

6.4.5 Properties of Silicon Dioxide

Table 3 summarizes properties of silicon dioxide deposited by different techniques, including the plasma-assisted deposition of silicon dioxide. In general, oxides deposited at higher temperatures resemble thermally grown silicon dioxide. However, high-temperature oxides cannot be deposited over aluminum and therefore cannot be used for the final device passivation or for insulation between aluminum levels. Consequently, the low-temperature, phosphorus-doped oxides are used in spite of their poor step coverage and somewhat inferior film properties.

Composition. Silicon dioxide deposited at low temperatures (400–500°C) contains hydrogen. This hydrogen is bonded within the silicon–oxygen network as

TABLE 3
Properties of silicon dioxide

Deposition	Plasma	SiH$_4$ + O$_2$	TEOS	SiCl$_2$H$_2$ + N$_2$O	Thermal
Temperature (°C)	200	450	700	900	1000
Composition	SiO$_{1.9}$(H)	SiO$_2$(H)	SiO$_2$	SiO$_2$(Cl)	SiO$_2$
Step coverage	nonconformal	nonconformal	conformal	conformal	conformal
Thermal stability	loses H	densifies	stable	loses Cl	stable
Density (g/cm^3)	2.3	2.1	2.2	2.2	2.2
Refractive index	1.47	1.44	1.46	1.46	1.46
Stress (10^9 dyne/cm^2)	3C–3T	3 T	1 C	3 C	3 C
Dielectric strength (10^6 V/cm)	3–6	8	10	10	11
Etch rate, nm/min (100:1 H$_2$O:HF)	40	6	3	3	2.5
Dielectric constant	4.9	4.3	4.0	—	3.9

silanol (SiOH), hydride (SiH), or water (H$_2$O).[39] The bonded hydrogen can be observed by infrared spectroscopy. Silicon dioxide deposited between 400 and 500°C typically contains 1 to 4 wt. % SiOH and less than 0.5 wt. % SiH. The amount of water in the film depends on the deposition temperature and increases with exposure to atmospheric moisture.

Silicon dioxide films deposited at 700°C by TEOS decomposition, or at 900°C by the dichlorosilane–nitrous oxide reaction, do not contain hydrogen that is detectable by infrared absorption. The films formed from dichlorosilane, however, contain chlorine. The chlorine can react with the silicon substrate or evolve from the film during high-temperature anneals.

Phosphorus concentrations in doped silicon dioxide are measured by infrared absorption, neutron activation, X-ray fluoresence, sheet resistance of diffused layers, etch-rate variation, refractive index, and electron microprobe.[40] Specialized X-ray fluoresence or infrared equipment is often used to measure the phosphorus concentration. Concentrations of other dopants, such as boron and arsenic, are also measured by infrared absorption.

Thickness. Film thickness is measured by a stylus instrument, reflectance spectroscopy, ellipsometry, or a prism coupler. Automated instruments suitable for routine use are available for all these techniques. While all four techniques are generally suitable for measuring silicon dioxide films, each has specific limitations. Stylus measurement requires etching a step or masking part of the substrate during deposition. Prism coupling cannot be used on oxide films less than 400 nm thick. Ellipsometry requires that the oxide thickness be known to within 250 nm, since the measured ellipsometric quantities are periodic functions of the thickness. Reflectance spectroscopy requires empirical calibration or accurate values for the film refractive index.

Structure. Deposited silicon dioxide has an amorphous structure consisting of SiO_4 tetrahedra.[41] Its structure is similar to that of fused silica. The film density ranges between 2.0 and 2.2 g/cm^3. The lower densities occur in films deposited below 500°C. Heating deposited silicon dioxide at temperatures between 600 and 1000°C causes densification; the oxide thickness decreases and the density increases to 2.2 g/cm^3. During densification the amorphous structure is maintained; however, the arrangement of the SiO_4 tetrahedra becomes more regular. Densification causes deposited silicon dioxide to take on many of the characteristic properties of thermally grown oxides.

Reactivity. Silicon dioxide deposited at a low temperature reacts with atmospheric moisture, especially if the oxide contains phosphorus. The phosphorus–oxygen double bond undergoes a reversible hydrolysis. This effect can be minimized by densification at 800 to 900°C.

   The etch rates of deposited oxides in a hydrofluoric acid solution depend on deposition temperature, annealing history, and dopant concentration.[39] These etch rates are important because fluoride solutions are frequently used for cleaning and etching. An etchant containing nitric acid, hydrofluoric acid, and water is useful for evaluating and comparing deposited oxides. Etch rates in this solution (often called P-etch) are sensitive to film density, porosity, and composition.

Refractive index and stress. The refractive index of silicon dioxide is 1.458 at a wavelength of 0.6328 μm.[42] Deposited oxides with refractive indices above 1.46 are usually silicon-rich. Oxides with lower indices are porous. An example is the oxide from the silane–oxygen deposition with a refractive index of about 1.44.

   Stress in silicon dioxide depends on deposition temperature, deposition rate, annealing treatments, dopant concentration, water content, and film porosity.[43] Undoped silicon dioxide deposited at a temperature between 400 and 500°C usually has a tensile stress of 1 to 4×10^9 $dyne/cm^2$. Undoped oxides deposited at a temperature between 650 and 750°C have a low compressive stress, 0 to 1×10^9 $dyne/cm^2$, and oxides deposited at 900°C have a slightly higher compressive stress, 2 to 3×10^9 $dyne/cm^2$. Stress in oxides is strongly influenced by impurities. The stress is usually more compressive when phosphorus is added, but is more tensile if the film contains water.

6.5   SILICON NITRIDE

Stoichiometric silicon nitride (Si_3N_4) is used for passivating silicon devices because it serves as an extremely good barrier to the diffusion of water and sodium. These impurities cause device metallization to corrode or devices to become unstable. Silicon nitride is also used as a mask for the selective oxidation of silicon. The silicon nitride is patterned and the exposed silicon substrate is oxidized. The silicon nitride oxidizes slowly and prevents the underlying silicon from oxidizing. This process of selective oxidation produces nearly planar device structures.[44]

Silicon nitride is chemically deposited by reacting silane and ammonia at atmospheric pressure at temperatures between 700 and 900°C or by reacting dichlorosilane and ammonia at reduced pressure at temperatures between 700 and 800°C.[45,46] The chemical reactions are

$$3SiH_4 + 4NH_3 \rightarrow Si_3N_4 + 12H_2 \tag{20}$$

$$3SiCl_2H_2 + 4NH_3 \rightarrow Si_3N_4 + 6HCl + 6H_2 \tag{21}$$

The reduced-pressure technique has the advantage of good uniformity and high wafer throughput. Thermal growth of silicon nitride by exposing silicon to ammonia at temperatures between 1000 and 1100°C has been investigated, but the resulting films contain oxygen and are very thin (≤ 10 nm).

6.5.1 Deposition Variables

Silicon nitride depositions at reduced pressure are controlled by temperature, total pressure, reactant concentrations, and temperature gradients in the furnace. The temperature dependence of the deposition rate is similar to that of polysilicon. The activation energy for the silicon nitride deposition is about 1.8 eV (41 kcal/mole). The deposition rate increases with increasing total pressure or dichlorosilane partial pressure, and decreases with an increasing ammonia to dichlorosilane ratio. A temperature ramp, with the furnace tube hotter at the exhaust end, is required for uniform depositions. (See Section 6.3.1 for a discussion of temperature ramps.)

6.5.2 Properties of Silicon Nitride

Silicon nitride, chemically deposited at temperatures between 700 and 900°C, is an amorphous dielectric containing up to 8 at. % hydrogen. The hydrogen is bonded to the nitrogen and the silicon. The amount of hydrogen depends on the deposition temperature and the ratio of reactants. More hydrogen is incorporated at low deposition temperatures or at high ammonia to dichlorosilane ratios. Silicon nitride deposited at low ammonia to dichlorosilane ratios contains excess silicon that decreases the electrical resistivity.

Silicon nitride films have a refractive index of 2.01 and an etch rate in buffered hydrofluoric acid of less than 1 nm/min. Both properties are used to check the quality of deposited nitrides. Oxygen impurities in the film cause a higher etch rate. High refractive indices suggest a silicon-rich film; low indices are caused by oxygen impurities. Figure 20 shows how the refractive index changes with composition. Silicon nitride films that contain oxygen are silicon oxynitride. The composition of silicon oxynitride can vary from silicon dioxide with a refractive index of 1.46 to silicon nitride with an index of 2.01. Films that are free of oxygen but have silicon to nitrogen ratios greater than 0.75 are silicon-rich silicon nitride and have a refractive index that increases as the amount of excess silicon increases.

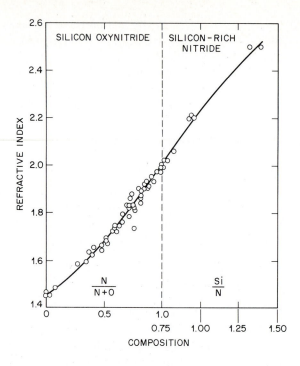

FIGURE 20
The effect of composition on the refractive index of silicon nitride and silicon oxynitride films deposited at temperatures between 700 and 900 °C.

Silicon nitride has a high tensile stress, about 1×10^{10} dyne/cm^2. Films thicker than 200 nm sometimes crack because of the high stress. The resistivity of silicon nitride at room temperature is about 10^{16} ohm-cm. The electrical resistivity depends on the deposition temperature, film stoichiometry, amount of hydrogen in the film, and the presence of oxygen impurities.

Silicon nitride is an excellent barrier to sodium diffusion. Its effectiveness is usually tested by evaporating radioactive sodium chloride (Na^{22}Cl) on the silicon nitride and then heating the samples at 600 °C for 22 h. The sodium that diffuses into the film is counted as the silicon nitride is removed by step etching. Typically, less than 10% of the original sodium diffuses more than 5 nm into the film. [47]

Table 4 summarizes the properties of LPCVD silicon nitride and plasma-deposited nitride.

6.6   PLASMA-ASSISTED DEPOSITIONS

Plasma-assisted depositions provide films at low sample temperatures. They do so by reacting the gases in a glow discharge. The discharge ionizes the gases, creating active species that react at the wafer surface. This technique is referred to as plasma deposition. [48,49]

Many inorganic and organic materials have been deposited by plasma deposition, but only two are useful in VLSI technology: plasma-deposited silicon

TABLE 4
Properties of silicon nitride

Deposition	LPCVD	Plasma
Temperature (°C)	700–800	250–350
Composition	$Si_3N_4(H)$	SiN_xH_y
Si/N ratio	0.75	0.8–1.2
Atom % H	4–8	20–25
Refractive index	2.01	1.8–2.5
Density (g/cm^3)	2.9–3.1	2.4–2.8
Dielectric constant	6–7	6–9
Resistivity (ohm-cm)	10^{16}	$10^6 - 10^{15}$
Dielectric strength $(10^6 V/cm)$	10	5
Energy gap (eV)	5	4–5
Stress $(10^9 \, dyne/cm^2)$	10 T	2C–5 T

nitride (SiNH) and plasma-deposited silicon dioxide. Plasma-deposited silicon nitride is used as the encapsulating material for the final passivation of devices. The plasma-deposited nitride provides excellent scratch protection, serves as a moisture barrier, and prevents sodium diffusion. Because of the low deposition temperature, 300 to 350°C, the nitride can be deposited over the final device metallization. Plasma-deposited nitride and oxide are both used as insulators between metallization levels. They are particularly useful when the bottom metal level is aluminum or gold.

6.6.1 Deposition Variables

Silicon dioxide films are deposited by reacting silane and nitrous oxide in an argon plasma. Silicon nitride is formed either by reacting silane and ammonia in an argon plasma or by reacting silane in a nitrogen discharge. The reactions are often assumed to be

$$SiH_4 + 4N_2O \rightarrow SiO_2 + 4N_2 + 2H_2O \qquad (22)$$

$$SiH_4 + NH_3 \rightarrow SiNH + 3H_2 \qquad (23)$$

$$2SiH_4 + N_2 \rightarrow 2SiNH + 3H_2 \qquad (24)$$

However, the products depend strongly on the deposition conditions. The radial-flow, parallel-plate reactor (Fig. 2a) and the hot-wall plasma reactor (Fig. 2b) are commonly used for device processing.

Many variables must be controlled during a plasma deposition, including discharge frequency and power, electrode spacing, total pressure, reactant partial pressures, pumping speed, sample temperature, electrode materials, and reactor geometry. Some variables have a predictable effect on the deposition. For instance, the deposition rate generally increases with increasing power or reactant pressure. Variables often interact to make measuring and interpreting the effect of a specific variable difficult. In other cases variables affect the deposition and film properties, but the effects are difficult to explain. The strong dependence on deposition conditions makes it difficult to compare films from different reactors. All deposition conditions must be carefully specified when discussing the properties of plasma-deposited films.

6.6.2   Properties of Plasma-Deposited Films

Plasma-deposited films contain large hydrogen concentrations that depend on the deposition conditions. Plasma silicon nitride may contain between 10 and 35 at. % hydrogen; however, most of the plasma nitride used in semiconductor processing contains 20 to 25 at. % hydrogen. The hydrogen is bonded to the silicon as Si—H and to the nitrogen as N—H. The plasma silicon nitride can contain up to 6 at. % oxygen as an impurity. Figure 21 shows how the plasma nitride composition varies with different deposition conditions.[50] The relative concentrations of Si—H and N—H change by large amounts and the silicon to nitrogen ratio varies between 0.7 and 1.7. These large variations in film composition and bonding cause large changes in film properties.

The empirical formula for plasma silicon nitride can be calculated from the silicon to nitrogen ratio and the at. % hydrogen. If the nitride contains 25 at. % H, has a Si/N ratio of 1.1, and the formula is written as SiN_xH_y, then

$$\frac{Si}{N} = \frac{1}{x} = 1.1 \tag{25}$$

$$at.\% \ H = \frac{100y}{1 + x + y} = 25 \tag{26}$$

Solving gives $x = 0.9$ and $y = 0.6$, so the formula is $SiN_{0.9}H_{0.6}$.

Figure 22 gives data for the hydrogen concentration in plasma-deposited silicon dioxide.[51] The hydrogen is bonded to silicon as Si—H and to oxygen as Si—OH and H_2O. The relative concentration of hydrogen in the three bonding sites strongly depends on deposition conditions; however, the total hydrogen varies only between 2 and 9 at. %. The changes in composition and bonding cause the film properties to depend on the specific deposition conditions. Compositions expressed as atom/cm^3, as in Fig. 22, can be converted to at. %, but the film density must be known. For instance, if plasma-deposited silicon dioxide contains 3×10^{21} H/cm^3 and the film density is 2.2 g/cm^3, then the atom density and at. % H are

$$\text{atom density (atom/cm}^3\text{)} = dn\frac{N_{AVO}}{MW} \tag{27}$$

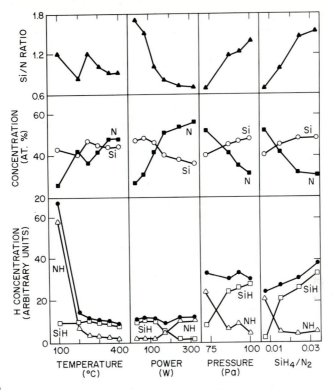

FIGURE 21

Composition of plasma nitride for different deposition conditions: solid circles denote total H, open squares denote SiH, open triangles denote NH, open circles denote Si, closed squares denote N, and closed triangles denote Si/N ratio. (*After Dun et al., Ref. 50.*)

$$\text{at. } \% \text{ H} = \frac{\text{atom density of H}}{\text{atom density}} \times 100\% \qquad (28)$$

where d is the film density, n is the number of atoms per molecule, N_{AVO} is Avogadro's number (6.023×10^{23}), and MW is the molecular weight. Substituting gives an atom density for silicon dioxide equal to 6.6×10^{22} atom/cm^3 and 4.5 at. % H.

Plasma-deposited silicon nitride has a large range of refractive indices (1.8 to 2.6), resistivities (10^5 to 10^{21} ohm-cm), and breakdown fields (10^6 to 10^7 V/cm). Figure 23 shows refractive index and resistivity data for different compositions of plasma nitride film. The silicon-rich films have high indices and low resistivities. The dielectric breakdown field, dielectric constant, and energy gap of plasma nitride also depend on composition.

Stress is one of the most important properties of plasma silicon nitride, since high stress can cause cracking of the nitride film. Crack-free films with low tensile stress, less than 2×10^9 dyne/cm^2, can be prepared, but the deposition

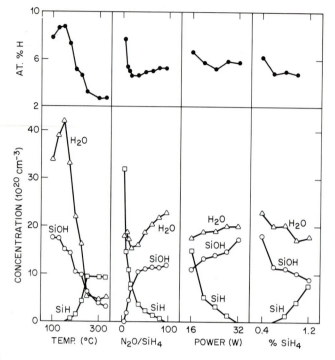

FIGURE 22

The concentration of hydrogen groups and the total at. % H in plasma SiO_2 for different deposition conditions: triangles denote H_2O, circles denote SiOH, squares denote SiH, and closed circles denote total hydrogen. (*After Adams et al., Ref. 51.*)

conditions must be carefully established and controlled. Stress depends not on the film composition but on the local bonding around individual atoms. Deposition variables such as temperature, rate, and ionic bombardment have large effects on the film stress.

Tables 3 and 4 list general properties of plasma-deposited silicon dioxide and silicon nitride. The properties of plasma nitride vary over large ranges because the film composition is variable. In contrast, the properties of plasma oxide are more nearly constant because the oxide composition is nearly stoichiometric.

6.7 OTHER MATERIALS

Several dielectric materials have been investigated for IC applications, primarily for passivation or for dual dielectric MOS devices. Silicon oxynitride is deposited by reacting silane, nitric oxide, and ammonia at temperatures between 700 and 800°C or in a plasma at temperatures between 200 and 350°C.[52,53] By adjusting the ratio of reactants, any film composition between SiO_2 and Si_3N_4 can be obtained. Since silicon dioxide has a compressive stress and silicon nitride is in tension, they form an intermediate composition of silicon oxynitride with zero stress. This composition is useful for passivation in some applications. The

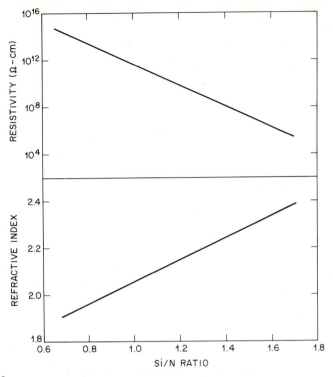

FIGURE 23
The effect of composition on resistivity and refractive index of plasma silicon nitride. The lines are least-squares fits of published data. (*After Adams, Ref. 48.*)

variation in refractive index for all compositions of silicon oxynitride is shown in Fig. 20.

Aluminum oxide, aluminum nitride, and titanium oxide have been evaluated as dielectrics for MOS applications. These films are chemically deposited between 800 and 1100°C and have high resistivities, high dielectric constants, and high breakdown fields.

Various organic compounds, usually polyimides, have been used as insulators between metal levels. These compounds are applied by spinning and then are cured at temperatures between 300 and 350°C. This process produces a planar surface that is ideal for metallization; however, the organic compounds have limited thermal stability and are porous to moisture penetration.

6.8   SUMMARY AND FUTURE TRENDS

Table 5 summarizes techniques for depositing dielectric and polysilicon films. The low-temperature processes for depositing P-glass and plasma nitride are particularly attractive for passivation, since the films can be deposited over aluminum or gold metallization. The poor step coverage, however, is a severe disadvantage if these films are used as insulating films between conducting layers.

TABLE 5
Comparison of different deposition methods

	Atmospheric pressure CVD	Low temperature LPCVD	Medium temperature LPCVD	Plasma assisted CVD
Temperature (°C)	300 – 500	300 – 500	500 – 900	100 – 350
Materials	SiO$_2$	SiO$_2$	Poly-Si	SiNH
	P-glass	P-glass BP-glass	SiO$_2$ P-glass BP-glass Si$_3$N$_4$ SiON SIPOS	SiO$_2$ SiON
Uses	Passivation, insulation	Passivation, insulation	Gate metal, insulation	Passivation, insulation
Throughput	High	High	High	Low
Step coverage	Poor	Poor	Conformal	Poor
Particles	Many	Few	Few	Many
Film properties	Good	Good	Excellent	Poor
Low temperature	Yes	Yes	No	Yes

Other deposition methods for dielectric and polysilicon films are available, such as evaporation, sputtering, anodization, and molecular beam techniques, but they are not widely used for VLSI processing. Their major problems include defects caused by excessive wafer handling, low throughput, poor step coverage, and nonuniform depositions over many wafers.

VLSI devices with small dimensions require precise lithography, pattern transfer with anisotropic etching, and shallow junctions. These conditions impose new requirements on the film deposition process. The major requirements are low processing temperatures to prevent movement of the shallow junctions, uniform step coverage over anisotropically etched features, few process-induced defects (mainly particles generated during wafer handling and loading), and high wafer throughput to reduce cost. These requirements are met by hot-wall, low-pressure depositions (chemical or plasma). Reactors designed to deposit at high rates on a few wafers at a time are becoming available. The advantages of these reactors are improved uniformity, automation, and reproducibility. The major disadvantage is low wafer throughput. Some of these new reactors have the added advantages of load-locked chambers to prevent contamination and to provide in situ wafer cleaning to give precisely controlled and reproducible interfaces.

Low-temperature depositions will become more important because the maximum processing temperature for devices with shallow junctions is about 800 to 900°C and decreases to less than 350°C for devices with metallization. Low deposition temperatures are achieved by reducing the amount of thermal energy supplied to the reaction and substituting energy in another form. These energy-enhanced depositions include plasma-assisted depositions, photo-induced depositions, and depositions aided by ultrasonic energy. Energy-enhanced depositions at low temperatures, 30 to 200°C, have been investigated and will probably find applications in new device technologies, especially with compound semiconductors.[48]

PROBLEMS

1 Find the empirical formula for LPCVD silicon nitride containing 5 at. % H and having a Si/N ratio of 0.75.

2 If LPCVD SiO_2 contains 5×10^{20} H/cm^3 and the density is 2.1 g/cm^3, find the at. % H and the empirical formula.

3 If the average chlorine concentration within the first 100 nm of a deposited SiO_2 is 1×10^{19} Cl/cm^3 and the film density is 2.2 g/cm^3, what is the at. % Cl in this region?

4 If polysilicon deposition has an activation energy of 1.7 eV and a deposition rate of 10 nm/min at 600°C, what is the deposition rate at 625°C?

5 Consider SiO_2 deposited by reacting silane and oxygen at 450°C. The deposition rate is 10 nm/min and the activation energy is 0.43 eV. How much must the temperature be increased to double the rate? Repeat the calculation for a TEOS deposition at 700°C and an activation energy of 1.95 eV.

6 A polysilicon deposition uses 30% silane in nitrogen at 625°C and 53.3 Pa (0.4 Torr). The total gas flow is 500 std. cm^3/min. The volume of the LPCVD reactor is 20 l, its length is 150 cm, and the cross-sectional area between the wafers and the walls is 45 cm^2. What are the partial pressure of the silane, and the linear velocity and residence time of the gas?

7 If the reactor in problem 6 has an effective area of 4000 cm^2 and 100 wafers have a total area of 15,000 cm^2, how many cm^3 of pure silane are required to deposit 0.5 μm of polysilicon if the reaction efficiency is 20%?

8 Sketch the step coverage expected for a conformal coating over a window 1 μm deep and 2 μm wide. Use film thicknesses of 0.5, 1.0, 1.5, and 2.0 μm. Repeat the calculation for a deposition with no surface migration, such as plasma oxide.

REFERENCES

1 W. Kern and V. S. Ban, "Chemical Vapor Deposition of Inorganic Thin Films," in J. L. Vossen and W. Kern, Eds., *Thin Film Processes*, Academic, New York, 1978, pp. 257–331.

2 W. Kern and G. L. Schnable, "Low-Pressure Chemical Vapor Deposition for Very Large-Scale Integration Processing—A Review," *IEEE Trans. Electron Devices*, **ED-26**, 647 (1979).

3 M. L. Hitchman and W. Ahmed, "Some Recent Trends in the Preparation of Thin Layers by Low Pressure Chemical Vapor Deposition," *Vacuum*, **34**, 979 (1984).

4 J. L. Vossen and W. Kern, "Thin-Film Formation," *Phys. Today*, **33**, 26 (1980).

5 M. L. Hammond, "Safety in Chemical Vapor Deposition," *Solid State Technol.*, **23**, 104 (1980).

6 W. A. P. Claassen, J. Bloem, W. G. J. N. Valkenburg, and C. H. J. Van den Brekel, "The Deposition of Silicon From Silane in a Low-Pressure Hot-Wall System," *J. Cryst. Growth*, **57**, 259 (1982).

7 W. A. Bryant, "The Kinetics of the Deposition of Silicon by Silane Pyrolysis at Low Temperatures and Atmospheric Pressure," *Thin Solid Films*, **60**, 19 (1979).

8 G. Harbeke, L. Krausbauer, E. F. Steigmeier, A. E. Widmer, H. F. Kappert, and G. Neugebauer, "Growth and Physical Properties of LPCVD Polycrystalline Silicon Films," *J. Electrochem. Soc.*, **131,** 675 (1984).

9 E. Kinsbron, M. Sternheim, and R. Knoell, "Crystallization of Amorphous Silicon Films During Low Pressure Chemical Vapor Deposition," *Appl. Phys. Lett.*, **42,** 835 (1983).

10 T. I. Kamins, "Structure and Properties of LPCVD Silicon Films," *J. Electrochem. Soc.*, **127,** 686 (1980).

11 T. I. Kamins, "Resistivity of LPCVD Polycrystalline-Silicon Films," *J. Electrochem. Soc.*, **126,** 833 (1979).

12 M. M. Mandurah, K. C. Saraswat, and T. I. Kamins, "Phosphorus Doping of Low Pressure Chemically Vapor-Deposited Silicon Films," *J. Electrochem. Soc.*, **126,** 1019 (1979).

13 R. E. Jones, Jr. and S. P. Wesolowski, "Electrical, Thermoelectric, and Optical Properties of Strongly Degenerate Polycrystalline Silicon Films," *J. Appl. Phys.*, **56,** 1701 (1984).

14 A. Baudrant and M. Sacilotti, "The LPCVD Polysilicon Phosphorus Doped In Situ as an Industrial Process," *J. Electrochem. Soc.*, **129,** 1109 (1982).

15 H. Kurokawa, "P-Doped Polysilicon Film Growth Technology," *J. Electrochem. Soc.*, **129,** 2620 (1982).

16 B. S. Meyerson and W. Olbricht, "Phosphorus-Doped Polycrystalline Silicon via LPCVD. 1. Process Characterization," *J. Electrochem. Soc.*, **131,** 2361 (1984).

17 C. Y. Lu and N. S. Tsai, "Thermal Oxidation of Undoped LPCVD Polycrystalline-Silicon Films," *J. Electrochem. Soc.*, **133,** 446 (1986).

18 T. I. Kamins, "Oxidation of Phosphorus-Doped Low Pressure and Atmospheric Pressure CVD Polycrystalline-Silicon Films," *J. Electrochem. Soc.*, **126,** 838 (1979).

19 K. C. Saraswat and H. Singh, "Thermal Oxidation of Heavily Phosphorus-Doped Thin Films of Polycrystalline Silicon," *J. Electrochem. Soc.*, **129,** 2321 (1982).

20 L. Faraone, R. D. Vibronek, and J. T. McGinn, "Characterization of Thermally Oxidized n^+ Polycrystalline Silicon," *IEEE Trans. Electron Devices,* **ED-32,** 577 (1985).

21 M. Sternheim, E. Kinsbron, J. Alspector, and P. A. Heimann, "Properties of Thermal Oxides Grown on Phosphorus In Situ Doped Polysilicon," *J. Electrochem. Soc.*, **130,** 1735 (1983).

22 L. Faraone and G. Harbeke, "Surface Roughness and Electrical Conduction of Oxide/Polysilicon Interfaces," *J. Electrochem. Soc.*, **133,** 1410 (1986).

23 B. G. Bagley, D. E. Aspnes, A. C. Adams, and C. J. Mogab, "Optical Properties of Low-Pressure Chemically Vapor Deposited Silicon Over the Energy Range 3.0-6.0 eV," *Appl. Phys. Lett.*, **38,** 56 (1981).

24 Ch. Kuhl, H. Schlotterer, and F. Schwidefsky, "Optical Investigation of Different Silicon Films," *J. Electrochem. Soc.*, **121,** 1496 (1974).

25 H. Mochizuki, T. Aoki, H. Yamoto, M. Okayama, M. Abe, and T. Ando, "Semi-Insulating Polycrystalline-Silicon (SIPOS) Films Applied to MOS Integrated Circuits," *Jap. J. Appl. Phys.*, **15,** 41 (1976).

26 W. R. Knolle and H. R. Maxwell, Jr., "A Model of SIPOS Deposition Based on Infrared Spectroscopic Analysis," *J. Electrochem. Soc.*, **127,** 2254 (1980).

27 C. Cobianu and C. Pavelescu, "A Theoretical Study of the Low-Temperature Chemical Vapor Deposition of SiO_2 Films," *J. Electrochem. Soc.*, **130,** 1888 (1983).

28 C. Cobianu and C. Pavelescu, "Silane Oxidation Study: Analysis of Data for SiO_2 Films Deposited by Low Temperature Chemical Vapour Deposition," *Thin Solid Films*, **117,** 211 (1984).

29 A. J. Learn, "Phosphorus Incorporation Effects in Silicon Dioxide Grown at Low Pressure and Temperature," *J. Electrochem. Soc.*, **132,** 405 (1985).

30 A. C. Adams and C. D. Capio, "The Deposition of Silicon Dioxide Films at Reduced Pressure," *J. Electrochem. Soc.*, **126,** 1042 (1979).

31 R. M. Levin and A. C. Adams, "Low Pressure Deposition of Phosphosilicate Glass Films," *J. Electrochem. Soc.*, **129,** 1588 (1982).

32 H. Huppertz and W. L. Engl, "Modeling of Low-Pressure Deposition of SiO_2 by Decomposition of TEOS," *IEEE Trans. Electron Devices*, **ED-26,** 658 (1979).

33 K. Watanabe, T. Tanigaki, and S. Wakayama, "The Properties of LPCVD SiO_2 Film Deposited by SiH_2Cl_2 and N_2O Mixtures," *J. Electrochem. Soc.*, **128**, 2630 (1981).

34 A. C. Adams and C. D. Capio, "Planarization of Phosphorus-Doped Silicon Dioxide," *J. Electrochem. Soc.*, **128**, 423 (1981).

35 R. A. Levy and K. Nassau, "Reflow Mechanisms of Contact Bias in VLSI Processing," *J. Electrochem. Soc.*, **133**, 1417 (1986).

36 W. Kern and R. K. Smeltzer, "Borophosphosilicate Glasses for Integrated Circuits," *Solid State Technol.*, **28**, 171 (1985).

37 T. Foster, G. Hoeye, and J. Goldman, "A Low Pressure BPSG Deposition Process," *J. Electrochem. Soc.*, **132**, 505 (1985).

38 A. C. Sharp and J. Patel, "Borophosphosilicate Glass for VLSI Device Fabrication," *Vacuum*, **35**, 441 (1985).

39 W. A. Pliskin, "Comparison of Properties of Dielectric Films Deposited by Various Methods," *J. Vac. Sci. Technol.*, **14**, 1064 (1977).

40 A. C. Adams and S. P. Murarka, "Measuring the Phosphorus Concentration in Deposited Phosphosilicate Films," *J. Electrochem. Soc.*, **126**, 334 (1979).

41 N. Nagasima, "Structure Analysis of Silicon Dioxide Films Formed By Oxidation of Silane," *J. Appl. Phys.*, **43**, 3378 (1972).

42 A. C. Adams, D. P. Schinke, and C. D. Capio, "An Evaluation of the Prism Coupler for Measuring the Thickness and the Refractive Index of Dielectric Films on Silicon Substrates," *J. Electrochem. Soc.*, **126**, 1539 (1979).

43 G. Smolinsky and T. P. H. F. Wendling, "Measurements of Temperature Dependent Stress of Silicon Oxide Films Prepared by a Variety of CVD Methods," *J. Electrochem. Soc.*, **132**, 950 (1985).

44 J. A. Appels, E. Kooi, M. M. Paffen, J. J. H. Schatorje, and W. H. C. G. Verkuylen, "Local Oxidation of Silicon and its Application in Semiconductor Device Technology," *Philips Res. Rep.*, **25**, 118 (1970).

45 T. Makino, "Composition and Structure Control by Source Gas Ratio in LPCVD SiN_x," *J. Electrochem. Soc.*, **130**, 450 (1983).

46 P. Pan and W. Berry, "The Compositional and Physical Properties of LPCVD Silicon Nitride Deposited with Different NH_3/SiH_2Cl_2 Gas Ratios," *J. Electrochem. Soc.*, **132**, 3001 (1985).

47 J. V. Dalton and J. Drobek, "Structure and Sodium Migration in Silicon Nitride Films," *J. Electrochem. Soc.*, **115**, 865 (1968).

48 A. C. Adams, "Plasma-Assisted Deposition of Dielectric Films," in R. Reif and G. R. Srinivasan, Eds., *Reduced Temperature Processing for VLSI,* The Electrochemical Society, Pennington, New Jersey, 1986, p. 111.

49 S. Veprek, "Plasma-Induced and Plasma-Assisted Chemical Vapour Deposition," *Thin Solid Films,* **130**, 135 (1985).

50 H. Dun, P. Pan, F. R. White, and R. W. Douse, "Mechanisms of Plasma-Enhanced Silicon Nitride Deposition Using SiH_4/N_2 Mixture," *J. Electrochem. Soc.*, **128**, 1555 (1981).

51 A. C. Adams, F. B. Alexander, C. D. Capio, and T. E. Smith, "Characterization of Plasma-Deposited Silicon Dioxide," *J. Electrochem. Soc.*, **128**, 1545 (1981).

52 M. J. Rand and J. F. Roberts, "Silicon Oxynitride Films from the $NO-NH_3-SiH_4$ Reaction," *J. Electrochem. Soc.*, **120**, 446 (1973).

53 W. A. P. Claassen, H. A. J. Th. v.d. Pol, A. H. Goemans, and A. E. T. Kuiper, "Characterization of Silicon-Oxynitride Films Deposited by Plasma-Enhanced CVD," *J. Electrochem. Soc.*, **133**, 1458 (1986).

CHAPTER
7

DIFFUSION

J. C. C. TSAI

7.1 INTRODUCTION

Diffusion of impurity atoms in silicon is important in silicon integrated circuit (SIC) processing. The idea of using diffusion techniques to alter the type of conductivity in silicon or germanium was disclosed in a patent by Pfann in 1952. Since then, various ways of introducing dopants into silicon by diffusion have been studied with the goals of controlling dopant distribution, total dopant concentration, its uniformity, and reproducibility; and of processing a large number of device wafers in a batch to reduce the manufacturing costs. Diffusion is used to form bases, emitters, and resistors in bipolar device technology, to form source and drain regions, and to dope polysilicon in MOS device technology. Dopant atoms that span a wide range of concentrations can be introduced into silicon in many ways. The most commonly used methods are (1) diffusion from a chemical source in a vapor form at high temperatures, (2) diffusion from a doped-oxide source, and (3) diffusion and annealing from an ion-implanted layer. For damage-free very shallow junction formation, dopant diffusion in silicides has become a subject of intense study.

Since ion implantation provides more precise control of total dopants from 10^{11} cm^{-2} to greater than 10^{16} cm^{-2}, it is used to replace the chemical or doped-oxide source wherever possible and is extensively applied in VLSI device fabrication. With the reduction of device sizes to the submicron range, the electrical activation of ion-implanted species relies on a rapid thermal annealing technique (RTA), resulting in as little movement of impurity atoms as possible. Thus, in the most advanced VLSI device technology, the study of diffusion in silicon becomes less important than methods for introducing impurity atoms into silicon for forming very shallow junctions. An example is the study of diffusion from a silicide source at low temperatures. The diffusion problem from silicide sources is discussed in Chapter 9. Atomic movements in the submicron range and interactions of impurity atoms at interfaces between the silicide or polysilicon layer and the silicon surface become very important.

However, many SICs and some VLSI circuits are still fabricated using diffusion techniques, and paradoxically, considerable advancement in the understanding of impurity diffusion in silicon has been made in the past few years. This is still an active area of research at many universities and research laboratories. An understanding of impurity diffusion will aid the understanding of the evolution of SIC and VLSI technology.

One aspect of the study of diffusion attempts to develop improved models from experimental data to predict diffusion results from theoretical analysis. The ultimate goal of diffusion studies is to calculate the electrical characteristics of a semiconductor device from the processing parameters. Diffusion theories have been developed from two major approaches: the continuum theory of Fick's simple diffusion equation (a second-order partial differential equation) and the atomistic theory, which involves interactions between point defects, vacancies and interstitial atoms, and impurity atoms.[1,2] The continuum theory describes the diffusion phenomenon from the solution of Fick's diffusion equation, with appropriate boundary conditions and the diffusivity of the impurity atoms. The diffusivity of a dopant element can be determined from experimental measurements (such as the surface concentration, junction depth, or concentration profiles) and the solutions of Fick's diffusion equation. In silicon, when impurity concentrations are low, the measured diffusion profiles are well behaved and agree with Fick's diffusion equation with a constant diffusivity, which can be calculated readily. In these cases the detailed atomic movements do not have to be known. When impurity concentrations are high, however, the diffusion profiles deviate from the predictions of the simple diffusion theory, and the impurity diffusion is affected by factors not considered in Fick's simple diffusion laws. Since the diffusion profile measurements reveal concentration-dependent diffusion effects, Fick's diffusion equation with concentration-dependent diffusivities is applied to the high-concentration diffusions. The concentration-dependent diffusivities are determined by a Boltzmann-Matano analysis or other formulations of profile analysis.

Various atomic diffusion models, based on defect-impurity interactions, have been proposed to explain the experimental results of concentration-dependent diffusivities and other anomalous diffusions. The atomistic diffusion theory is still undergoing active development. Theoretical and experimental results of the diffusion of Group III and V elements in silicon have been incorporated into various process models. Chapter 10 discusses the process models in detail. Because process modeling is still developing, we have to be aware of the model's limitations. The computer program, SUPREM-IV, simulates diffusion and oxidation by solving simultaneous continuity equations including point-defect equations developed at Stanford University. It was released in July 1986. Other simulation models are actively being developed at a few research centers. PREDICT, which has been developed by the MCNC (Microelectronic Center of North Carolina), is a process simulator that relies on verified empirical models and coupled diffusion equations.

7.2   MODELS OF DIFFUSION IN SOLIDS

At high temperatures, point defects such as vacancies and self-interstitial atoms are generated in a single-crystal solid. When a concentration gradient of the host or impurity atoms exists, such point defects affect atom movement (diffusion). Diffusion in a solid can be visualized as atomic movement of the diffusant in the crystal lattice by vacancies or self-interstitials. Figure 1 shows some common atomic diffusion models in a solid, using a simplified two-dimensional crystal structure with lattice constant a. The open circles represent the host atoms occupying the low-temperature lattice positions. The solid circles represent either host or impurity atoms. At elevated temperatures the lattice atoms vibrate around the equilibrium lattice sites. Occasionally a host atom acquires sufficient energy to leave the lattice site, becoming a self-interstitial atom and creating a vacancy. When a neighboring atom (either the host or the impurity atom) migrates to the vacancy site, it is called diffusion by a vacancy (Fig. 1a). If the migrating atom is a host atom, the diffusion is referred to as self-diffusion; if it is an impurity atom, the diffusion is impurity diffusion. To produce impurity atom movement other than oscillating between two lattice sites by exchanging positions with a vacancy, the vacancy has to diffuse away from the site that the impurity atom had just occupied, or the impurity atom has to move to a second vacancy that is at the nearest neighbor of the original vacancy site. The latter can lead to the concept of *diffusion assisted by a double vacancy,* or *divacancy.*

An interstitial atom moving from one place to another without occupying a lattice site (Fig. 1a) is called the interstitial diffusion mechanism. An atom smaller than the host atom that does not form covalent bonds with silicon often moves

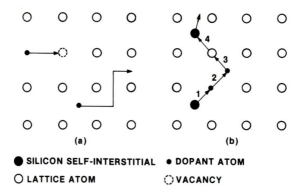

(a)

(b)

● SILICON SELF-INTERSTITIAL   • DOPANT ATOM

○ LATTICE ATOM   ◌ VACANCY

FIGURE 1
Models of atomic diffusion mechanism for a two-dimensional lattice. (a) Vacancy and interstitital mechanisms. (b) Interstitialcy mechanism.

interstitially. The activation energies required for the diffusion of interstitial atoms are lower than those for diffusion of lattice atoms by a vacancy mechanism, since no energy is required to form a vacancy.

Figure 1b shows a crude 2-dimensional picture of the atomic movement of a self-interstitial atom displacing an impurity atom (step 1), which in turn becomes an interstitial atom (step 2). Subsequently, the impurity atom displaces another host atom (step 3), and the second host atom becomes a self-interstitial (step 4). This is an example of the extended interstitial mechanism, sometimes called the *interstitialcy* mechanism.[2] The vacancy and the interstitialcy mechanisms are considered the dominating mechanisms for dopant (P, B, As, Sb) impurity diffusion in silicon. Boron and phosphorus diffuse via a dual (vacancy and interstitialcy) mechanism, with the interstitialcy component dominating.[2] Arsenic and antimony diffuse predominately via a vacancy mechanism.

Group I and VIII elements have small ionic radii, and they are fast diffusers in silicon. They are usually considered to diffuse by an interstitial mechanism. The two-dimensional lattice diagrams in Figure 1 illustrate the concepts of possible diffusion models. However, atomic movements are three-dimensional, and they are difficult to illustrate on a two-dimensional drawing.

7.3 FICK'S ONE-DIMENSIONAL DIFFUSION EQUATIONS

In 1855, Fick published his theory on diffusion. He based his theory on the analogy between material transfer in a solution and heat transfer by conduction. Fick assumed that in a dilute liquid or gaseous solution, in the absence of convection, the transfer of solute atoms per unit area in a one-dimensional flow can be described by the following equation:

$$J = -D \frac{\partial C(x, t)}{\partial x} \tag{1}$$

where J is the rate of transfer of solute per unit area or the diffusion flux, C is the concentration of solute (which is assumed to be a function of x and t only), x is the coordinate axis in the direction of the solute flow, t is the diffusion time, and D is the diffusivity (sometimes called the diffusion coefficient or the diffusion constant).

Equation 1 states that the local rate of transfer (local diffusion rate) of solute per unit area per unit time is proportional to the concentration gradient of the solute, and defines the proportionality constant as the diffusivity of the solute. The negative sign on the right-hand side of Eq. 1 states that the matter flows in the direction of decreasing solute concentration (i.e., the gradient is negative). Equation 1 is called Fick's first law of diffusion.

From the law of conservation of matter, the change of solute concentration with time must be the same as the local decrease of the diffusion flux, in the absence of a source or a sink; that is

$$\frac{\partial C(x, t)}{\partial t} = -\frac{\partial J(x, t)}{\partial x} \tag{2}$$

Substituting Eq. 1 into Eq. 2, yields Fick's second law of diffusion in one-dimensional form:

$$\frac{\partial C(x,t)}{\partial t} = \frac{\partial}{\partial x}\left[D\frac{\partial C(x,t)}{\partial x}\right] \tag{3}$$

When the concentration of the solute is low, the diffusivity at a given temperature can be considered as a constant, and Eq. 3 becomes

$$\frac{\partial C(x,t)}{\partial t} = D\frac{\partial^2 C(x,t)}{\partial x^2} \tag{4}$$

Equation 4 is often referred to as Fick's second law of diffusion. In Eq. 4, D is given in units of cm^2/s and $C(x,t)$ is in units of atoms/cm^3. Sometimes D is also expressed in μm^2/h. Solutions for Eq. 4 with various simple initial and boundary conditions have been obtained.[3] The most commonly used solutions are given in the following section.

7.3.1 Constant Diffusivities

Constant Surface Concentration. The initial condition at $t = 0$ is $C(x,0) = 0$, and the boundary conditions are $C(0,t) = C_s$ and $C(\infty,t) = 0$.

The solution of Eq. 4 that satisfies the initial and boundary conditions is given by

$$C(x,t) = C_s\,\text{erfc}\left(\frac{x}{2\sqrt{Dt}}\right) \tag{5}$$

where C_s is the concentration (in atoms/cm^3), D is the constant diffusivity (in cm^2/s), x is the distance (in cm), t is the diffusion time (in s), and erfc is the complementary error function.

Constant Total Dopant. Suppose that a thin layer of dopant is deposited onto the silicon surface with a fixed (or constant) total amount of dopant Q_T per unit area, and that the dopant diffuses only into the silicon; that is, all of the dopant atoms remain in the silicon. The initial and boundary conditions, and the solution of the diffusion equation (Eq. 4) that satisfy these conditions are:

Initial condition $C(x,0) = 0$, boundary conditions $\int_0^\infty C(x,t)dx = Q_T$ and $C(\infty,t) = 0$.

The solution of the diffusion equation Eq. 4 that satisfies the above conditions is

$$C(x,t) = \frac{Q_T}{\sqrt{\pi Dt}}\exp\left(-\frac{x^2}{4Dt}\right) \tag{6}$$

By setting $x = 0$ we obtain the surface concentration

$$C_s = C(0,t) = \frac{Q_T}{\sqrt{\pi Dt}} \tag{7}$$

where Q_T is the total impurity in atoms/cm^2.

Equation 6 is often called the Gaussian distribution, and the diffusion condition is referred to as the predeposition from a thin layer source or drive-in diffusion from a fixed total dopant concentration. Impurity atom distribution from ion implantation into an amorphous material can also be approximated by a Gaussian function (See Chapter 8).

Problem

Find the diffusivity from a known impurity profile. Assume that boron is diffused into an n-type Si single-crystal substrate with a doping concentration of 10^{15} atoms/cm^3, and also that the diffusion profile can be described by a Gaussian function. Using a diffusion time of 60 minutes, one obtains a measured junction depth of 2 microns (2×10^{-4}cm) and a surface concentration of 1×10^{18} cm^{-3}. (This is close to the surface concentration of a diffused base.)

Solution

Equation 6 can be used to calculate the diffusivity and the total dopant in the diffused layer.

$$C_s = 10^{15} \exp\left[\frac{4 \times 10^{-8}}{4 \times 3600D}\right] = 10^{18} \text{ and } D = 4 \times 10^{-13} \text{ cm}^2/\text{s}$$

From Eq. 7, the total dopant in the diffused layer is 6.8×10^{13} atoms/cm^2. For transistors made by the double diffusion process (often called the planar diffused transistor process, which is the basic technology for all of the silicon integrated circuits), the total base dopant is in the range of 1 to 3×10^{14} atoms /cm^2 and the net total doping concentration in the base region under the emitter of a transistor is 7×10^{11} to 1×10^{12} atoms/cm^2. A monolayer of atoms is close to 10^{15} atoms/cm^2. Thus, we are dealing with a very small number of atoms in the active region of modern silicon devices.

Since the solutions of Fick's diffusion equation depend on the initial and the boundary conditions for a given diffusion process, aside from the above-mentioned simple cases, various other diffusion conditions lead to complex mathematical expressions. A few commonly used expressions and results from practical applications are given in Reference 4.

Sheet Resistance of a Diffused Layer. For a diffused layer that forms a *pn* junction, an average sheet resistance R_s is defined and related to the junction depth x_j, the carrier mobility μ, and the impurity distribution $C(x)$ by the following expression:

$$R_s = \frac{1}{q\int_0^{x_j} \mu C(x)dx} = \frac{1}{q\mu_{\text{eff}}\int_0^{x_j} C(x)dx} \tag{8}$$

where μ is the majority carrier mobility in cm^2/V-s and is a function of the majority carrier concentration when $C(x)$ is greater than 10^{16} atom/cm^3; and μ_{eff} is an effective mobility in cm^2/V-s, which is defined as $\mu_{\text{eff}}\int_0^{x_j} C(x)dx = \int_0^{x_j} \mu C(x)dx$

Empirical expressions of μ versus C have been determined and the latest data are given as follows.[4] A word of caution: at concentrations above 10^{16} atoms/cm^3, the minority carrier mobility is different from the majority carrier mobility.[5,6] For n-type silicon[7]

$$\mu = (\mu_n)_n = \frac{1360 - 92}{1 + (C/1.3 \times 10^{17})^{0.91}} + 92 \ \text{cm}^2/\text{V-s} \tag{9}$$

where $(\mu_n)_n$ is the electron majority carrier mobility in n-type Si. The NBS (National Bureau of Standards) recommends a polynomial expression for the electron mobility in phosphorus-doped silicon. The calculated electron mobilities from Eq. 9 are comparable to those obtained using the polynomial function. For p-type silicon[8]

$$\mu = (\mu_p)_p = \frac{468 - 49.7}{1 + (C/1.6 \times 10^{17})^{0.7}} + 49.7 \ \text{cm}^2/\text{V-s} \tag{10}$$

where $(\mu_p)_p$ is the hole majority carrier mobility in p-type Si.

The average resistivity, ρ, of the diffused layer, is

$$\rho = R_s x_j \tag{11}$$

The average resistivity, ρ, is uniquely related to the surface concentration of the diffused layer and the substrate dopant concentration for an assumed diffusion profile. Using Eqs. 8 and 11, design curves relating to the surface concentration and the average conductivity $1/\rho$ (also called the effective conductivity), have been calculated for simple diffusion profiles, such as exponential, Gaussian, or erfc distributions. They are often called Irvin's curves.[9] Often, when the diffusion source is present during the diffusion period, it is assumed that a constant surface concentration prevails and an erfc profile is assumed. From the measured sheet resistance and the junction depth, the surface concentration is read from the Irvin's curve for erfc distributions. From the measured substrate concentration, the junction depth, and the known diffusion time, the diffusivity can be calculated from Eq. 5. However, when the actual diffusion profile deviates from the assumed erfc function, a fictitiously high surface concentration is often obtained and the calculated diffusivity is also in error. Many early published diffusivity data (1960–1970) were incorrect because of this error in assuming that the impurity distribution is an erfc function. When the calculated surface concentration from the Irvin's curve is above $(2 - 4) \times 10^{20} \ \text{cm}^3$, one should be suspicious about the use of the Irvin's curves for estimating the surface concentration of a diffused layer. When the diffusivities have been determined from an assumed constant surface concentration (i.e., an erfc profile) and using Irvin's curve with $C_s > 10^{19}$ atoms/cm^3, one should question the validity of such data. Since impurity profiles can be measured readily for junctions greater than 0.5 micron with the available analytical techniques (except extremely shallow junctions), Irvin's curves are not widely used for calculating the diffusivities. When a diffusion profile is truly a Gaussian or an erfc function, Irvin's curves can be used for estimating the surface concentrations. However, in Reference 9, the hole-mobility values are in error by as much as 30% in the concentration range 10^{17} to 10^{18} atoms/cm^3.

Because both the junction-depth and the sheet-resistance measurements are simple and give important information about a diffused layer without elaborate profile measurements, they are used routinely for monitoring diffusion processes. For ion-implanted samples, sheet-resistance measurement is a simple method used to check the electrical activity (the combined effects of mobilities and carrier concentrations) after the sample is annealed or diffused.

7.3.2 Concentration-Dependent Diffusivities

At high concentrations, when the diffusion conditions are close to the constant surface concentration case or to the constant total dopant case, the measured impurity profiles deviate from Eqs. 5 and 6, respectively. In these high-concentration regions, the impurity profile can often be represented by concentration-dependent diffusivities. Anderson and Lisak obtained approximate solutions to the nonlinear diffusion equations[10] (Eq. 3). Because of the rather lengthy derivations, the results of their work are summarized briefly in this section. Major procedures used in finding the diffusivity from the measured impurity profiles are also discussed in this section.

However, with the aid of a computer, diffusivities are frequently determined from both the measured profiles and use of numerical methods for solving Eq. 3, with appropriate initial and boundary conditions.

Approximate Solutions. Equation 3 with a concentration-dependent diffusivity as given in Eq. 12 can be solved by a similarity method,[10] by assuming

$$D = 2D_i \left(\frac{C}{n_i} \right)^r \tag{12}$$

where D_i is the constant diffusivity at low concentrations, sometimes called the intrinsic diffusivity; n_i is the intrinsic carrier concentration, and r is a constant. Equation 12 can be derived from atomic diffusion models that consider point defect and dopant interactions.[2]

The solution of Eqs. 3 and 12 can be approximated as

$$C(x, t) \approx t^\alpha C(y) \tag{13}$$

and

$$y \equiv \frac{x}{t^\beta} \tag{14}$$

The values of α, β, and the function $C(y)$ are determined from the initial and boundary conditions.

When $\alpha = 0$ and $\beta = 1/2$, Eq. 14 becomes Boltzmann's transformation, which has been used to determine the concentration-dependent diffusivities by the following equation:

$$D(C) = \frac{-1/2 \int_{C_o}^{C} y \, dC}{\frac{dC}{dy}} \tag{15}$$

where

$$y = \frac{x}{t^{1/2}} \tag{16}$$

and C_o is the constant surface concentration. Equation 15 can be obtained from Eq. 3 by substituting y for x and t.

Constant Total Dopants.[10] Assuming that the impurity atoms are introduced into silicon in a thin sheet (i.e., a delta function) with a total dopant Q_T and that there is no loss of the dopant during subsequent diffusion, the solution of Eq. 13 with the variable y takes a simple form:

$$C(x,t) = C_s(t)\left[1 - \frac{x^2}{x_F^2(t)}\right]^{1/r} \tag{17}$$

where

$$C_s(t) = \text{the surface concentration} = \left[\frac{r}{2(r+2)\Gamma_r^2} \cdot \frac{Q_T^2 n_i^r}{D_i t}\right]^{\frac{1}{r+2}} \tag{18}$$

and

$$x_F(t) = \left[\frac{2(r+2)}{r\Gamma_r^r} \cdot \frac{Q_T^r D_i t}{n_i^r}\right]^{\frac{1}{r+2}} \tag{19}$$

Γ_r is related to the Gamma integral; for our interest, $\Gamma_1 = 2/3$ and $\Gamma_2 = \frac{\pi}{4}$, for $r = 1$ and 2, respectively. $x_F(t)$ corresponds to the depth, x, at $C = 0$ from Eq. 17. Figure 2 shows the plot of C/C_s versus x/x_F. For comparison, the Gaussian distribution is also shown in Fig. 2. The meanings of Eqs. 18 and 19 will become clear when experimental profiles are discussed in Section 7.5.5.

7.3.3   Temperature Dependence of the Diffusivities

The diffusivities determined experimentally over a range of diffusion temperatures can often be expressed as

$$D = D_o \exp\left(-\frac{E}{kT}\right) \tag{20}$$

where D_o is the frequency factor (in cm^2/s), E is the activation energy (in eV), T is temperature (in K), and k is the Boltzmann constant (in eV/K). Thus, when D is plotted versus $1/T$ in semilogarithmic coordinates, D is a straight line with slope E/kT. From the atomic diffusion theories involving the defect-impurity interactions, D_o is related to the atomic jumping frequency or the lattice vibration frequency (typically 10^{13} Hz) and a jumping distance of an impurity, a defect, or defect-impurity pairs. At the diffusion temperatures, D_o can often be considered temperature independent. The activation energy E is related to the energies of motion and the energies of formation of defect-impurity complexes. Both D_o and

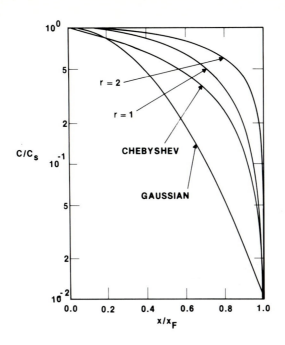

FIGURE 2
Normalized doping profiles for $D \propto C(r = 1)$, $D \propto C^2(r = 2)$, a Gaussian profile, and the Chebyshev polynominal approximation. (*After Anderson and Lisak, Ref. 10.*)

E can be derived from the atomic diffusion models. A thorough discussion on this topic is given in References 2 and 17.

In metals and for some elements with a simple vacancy diffusion model in silicon, E is between 3 and 4 eV, while for the interstitial diffusion model, E is between 0.6 and 1.2 eV. Thus, by measuring the diffusivity as a function of temperature, we can speculate whether the diffusion is dominated by an interstitial or vacancy mechanism. The validity of a diffusion mechanism, however, has to be verified from other experiments and theoretical considerations.

7.4 ATOMIC DIFFUSION MECHANISMS

The preceding section discussed solutions of diffusion problems from the phenomenological approach by solving Fick's diffusion equations without looking into the mechanisms of the atomic movements in the silicon crystal. Dopant impurity atoms occupying a substitutional site in silicon cannot move without the presence of point defects. The generation, annihilation, and movement of these point defects and their interactions with the impurity atoms affect the diffusion results and the measured diffusivities. To understand these effects one must look deeper into the diffusion mechanisms involving point defects and impurity atoms.

Point-defect impurity interactions and their effects on impurity diffusion are further explored in this section. Experimental results for impurity diffusion at low concentrations follow the phenomenological description of the diffusion process defined by Fick's diffusion law, Eq. 4. The upper limit of the dopant concentration for which the diffusivity is a constant can be estimated from the intrinsic carrier concentrations, n_i, at the diffusion temperature. When the impurity concentration, $C(x)$, is less than n_i, the diffusion results can be described by a concentration-independent diffusivity, which can be determined from the solution of Eq. 4 in Section 7.3 given the appropriate boundary conditions, and the measured diffusion profiles. The diffusivity at low concentrations is often referred to as the intrinsic diffusivity, D_i. When the impurity concentration, including both the substrate doping and the dopant impurity, is greater than $n_i(T)$, the silicon is considered as extrinsic silicon and the diffusivity is considered as the extrinsic diffusivity D_e. Experimentally measured values of D_i and D_e for boron, phosphorus, arsenic, and antimony are summarized in Section 7.5.

To understand the diffusion process at high concentration levels and the physical mechanisms for the impurity diffusion at various concentration levels, atomic models of solid-state diffusion have been proposed and compared with experimental measurements. Diffusion in silicon can be described by mechanisms involving impurity and point-defect interactions, such as vacancies and interstitials, at different charge states.

Point defects can become electrically active when they accept or lose electrons. A vacancy can act as an acceptor by acquiring a negative charge, V^-.

$$V + e \rightleftharpoons V^- \tag{21}$$

Similarly, a self-interstitial atom can act as an acceptor by acquiring a negative charge, I^-:

$$I + e \rightleftharpoons I^- \tag{22}$$

where V represents a vacancy and I represents a Si self-interstitial. A vacancy represents a lattice site where the silicon atom is missing. A self-interstitial is a silicon atom that is not on a lattice site (see Fig. 1). These concepts of ionized point defects have been applied to impurity diffusion in silicon with varied success. It has been found that both vacancy and interstitial atoms can be neutral, singly charged, or doubly charged. The probability of a charge state higher than two is very small. The exact mechanisms that dominate a diffusion process depend on the species under consideration; in many cases a consensus about those mechanisms has not been reached. The current prevailing view on the influences of point defects on impurity diffusion in silicon has been reviewed.[11]

If an impurity atom occupying a substitutional site is "kicked out" by a silicon atom from an interstitial site (described as a silicon self-interstitial atom), the impurity atom becomes an interstitial atom. It could move to another vacancy site, kick out another lattice silicon atom some distance away from the original lattice site, or diffuse interstitially for some distance. The first reaction can be described as

$$A_S + I_{Si} \rightleftharpoons A_I \qquad (23)$$

where A_S is an impurity atom occupying a lattice site, I_{Si} is the silicon self-interstitial atom, and A_I is an impurity atom which is not occupying a lattice site. Equation 23 has been successfully applied to gold diffusion in silicon under steady-state conditions, and it is called the kick out mechanism.[12]

On the other hand, if an impurity atom occupying a substitutional site has left that site, and if it becomes an interstitial atom and a vacancy is left behind, the reaction can be described by the following equation, which is called the dissociative mechanism.

$$A_S \rightleftharpoons A_I + V \qquad (24)$$

Equation 24 has been invoked to describe the Cu diffusion in Ge and also the Au diffusion[13] in Si. Both Eqs. 23 and 24 have been proposed as the mechanisms of Au diffusion in Si.

If the impurity atom in Eq. 24 is replaced by a silicon atom, Eq. 24 describes the creation of a Frenkel pair, which is a vacancy and a self-interstitial atom pair (See Chapter 1). For a perfect crystal at thermal equilibrium, the Frenkel pair implies that the numbers of vacancies and interstitials are equal. When other sources of vacancies or interstitials are present, the above statement is no longer true.

Equations 21 through 24 express equilibrium reactions, so the law of mass action can be applied to determine the equilibrium constants. The law of mass action states that the equilibrium constant of a chemical reaction in the gas phase can be expressed in terms of the chemical activity of the reactants and products. Consider a simple reversible chemical reaction

$$\alpha A + \beta B \rightleftharpoons \gamma C \qquad (25)$$

The equilibrium constant of the reaction toward the right-hand side is

$$K_c = \frac{a_A^\alpha a_B^\beta}{a_C^\gamma} \qquad (26)$$

where K_c is the equilibrium constant, a_A is the chemical activity of element A, a_B is the chemical activity of element B, a_C is the chemical activity of the product C, and α, β, and γ represent the mole concentrations of elements A, B, and C of the reaction shown in Eq. 25, respectively. For a dilute solution (a near ideal solution), the activities can be replaced by the concentrations of the reactants and products according to Raoult's law, and Eq. 26 becomes

$$K_c = \frac{[A]^\alpha [B]^\beta}{[C]^\gamma} \qquad (27)$$

where $[A]$ is the concentration of element A, $[B]$ is the concentration of element B, and $[C]$ is the concentration of element C.

The law of mass action has been applied to dilute solid solutions where point defects in a solid are considered as dilute solid solutions of defects in the crystal lattice. The law of mass action is applicable to a dilute solid solution when the reactions are in thermal equilibrium, and sometimes applicable for the reactions that are in quasi-thermal equilibrium.

Vacancy and self-interstitial concentrations can be determined from statistical thermodynamics. They are expressed in terms of entropies of formation ΔS and formation energies ΔH. For a neutral monovacancy in silicon, the concentration C_V^x can be expressed as

$$C_V^x = 5.5 \times 10^{20} \exp\left(\frac{\Delta S_V^x}{k}\right) \exp\left(\frac{-\Delta H_V^x}{kT}\right) \tag{28}$$

where ΔS_V^x is the entropy of formation of a neutral monovacancy and ΔH_V^x is the formation energy of a neutral monovacancy (expressed in eV). The superscript x represents a neutral charge state of the defect, and the subscript V denotes a vacancy defect. For silicon, ΔH_V^x is estimated to be greater than or equal to 2.5 eV, and ΔS_V^x is estimated to equal $1.1\ k$. Thus, the intrinsic concentration of monovacancies at the diffusion temperatures of interest is rather low for silicon; for example, $C_v^x = 2.1 \times 10^{11} \mathrm{cm}^{-3}$ at 1000°C.

For extrinsic silicon, the acceptor-type vacancy concentration can be expressed as[14]

$$
\begin{aligned}
C_V^- &= \frac{1 + 1/2 \exp\left(\frac{E_V - E_{Fi}}{kT}\right)}{1 + 1/2 \exp\left(\frac{E_V - E_F}{kT}\right)} C_i(V^-) \\
&\cong \frac{\exp\left(\frac{E_V - E_{Fi}}{kT}\right)}{\exp\left(\frac{E_V - E_F}{kT}\right)} C_i(V^-)
\end{aligned}
\tag{29}
$$

for $(E_V - E_F) \gg kT$ and $(E_V - E_{Fi}) \gg kT$. C_V^- is the acceptor-vacancy concentration in the extrinsic silicon; $C_i(V^-)$ is the acceptor-vacancy concentration in the intrinsic silicon; E_V is the acceptor-vacancy energy level (in eV); E_{Fi} is the intrinsic Fermi level (in eV), and E_F is the Fermi level of the extrinsic silicon (in eV). Thus,

$$\frac{C_V^-}{C_i(V^-)} = \exp\left(\frac{E_F - E_{Fi}}{kT}\right) \tag{30}$$

For n-type silicon and for the nondegenerate case, we obtain

$$n = n_i \exp\left(\frac{E_F - E_{Fi}}{kT}\right)$$

Thus, Eq. 30 becomes

$$\frac{C_V^-}{C_i(V^-)} = \frac{n}{n_i} \tag{31}$$

If the impurity diffusion is dominated by the acceptor monovacancy mechanism, the diffusivity is approximately proportional to the acceptor monovacancy concentration. If the impurity atoms interact with singly-charged acceptor silicon interstitial atoms, one can replace C_V^- with C_I^- and $C_i(V^-)$ with $C_i(I^-)$ in Eq. 31.

Thus, we have

$$\frac{D}{D_i} = \frac{n}{n_i} \tag{32}$$

where D is the diffusivity in extrinsic Si, and D_i is the diffusivity in intrinsic Si, assuming D is proportional to the defect concentrations and thus proportional to the electron concentrations.

The intrinsic carrier concentration n_i can be calculated using the following empirical formula:[15]

$$n_i^2 = 1.5 \times 10^{33} T^3 \exp\left[(-1.21 + \Delta E_g)/kT\right] \qquad (\mathrm{cm}^{-3})^2 \tag{33}$$

where

$$\Delta E_g = -7.1 \times 10^{-10} \cdot \left(\frac{n_i}{T}\right)^{1/2} \tag{34}$$

and an assumed $E_g = 1.21$ eV.

Equation 32 states that the interaction of the impurity atoms with charged acceptor vacancies leads to the diffusivity's dependence on the Fermi level at the diffusion temperature. Since vacancies and interstitials can have various charge states, Eq. 32 can be generalized to include all possible combinations of impurity-point-defect interactions.[16]

$$D = D^x + D^-\left(\frac{n}{n_i}\right) + D^=\left(\frac{n}{n_i}\right)^2 + D^+\left(\frac{n_i}{n}\right) + \cdots \tag{35}$$

where D^x represents the intrinsic diffusivity of impurity interaction with a neutral point defect; D^- represents the intrinsic diffusivity of impurity interaction with a singly charged acceptor point defect; D^+ represents the intrinsic diffusivity of impurity interaction with a singly charged donor point defect; and $D^=$ represents the intrinsic diffusivity of impurity interaction with doubly charged acceptor defects. Thus, when Eq. 35 is substituted into Eq. 3 to fit experimental profiles with defects of different charge states, it does not specify the dominating diffusion mechanism or mechanisms. The exact mechanisms involved in the impurity-defect interaction during the diffusion process, either vacancy or self-interstitial type, have to be determined from other experimental evidence and/or theoretical considerations. Therefore, we can consider Eq. 35 as a phenomenological expression of the concentration dependence of the diffusivity, and it provides a description of diffusion phenomena by expanding Fick's diffusion equation, Eq. 3, into components of $D(n)$ or $D(p)$. However, n or p is related to the dopant concentration, C. Derivation of Eq. 35 from different assumed dopant and point-defect interactions is given in Reference 2.

The concentration-dependent diffusivities can be determined from the experimental diffusion profiles without knowing the details of the atomic-diffusion mechanisms. However, the measured diffusivities as a function of diffusion temperature can sometimes be fitted to appropriate impurity-point-defect interaction models.

Isolated point defects in silicon are generated at or below room temperatures by high energy (≥ 1 MeV) electron, X-ray, or neutron irradiations. When these defects are in various charge states, their electronic states and annealing properties can be studied by electron paramagnetic resonance (EPR) measurements, by infrared absorption spectra analysis for neutral defects, and by other techniques. The deep-level transient spectroscopy (DLTS) method has also been used to study electrically active defects in proton-bombarded silicon crystals. Theoretical calculations of these defects have also been made, using various models of the charge states of these point defects and their annealing properties to explain the experimental observations. For vacancies in silicon, the EPR and optical absorption studies have identified four charge states: V^+ is a donor vacancy, V^x a neutral vacancy, V^- an acceptor vacancy, and $V^=$ a doubly charged acceptor vacancy. Three kinds of interstitials have been used to calculate the characteristics of the observed defect configurations theoretically: the simple tetrahedron, the bond-centered, and the $<110>$ split interstitial. In a unit cell, the positions for the five interstitial sites are (1/2, 1/2, 1/2), (1/4, 1/4, 1/4), (1/4, 3/4, 3/4), (3/4, 1/3, 3/4), and (3/4, 3/4, 1/4).

The investigations of point-defect formation made through study of the radiation effects in Si have provided fundamental information on the defect configurations, energies, and entropies of formation and migration. However, these were determined at temperatures well below room temperatures. This information is used to analyze atomic diffusion mechanisms in silicon from the measured diffusivities, as functions of temperature, by extrapolations to the diffusion temperatures.

7.5   DIFFUSIVITIES OF B, P, As, AND Sb

In VLSI technology, boron, phosphorus, arsenic, and sometimes antimony are used as dopant elements for junction formation. Hence, the diffusivities of these elements are of interest, and they are summarized in this section. We give both the intrinsic and extrinsic diffusivities. By applying the multiple-charge-state point-defect-impurity interaction model for diffusion, we can tentatively identify the species contributing to the diffusivities. Since this diffusion theory is still being developed, the identification of these species has not been confirmed. The following identification of the point defects responsible for each element is based on the proposal given in References 2 and 17. A detailed analysis of the point-defect-dopant diffusion problem through examination of the energies related to the diffusion process via vacancy, interstitialcy, and kick-out mechanisms is given in Reference 2. However, there are insufficient experimental or theoretical values for the entropies and enthalpies of the formation and movement of these defects.

Consequently, quantitative data are difficult to extract from the general analysis, which considers all the possibilities of impurity-point-defect interactions through coupled differential equations.

7.5.1   Low-Impurity Concentration Diffusion Into Intrinsic Silicon

Table 1 shows the intrinsic diffusivities[17] of boron, phosphorus, arsenic, and antimony in terms of a frequency factor D_o and an activation energy E. The expression of the diffusivity as a function of temperature was given in Eq. 20. For arsenic, three sets of D_o and E are given in Table 1. Since each set of data represents the measured values for the experimental conditions studied, and all of them are within the scattering of the measurement, no attempt is made to express preferences for any of them. In Table 1, D_i represents the impurity's intrinsic diffusivity: $(D_i)_B$ for boron, $(D_i)_P$ for phosphorus, and so on. Figure 3 summarizes the diffusivities of boron, phosphorus, arsenic, and antimony as functions of diffusion temperatures. Detailed descriptions of the experimental data on which these figures are based are given in Reference 17.

7.5.2   The Electric-Field Effect

When impurity atoms are ionized at the diffusion temperature, a local electric field is set up between the ionized impurity atoms and the electrons, or holes. The concentration gradient of these ionized impurity atoms (donors or acceptors) produces an internal electric field that enhances the diffusivity of the ionized impurity atoms. This internal electric field is related to the electrical potential $\phi(x)$ as

$$E_x = -\frac{\partial}{\partial x}\phi(x,t) \tag{36}$$

For a donor impurity, $\phi(x,t)$ can be expressed as

$$\phi(x,t) = (E_C - E_F)/q \tag{37}$$

TABLE 1
Intrinsic diffusivity of B, P, As, and Sb

	Unit	Boron	Phosphorus	Arsenic* CS	Arsenic* PD	Arsenic* IS	Antimony
		$(D_i)_B$	$(D_i)_P$		$(D_i)_{As}$		$(D_i)_{Sb}$
D_o	cm^2/s	0.76	3.85	24	22.9	60	0.214
E	eV	3.46	3.66	4.08	4.1	4.2	3.65

*CS represents results from a chemical source and PD results from predeposition diffusion of ion-implanted ^{75}As$^+$ and low-concentration predeposited layers.[78] IS represents results from isoconcentration diffusion experiments.[79]

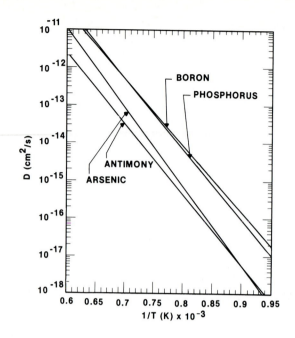

FIGURE 3
Intrinsic Diffusivity versus $1/T$ for Sb, As, B, and P in Silicon (calculated using D_o and E from Table 1 and for As the values are for ion implant and predeposition). *(After Fair, Ref. 17.)*

where E_C is the energy at the bottom of the conduction band, E_F is the Fermi level, and q is the magnitude of the electronic charge. Assuming that a charge neutrality exists between the ionized donor and the electron and that all donor atoms N_D are ionized, we have $np = n_i^2$ and $N_D \cong n$. It can be shown that

$$E_x = \frac{kT}{q} \frac{\partial}{\partial x} \ln \left(\frac{N_D}{n_i} \right) \tag{38}$$

The diffusion flux in an electric field can be expressed as

$$J = -qD \frac{\partial N_D}{\partial x} - qZD \frac{q}{kT} N_D E_x \tag{39}$$

where Z is the charge state of the donor atoms. For a singly charged donor atom, $Z = 1$. Substituting Eq. 38 into Eq. 39 and changing variables from $\partial/\partial x$ to $(\partial/\partial N_D)(\partial N_D/\partial x)$, Eq. 39 becomes

$$J = -qDh \frac{\partial N_D}{\partial x}$$

and

$$h \equiv 1 + ZN_D \frac{\partial}{\partial N_D} \ln \left(\frac{N_D}{n_i} \right) \tag{40}$$

where h is the electric-field enhancement factor. It can be shown that

$$h = 1 + \frac{N_D}{2n_i} \cdot \frac{1}{\left[\left(\frac{N_D}{2n_i}\right)^2 + 1\right]^{1/2}} \tag{41}$$

When $N_D/2n_i \gg 1$, h equals 2, which means that the maximum enhancement of the diffusivity from the electric-field effect is 2. For an acceptor diffusion with the electric-field enhancement, N_A should be substituted for N_D in Eq. 41.

For phosphorus-diffused samples at temperatures below 900°C, an electric-field enhancement of the diffusivity has been observed, in which neutral vacancies V^x dominate the diffusion and the measured D_e/D_i^x resembles h, as shown in Eq. 41. Figure 4 shows this electric-field enhancement for phosphorus.[17]

7.5.3 The Bandgap-Narrowing Effect

The previous section presented the ideas that the mechanisms of impurity atom diffusion are related to the concentration of the point defects (vacancies and interstitials), which can be either electrically active or neutral. The concentration of the charged point defects is related to the Fermi level at the diffusion temperature. Since the relative location of the Fermi level with respect to the conduction, or the valence, band edge is related to the energy bandgap, effects on the bandgap will affect the concentrations of the electrically active point defects. At high impurity doping concentrations, various factors can cause the energy bandgap to deviate from the value 1.12 eV at low doping levels. A bandgap narrowing from the misfit

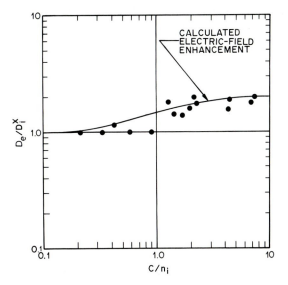

FIGURE 4
Electric-field-enhanced diffusion of phosphorus in silicon at 900°C. ● represents data from diffusion of phosphorus in silicon in the temperature range 875-900°C. (*After Fair, Ref. 17.*)

strain effect was attributed to the phosphorus diffusion in Si at high phosphorus concentrations.[18] The strain is the result of the mismatch between the phosphorus atom and the silicon lattice, as the tetrahedral covalent radius of phosphorus is smaller than that of silicon. Such lattice strain caused a bandgap narrowing for a high concentration of phosphorus diffused from a chemical source. For ion-implanted layers, an additional cause of stress in the silicon lattice is from the ion-implant damage. This also causes bandgap narrowing.[18] Due to the bandgap-narrowing effect, the effective intrinsic electron density is reduced from the intrinsic electron density by the following expression:

$$n_{ie} = n_i \exp\left(-\frac{\Delta E_g}{2kT}\right) \tag{42}$$

where ΔE_g is the change of the bandgap energy from lattice strains, in eV.

Because the theoretical analysis of bandgap narrowing is for room temperatures, the relationship between the ΔE_g and impurity concentration at the diffusion temperature is unknown. Empirically, a good fit to the experimental data for phosphorus diffusions from a chemical source is

$$\Delta E_g = -1.3 \times 10^{-22}(C_s - C_z) \tag{43}$$

where C_s is the surface concentration of phosphorus in cm^{-3} and C_z is an empirical constant. Various values of C_s, C_z, and ΔE_g are given in Table 2. C_s is close to the solid solubility at the diffusion temperature. The net effect of the bandgap narrowing is a reduction of the diffusivity of the doubly charged specie, which affects both the diffusion profile in the surface and the tail region for phosphorus.

For samples that were ion implanted and diffused with phosphorus, the estimated value of ΔE_g is close to 0.03 eV.[18] Since the covalent radius of arsenic is close to that of silicon, the diffusion profile of arsenic can be modeled without the inclusion of the bandgap narrowing at the diffusion temperatures.

TABLE 2
Energy bandgap narrowing due to strain-effect from high-concentration phosphorus diffusion from a $POCl_3$ source in <111> Si and a P ion-implanted-diffused layer[18]

	POCl₃				P Implant†
Temperature (°C)	900	1000	1100	1200*	1040
Time (min)	30	60	25	15	55
C_s (10^{20} cm^{-3})	7	9	5.6	4	—
C_z (10^{20} cm^{-3})	3	2.85	2.6	1.5	—
ΔE_g (eV)	0.052	0.08	0.039	0.033	0.03

*C_s and the diffusion time are the assumed values.

† Phosphorus implantation at 30 keV with an ion dose of 3×10^{16} ions/cm^2. Diffusion in nitrogen.

The values for ΔE_g shown in Table 2 were determined from fitting curves to the experimental phosphorus profiles. They are useful only for diffusion conditions comparable to the reported conditions. One should be careful in extrapolating these results to other diffusion conditions. However, these values can be used for estimations. These values are different from those measured at room temperatures from diodes or transistors because those values are related to "band tails" and impurity band formation in heavily doped silicon. For example, for an As emitter with a flat surface concentration at 1×10^{20} cm^{-3}, the calculated ΔE_g at 300–360 K is -0.133 to -0.149 eV for an electron effective mass of $1.45m_0$ to $1.1m_0$, respectively.[19] Whether the ΔE_g of Eq. 43 is really a bandgap narrowing has not been accepted universally.[2]

7.5.4 High-Concentration Effects

This section summarizes diffusion results of arsenic, boron, and phosphorus at high concentrations, when the surface concentrations are greater than n_i. For high-concentration As, a model for the cluster formation of As atoms explains the observation that only a portion of the diffused arsenic atoms is electrically active at room temperature. Electrically inactive boron has also been observed for high-concentration boron-diffused layers. However, no physical model that takes account of this reduction of the electrical activity for boron has been established. Results from a phosphorus diffusion model are also given.

Arsenic. When the As surface concentration is above 10^{20} cm^{-3} (approximately), the measured profile has a nearly flat portion near the surface region. The arsenic diffusivity in this high-concentration region can be expressed as[17]

$$D_{As} = (2\frac{n}{n_i})(D_i)_{As} \tag{44}$$

Equation 44 is similar to Eq. 32; the factor 2 represents the electric-field effect, h (Eq. 41). A similar expression based on interactions of charge vacancies with arsenic is[20]

$$D_{As} = \left(\frac{1 + \gamma n/n_i}{1 + \gamma} \right)(D_i)_{As} \tag{45}$$

with $\gamma \cong 100$ for donor-impurity diffusions. Thus, D_{As} from Eq. 44 is almost twice that of Eq. 45.

The electric activity of arsenic from ion-implanted samples depends on the ion dose and the annealing or diffusion temperatures. For arsenic ion doses below 1×10^{16} cm^{-2} and diffusion temperatures greater than 1000°C, nearly all of the arsenic atoms are ionized and contribute to the electrical activities. However, for diffusion temperatures below 1000°C and an arsenic ion dose greater than 10^{16} cm^{-2}, the concentration of ionized arsenic is a fraction of the total arsenic, and the differences increase[21] as the diffusion temperature is decreased below 900°C.

The difference between the ionized and the total arsenic can be explained by an arsenic clustering model. Arsenic atoms form clusters when their concentration is above 10^{20} cm^{-3}. The most recent clustering model consists of three arsenic atoms and one electron that are electrically active at the diffusion (or annealing) temperature and electrically neutral at room temperature. [21] The model is expressed as

$$3As^+ + e^- \underset{\rightleftharpoons}{\overset{\text{high temp.}}{}} As_3^{+2} \overset{25°C}{\rightleftharpoons} As_3 \tag{46}$$

Applying the law of mass action to the high-temperature region, the equilibrium constant is

$$K_{eq} = \frac{[As_3^{+2}]}{[As^+]^3 n} \tag{47}$$

and the carrier concentration at the annealing/diffusion temperature is

$$n = [As^+] + 2[As_3^{+2}] \tag{48}$$

where $[As^+]$ is the carrier concentration from isolated arsenic atoms and $2[As_3^{+2}]$ is the carrier concentration due to arsenic clusters $[As_3^{+2}]$ at high temperatures. At room temperature, $[As_3^{+2}]$ is electrically neutral and the carrier concentration is $[As^+] = C$. Thus, the total arsenic can be expressed as the sum of the unclustered arsenic $[As^+]$ and the clustered arsenic, which has three arsenic atoms per cluster:

$$C = [As^+] + 3[As_3^{+2}]$$

$$= C + \frac{K_{eq}C^4}{1 - 2K_{eq}C^3} \tag{49}$$

The second term on the right-hand side of Eq. 49 can be obtained from Eqs. 47 and 48. Limiting values for the electrically active arsenic are determined by letting $1 - 2K_{eq}C^3 = 0$. Then

$$C_{max} = (2K_{eq})^{-1/3} = 1.584 \times 10^{23} \exp\left(-\frac{0.687}{kT}\right) \tag{50}$$

A generalized model for cluster formation of arsenic atoms has been derived. [22] The model considered m arsenic atoms interacting with k electrons and analyzed all the possibilities. The conclusions supported the model shown in Eq. 46. The expression for C_{max} is[22]

$$C_{max} = 1.896 \times 10^{22} \exp\left(-\frac{0.453}{kT}\right) \tag{51}$$

Equations 50 and 51 give comparable values at temperatures above 900°C, but Eq. 51 gives a better fit to experimental data at temperatures below 900°C.

Figure 5 shows the maximum carrier concentration C_{max} as a function of annealing/diffusion temperature for arsenic at high concentrations. Experimental results agree with this model rather well.

At temperatures below 1000°C, the arsenic clusters are nearly immobile.

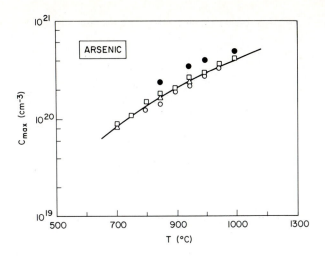

FIGURE 5
Maximum carrier concentration of arsenic in silicon versus temperature curve fits Eq. 51. $\triangle$, $\square$, $\bigcirc$, and $\bullet$ represent experimental data. (*After Guerrero et al., Ref. 22.*)

Eq. 44 or 45 could be considered to be the diffusivity of the mobile arsenic atoms that did not form clusters. At higher temperatures these clusters separate first (decluster) and diffuse as separate arsenic species, with the diffusivities given in Eq. 44 or 45.

Boron. When the multi-charge-state impurity-defect interaction mechanism is applied to the experimental profiles, the diffusivity of boron at high concentrations can be expressed as[17]

$$D_B = (D_i^+)\frac{p}{n_i} \tag{52}$$

by using Eq. 35 with $D^x = D^- = D^= = 0$ and $n = n_i^2/p$. In ion-implanted samples, when the boron concentration is above $10^{20}\,\mathrm{cm}^{-3}$, the concentration of the electrically active boron is also less than that of the total boron in the high-concentration region.[23] The diffusivity of boron in the high-concentration region is reduced considerably—almost to zero. The limiting values for the electrically active boron have been obtained experimentally; however, a physical model has not been developed. Figure 6 shows the experimental activity limits for boron at different temperatures.[23]

Phosphorus. Phosphorus is useful not only as an emitter and base dopant, but also for gettering fast-diffusing metallic contaminants, such as Cu and Au, which cause junction-leakage-current problems. Thus, phosphorus is indispensable in VLSI technology. However, npn transistors made with arsenic-diffused emitters have better low-current gain characteristics and better control of narrow base

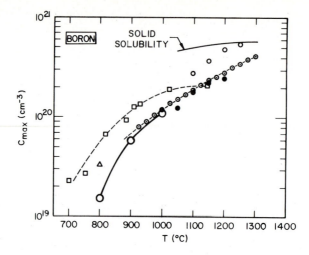

FIGURE 6

Maximum carrier concentration of boron in silicon versus temperature. ○: TEM data; o, ⊙ : nuclear reaction data, △, ☐, ●: electrical data (*After Ryssel et al., Ref. 23.*)

widths than those made with phosphorus-diffused emitters. Therefore, in VLSI, the use of phosphorus as an active dopant in small, shallow junctions and low-temperature processing will be limited to its use as the base dopant of pnp transistors and as a gettering agent. Arsenic is the most frequently used dopant for the source and drain regions in n-channel MOSFETs.

For completeness, the diffusion model for phosphorus diffusion via a vacancy mechanism is discussed briefly.[24] The characteristic profile of phosphorus can be described as consisting of three regions (see Fig. 7): the high-concentration region, the transition region (often called the "kink" of the profile), and the low-concentration region (the tail region).

In the high-concentration region, a fraction of the phosphorus ion (P^+) concentration pairs with $V^=$ vacancies as $(PV)^-$. The concentration of $(PV)^-$ is proportional to n_s^3, the surface electron concentration or the peak concentration for a Gaussian implanted profile. The n_s has to be determined experimentally. The diffusivity of phosphorus, D_P, in this region has two components: D_i^x corresponds to phosphorus-neutral-vacancy interaction and $D_i^=$ corresponds to $(PV)^-$ interaction $D_i^=$ is proportional to n^2, the electron concentration. Thus,

$$D_P = [D_i^x + D_i^= (n/n_i)^2] \tag{53}$$

where

$$D_i^x = 3.85 \exp(-3.66/kT) \tag{54}$$

and

$$D_i^= = 44.2 \exp(-4.37/kT) \tag{55}$$

Near the transition region, the electron concentration decreases, and when the

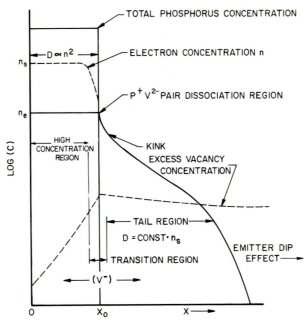

FIGURE 7
A model for phosphorus diffusion in silicon. (*After Fair, Ref. 17.*)

Fermi level is close to 0.11 eV below the conduction band, the $(PV)^-$ pairs show significant dissociations. The electron concentration in this transition region is

$$n_e = 4.65 \times 10^{21} \exp(0.39/kT) \tag{56}$$

The dissociation of $(PV)^-$ increases the vacancy concentration in the tail region, which can be expressed as

$$(PV)^- \rightarrow (PV)^x + e^- \tag{57}$$

and

$$(PV)^x \rightleftharpoons P^+ + V^- \tag{58}$$

The arrows shown in Fig. 7 next to (V^-) signify that at $n = n_e$ the excess vacancies diffuse in both directions from $x = x_o$. The diffusivity in the tail region increases as V^- is increased and is

$$D_{tail} = D_i^x + D_i^- \cdot \frac{n_s^3}{n_e^2 n_i} [1 + \exp(0.3 \text{ eV}/kT)] \tag{59}$$

where

$$D_i^- = 4.44 \exp(-4 \text{ eV}/kT) \tag{60}$$

The expression for the total phosphorus concentration and the electrically active phosphorus is

$$C_T = n + 2.4 \times 10^{-41} n^3 \qquad\qquad (61)$$

for temperatures between 900 and 1050°C.

This model for phosphorus diffusion was derived from experimental data for phosphorus concentrations at or slightly below the solid solubility limit at diffusion temperatures from 890 to 1050°C for diffusion from a $POCl_3$ source, or an ion implant and diffusion source. Both diffusions were in an ambient with low concentration of oxygen. When the diffusion conditions differ from the experimental conditions used in the study, the results must be verified with the measured experimental profiles. Some adjustments of the empirical constants could result. This seeming lack of universality of the diffusion model is really a manifestation of the sensitivity of the diffusion process to interactions with point defects; that is, variations in these point-defect concentrations and distributions will affect the initial and boundary conditions associated with the diffusion problem. Any change of the concentrations of these charged point defects should alter the boundary conditions and hence the detail profile calculations. Furthermore, this model does not apply to diffusion in oxygen or a subsequent oxidation in a steam ambient after the initial diffusion; oxidation enhancement of diffusion was not included in this model.

For the purpose of passivating the emitter–base junction, an oxidation step is generally used following the emitter diffusion (except for the high-frequency npn transistors which use a wash-out emitter process, the junction is passivated by the lateral diffusion of the emitter dopant underneath the masking oxide). The emitter-push effect is enhanced at this oxidation step. It has been established that oxidation of silicon injects silicon self-interstitials from the oxide–silicon interface into the silicon crystal. The emitter-push effect and the diffusion of phosphorus are also explained by an interstitialcy mechanism.

The phosphorus diffusion model was derived with the assumption that PV^- pairs coexisted with neutral vacancies. However, the model will not change much if PI^- (a phosphorus and a doubly charged self-interstitial) is the assumed defect pair. The PI^- model comes from the fact that during high-concentration phosphorus diffusion (not necessarily for the case above the solid solubility), in an oxidizing or slightly oxidizing ambient, extrinsic stacking faults (SF) are formed in the region beyond the phosphorus-diffused layer. Extrinsic SFs are formed from silicon interstitial condensation at defect regions.[25] Hence, it was also proposed that phosphorus diffusion is related to the interaction with silicon interstitials. Expressions based on interactions of P or B and silicon self-interstitials have been derived to explain the diffusion in the tail region.[26] However, the PV^- model is being used successfully in process-modeling programs, such as the SUPREM-III from Stanford University[27] or PREDICT[28] from the Microelectronics Center of North Carolina. A program, SUPREM-IV, for modeling oxidation/diffusion in two-dimensions, has been released.[29] In SUPREM-IV, the dopant and point defect interactions are modeled through 10 coupled-diffusion equations. The calculation of a diffusion profile at one temperature can take 8 hours on a Convex C-1 computer. Thus, effective use of the simulation program is better suited for fast systems of the supercomputer class. Simulation of diffusion is discussed in Chapter 10.

When phosphorus concentrations are above the solid solubility, using a chemical diffusion source such as $POCl_3$ (or PBr_3) + O_2 (a few %) or an ion-implanted phosphorus at ion doses greater than $10^{16} cm^{-2}$, the diffusivity is reduced by an empirical bandgap-narrowing effect, as discussed in a previous section. In calculating the diffusion profiles under these conditions, this empirical bandgap-narrowing effect due to lattice strain should be included. The magnitude of the bandgap narrowing, ΔE_g, depends on the diffusion sources and the phosphorus concentrations.

Emitter-push Effect. In npn narrow-base transistors using phosphorus-diffused emitter and boron-diffused base, the base region under the emitter (phosphorus) region is deeper than that outside the emitter region by 0.2 to 0.6 μm. This phenomenon is called the emitter-push effect. Since the discovery of this phenomenon, researchers have proposed various physical mechanisms to explain it. However, a bandgap-narrowing effect together with the phosphorus diffusion model shown in Fig. 7 explain the emitter-push effect adequately. Detailed derivations of the emitter-push effect can be found in Reference 24.

7.5.5 Analytical Expressions at High Concentrations

Although numerical solutions to nonlinear diffusion equations are widely used with the help of fast computers, the quest for analytical solutions, even approximate solutions, continues. Such analytical approximations provide not only elegance but also functional relations that can be readily visualized. Furthermore, the calculations take less time, present no problem for the converging of solutions that frequently occurs in numerical methods, and can be performed with a portable calculator. As an example, an application of Eqs. 17–19 in Section 7.3.2 is illustrated in the following problem.

Problem
Assume that a thin layer of phosphorus has been introduced into silicon at concentrations above n_i. Assume also that the diffusivity is proportional to the square of the local phosphorus concentration: $D = D_o(C/n_i)^2$. Find the expressions for the surface concentration and the depth where the high-concentration effect becomes negligible, as a function of the diffusion time.

Solution[30]
From Eq. 12, we find that $D_o = 2D_i$, $r = 2$ and $\Gamma_2 = \frac{\pi}{4}$. Equations 17 through 19 with $r = 2$ become

$$C(x,t) = C_s(t)\left[1 - \frac{x^2}{x_F^2(t)}\right]^{1/2} \tag{62}$$

$$C_s(t) = \left(\frac{4}{\pi^2} \frac{Q_o^2 n_i^2}{D_o t} \right)^{1/4} \tag{63}$$

$$X_F(t) = \left(\frac{64}{\pi^2} \frac{Q_o^2 D_o t}{n_i^2} \right)^{1/4} \tag{64}$$

The normalized profile for Eq. 62 is shown in Fig. 2, with $r = 2$. Equation 62 gives the expression of the phosphorus profile in the high-concentration region, and Eq. 63 gives the surface concentration as $t^{-1/4}$. However, for a Gaussian profile, $C_s \alpha t^{-1/2}$ (see Eq. 7). Equation 64 can be used for estimating the approximate location of the kink in the phosphorus profile, as shown in Fig. 7. The location for the local diffusivity that ceased to be proportional to C^2 can be estimated as x_F, which is proportional to $t^{1/4}$. For an ion-implanted phosphorus, Q_o can be assumed to be approximately the ion dose. Equations 8 and 62 can be used to estimate the sheet resistance and Eq. 63 to estimate C_s. The expressions for the diffusion in the tail region are given in Reference 30.

Arsenic. Although the As concentration-dependent diffusivities can be determined from the experimental diffusion profiles by numerical analysis, approximate analytical expressions represent the experimental data rather well. These expressions are useful for estimating R_s, C_s, and x_j. Chebyshev orthogonal polynomials were used to represent ion-implanted-diffused As profiles.[17] The expressions are

$$C/C_s = 1 - 0.87Y - 0.45Y^2 \tag{65}$$

$$Y = x \left[8 \frac{C_s}{n_i} (D_i)_{As} t \right]^{-1/2} \tag{66}$$

Expressions for x_j, R_s, C_s, and Q_T can be derived from Eqs. 65 and 66.

$$x_j = 2 \left[\left(\frac{Q_T}{n_i} \right) (D_i)_{As} t \right]^{1/3} \tag{67}$$

$$R_s = \frac{1.76 \times 10^{10}}{Q_T^{7/9}} \left[\frac{n_i}{(D_i)_{As} t} \right]^{1/9} \tag{68}$$

$$C_s = 0.91 \left[\frac{Q_T^2 n_i}{(D_i)_{As} t} \right]^{1/3} \tag{69}$$

$$Q_T = 0.55 C_s x_j \tag{70}$$

where x_j is the junction location at a concentration equal to $0.01 C_s$, Q_T is the ion dose in cm^{-2}, $(D_i)_{As}$ is the intrinsic diffusivity of arsenic in cm^2/s, R_s is the sheet resistance in $\Omega / \square$, and C_s is the surface concentration in cm^{-3}. In order to calculate R_s and Q_T, we arbitrarily select $0.01 C_s$ for the location of x_j. Since the arsenic profiles have a steep gradient for concentrations below $0.1 C_s$, assuming the junction depth to be $0.01C_s$ introduces small errors in the estimation of sheet resistance R_s and Q_T but not in x_j.

Equation 67 gives only an estimate of the junction depth; the exact junction depth depends on the dopant concentration level where the junction is formed. Calculated results for As from Eqs. 65 through 70 are given as families of design curves in Reference 4.

Using the similarity solution, Eqs. 17–19, one can derive expressions [10] similar to those given in Eqs. 67 through 70. In this case, $r = 1$ is assumed, and the expressions are

$$D \simeq D_o \frac{n}{n_i} \text{ and } D_o \simeq 2(D_i)_{As}$$

$$C(x, t) = C_S \left(1 - \frac{x^2}{x_F^2} \right) \tag{71}$$

$$C_s(t) \simeq 0.57 \left(\frac{Q_T^2 n_i}{(D_i)_{As} t} \right)^{1/3} \tag{72}$$

$$x_F(t) \simeq 2.62 \left(\frac{Q_T (D_i)_{As} t}{n_i} \right)^{1/3} \tag{73}$$

Note the similarity between Eqs. 69 and 72 and Eqs. 67 and 73. In using Eq. 65, the Chebyshev polynominal, to find $C_s(t)$, it was implicitly assumed that C_s is either a constant or a slowly varying function of the diffusion time. Such an assumption is a weakness in the derivation of Eqs. 67–70. The derivation of Eq. 71 does not have this weakness.

7.6 MEASUREMENT TECHNIQUES

Diffusivity, an important parameter in diffusion study, must be determined experimentally. This section discusses a few measurement techniques for determining diffusivities in diffusion study.

7.6.1 Junction Depth and Sheet Resistance

Junction Depth Measurement. Diffusion results can be checked by two simple measurements: the junction depth and the sheet resistance of the diffused layer. The junction depth is commonly measured on an angle-lapped (1° to 5°) sample chemical staining by a mixture of 100 cc HF (49%) and a few drops of HNO_3. Al_2O_3 powder, in water, is often used as the lapping mixture. Often, HF alone is sufficient. If the sample is put under a strong illumination for 1–2 minutes , the p-type region will be stained darker than the n-type region. With the aid of the interference-fringe techniques of Tolansky, the junction depths can be measured accurately from 0.5 micron to over 100 microns. In a production environment the angle lapping is replaced by a grooving method, which uses a commercially available groover. Various mixtures of staining solutions and solutions for Cu or Ag electrodeless-plating have also been used to delineate diffused junctions. For very high-resistivity substrates or low surface concentrations, difficulties in staining the junction can occur. This is why many staining solutions are reported. The HF–HNO$_3$ or HF alone is the method used most often and is successful for junctions over a wide range of concentrations.

The mechanism of staining has been established; the resulting stain film is a sub-oxide of silicon.[31] Due to the majority hole carriers in the p-type region, the surface silicon atoms are oxidized into some form of SiO_x ($x < 2$), which changes the surface reflectivity and gives a dark stain film of this sub-oxide. The location of the stained junction depends on the p-type concentration level and sometimes on the concentration gradient. In general, the stain boundary corresponds to a concentration level in the range of mid-10^{17} atoms/cm^3.

The Four-Point Probe Technique. The sheet resistance (also often called the sheet resistivity) of a diffused layer can be measured by a four-point probe technique (see Chapter 1). The sheet resistance R_s is given by

$$R_s = \frac{V}{I} \text{C.F.} \tag{74}$$

where R_s is the sheet resistance (in $\Omega/\square$); V is the dc voltage across the voltage probes (in volts); I is the constant dc current passing through the current probes (in amperes); and C.F. is the correction factor.[4] The most often used probe set has four probe points in line with the dc current passing through the outer two probes and the voltage measured across the inner two probes. $R_s \cdot x_j$ is the average resistivity of a diffused layer.

The correction factors for a circular sample (with a diameter d) and a rectangular sample (with the side parallel to the probe line as a and that perpendicular to the probe line as d) are given in Table 3 (where s is the probe spacing). Note that for a large d/s, the correction factor approaches that of a two-dimensional sheet extending to infinity in both directions, that is, C.F. = 4.53. For the correction factors to be insensitive to the sample size and the positions of the probe points with respect to the sample edge, a large d/s is desirable. Equation 74 and the correction factors in Table 3 are valid only for junctions, which are diffused only on the front side of the sample. Diffusion from a chemical source will have the diffused region wrapped around the sample. The back side of the diffused layer has to be either removed or isolated from the front side; otherwise, a different correction factor should be used (see Section 4 of Reference 4).

Van der Pauw Technique. The sheet resistance of a sample with an irregular shape can be determined using the Van der Pauw technique.[32] The measurement consists of attaching four contact points along the periphery of a sample, as shown in Figure 8a. A current is forced to flow between two adjacent contacts,

TABLE 3
Correction factor C.F. for the measurement of sheet resistances with the four-point probe[41]

		Square	Rectangle		
d/s	Circle	$a/d = 1$	$a/d = 2$	$a/d = 3$	$a/d \geq 4$
1.0				0.9988	0.9994
1.25				1.2467	1.2248
1.5			1.4788	1.4893	1.4893
1.75			1.7196	1.7238	1.7238
2.0			1.9475	1.9475	1.9475
2.5			2.3532	2.3541	2.3541
3.0	2.2662	2.4575	2.7000	2.7005	2.7005
4.0	2.9289	3.1127	3.2246	3.2248	3.2248
5.0	3.3625	3.5098	3.5749	3.5750	3.5750
7.5	3.9273	4.0095	4.0361	4.0362	4.0362
10.0	4.1716	4.2209	4.2357	4.2357	4.2357
15.0	4.3646	4.3882	4.3947	4.3947	4.3947
20.0	4.4364	4.4516	4.4553	4.4553	4.4553
40.0	4.5076	4.5120	4.5129	4.5129	4.5129
∞	4.5324	4.5324	4.5325	4.5325	4.5324

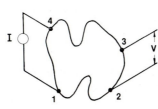
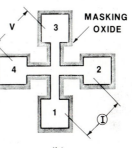

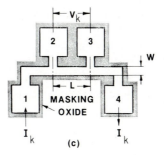

FIGURE 8
Van der Pauw and Kelvin resistor structures: (a) sample with an arbitrary shape; (b) an in-process tester design; (c) a Kelvin resistor design for measuring the effective window width.

and the voltage across the other pair of contacts is measured with a high-input-impedance voltmeter. Since no current flows through the voltage contacts, a contact resistance problem does not affect the results. The resistivity of the sample is

$$\rho = \frac{\pi d}{\ln 2}\left(\frac{R_{12,34} + R_{23,41}}{2}\right)F(Q) \tag{75}$$

$$\text{where } R_{12,34} = \frac{V_{12}}{I_{34}} \text{ and } R_{23,41} = \frac{V_{23}}{I_{41}}$$

$F(Q)$ is a correction factor and d is the sample thickness;* $Q = R_{12,34}/R_{23,41}$. The sheet resistance of a diffused layer can also be measured by this method. Various configurations of diffused layers have been designed for this measurement on SIC wafers in a testing area for in-process monitoring. An example is shown in Figure 8b, in which an oxide window is defined during the photolithographic operation. Impurity atoms are diffused into silicon in the region defined by the oxide window. After diffusion the sheet resistance is measured with four probe contacts connected to the large contact areas. Current flows through two adjacent probes and the voltage is measured across the remaining two contacts. An average resistance is obtained by rotating the current and the voltage probes;

$$R = \frac{1}{4}\left(\frac{V_{12}}{I_{34}} + \frac{V_{23}}{I_{41}} + \frac{V_{34}}{I_{12}} + \frac{V_{41}}{I_{23}}\right) \Omega/\square \tag{76}$$

Since the structure shown in Fig. 8b is symmetrical, $F = Q = 1$, and

$$R_s = \frac{\pi}{\ln 2}R = 4.53R \tag{77}$$

In this measurement, the measured sheet resistance does not depend on the linewidth of the diffused region. When the resistance from a Kelvin resistor is measured (Fig. 8c), the effective linewidth of a diffused resistor can be determined from these two measurements. In a Kelvin resistor structure, a constant dc current flows through the two end contacts, and the voltage across the middle two contacts is measured with a high-input-impedance electronmeter, which measures the potential difference across the center probes without disturbing the equal potential lines or the current flow. The measured resistance from the Kelvin resistor, R_k, is V_k/I_k, with V_k the measured voltage across the voltage probes and I_k the current. The ratio of R_s/R_k can be used to find an effective window width. The effective window width includes errors from the oxide window etching and the lateral diffusion. Assume the Kelvin resistor has a designed length, L, to width, W, ratio of 10/1; then the effective window width is

$$\frac{W_{\text{eff}}}{W} = \frac{L}{W}\frac{R_s}{R_k} = 10\frac{R_s}{R_k} \tag{78}$$

---

*This structure is often used in the Hall effect measurement on a bulk crystal with thickness d, to determine the carrier concentration and the carrier mobility of a semiconductor.

where R_s is the sheet resistance measured from the Van der Pauw pattern (Eq. 77) and R_k is the resistance measured from the Kelvin resistor.

$$R_k = \frac{V_{23}}{I_{14}} = R_s \frac{L}{W_{\text{eff}}} \text{ ohms} \qquad (79)$$

where W_{eff} is the actual width of the Kelvin resistor including the lateral diffusion effect and the oxide window etching errors.

7.6.2 Profile Measurements

The accuracy of the diffusion model and its associated diffusivities depend on the correctness of the diffusion profile measurements. Hence, the profile measurements are indispensable in diffusion studies. The simple measurements of the junction depth and the sheet resistance of a diffused layer, although useful for process monitoring, are grossly inadequate for diffusion study when the diffusivity depends on the impurity concentration. A few commonly used techniques for diffusion profile measurements and their limitations are discussed in the following sections.

C-V Technique. From the pn junction theory, the space-charge capacitance is a function of the reverse-bias voltage. For the depletion approximation, this capacitance can be treated as a parallel-plate capacitor. For an abrupt junction (Fig. 9a), where the impurity concentration is very high on one side of the junction

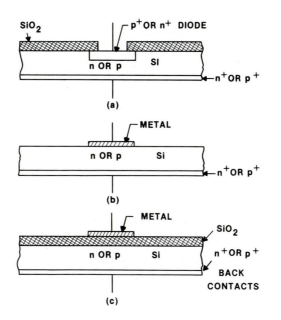

FIGURE 9
(a) A shallow p^+n or n^+p diode C-V structure. (b) A Schottky diode C-V structure. (c) A MOS capacitor C-V structure.

and decreases to a low value abruptly on the other side (i.e., an n$^+$p or p$^+$n junction), the following expressions can be derived:[33]

$$C(x) = \frac{2}{q\epsilon_s \frac{d}{dV}\left[\frac{1}{C(V)}\right]^2} = \frac{C^3(V)}{q\epsilon_s} \cdot \frac{1}{\left[\frac{dC(V)}{dV}\right]} \tag{80}$$

$$x = \frac{\epsilon_s}{C(V)} \tag{81}$$

where $C(x)$ is the impurity concentration at the space-charge layer edge, $C(V)$ is the junction reverse-bias capacitance per unit area at a reverse voltage V, and ϵ_s is the dielectric permittivity of silicon. $C(x)$, C_A, and C_N are concentrations and $C(V)$ means junction capacitance. Now

$$V = V_R + V_{bi} \tag{82}$$

where V_R is the applied reverse bias and V_{bi} is the built-in potential of the pn junction.

$$V_{bi} \cong \frac{kT}{q} \ln\left(\frac{C_A C_D}{n_i^2}\right) \equiv \frac{kT}{q} \ln\left(N_A \frac{N_D}{n_i^2}\right) \tag{83}$$

where C_A (or N_A) is the acceptor concentration, and C_D (or N_D) is the donor concentration. Thus,

$$C(V) = \left(\frac{q\epsilon_s}{2} C_B\right)^{1/2} \left(V_{bi} \pm V_R - \frac{2kT}{q}\right)^{-1/2}$$

$$= \frac{\epsilon_s}{\sqrt{2}L_D}(\beta V_{bi} \pm \beta V - 2)^{-1/2} \tag{84}$$

where C_B is the substrate doping concentration, $\beta \equiv q/kT$, and

$$L_D = \text{the Debye length} = \left(\frac{\epsilon_s}{qC_B} \cdot \frac{kT}{q}\right)^{1/2} \tag{85}$$

Thus, V_{bi} can be determined from the junction capacitance at zero reverse bias from Eq. 84. The C-V method is limited to a few L_Ds away from the depletion layer edge at zero bias, and it cannot resolve the concentration distribution within a few L_Ds. The impurity profile can be determined by measuring the reverse-bias capacitance as a function of the applied voltage from Eqs. 80, 81, and 84. The C-V method can also be used for a Schottky diode (Fig. 9b) to determine impurity profiles. In this case, a true high-low junction is formed. Both the junction and the Schottky diodes are limited to a maximum concentration of 10^{17} atoms/cm^3, or a total concentration close to 10^{12} atoms/cm^2. Schottky diodes are easy to make on n-type Si, but leakage current can be a problem on p-type Si. MOS capacitor structures have also been used for measuring impurity profiles near the surface (Fig. 9c), using a voltage pulse to cause a deep depletion of carriers near the oxide–Si interface and measure the transient capacitance changes. In C-V methods, the concentration near the junction space-charge region at a zero bias cannot be measured.

Differential Conductivity Technique. This is one of the oldest techniques for measuring the diffusion profiles in silicon by the electrical method.[34] This technique involves repeated measurement of the sheet resistance of a diffused layer by the four-point probe technique after removing a thin layer of silicon by anodic oxidation and etching the oxide off in a HF solution. Because the anodic oxidation is at room temperature, the impurity atoms do not move during the anodic oxidation and there is no segregation effect; hence, a true distribution profile can be determined. To use this technique, either the carrier mobility is measured by the Hall effect measurement, or the resistivity versus impurity concentration curves are used.[4] Figure 20 of Chapter 1 gives the composite curves of the resistivities for boron- and phosphorus-doped silicon over a wide range of concentrations. The differential conductivity technique is not suitable for diffusion studies in VLSI process development when shallow junction profiles are of interest. Normally the four-point probe penetrates the silicon surface and measures an average value over a volume under the probe points. Measurement errors can be significant for shallow junctions.

Spreading Resistance Technique. The C-V technique has a limited range of junction depth and dopant concentration that can be used for profile measurement, and the differential conductivity technique is a time-consuming method for profiling diffused layers. Various other techniques have been investigated to improve the spatial resolution and to reduce the measurement time. The two-point probe spreading resistance technique has been developed for diffusion profile measurement.[35] Because a refined and improved instrument is commercially available, the spreading resistance technique for diffusion profile measurement is becoming a routine evaluation technique.

For a two-point probe arrangement, the total spreading resistance is given by

$$R_{sr} = \frac{\rho}{2a} \tag{86}$$

where R_{sr} is the spreading resistance, ρ is the average resistivity near the probe points, and a is the probe radius. The spreading resistance technique is very sensitive to local impurity concentration variations; that is, it has high spatial resolution. However, measurements are also sensitive to the sample surface and the conditions of the probe points. Unless very elaborate measuring and checking procedures are conducted, this technique is best used to compare an unknown sample with a sample of known profile. For profile comparison of the spreading resistance versus distance, this technique is often sufficient. Concentration profiles, however, should be checked with another method, such as the differential conductivity method or the Secondary Ion Mass Spectrometry (SIMS) method.

To convert spreading resistance into concentration, various correction factors have been derived for different boundary conditions. Because we have imprecise knowledge of these correction factors and varying probe conditions, empirical calibration curves have to be used. Often, only the spreading resistance profiles are used for comparing different treatment results. Figure 10 shows the spreading resistance profile of a transistor structure; the collector–base junction x_{cb} and the emitter–base junction x_{eb} are clearly shown. The emitter region n^+ is phosphorus-diffused and shows a kink in the profile about 1.2 μm from the surface. This kink in the phosphorus profile was discussed in Section 7.5.4.

SIMS Technique.[36] This technique has been extensively used in measuring impurity diffusion profiles and is suitable for studying dopant profiles in VLSI technology. It will be discussed in Chapter 12.

Summary of Profiling Techniques. Various other techniques have also been used for impurity profile measurements. These techniques often require special laboratory setups or special equipment, but are useful for independently determining the total impurity concentration profile and for verifying the results from the electrical or the SIMS measurement. Table 4 summarizes the measurement techniques discussed in the previous sections, and others that were not discussed in detail but are mentioned here.

The Rutherford backscattering (RBS) technique has been used for measur-

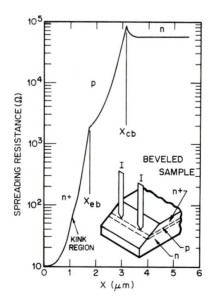

FIGURE 10
The spreading resistance profile of an npn transistor structure: x_{eb} = the emitter–base junction depth = 1.7 μm; x_{cb} = the collector–base junction depth = 3.2 μm.

TABLE 4
Commonly used diffusion profile measurement techniques

Profile techniques	Characteristics	Ref.
Capacitance-Voltage	Carrier concentration at the edge of the depletion layer of a pn junction. Maximum total dopants 2×10^{12} atoms/cm^2.	33
Differential conductance	Resistivity and Hall effect mobility of net electrically active species. Requires thin-layer removal, concentration range from 10^{20} to 10^{18} atoms/cm^3.	34
Spreading resistance	Resistance on angle-beveled sample. Good for comparison with known profiles and quick semi quantitative evaluation. $x_j \geq 1\mu$m.	35
SIMS	High sensitivity on many elements; for B and As detection limit is 5×10^{15}cm^{-3}. Capable of measuring total dopant profiles in 1000Å range. Needs standards.	36
Radioactive tracer analysis	Total concentration. Lower limit is 10^{15}cm^{-3}. Limited to radioactive elements with suitable half-life times: P, As, Sb, Na Cu, Au, etc.	37
Rutherford backscattering	Applicable only for elements heavier than Si.	38
Nuclear reaction	Measures total boron through ^{10}B(n,^{4}He)^{7}Li, or ^{11}B(p,α). Needs Van de Graaff generator.	39 40

ing distributions of heavy elements (arsenic, platinum, gold, etc.) in silicon but cannot be used for measuring boron or phosphorus profiles.[38] In this technique, high-energy helium ions (1 to 3 MeV) are used as the incident ion beam, and the backscattered He ion energy-loss spectra are analyzed. A few nuclear reaction processes have been used for measuring the boron atom distribution nondestructively. For example, thermal neutrons interreact with ^{10}B, causing the emission of monoenergetic ^{4}He ions at 1471 keV.[39] By analyzing the energy losses of the helium ions, one can determine the depth of boron atoms from the experimentally measured specific-energy-loss spectra of ^{4}He ions in silicon. The boron concentration can be related to the ^{7}Li particle signals at 839 keV (94%) and 1014 keV (6%) that are generated in the nuclear reaction of ^{10}B (n,^{4}He)^{7}Li. Another nuclear reaction for measuring boron profiles involves the use of a proton beam at 400 keV, which reacts with ^{11}B in silicon.[40] The energy spectra of the α-particles from their reaction have been analyzed. This reaction is expressed as ^{11}B(p,α). For boron-implanted profiles, the results of this method and of the SIMS method agree.

7.7 FAST DIFFUSANTS IN SILICON

Group I and VIII elements are fast diffusants in silicon. They form deep-level traps and affect the minority-carrier lifetime and the junction-leakage currents. For example, gold and platinum were used to reduce the storage time of switching transistors in the 1960s. These elements diffuse mainly through an interstitial mechanism that is modified to account for the experimental results. For instance, the distribution of gold throughout a silicon wafer resembles a U-shape, with high concentrations near the front and the back surfaces of a silicon wafer and a nearly uniform low-concentration distribution in the center of the wafer. A kick-out mechanism has been proposed for the diffusion of gold in Si[11,12,13]. Many factors affect the distribution and diffusion rates of these elements. These factors include the dislocation concentration, the precipitation and clustering of these elements near dislocations and point defects, the cooling rates, the presence of high concentrations of dopant elements such as phosphorus and boron, and the heat treatment history of the substrate silicon crystal. It is almost impossible to measure the diffusivities of these elements with any consistency, and unusual impurity profiles have been observed. Copper, gold, iron, and so forth are undesirable contaminants in VLSI and are gettered away from the active regions by high-concentration P diffusion.

Table 5 shows the diffusivities, solubilities, and the distribution coefficients at melting temperature of the fast diffusants in silicon.[41] Diffusivities of hydrogen, oxygen, and recent values for Pt, Cr, and Co are also given.

7.8   DIFFUSION IN POLYCRYSTALLINE SILICON

Polysilicon films are used in VLSI for two major purposes: (1) as a polysilicon gate in a self-aligned structure and (2) as an intermediate conductor in two-level structures. Recently, the use of doped poly-Si as thin-film, high-value resistors and as an emitter dopant source have been explored. Results from these studies are discussed briefly. The resistivity of polysilicon is often reduced by doping with boron, phosphorus, or arsenic. Since the gate electrode is over a thin oxide, typically 150 to 500 Å, it is very important that the dopant atoms in the polysilicon film not diffuse through the gate oxide or cause degradation of the gate oxide. To minimize this problem, the polysilicon film is deposited at a low temperature without doping elements. After the gate region is defined, the polysilicon film is doped. Dopant atoms are introduced by diffusion from a doped-oxide source, from a chemical source, or by ion implantation.

Impurity diffusion in polysilicon film can be explained qualitatively by a grain-boundary diffusion model.[46] The diffusivity of impurity atoms that diffuse along grain boundaries can be up to 100 times larger than the diffusivities in a single crystal lattice. The polycrystal film is considered to be composed of single crystallites of varying sizes that are separated by grain boundaries. For thin films, the grain sizes will be limited by the film thicknesses. Experimental results indicate that the impurity atoms inside each crystallite have diffusivities either comparable to or a factor of 10 larger than those found in the single crystal. Impurity atoms also diffuse along grain boundaries, so the diffusivity in a polysilicon film depends strongly on the texture of the film. The textures of the films are functions of the film deposition temperature, rate of deposition, thickness of the film, and composition of the substrate film, which can be an

TABLE 5
The diffusivity, solubility, and distribution coefficient at melting temperature of the fast diffusants in silicon

Element	Ref.	Diffusivity D_o (cm^2/s)	E (eV)	Solubility (cm^{-3})	Distribution coefficient
Li (25–1350°C)	41	2.3×10^{-3} to 9.4×10^{-4}	0.63– 0.78	Max. 7×10^{19} (1200°C)	10^{-2}
Na (800–1100°C)	41	1.6×10^{-3}	0.76	$(1–9) \times 10^{18}$ (600–1200°C)	
K (800–1100°C)	41	1.1×10^{-3}	0.76	$9 \times 10^{17} – 7 \times 10^{18}$ (600–1200°C)	
Cu (800–1100°C)	41	4.0×10^{-2}	1.0	$5 \times 10^{15} – 3 \times 10^{18}$ (600–1300°C)	4×10^{-4}
(Cu)$_i$ (300–700°C)	41	4.7×10^{-3}	0.43		
Ag (1100–1350°C)	41	2×10^{-3}	1.6	$6.5 \times 10^{15} – 2 \times 10^{17}$ (1200–1350°C)	1.1×10^{-4}
Au (800–1200°C) (Au)$_i$ (Au)$_s$ (700–1300°C)	41	1.1×10^{-3} 2.4×10^{-4} 2.8×10^{-3}	1.12 0.39 2.04	$5 \times 10^{14} – 5 \times 10^{16}$ (900–1300°C)	2.5×10^{-5}
Pt (800–1000°C)	42	$1.5–$ 1.7×10^{2}	2.22– 2.15	$4 \times 10^{16} – 5 \times 10^{17}$ (800–1000°C)	
Fe (1100–1250°C)	41	6.2×10^{-3}	0.87	$10^{13} – 5 \times 10^{16}$ (900–1300°C)	8×10^{-6}
Ni (450–800°C)	41	0.1	1.9	6×10^{18} (1200–1300°C)	$\sim 10^{-4}$
Cr (1100–1250°C)	43	0.01	1.0	$2 \times 10^{13} – 2.5 \times 10^{15}$ (900–1280°C)	
Co (900–1200°C)	44	9.2×10^{4}	2.8	Max. 2.5×10^{16} (1300°C)	8×10^{-6}
O$_2$ (700–1240°C)	45	7×10^{-2}	2.44	$1.5 \times 10^{17} – 2 \times 10^{18}$ (1000–1400°C)	5×10^{-1}
O$_i$ (330–1250°C)	45	0.17	2.54		
H$_2$	41	9.4×10^{-3}	0.48	$2.4 \times 10^{21} \exp \frac{-1.86}{kT}$ at 1 atm	

Note: (Cu)$_i$ = interstitial copper, (Au)$_i$ = interstitial gold, (Au)$_s$ = substitutional gold, (O)$_i$ = interstitial (or isolated) oxygen.

oxide layer, a silicon nitride film, or a single-crystal silicon surface, depending on the structure being fabricated.

Although diffusion results that are universally useful are difficult to summarize, some general observations can be made. Experimental profiles in polysilicon films resemble simple diffusion results, such as a complementary error function or a Gaussian function, which depend on the applicable diffusion conditions. Because of this resemblance, the diffusivities have been estimated from the measured junction depth and the surface concentration[46] using Eq. 5 or 6.

Recently, large-value polysilicon film resistors have been evaluated from arsenic implanted at a dose $\leq 3 \times 10^{14}$ atoms/cm^2 and diffused in N$_2$ at temperatures below 1100°C.[47] Figure 11 shows examples of the sheet resistance versus the annealing temperature for a few ion doses. The polysilicon film was deposited at 640°C in a low pressure CVD system on top of an oxide 400 Å thick. The polysilicon was 6500 Å thick, and the implant energy was 150 keV. The profiles were measured by the SIMS and the spreading resistance techniques. Diffusivities for As inside the grains and along the grain boundaries are determined by solving the Fick's diffusion equations, using an implicit, finite-difference method. Diffusion inside the grain (D_g) is assumed to proceed only in the horizontal direction (i.e., parallel to the sample surface), and diffusion along the grain boundary (D_{gb}) only in the vertical direction. The values of the diffusivities decrease with the diffusion

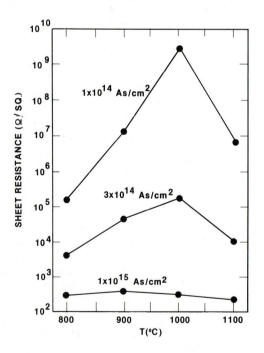

FIGURE 11

Sheet resistance versus annealing temperature for As in polycrystalline silicon exhibits reverse annealing effect at low doses. (*After Schubert, Ref. 47.*)

time. For example, at 800°C, $D_g = 9 \times 10^{-17}$ cm²/s and $D_{gb} = 6.2 \times 10^{-15}$ cm²/s for a diffusion time of 4.9 h; however, the values for an 18.5 h diffusion are $D_g = 3.5 \times 10^{-17}$ cm²/s and $D_{gb} = 4.5 \times 10^{-15}$ cm²/s.

Another application of a doped polysilicon film is to use the polysilicon film as the diffusion source for emitters in high-frequency (shallow emitter depth), bipolar transistor integrated circuits.[48] In this case, the emitter dopants are introduced into the polysilicon film by ion implantation or diffusion from a chemical source. This process is used for forming emitters with junction depths less than 0.5 micron. The diffusion depth is strongly affected by the presence or absence of a thin oxide layer (20 to 30 Å thick) at the polysilicon and single-crystal interface. Figure 12a shows SIMS profiles of As after ion implantation and after diffusion at 950°C for 30 min. The dashed curve is the As profile after implantation inside the polycrystalline silicon (poly-Si), and the solid curve is the profile after diffusion inside the poly-Si and the single-crystalline silicon (mono-Si). A pileup of arsenic at the poly-Si and the mono-Si interface was observed. The arsenic at the interface is not electrically active. C_{INT}^{POLY} and C_{INT}^{MONO} represent the extrapolated interface concentration from the polysilicon and the single-crystal side, respectively. For arsenic, the diffusion across the interface is continuous (Fig. 12a) and for boron and C_{INT}^{MONO} equals the saturated value at the diffusion temperature (Fig. 12b). In Fig. 12b, SIMS profiles of boron after the ion implantation are shown by the dashed curve, and the profiles after diffusion at 950°C

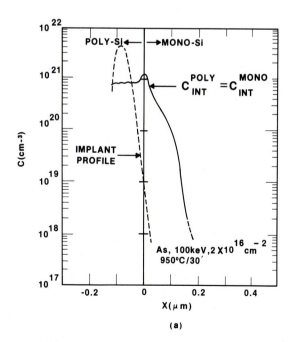

(a)

FIGURE 12
Arsenic and boron profiles in polycrystalline and single-crystal silicon: (a) Arsenic profile.

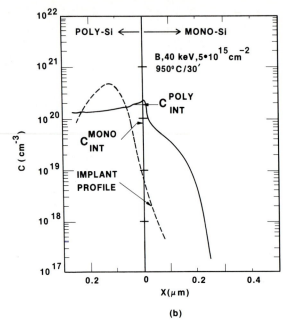

FIGURE 12
(b) Boron profile. (*After Schaber et al., Ref. 48.*)

for 30 min are shown by the solid curves. A pileup of boron at the interface was also observed. When polysilicon film deposited at a low temperature (for example, <650°C) is annealed at temperatures above 1000°C, epitaxial realignment of the polysilicon grains with the underlying single crystal occurred. Fast diffusion along grain boundaries becomes important, and the diffusions of high-concentration dopants are much too deep for shallow-junction applications.

For boron, an empirical resistivity versus concentration curve for a polysilicon film has been determined.[49] The electrical property of the As- and P-doped polysilicon films indicates that these elements segregate at the grain boundaries. Heat treatment of these films between 800 and 900°C showed reversible change of the resistivities.[51] The resistivity of As- and P-doped polyfilms is influenced by both carrier trapping (electrons) and atom trapping (P or As) at the grain boundaries. The resistivity increases when the dopant atoms are trapped at the grain boundaries. However, boron atoms do not appear to segregate at the grain boundaries.

Table 6 gives a few examples of the diffusivities of As, B, P, and Sb in polysilicon films. For VLSI circuits using CMOS transistors, thin silicide films, which are more conductive than doped poly-Si films, are being used as the low-resistivity interconnection conductors, and dopant diffusion from a silicide source for shallow junction formation is under active development (see Chapter 9).

TABLE 6
Examples of diffusivities in polysilicon films

Elements	D_o (cm^2/s)	E (eV)	D (cm^2/s)	T (°C)	Ref.
As	8.6×10^4	3.9	2.4×10^{-14}	800	50
B	$(1.5\text{–}6) \times 10^{-3}$	2.4–2.5	9×10^{-14}	800	52
			4×10^{-14}	925	49
P			$(6.9\text{–}63) \times 10^{-13}$	1000	46
			$(1\text{–}7) \times 10^{-11}$	1200	46
Sb*	$D_g = 13.6 \pm \begin{smallmatrix}60\\11\end{smallmatrix}$	3.9 ± 0.2			53
	$D_{gb} = 812 \pm \begin{smallmatrix}4000\\670\end{smallmatrix}$	2.9 ± 0.2			53

* For the Sb diffusion, the diffusivities were determined from an assumed model. The D_g inside the grain and the D_{gb} along the grain boundary are given for temperatures between 930 and 1150 °C for D_g, and between 1050 and 1150 °C for D_{gb}.[53]

7.9 DIFFUSION IN SiO$_2$

VLSI and silicon planar device fabrication rely on the thermal oxide of silicon as a mask to prevent diffusion of impurity atoms into silicon. Therefore, understanding diffusion in SiO$_2$ films is important. The diffusivities in SiO$_2$ were frequently deduced by measuring the C_s of the dopants in silicon that diffused through the oxide, and by using the solutions of the diffusion equations from Fick's law with an assumed set of initial and boundary conditions. The impurity distribution at the Si–SiO$_2$ interface is assumed to be in equilibrium, and the concentration ratio is described by a segregation coefficient, which was discussed in Section 3.6 of Chapter 3. Both the diffusivity and the segregation coefficient are unknown. The diffusivities are calculated, and the segregation coefficient is either assumed or deduced. For masking diffusants, a minimum oxide thickness can be determined empirically for the diffusion source and conditions in use. Hence, there is little practical value in measuring the diffusivities in SiO$_2$ films, and the diffusion of dopant in SiO$_2$ has not been extensively investigated. Besides, very few techniques were available for direct measurement of impurity profiles in SiO$_2$ films. Only recently has the SIMS technique been more extensively used for thin SiO$_2$ films, where charging of the oxide during the measurement is not a problem.

Since Group III and Group V elements are glass formers in SiO$_2$, they lower the melting temperature of the oxide film. The diffusivities of these elements depend strongly on their concentrations. For example, phosphorus at 3 to 6 at. % forms a thin, viscous film on SiO$_2$ that flows at 800 to 900°C. (Phosphosilicate, PSG, or borophosphosilicate, BPSG, glass is used for planarization in VLSI circuits, as discussed in Chapter 6.) However, outside the liquid–solid boundary, the phosphorus concentration becomes too low to show any diffusion. When P_2O_5 is used as the diffusion source, a very high-concentration layer is present on the thermal oxide. This phosphorus layer can be considered as a liquid; and the diffusion is from the liquid–solid interface into SiO_2 (the thermal oxide layer).

As a general rule, over the temperature range used in VLSI, diffusivities of these dopant elements (B, As, P, and Sb) are very low when their concentrations are below that which causes the glass to melt. Hydrogen, He, OH, Na, O_2, and Ga are fast diffusants in SiO_2. At 900°C the diffusivities of these elements are greater than 10^{-13} cm^2/s. Table 7 shows the diffusivities of some elements used in VLSI technology.[54] These values represent the magnitudes of the deduced diffusivities for the diffusion conditions listed. The estimated diffusivities at 900°C are calculated from the D_o and E, given in Table 7 from Eq. 20. The values of arsenic diffusivities in SiO_2 are calculated from measured profiles of arsenic in SiO_2 films, rather than being deduced from measurements in silicon.[56] Most of the deduced values of diffusivities in SiO_2 are rather good estimates.

SIMS and RBS techniques have recently been used to study diffusion of As and Sb in SiO_2.[56–58] The diffusivity of As is a strong function of concentration and the annealing ambient. For concentrations below 5×10^{20} cm^{-3}, there is some difference in activation energies and diffusivities between diffusion in oxidizing and inert ambients, as shown in Table 7 under columns E and D (900°C). At concentrations exceeding 5×10^{20} cm^{-3}, the behavior of As in SiO_2 is more complicated. In an oxidizing ambient, As diffuses more rapidly than at low concentrations, because its presence has reduced the oxide viscosity.[58] The presence of hydroxyl groups (OH or O_2/H_2O) further enhances diffusion.[57] In an inert ambient As precipitates into spherical As-rich inclusions of 50 to 500 Å diameter, which are essentially immobile under isothermal conditions, even after many hours at 1400°C.[59] In addition, the entire distribution of arsenic can be swept in one direction without significant broadening of the distribution by applying a temperature gradient. The unidirectional drift of As-rich droplets is caused by the process of thermomigration, which is better known from the studies of metallic droplets propagating in semiconductors. For Sb diffusion toward the surface, the calculated diffusivities at 900°C are given in Table 7 for diffusion in N_2 or O_2 and O_2/H_2O.[57] No diffusion of Sb inside the SiO_2 film was observed up to 1200°C, suggesting that the Sb atoms are immobile. Diffusion of metal elements (Ag, Cu, Au, Pd, and Ti) in SiO_2 films under an applied electric field was investigated using MOS structures.[60] This method is often called the bias-temperature stress (BTS) test. It is frequently used for evaluating the reliability of the gate oxide for a MOS transistor. Figure 9c shows the test structure. Au and Pd showed no diffusion in SiO_2 for temperatures of 150–600°C and at electric fields between $\pm 10^6$ V/cm. Ti forms TiO_2. For Ag, the diffusivity is approximately 4.5×10^{-13} cm^2/s for diffusion in N_2 at 300°C under an electric field of 10^5 V/cm. The activation energy is $\cong 1.24$ eV in the temperature range between 275° and 365°C. Sodium and potassium are fast diffusants at temperatures above 100°C and under an applied electric field.

TABLE 7
Diffusivities in SiO$_2$

Element	Ref.	D_o (cm^2/s)	E (eV)	$D(900°C)$ (cm^2/s)	C_s (cm^{-3})	Source and ambient
Boron	54	7.23×10^{-6}	2.38	4.4×10^{-16}	10^{19}– 2×10^{20}	B$_2$O$_3$ vapor, O$_2$ + N$_2$
	54	1.23×10^{-4}	3.39	3.4×10^{-19}	6×10^{18}	B$_2$O$_3$ vapor, Ar
	54	3.16×10^{-4}	3.53	2.2×10^{-19}	$<3 \times 10^{20}$	Borosilicate
Gallium	54	1.04×10^5	4.17	1.3×10^{-13}	$\cdots$	Ga$_2$O$_3$ vapor, H$_2$ + N$_2$ + H$_2$O
Phosphorus	54	5.73×10^{-5}	2.30	7.7×10^{-15}	8×10^{20}– 10^{21}	P$_2$O$_5$ vapor, N$_2$
	55	1.86×10^{-1}	4.03	9.3×10^{-19}	8×10^{17}– 8×10^{19}	Phosphosilicate, N$_2$
Arsenic	56	67.25	4.7	4.5×10^{-19}	$<5 \times 10^{20}$	Ion implant, N$_2$
	56	3.7×10^2	3.7	4.8×10^{-18}	$<5 \times 10^{20}$	Ion implant, O$_2$
Antimony	57	3.7×10^{-11}	1.32	7.9×10^{-19}		Ion implant, N$_2$ or O$_2$
		1.2×10^{-7}	2.25	2.6×10^{-17}		Ion implant, O$_2$/H$_2$O
Hydrogen (H$_2$)	4	5.65×10^{-4}	0.446	7×10^{-6}		
Helium	4	3×10^{-4}	0.24	2.8×10^{-5}		
Water	4	10^{-6}	0.79	4×10^{-10}		
Oxygen	4	2.7×10^{-4}	1.16	2.8×10^{-9}		
Gold	4	8.2×10^{-10}	0.8	3×10^{-13}		
	4	1.52×10^{-7}	2.14	10^{-16}		
Platinum	4	1.2×10^{-13}	0.75	7.2×10^{-17}		
Sodium	4	6.9	1.3	1.8×10^{-5}		

Note: C_s = Surface concentration on silicon after diffusion from the specified source and ambient in the absence of an oxide barrier.

7.10 DIFFUSION ENHANCEMENTS AND RETARDATIONS

Diffusion study is complicated by not only the presence of defects or high-concentration effects, but also other processing factors. Diffusion in an oxidizing ambient and the lateral enhancement of diffusivity can significantly affect VLSI structures. In the study of diffusion in an NH_3 ambient, it was discovered that the exposure of a Si surface to NH_3 caused vacancy injection and the SiO_2 surface caused interstitial injection into the silicon underneath.[2] The results from these studies are discussed briefly in this section.

7.10.1 Effect of an Oxidizing Ambient

A few processing conditions were shown to enhance or retard diffusion in silicon. Among these, diffusion in an oxidizing ambient of boron, phosphorus, arsenic, and antimony were investigated extensively. Most of the experimental data were obtained from samples that had been processed under conditions similar to those under which self-aligned gate MOS devices and circuits are fabricated.

The oxidation-enhanced diffusion (OED) of boron was first observed in high-concentration diffusions into both <100> and <111> oriented silicon wafers.[61] Some experiments attempted to separate the oxidation effect from the high-concentration effect by diffusing dopants at concentration levels below n_i at the diffusion temperature. Dopants were introduced at low concentrations to form a prediffused layer from a chemical source or an ion-implanted source. A thin oxide layer (100 to 500 Å) is grown at low temperatures to protect the silicon surface, and is then covered by the deposition of a silicon nitride film 0.1 to 0.2 μm thick. The thin oxide layer between the silicon nitride and the silicon surface also serves to adjust the interface properties. Without the thin oxide layer, the interface between a Si_3N_4 film and a silicon surface exhibits a charge storage effect, which causes surface leakage current and instabilities. Strips of silicon nitride and oxide films are removed by a selective photolithography and etching technique. These samples have alternating regions of free silicon surface and nitride–oxide-protected surface, and are oxidized at different oxidation temperatures, in different ambients, for different time periods, and sometimes with both <100> and <111> oriented wafers. The enhancement or retardation is evaluated by measuring the junction depths, by spreading resistance profiles, or by concentration profiles using the differential conductivity method.

Figure 13 shows a cross section of the diffusion structures, with adjacent oxidized and masked regions.[62] The junction depth on the right side under the silicon nitride mask is shallower than the one on the left side. All the junction depths are measured from the original sample surface, prior to the oxidation but after the silicon nitride deposition. The enhancement or retardation depth Δx_j can be expressed as

$$\Delta x_j = (x_j)_{fo} - (x_j)_f \tag{87}$$

where $(x_j)_{fo}$ is the final junction depth under the oxide region and $(x_j)_f$ is the final

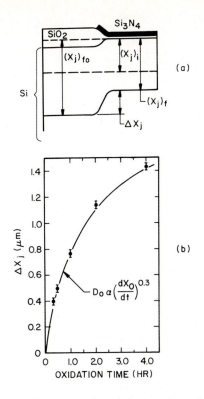

FIGURE 13
Oxidation-enhanced diffusion. (a) Cross section of the experimental structure. (b) Δx_j versus oxidation time. Boron diffusion at 1100°C in wet oxygen. (*After Taniguchi et al., Ref. 62.*)

junction depth under the silicon oxide–nitride mask. $(x_j)_i$ is the initial junction depth shown in Fig. 13a. Figure 13b shows an example of the measured Δx_j as a function of the oxidation time for boron at 1100°C. Since the concentration levels are below n_i, the diffusion under the masking oxide–nitride film is due to the intrinsic diffusivity, and the diffusivity under the oxide can be expressed as

$$D_A = D_i + \Delta D(T, t, P_{O_2}, \text{orientation, etc.)} \tag{88}$$

where D_A is the diffusivity for oxidation-enhanced diffusion; D_i is the intrinsic diffusivity or the diffusivity in a nonoxidizing ambient; and ΔD is the enhancement diffusivity that can depend on diffusion temperature, time, partial pressure of oxygen (P_{O_2}), crystal orientations, crystal type (Czochralski or floating-zone-grown crystal), and so forth.

Since the observed enhancement showed a strong dependence on the diffusion time but the measured results were from a given diffusion-oxidation period, an effective diffusivity, sometimes called the diffusion-time average diffusivity, is used.[63] In these cases, the time-dependence of enhancement is approximated by

$$< D_A >_{\text{eff}} = \frac{1}{t} \int_0^t D_A dt \qquad (89)$$

where D_A is the diffusivity of the dopant at time t. The ΔD is proportional to a fractional power, n, of the oxidation rate.[64]

$$\Delta D = \alpha \left[\frac{dX_{ox}}{dt} \right]^n \qquad (90)$$

where α is a proportional constant that can be estimated from an assumed diffusion model, n is between 0.3 and 0.6, and dX_{ox}/dt is the oxidation rate. Results on arsenic and phosphorus diffusion in dry oxygen show that ΔD decreases as the oxidation temperature increases.[65]

Oxidation enhancement of As and P have been investigated using prediffused samples.[66] In this case, the diffusion equation, with a moving boundary during the oxidation, was solved by the numerical methods with a measured initial profile after the prediffusion and an assumed parabolic oxidation-rate relationship; that is, $dX_{ox}/dt = Bt^{-1/2}$. The results can be summarized as follows:

1. In dry N_2, the diffusivities are the same in $<100>$ and $<111>$ oriented wafers for both arsenic and phosphorus.
2. In dry O_2, the diffusivities are enhanced for As and P in $<100>$ oriented wafers and for P in $<111>$ wafers, but little enhancement was observed for As in $<111>$ silicon.
3. The enhancement in $<100>$ Si is greater than that in $<111>$ Si.
4. Since the oxidation rates are higher at shorter oxidation times, the enhancement is larger for short oxidation time and decreases with increasing oxidation time.
5. The diffusivity enhancement $\Delta D = (D)_{O_2} - (D)_{N_2}$ can be expressed in terms of an effective oxidation rate $(X_{ox}/t)^n$.

A retardation of antimony diffusion in silicon during oxidation was observed.[67] In addition, the stress at the Si–Si$_3$N$_4$ edge caused junction retardation under the silicon nitride film, laterally, to 20–30 μm inside the nitride film edge. The retardation is a fraction of a micron for junction depths of 3 to 6 μm and diffusion temperatures between 1000° and 1200°C.

Diffusion effects from diffusion ambients and from variations of the capping films have been investigated, as have the effects of oxidation from the back side of a wafer on the impurity diffusion in the front surface under a silicon nitride cap or an oxide–nitride cap.[68] Heat treatment of SiO_2 film in N_2, NH_3, or $H_2 + N_2$ will convert the surface into a thin film called nitroxide (or oxynitride). When a silicon surface is exposed to N_2, NH_3, or $H_2 + N_2$, a thin nitride film is formed, and this process is called "direct nitridation." The diffusions of Sb, As, B, and P under these conditions showed different results.[2,69,70] Table 8 summarizes some of the results from NH_3 and oxygen treatments on the dopant diffusion, including the growth or shrinkage of oxidation-induced stacking faults (OSF).[70]

Observation of the oxidation-induced stacking fault (OSF) and oxidation-

TABLE 8
Diffusion enhancement/retardation[70]

Elements	Treatments		
	Oxidation	Oxynitridation	Direct nitridation
P,B			
intrinsic	enhanced	enhanced	retarded
extrinsic	enhanced	enhanced	retarded
As			
intrinsic	enhanced	enhanced	enhanced
extrinsic	retarded or no effect	enhanced	enhanced
Sb*			
intrinsic	enhancement precedes retardation	enhancement precedes retardation	enhanced
Stacking faults	grow	grow	shrink

*The solid solubility of Sb is below 5×10^{19} cm^{-3}; all diffusions were under intrinsic conditions.

enhanced diffusion (OED) has led to the proposal of a dual diffusion mechanism. The diffusion of an impurity under the nitride layer is considered to be dominated by the vacancy mechanism, and the oxidation enhancement of diffusion is attributed to the presence of silicon self-interstitials that also cause the extrinsic stacking faults to grow.[64] Interstitials are generated at the Si–SiO$_2$ interface during oxidation. By assuming that the vacancy concentration is constant during oxidation, one can conclude that the enhancement of boron and phosphorus diffusivities during oxidation is due to the excess of interstitials diffusing away from the oxide–silicon interface, and that these elements are governed by a dual mechanism: vacancy and interstitialcy.[71]

The observation of oxidation-retarded diffusion of antimony suggests that during diffusion and oxidation, thermal equilibrium between vacancies and interstitials exists. The generation of interstitials at the oxide–silicon interface causes the depression of vacancy concentrations.[67] The diffusion retardation of antimony could be due to the reduction of vacancy concentrations; thus, antimony diffusion could be governed by a vacancy mechanism. By similar reasoning, since silicon interstitials are enhanced at the oxide–silicon interface, it has been suggested that both boron and phosphorus diffuse mainly via the interstitialcy mechanism in either an oxidizing or neutral ambient.[2]

7.10.2 Lateral Enhancement of Diffusivity

Another important enhancement effect in VLSI devices is the enhanced diffusion in narrow oxide or silicon–nitride windows. Diffusion into narrow windows of silicon oxide can result in anomalous junction depths.[72] Various enhancements and retardations of the junction depth near the oxide window edges have been observed; these are the results of elastic strain fields near the window edges.

For boron diffusion in a structure similar to that shown in Fig. 14, laterally enhanced diffusion extends under the nitride layer up to 30 μm.[73] Strips of silicon nitride layers with widths varying from 2.5 to 100 μm were separated with 100 μm windows without oxide. The samples were oxidized after boron implantation and annealing at 900°C. The junction depth at the center of the nitride–oxide strip was measured as a function of the widths of the strips. For narrow strips, the lateral enhancement of the diffusivity is significant for VLSI device designs. In a narrow structure, the junction depths under the silicon nitride film are enhanced and nonuniform. These results suggest the importance of the study of two-dimensional (2D) or 3D diffusion for small device simulations, which is being actively pursued at the Stanford Electronics Laboratories.[74]

7.11 SUMMARY AND FUTURE TRENDS

This chapter has discussed various diffusion effects in silicon with emphasis on VLSI applications. Various factors affecting diffusion control are presented. Fick's classical diffusion laws with constant diffusivities are obeyed for Group III and V elements when the concentrations are below the intrinsic carrier concentration. At high concentrations, $C > n_i$, Fick's generalized diffusion equation with concentration-dependent diffusivities can be solved by numerical methods or by a similarity function for some simple cases. The concen-

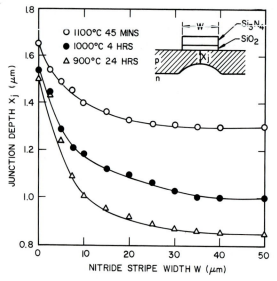

FIGURE 14
Lateral enhancement of x_j under a Si_3N_4 film during oxidation. (*After Lin et al., Ref. 73.*)

tration-dependent diffusivities can be determined from the measured profiles using mathematical formulations of a Boltzmann transformation or modification of it.

Atomic diffusion mechanisms are being developed to relate the interaction of impurity diffusion with charged point defects. Attempts were made to construct diffusion models based on defect-impurity interactions. The diffusivities are functions of the concentration of the ionized point defects, vacancies, or self-interstitials. This approach is successful in explaining the high concentration diffusions of the Group III and Group V elements. Various other models have also been proposed and tested. Coupled differential equations for diffusion of dopant and point defects are used to address the complex diffusion problems.

Diffusion in an oxidizing ambient also exhibits a time dependency because of the parabolic oxidation rate of silicon. Observations of the coexistence of the oxidation-enhanced diffusion and the formation of oxidation-induced stacking faults suggest that an extrinsic mechanism for generating silicon self-interstitials near the silicon–oxide interface may be applied to explain the impurity diffusions. The oxidation-induced stacking faults are extrinsic in nature; that is, they grow by absorbing silicon self-interstitials. These observations have led to the proposal of a dual vacancy-interstitialcy diffusion mechanism. In theoretical modeling, the prevailing trend favors an interstitialcy mechanism; further quantitative analysis based on this mechanism is needed.

As the device size becomes smaller and smaller, the need for better measurement techniques becomes more urgent. At present, the spreading resistance technique is widely used for profile measurements. Unfortunately, it relies on a beveled sample technique that limits it to junction depths greater than 1 μm, and it is a comparative and semiquantitative method. Some efforts have been made to measure shallow junctions (x_j less than 0.5 micron) by this method. This appears to be the limits of this technique. Conversion of the spreading resistance into impurity (or carrier) concentration remains a problem.

SIMS analysis is a powerful tool for diffusion profile measurements. It can measure boron and arsenic concentrations as low as 5×10^{15} atoms/cm^3, and has a high depth resolution of a few tens of angstroms. Therefore, it is an ideal tool for measuring shallow diffusion profiles. This technique will provide the needed precision for profile measurements in VLSI structures. However, it measures the total impurity atom concentration, that is, both electrically active and inactive atoms.

Dopant segregation at the interface between the polycrystalline silicon and the single-crystal silicon, or at the interface between the SiO$_2$ and Si, could have profound effects on the electrical characteristics at these interfaces. The segregation of dopants at these interfaces has been observed by the SIMS, the Auger electron spectroscopic, the RBS, the scanning transmission electron microscopy (STEM), and the electron spin resonance (ESR) methods.[75-77] Little is known about the electrical properties of these dopants at the interfaces. For components in VLSI circuits, the pileup of inactive impurities near an interface could be important, and it is an area for future investigation.

PROBLEMS

1 (a) Derive expressions of concentration gradients for the erfc and Gaussian distributions. If the substrate doping density is C_{sub}, derive the expressions of the junction depths.

(b) Assuming $C_s = 10^{19}$ cm^{-3} for an erfc distribution and $Q_T = 1 \times 10^{13}$ atoms/cm^2 for a Gaussian distribution, $C_{sub} = 10^{15}$ atoms/cm^3, and $D = 1 \times 10^{-15}$ cm^2/s (which is close to the boron diffusivity at 900°C), calculate the junction depths and the concentration gradients for both distributions. Calculate the integrated dopants for the erfc distribution and the surface concentration for the Gaussian distribution for diffusion times of 10, 30, and 60 min.

(c) Compare and discuss the results of **(a)** and **(b)**.

2 Assume the measured phosphorus profile can be represented by a Gaussian function with a diffusivity $D = 2.3 \times 10^{-13}$ cm^2/s. The measured surface concentration is 1×10^{18} atoms/cm^3 and the measured junction depth is 1 micron at a substrate concentration of 1×10^{15} atoms/cm^3.

(a) Calculate the diffusion time and the total dopant in the diffused layer.

(b) If the diffusivity is 1.4×10^{-13} cm^2/s, what is the calculated junction depth?

3 Derive Eq. 15 from Eq. 3. (Hint: assume $y = \frac{1}{2}\frac{x}{t^{1/2}}$, and find $\frac{\partial C}{\partial y}$, etc.).

4 Use values given in Table 1 and Eq. 20 to calculate the intrinsic diffusivities of B, P, As, and Sb between 850 and 1200°C at 50°C intervals and plot D versus $1/T$ (K) on semilog graph paper. Use the PD values for Arsenic.

5 Derive Eq. 29, assuming that the ionized acceptor vacancy concentration can be expressed as a function of the Fermi level and E_V, the activation energy of the acceptor vacancy.

6 In order to determine if the intrinsic diffusivity of an impurity is applicable at a given diffusion temperature, one has to know the intrinsic carrier concentration, n_i. Thus, the plot of n_i versus temperature is a very useful curve. Using Eqs. 33 and 34, construct n_i versus T.

7 Using Eq. 36, derive the electric-field enhancement factor of Eq. 41.

8 A p-type <100> oriented silicon wafer with a substrate doping at 10^{16} atoms/cm^3 has been implanted and diffused with arsenic to an ion dose of 1×10^{15} cm^{-2} at 30 keV and diffusion at 850°C for 20 min in nitrogen.

(a) Calculate the sheet resistance from Eq. 68.

(b) Calculate the surface concentrations from Eq. 69.

(c) Find the surface concentration of the electrically active arsenic.

(d) Discuss the results.

9 (a) Using Eqs. 65 and 66 for an arsenic-implanted-diffused profile, derive the approximate expressions for Eqs. 67, 68, and 69.

(b) For an arsenic dose less than 10^{16} cm^{-2} and a diffusion temperature greater than 1000°C, assuming the electrically active arsenic equals the total arsenic (neglect As clusters), derive Eq. 68 using Eq. 8 and an effective electron mobility of

$$\mu_{eff} \cong \frac{28.2 \times 10^7}{C_A^{1/3}} \text{ cm}^2/\text{V-s}$$

for 10^{19} cm$^{-3} < C_A < 6 \times 10^{20}$ cm^{-3} where C_A is the concentration of the electrically active As.

10 For an acceptor-type impurity, the diffusion current including the electric-field term, is

$$J = -qD_A \frac{\partial C_A}{\partial x} + q\mu_A C_A E$$

where D_A is the diffusivity, C_A is the acceptor concentration, μ_A is the mobility, and E is the electric field. If

$$D_A = D_i \frac{p}{n_i}$$

show that

$$J = -q\frac{d}{dx}(D_A C_A)$$

The above expression reduces the computation time when it is used for analyzing diffusion profiles numerically.

REFERENCES

1 J. Crank, *The Mathematics of Diffusion,* Oxford University Press, London, 1957.
2 P. M. Fahey, "Point Defects and Dopant Diffusion Silicon," PhD. Thesis, Stanford University, 1985.
3 B. I. Boltaks, *Diffusion in Semiconductors,* Academic, New York, 1963.
4 W. E. Beadle, J. C. C. Tsai, and R. D. Plummer, *Quick Reference Manual for Silicon Integrated Circuit Technology,* Wiley, New York, 1985.
5 S. E. Swirhun, J. A. Del Alamo, and R. M. Swanson, "Measurement of Hole Mobility in Heavily Doped n-Type Silicon," *IEEE Electron Device Lett.,* **EDL-7**, 168 (1986).
6 S. E. Swirhun, Y. H. Kwark, and R. M. Swanson, "Measurement of Electron Lifetime, Electron Mobility and Band-Gap Narrowing in Heavily Doped p-Type Silicon," *IEDM Technical Digest,* 1986, p. 24.
7 G. Baccarani and P. Ostoja, "Electron Mobility Empirically Related to the Phosphorus Concentration in Silicon," *Solid State Electronics,* **18**, 579 (1975).
8 D. A. Antoniadis, A. G. Gonzalez, and R. W. Dutton, "Boron in Near-Intrinsic <100> and <111> Silicon Under Inert and Oxidizing Ambients—Diffusion and Segregation," *J. Electrochem. Soc.,* **125**, 813 (1978).
9 J. C. Irvin, "Resistivity of Bulk Silicon and Diffused Layers in Silicon," *Bell Syst. Tech. J.,* **41**, 387 (1962).
10 D. Anderson and M. Lisak, "Approximate Solutions of Some Nonlinear Diffusion Equations," *Phys. Rev.* **A22**, 2761 (1980). D. Anderson and K. O. Jeppson, "Nonlinear Two-Step Diffusion in Semiconductors," *J. Electrochem. Soc.,* **131**, 2675 (1984).
11 U. Gosele, "Current Understanding of Diffusion Mechanisms in Silicon," *Semiconductor Silicon 1986,* H. R. Huff, T. Abe, and B. Kolbesen, Eds., *Electrochem. Soc.,* Pennington, NJ, 1986, p. 541.
12 N. A. Stolwijk, B. Schuster, and J. Holzl, "Diffusion of Gold in Silicon Studied by Means of Neutron-Activation Analysis and Spreading Resistance Measurements," *Appl. Phys.,* **A33**, 133 (1984).
13 F. A. Huntley and A. F. W. Willoughby, "The Diffusion of Gold in Thin Silicon Slices," *Solid State Electronics,* **13**, 1231 (1970).
14 R. L. Longini and R. F. Green, "Ionization Interaction Between Impurities in Semiconductors and Insulators," *Phys. Rev.,* **102**, 992 (1956).
15 F. J. Morin and J. P. Maita, "Electrical Properties of Silicon Containing Arsenic and Boron," *Phys. Rev.,* **96**, 28 (1954).
16 D. Shaw, "Self and Impurity Diffusion in Ge and Si," *Phys. Status Solidi B,* **72**, 11 (1975).
17 R. B. Fair, "Concentration Profiles of Diffused Dopants in Silicon," in F. F. Y. Wang, Ed., *Impurity Doping Processes in Silicon,* North-Holland, New York, 1981, Chapter 7.

18 R. B. Fair, "The Effect of Strain-Induced Bandgap Narrowing on High Concentration Phosphorus Diffusion in Silicon," *J. Appl. Phys.,* **50**, 860 (1979).

19 J. S. Park, A. Neugroschel, and F. A. Lindholm, "Comments on Determination of Band Gap Narrowing from Activation Plots," *IEEE Trans. Electron Devices,* **ED-33**, 1077 (1986).

20 S. M. Hu and S. Schmidt, "Interactions in Sequential Diffusion Processes in Semiconductors," *J. Appl. Phys.,* **39**, 4272 (1968).

21 M. Y. Tsai, F. F. Morehead, and J. E. E. Baglin, "Shallow Junctions by High Dose As Implants in Si: Experiments and Modeling," *J. Appl. Phys.,* **51**, 3230 (1980).

22 E. Guerrero, H. Potzl, R. Tielert, M. Grasserbauer, and G. Stingeder, "Generalized Model for the Clustering of As Dopants in Si," *J. Electrochem. Soc.,* **129**, 1826 (1982).

23 H. Ryssel, K. Muller, K. Haberger, R. Henkelmann, and F. Jahael, "High Concentration Effects of Ion Implanted Boron in Silicon," *J. Appl. Phys.,* **22**, 35 (1980).

24 R. B. Fair and J. C. C. Tsai, "A Quantitative Model for the Diffusion of Phosphorus in Silicon and the Emitter Dip Effect," *J. Electrochem. Soc.,* **124**, 1107 (1978).

25 S. Mahajan, G. A. Rozgonyi, and D. Brasen, "A Model for the Formation of Stacking Faults in Silicon," *Appl. Phys. Lett.,* **30**, 73 (1977).

26 F. F. Morehead and R. F. Lever, "Enhanced Tail Diffusion of Phosphorus and Boron in Silicon: Self-Interstitial Phenomenon," *Appl. Phys. Lett.,* **48**, 151 (1986).

27 C. P. Ho, J. D. Plummer, S. E. Hansen, and R. W. Dutton, "VLSI Process Modeling— SUPREM-III," *IEEE Trans. Electron Devices,* **ED-30**, 1438 (1983).

28 R. B. Fair and R. Subrahmanyan, "PREDICT-A New Design Tool for Shallow Junction Process," proceedings of SPIE on Advanced Application of Ion Implantation, ed. M. I. Current and D. K. Sadana, Bellingham, WA, Vol. 530, 1985, p. 88.

29 J. D. Plummer, and R. W. Dutton, "Process Simulators for Silicon VLSI and High Speed GaAs Devices," SRC Technical Report No. T86085, Integrated Circuits Laboratory, Stanford University, CA., October 1986.

30 K. O. Jeppsen and D. A. Anderson, "Analytical Model for Phosphorus Diffusion in Silicon," *J. Electrochem. Soc.,* **133**, 388 (1986).

31 D. G. Schimmel and M. J. Elkind, "An Examination of the Chemical Staining of Silicon," *J. Electrochem. Soc.,* **125**, 152 (1978).

32 L. J. Van der Pauw, "A Method of Measuring Specific Resistivity and Hall Effect of Discs of Arbitrary Shape," *Phillips Res. Repts.,* **13**, 1 (1958).

33 C. P. Wu, E. C. Douglas, and C. W. Mueller, "Limitations of the CV Technique for Ion Implanted Profiles," *IEEE Trans. Electron Devices,* **ED-22**, 319 (1975).

34 E. Tannenbaum, "Detailed Analysis of Thin Phosphorus Diffused Layers in P-Type Silicon," *Solid State Electron.,* **3**, 123 (1961).

35 R. G. Mazur and D. H. Dickey, "A Spreading Resistance Technique for Resistivity Measurements on Silicon," *J. Electrochem. Soc.,* **113**, 255 (1966).

36 W. K. Hofker, "Implantation of Boron in Silicon," *Philips Res. Rept. Suppl.,* **8**, 1 (1975).

37 P. F. Kane and G. B. Larrabee, *Characterization of Semiconductor Materials,* McGraw-Hill, New York, 1970, Chapter 9, p. 278.

38 W. K. Chu, J. W. Mayer, M-A Nicolet, T. M. Buck, G. Amsel, and F. Eisen, "Microanalysis of Surface, Thin Films and Layered Structures by Nuclear Backscattering and Reactions," in H. R. Huft and R. R. Burgess, Eds., *Semiconductor Silicon 1973,* Electrochem. Soc., New York, 1973, p. 416.

39 J. F. Ziegler, G. W. Cole, and J. E. E. Baglin, "Technique for Determining Concentration Profiles of Boron Impurities in Substrates," *J. Appl. Phys.,* **43**, 3809 (1972).

40 J. L. Combasson, J. Bernard, G. Guernet, N. Hilleret, and M. Bruel, "Physical Profile Measurements in Insulating Layers Using the Ion Analyzer," in B. L. Crowder, Ed., *Ion Implantation in Semiconductors and Other Materials,* Plenum, New York, 1973, p. 285.

41 B. L. Sharma, "Diffusion in Semiconductors," *Trans. Tech. Pub. Germany,* 87 (1970).

42 R. F. Bailey and T. G. Mills, "Diffusion Parameters of Platinum in Silicon," in R. R. Habarecht and E. L. Kern, Eds., *Semiconductor Silicon 1969,* Electrochem. Soc., New York, 1969, p. 481.

43 W. Wurker, K. Roy, and J. Hesse, "Diffusion and Solid Solubility of Chromium in Silicon," *Mater. Res. Bull. (U.S.A.),* **9**, 971 (1974).

44 H. Kitagano and K. Hashimoto, "Diffusion Coefficient of Cobalt in Silicon," *J. Appl. Phys. Jpn.,* **16**, 173 (1977).

45 J. C. Mikkelsen, Jr., "Diffusivity of Oxygen in Silicon During Steam Oxidation," *Appl. Phys. Lett.,* **40**, 336 (1982). M. Stavola, J. R. Patel, L. C. Kimerling, and P. E. Freeland, "Diffusivity of Oxygen in Silicon at the Donor Formation Temperatures," *Appl. Phys. Lett.,* **42**, 73 (1983).

46 T. I. Kamins, J. Manolin, and R. N. Tucker, "Diffusion of Impurities in Polycrystalline Silicon," *J. Appl. Phys.,* **43**, 83 (1972).

47 W. K. Schubert, "Properties of Furnace-Annealed, High-Resistivity, Arsenic-Implanted Polycrystalline Silicon Films," *J. Mater. Res.,* **1**, 311 (1986).

48 H. Schaber, R. V. Criegerm, and I. Weitzel, "Analysis of Polycrystalline Silicon Diffusion Sources by Secondary Ion Mass Spectrometry," *J. Appl. Phys.,* **58**, 4036, (1985).

49 C. J. Coe, "The Lateral Diffusion of Boron in Polycrystalline Silicon and Its Influence on the Fabrication of Sub-Micron Mosts," *Solid State Electron.,* **20**, 985 (1977).

50 B. Swaminathan, K. C. Saraswat, R. W. Dutton, and T. I. Kamins, "Diffusion of Arsenic in Polycrystalline Silicon," *Appl. Phys. Lett.,* **40**, 795 (1982).

51 M. M. Mandurah, K. C. Saraswat, C. R. Helms, and T. I. Kamins, "Dopant Segregation in Polycrystalline Silicon," *J. Appl. Phys.,* **51**, 5755 (1980).

52 S. Horiuchi and R. Blanchard, "Boron Diffusion in Polycrystalline Silicon Layers," *Solid State Electron.,* **18**, 529 (1975).

53 F. H. M. Spit, H. Albers, A. Lubbes, Q. J. A. Rijke, L. J. v. Ruijven, J. P. A. Westerveld, H. Bakker, and S. Radelaar, "Diffusion of Antimony (^{125}Sb) in Polycrystalline Silicon," *Phys. Stat. Sol.,* **(a)89**, 105 (1985).

54 M. Ghezzo and D. M. Brown, "Diffusivity Summary of B, Ga, P, As, and Sb in SiO_2," *J. Electrochem Soc.,* **120**, 146 (1973).

55 R. N. Ghoshtagore, "Silicon Dioxide Masking of Phosphorus Diffusion in Silicon," *Solid State Electron.,* **18**, 399 (1975).

56 Y. Wada and D. A. Antoniadis, "Anomalous Arsenic Diffusion in Silicon Dioxide," *J. Electrochem. Soc.,* **128**, 1317 (1981).

57 A. H. van Ommen, "Diffusion of Ion-Implanted As in SiO_2," *J. Appl. Phys.,* **56**, 2708 (1984); "Diffusion of Ion-Implanted Sb in SiO_2," *J. Appl. Phys.,* **61**, 993 (1987).

58 R. Singh, M. Maier, H. Kraute, D. R. Young, and P. Balk, "Diffusion of Ion Implanted Arsenic in Thermally Grown SiO_2 Films," *J. Electrochem. Soc.,* **131**, 2645 (1984).

59 G. K. Celler, L. E. Trimble, K. W. West, L. Pfeiffer, and T. T. Sheng, "Segregation and Drift of Arsenic in SiO_2 Under the Influence of a Temperature Gradient," *Appl. Phys. Lett.,* **50**, (11), 664 (1987).

60 J. D. McBrayer, R. M. Swanson, and T. W. Sigman, "Diffusion of Metals in Silicon Dioxide," *J. Electrochem. Soc.,* **133**, 1242 (1986).

61 W. G. Allen and K. V. Anand, "Orientation Dependence of the Diffusion of Boron in Silicon," *Solid State Electron.,* **14**, 397 (1971).

62 K. Taniguchi, K. Kurosawa, and M. Kashiwagi, "Oxidation Enhanced Diffusion of Boron and Phosphorus in (100) Silicon," *J. Electrochem. Soc.,* **127**, 2243 (1980).

63 S. M. Hu, "Formation of Stacking Faults and Enhanced Diffusion in the Oxidation of Silicon," *J. Appl. Phys.,* **45**, 1567 (1974).

64 A. M. R. Lin, D. A. Antoniadis, and R. W. Dutton, "The Oxidation Rate Dependence of Oxidation-Enhanced Diffusion of Boron and Phosphorus in Silicon," *J. Electrochem. Soc.,* **128**, 1131 (1981).

65 D. A. Antoniadis, A. M. Lin, and R. W. Dutton, "Oxidation-Enhanced Diffusion of Arsenic and Phosphorus in Neat Intrinsic (100) Silicon," *Appl. Phys. Lett.,* **33**, 1030 (1978).

66 Y. Ishikawa, Y. Skaina, H. Tanaka, S. Matsumoto, and T. Niimi, "The Enhanced Diffusion of Arsenic and Phosphorus in Silicon by Thermal Oxidation," *J. Electrochem. Soc.,* **129**, 644 (1982).

67 S. Mizuo and H. Higuchi, "Retardation of Sb Diffusion in Si During Thermal Oxidation," *J. Appl. Phys. Jpn.,* **20**, 739 (1981).

68 S. Mizuo and H. Higuchi, "Investigation of Point Defects in Si by Impurity Diffusion," Impurity Diffusion and Gettering in Silicon, in R. B. Fair, C. W. Pearce and J. Washburn, Eds., *Symposia Proceedings,* Vol. 36, Material Research Society, Pittsburgh, PA, 1985, p. 125.

69 S. Mizuo, T. Kusaka, A. Shintani, M. Nanba, and H. Higuchi, "Effect of Si and SiO_2 Thermal Nitridation on Impurity Diffusion and Oxidation Induced Stacking Fault Size in Si," *J. Appl. Phys.,* **54**, 3860 (1983).

70 P. Fahey and R. W. Dutton, "Dopant Diffusion Under Conditions of Thermal Nitridation of Si and SiO_2," *Semiconductor Silicon 1986* (see Reference 11), 1986, p. 571.

71 T. Y. Tan and U. Gosele, "Oxidation-Enhanced or Retarded Diffusion and the Growth or Shrinkage of Oxidation-Induced Stacking Faults in Silicon," *Appl. Phys. Lett.,* **40**, 616 (1982).

72 C. F. Gibbon, E. I. Povilonis, and D. R. Ketchow, "The Effect of Mask Edges on Dopant Diffusion into Semiconductors," *J. Electrochem. Soc.,* **119**, 767 (1972).

73 A. M. Lin, R. W. Dutton, and D. A. Antoniadis, "The Lateral Effect of Oxidation on Boron Diffusion in <100> Silicon," *Appl. Phys. Lett.,* **35**, 799 (1979).

74 P. B. Griffin and J. D. Plummer, "Process Physics Determining 2-D Impurity Profiles in VLSI Devices," *IEDM Technical Digest,* December (1986), p. 522.

75 V. Probst, H. J. Bohm, H. Schaber, H. Oppolzer, and I. Weitzel, "Analysis of Polysilicon Diffusion Sources," *Semiconductor Silicon 1986,* 1986, p. 594.

76 A. Stesmans, "ESR Observation of Phosphorus Pile up at the Si/SiO_2 Interface in Boron Doped Si," *Solid State Comm.,* **58**, 299 (1986).

77 C. Y. Wong, C. R. M. Grovenor, P. E. Batson, and R. D. Isaac, "Arsenic Segregation to Silicon/Silicon Oxide Interfaces," *J. Appl. Phys.,* **58**, 1259 (1985).

78 R. B. Fair and J. C. C. Tsai, "The Diffusion of Ion Implanted Arsenic in Silicon," *J. Electrochem. Soc.,* **122**, 1689 (1975).

79 B. J. Masters and J. M. Fairfield, "Arsenic Isoconcentration Diffusion Studies in Silicon," *J. Appl. Phys.,* **40**, 2390 (1969).

CHAPTER
8

ION
IMPLANTATION

M. D. GILES

8.1   INTRODUCTION

The feature of semiconductors that makes them so useful for electronic devices is that their conduction properties can be changed by introducing small quantities of dopant atoms. We need a method of introducing the dopant that is controllable, reproducible, and free from undesirable side effects. Initially, dopant was diffused into the wafer from a surface source such as a doped glass. To achieve reproducibility, the dopant concentration at the surface was maintained at solid solubility. This seriously limited the possible dopant distributions that could be obtained. During the 1960s, the new method of ion implantation was developed, which largely satisfies the conditions of controllability and reproducibility, and over the past 20 years this has become the method of choice for fabricating integrated circuits.

During ion implantation, dopant atoms are vaporized, accelerated, and directed at a silicon substrate. They enter the crystal lattice, collide with silicon atoms, and gradually lose energy, finally coming to rest at some depth within the lattice. The average depth can be controlled by adjusting the acceleration energy. The dopant dose can be controlled by monitoring the ion current during implantation. The principle side effect—disruption of the silicon lattice caused by the ion collisions—is removed by subsequent heat treatments. Ion implantation therefore satisfies the conditions for a generally useful doping process. Implantation energies range from 1 keV to 1 MeV, resulting in ion distributions with average depths ranging from 100 Å to 10 μm. Doses range from 10^{12} ions/cm^2 for threshold adjustment to 10^{18} ions/cm^2 for buried insulators.

The idea of doping semiconductors using ion implantation was patented by Shockley[1] at Bell Labs in 1954. Early work has been summarized by Gibbons,[2] covering developments up to 1968. An overview of work since then can be found in the proceedings of the International Conferences on Ion Implantation[3] and the Conferences on Ion Beam Modification of Materials,[4] and in various other books.[5] There are still many active areas of research on ion implantation in silicon, with over 3000 papers published during the past 5 years. High-dose and high-energy implants have been used to form buried layers of different materials. New, low-energy machines have been designed to allow very shallow dopant layers to be formed, leading, in the limit, to ion beam deposition systems. Focused ion beams can be used directly to write dopant patterns without the need for a masking layer. These new directions provide challenges and opportunities for equipment manufacture, process design, and process simulation.

This chapter describes the implantation process and its application to VLSI integrated circuit fabrication. First, the theory of ion penetration in crystalline solids will be covered. This provides the basis for a discussion of ion implantation equipment, followed by a description of annealing and its associated effects. The remainder of the chapter covers the application to VLSI processing: shallow junctions, buried layers, and other issues. Finally, we will summarize and mention areas for future work. We will concentrate on implantation into silicon because it is the only substrate used for VLSI devices at present—other materials may differ significantly.

8.2   RANGE THEORY

This section covers the basic theory of ion implantation: the physics of ion collisions, its application to calculating range distributions, the effect of a crystal lattice, damage, and recoil distributions.

8.2.1   Ion Stopping

As each implanted ion enters the target, it undergoes a series of collisions with target atoms until it finally comes to rest at some depth, as shown in Fig. 1. The initial ion energy, typically 100 keV, is much higher than lattice binding energies, 10 to 20 eV, so the ion must come very close to a lattice atom if it is to be significantly deflected. This allows us to develop a theory for ion scattering based on elastic collisions between pairs of nuclei, ignoring the relatively weak lattice forces. A second component of scattering comes from inelastic collisions with electrons in the target. The total stopping power S of the target, defined as the energy (E) loss per unit path length of the ion (x), is the sum of these two terms:

$$S = \left(\frac{dE}{dx}\right)_{\text{nuclear}} + \left(\frac{dE}{dx}\right)_{\text{electronic}} \tag{1}$$

Once we have evaluated S, we can integrate it to sum the energy losses

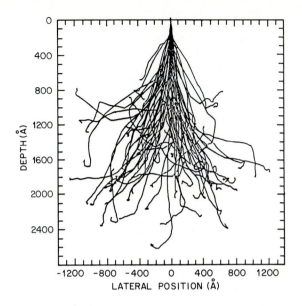

FIGURE 1
Monte Carlo calculation of 128 ion tracks for 50 keV boron implanted into silicon.

to find how far an ion will travel before all of its energy has gone. Figure 2 shows the relative contributions to S of each of the terms over a wide energy range.[6] Energies typical for implantation, 10 to 200 keV, fall at the far left of the figure, dominated by nuclear stopping or in the "low-energy" electronic stopping regime.

Nuclear stopping is caused by a collision between two atoms, and can be

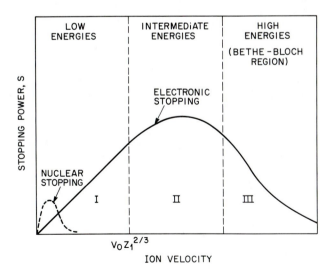

FIGURE 2
Nuclear and electronic components of the ion stopping power as a function of ion velocity. The quantity v_o is the Bohr velocity, $q^2/4\pi\epsilon_0\hbar$, and Z_1 is the ion atomic number. (*After Cruz, Ref. 6.*)

well described by classical kinematics. If the atoms were bare nuclei, then at a separation r there would be a coulombic potential between them, given by

$$V_c(r) = \frac{q^2 Z_1 Z_2}{4\pi\epsilon_0 r} \tag{2}$$

where Z_1 and Z_2 are the atomic charge numbers of ion and target atoms respectively, and q is the charge magnitude of a single electron. Electrons screen the nuclear charge, so a more realistic potential contains extra terms that reduce the interaction. A screening function $f_s(r)$ is usually written as the sum of exponential terms

$$f_s(r) = a_1 e^{-r/b_1} + a_2 e^{-r/b_2} + a_3 e^{-r/b_3} + \ldots \tag{3}$$

and

$$V(r) = V_c(r) f_s(r) \tag{4}$$

Given this description of the interaction potential, the equations of motion of the atoms can be integrated to yield the scattering angle for any incident ion trajectory, although this must be done numerically for realistic potentials. Working in the center-of-mass (CM) frame simplifies the form of the equations involved. The results of such calculations are available in the form of fitted functions that are easy to use,[7] giving the scattering angle θ in the CM frame as a function of the impact parameter p in Fig. 3. Classical dynamics allows us to relate θ to the energy lost by the ion, T, independent of the particular potential used:

$$T(p) = \frac{4M_1 M_2}{(M_1 + M_2)^2} E \sin^2\left[\frac{\theta(p)}{2}\right] \tag{5}$$

M_1 and M_2 are the atomic mass numbers of ion and target atoms respectively. The probability of having an impact parameter between p and $p + dp$ is $2\pi p\, dp$, also known as the differential scattering cross section $d\sigma$. We can find the rate of energy loss to nuclear collisions per unit path length by summing the energy loss

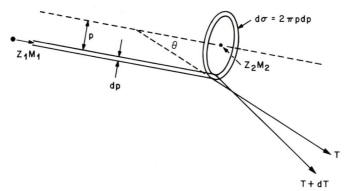

FIGURE 3
Schematic view of ion scattering, showing the relation between impact parameter and scattering cross section.

for each possible impact parameter multiplied by the probability of that collision occurring. If the maximum possible energy transfer in a collision is $T_{\max}$ and there are N target atoms per unit volume, then

$$S_n = \left(\frac{dE}{dx}\right)_{\text{nuclear}} = N \int_0^{T_{\max}} T\,d\sigma \tag{6}$$

Nuclear stopping is elastic, so energy lost by the incoming ion is transferred to the target atom, which is recoiled away from its lattice site.

Problem
A 100 keV boron atom makes a head-on collision with a silicon atom. Use an unscreened coulomb potential to estimate the closest distance the atoms come to each other during the collision. How much energy does the boron atom lose during the collision?

Solution
If we write the hydrogen mass as m_H and the boron atom velocity as v_{imp}, then the CM frame (marked with primes) is defined by

$$v_1 = v_1' + v_2' \quad \text{and} \quad M_1 v_1' = M_2 v_2'$$

Given that $M_1 = 11m_H$, $M_2 = 28m_H$, and $v_1 = v_{\text{imp}}$, we can solve for v_1' and v_2':

$$v_1' = 0.72 v_{\text{imp}} \quad \text{and} \quad v_2' = 0.28 v_{\text{imp}}$$

so in the CM frame, boron has energy 52 keV and silicon 20 keV. At minimum separation, the combined kinetic energy of 72 keV is changed to potential energy. Using Eq. 2, we can write

$$r = \frac{q^2 Z_1 Z_2}{4\pi\epsilon_0 V} = 0.014 \text{ Å}$$

To find the energy loss, use Eq. 5 and set $\theta = 180°$:

$$T_{\max} = \frac{4 M_1 M_2}{(M_1 + M_2)^2} E = 81 \text{ keV}$$

Electronic stopping is caused by interaction with the electrons of the target. Detailed modeling is very complex, but in the low-energy regime (see Fig. 2) the stopping is similar to a viscous drag force and is proportional to the ion velocity v. Lindhard and Scharff proposed the following expression, which is commonly used:[8]

$$S_e = \left[\frac{2q^2 a_0 Z_1^{7/6} Z_2 N}{\epsilon_0 (Z_1^{2/3} + Z_2^{2/3})^{3/2}}\right]\left(\frac{v}{v_0}\right) = \left[\frac{Z_1^{7/6} Z_2}{(Z_1^{2/3} + Z_2^{2/3})^{3/2}}\right] 8\pi a_0 \hbar N v \tag{7}$$

where a_0 and v_0 are the Bohr radius and velocity, and N is the target atom number density. Experimental measurements have shown that, although this expression predicts the correct general trend, there are oscillations in the stopping power as the ion number increases. One solution has been the use of empirical correction factors to bring S_e into agreement with experiment, writing $S_e = k_e \sqrt{E}$. More recently, several workers have published expressions whose coefficients have been fitted to the available experimental data. Figure 4 compares a four-parameter formula with experiment and with Eq. 7 for ions traveling in a carbon target.[9] The experimental oscillations are closely reproduced, giving much better agreement than the simpler Lindhard-Scharff formula. Electronic stopping is inelastic—the energy lost by incident ions is dissipated through the electron cloud into thermal vibrations of the target.

8.2.2 Range Distributions

Each implanted ion has a random path as it moves through the target, losing energy by nuclear and electronic stopping. Since each implantation dose contains more than 10^{12} ions/cm^2, their average behavior can be well predicted. We will consider an ion descending vertically and entering a horizontal silicon surface. The average total path length in silicon is called the range R, and is composed of a mixture of vertical and lateral motion. The average depth of the implanted ions is called the projected range R_p, and the distribution of ions about that depth can be approximated as Gaussian with standard deviation σ_p. The lateral motion

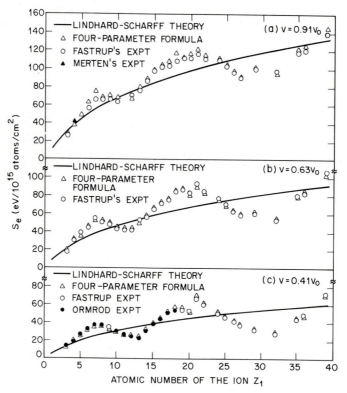

FIGURE 4
Electronic stopping powers for various ions in carbon, comparing theory with experiment. (*After Xia and Tan, Ref. 9.*)

of the ions leads to a lateral Gaussian distribution with standard deviation $\sigma_\perp$. The ion range is shown schematically[10] in Fig. 5. Far from a mask edge, we can neglect the lateral motion and write the ion concentration $n(x)$ as

$$n(x) = n_0 \exp\left[\frac{-(x - R_p)^2}{2\sigma_p^2}\right] \tag{8}$$

If the total dose is Φ, then integrating Eq. 8 gives us an expression for the peak concentration n_0:

$$n_0 = \frac{\Phi}{\sqrt{2\pi}\sigma_p} \cong \frac{0.4\Phi}{\sigma_p} \tag{9}$$

In general, an arbitrary distribution $n(x)$ can be characterized in terms of its moments. The normalized first moment of an ion distribution is the projected range. For convenience, higher moments are usually taken about R_p.

$$m_i = \frac{1}{\Phi} \int_{-\infty}^{\infty} (x - R_p)^i n(x) dx \tag{10}$$

The second, third, and fourth moments are typically expressed in terms of the following parameters:

$$\text{(standard deviation)} \quad \sigma_p = \sqrt{\frac{m_2}{\Phi}}$$

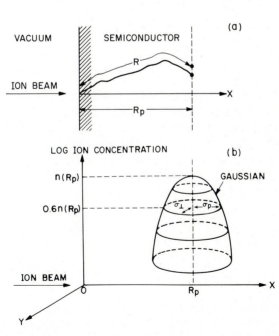

FIGURE 5

Schematic views of the ion range. (a) The total path length R is longer than the projected range R_p. (b) The stopped atom distribution is two-dimensional Gaussian. (*After Sze, Ref. 10.*)

$$(\text{skewness}) \; \gamma = \frac{m_3}{\sigma_p^3}$$

$$(\text{kurtosis}) \; \beta = \frac{m_4}{\sigma_p^4}$$

Qualitatively, skewness measures the asymmetry of the distribution—positive skewness places the peak of the distribution closer to the surface than R_p. Kurtosis measures how flat the top of a distribution is. Gaussian distributions have a skewness of 0 and a kurtosis of 3. In cases where the kurtosis is not available, a universal expression is often used:

$$\beta \cong 2.8 + 2.4\gamma^2 \tag{11}$$

Several different distributions have been used to give a more accurate fit to the moments of an ion distribution than is possible using a Gaussian. The most popular of these is the Pearson family of distributions, used in statistics, which is given by the solutions of

$$\frac{df(s)}{ds} = \frac{(s-a)f(s)}{b_0 + b_1 s + b_2 s^2} \tag{12}$$

where $s = x - R_p$. The Pearson coefficients can be written in terms of the first four moments of the distribution:

$$a = -\gamma\sigma_p(\beta + 3)/A \tag{13a}$$

$$b_0 = -\sigma_p^2(4\beta - 3\gamma^2)/A \tag{13b}$$

$$b_1 = a \tag{13c}$$

$$b_2 = -(2\beta - 3\gamma^2 - 6)/A \tag{13d}$$

where $A = 10\beta - 12\gamma^2 - 18$. There are seven different types of solution to Pearson's equation, depending on the values of the coefficients. For implantation profiles, the most commonly used is Pearson IV, which applies when the coefficients satisfy $0 < b_1^2/4b_0 b_2 < 1$. An advantage of this solution is that it is defined for $-\infty < s < \infty$, whereas other types have a restricted range. The solution has a single maximum at $x = R_p + a$ and decays smoothly to zero on both sides. If Eq. 11 is used for the kurtosis, this condition will always be satisfied. Integrating Eq. 12 for a Pearson IV solution gives

$$\ln\left[\frac{f(s)}{f_0}\right] = \frac{1}{2b_2}\ln(b_0 + b_1 s + b_2 s^2) - \frac{b_1/b_2 + 2b_1}{\sqrt{4b_0 b_2 - b_1^2}}\tan^{-1}\left(\frac{2b_2 s + b_1}{\sqrt{4b_0 b_2 - b_1^2}}\right) \tag{14}$$

Figure 6 compares measured boron profiles[11] with fitted distributions for energies between 30 keV and 800 keV. Ions were implanted into fine-grained polycrystalline silicon to avoid channeling effects (described in Section 8.2.4). As the energy increases, the profiles become more and more negatively skewed

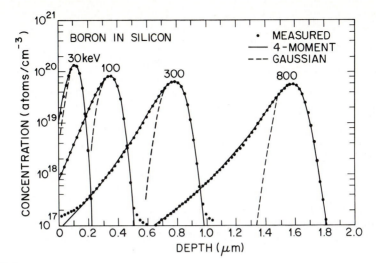

FIGURE 6
Boron implanted atom distributions, comparing measured data points with four-moment (Pearson IV) and Gaussian fitted distributions. The boron was implanted into amorphous silicon without annealing. (*After Hofker et al., Ref. 11.*)

and Gaussian distributions are not able to give a good fit to the measurements. For heavier ions, the profiles at low energies have positive skew which decreases more slowly but can also become negative for sufficiently high energies. Figure 7 shows the first four moments for some common ions in silicon, obtained from Monte Carlo calculations.[12] Range and standard deviation data[13,14,15] for boron in a variety of materials is given in Table 1. There are differences on the order of 10% between published range values, due to differences in stopping powers and to the approximations used.

Problem
What is the position of the peak of a 150 keV boron implant into silicon?

Solution
From Fig. 7 we find the following range parameters: $R_p = 4300$ Å, $\sigma_p = 800$ Å, $\gamma = -1.4$, and $\beta = 6$. The distribution is significantly skewed, so we should use a Pearson distribution, which has the peak at $x = R_p + a$. From Eq. 13,

$$A = 18.4 \qquad a = 550 \text{ Å}$$

so the peak is at 4850 Å.

If the target is composed of multiple layers of different materials, it is still possible to estimate the ion distribution from the moments of the ion in each material separately. The simplest method is to consider a modified target where each layer is scaled in thickness according to the relative ranges of the ion in that layer and the substrate, $t_{\text{layer}}^{\text{scaled}} = t_{\text{layer}}^{\text{real}} \times R_p^{\text{substrate}}/R_p^{\text{layer}}$. This makes the implantation profile in the scaled target the same as the implantation profile in an unlayered

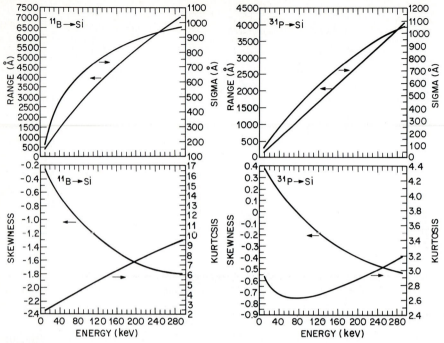

FIGURE 7

Monte Carlo – calculated moments of implanted atom distributions for common dopants in silicon. (*After Petersen, Fichtner, and Grosse, Ref. 12.*)

substrate target. The true implantation profile is obtained by reversing the scaling, multiplying the concentration at any point in each layer by $R_p^{\text{substrate}}/R_p^{\text{layer}}$. A more accurate method[16] combines scaling with dose matching: suppose there is a single surface layer of thickness t and that the profiles in a thick surface film and in the substrate are given by $n_1(x)$ and $n_2(x)$, respectively. The resulting profile is given by

TABLE 1
Boron ranges in various materials[13,14,15]

100 keV boron implantation				
Material	**Symbol**	**Density (g/cm^3)**	**R_p(Å)**	**σ_p(Å)**
Silicon	Si	2.33	2968	735
Silicon dioxide	SiO$_2$	2.23	3068	666
Silicon nitride	Si$_3$N$_4$	3.45	1883	408
Photoresist AZ111	C$_8$H$_{12}$O	1.37	10569	1202
Titanium	Ti	4.52	2546	951
Titanium silicide	TiSi$_2$	4.04	2154	563
Tungsten	W	19.3	824	618
Tungsten silicide	WSi$_2$	9.86	1440	555

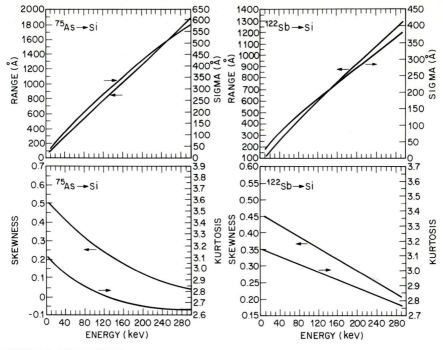

FIGURE 7 (CONTINUED)

$$n(x) = \begin{cases} n_1(x) & 0 \le x \le t \\ \alpha n_2\left[x - t\left(1 - \dfrac{R_{p2}}{R_{p1}}\right)\right] & x \ge t \end{cases} \tag{15}$$

where α is chosen to satisfy $\Phi = \int_{\infty}^{-\infty} n(x)dx$.

Experimentally, it is comparatively easy to measure accurately vertical atom profiles to improve our modeling of implanted depth distributions. Very little progress has been made in measuring lateral atom profiles, so simple Gaussian distributions have been used almost universally. A two-dimensional distribution is written as the product of vertical and lateral distributions

$$n(x, y) = \frac{n_{\text{vert}}(x)}{\sqrt{2\pi}\sigma_\perp}\exp(-y^2/2\sigma_\perp^2) \tag{16}$$

This equation describes the result of implanting at a single point on the surface. To find the result of implanting through a mask window, we must integrate Eq. 16 over the open areas where the ion beam can enter. For a mask with an opening from $y = -a$ to $y = a$, this gives a density $d(x, y)$ of

$$d(x, y) = \int_{-a}^{a} n(x, y - y')dy' = \frac{n(x)}{2}\left[\text{erfc}\left(\frac{y - a}{\sqrt{2}\sigma_\perp}\right) - \text{erfc}\left(\frac{y + a}{\sqrt{2}\sigma_\perp}\right)\right] \tag{17}$$

Figure 8 illustrates this for a 70 keV boron implant through a 1 μm slit

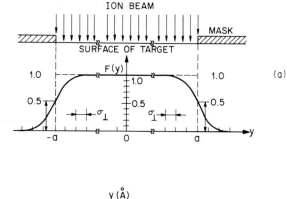

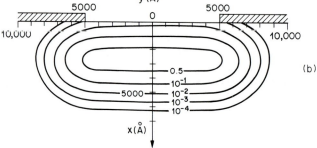

FIGURE 8
Two-dimensional implantation profiles. (a) Fraction of total dose as a function of lateral position for an opaque mask. (b) Equi-concentration contours for a 70 keV boron implant through a 1 μm slit. (*After Furukawa, Matsumura, and Ishiwara, Ref. 17.*)

in a thick mask, showing that ions scatter well outside the area indicated by the masking layer.[17]

A more accurate treatment of two-dimensional ion profiles requires three other factors to be taken into account. First, masking layers are often tapered at the edge rather than perfectly abrupt, so that ions are gradually prevented from entering the silicon. If the masking layer has thickness $m_{mask}(y)$ then it can be considered as an equivalent silicon film of thickness $m_{Si}(y)$ by applying the principles of multilayer target scaling described previously. The superposition integral shifts each surface element according to the scaled mask thickness.

$$d(x, y) = \int_{-\infty}^{\infty} n[x + m_{Si}(y'), y - y']dy' \qquad (18)$$

Figure 9a shows the effect of implanting through a 60° sloping mask edge with a stopping power similar to that of silicon.[18]

Second, more advanced modeling (described in Chapter 10) has shown that the lateral standard deviation $\sigma_{\perp}$ varies with depth. This effect is most significant for high-energy, light-ion profiles such as 200 keV boron in silicon, where the profiles have large negative skewness. In this case, the peak of the lateral distribution is shifted closer to the surface than the projected range[19] by

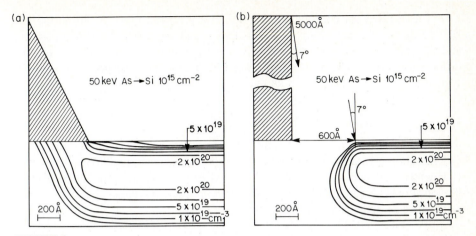

FIGURE 9
Modified masked ion profiles for 50 keV arsenic. Contours show concentration of arsenic after implanting a dose of 10^{15} cm^{-2}. (a) Ions penetrate through the thinner regions of a sloping mask edge, increasing the doping near the corner. (b) A thick mask casts a shadow at the base which is 12% of the mask height for a 7° beam tilt. Note also the variation in lateral standard deviation with depth. (*After Giles, Ref. 18.*)

1500 Å. The effect can also be seen for low-energy arsenic, Fig. 9b, where the lateral peak is pushed deeper.

Finally, we must consider the effect of the 7° beam tilt used to minimize channeling (see Section 8.2.4). This tilt reduces the profile peak depth less than 1% and shifts the peak laterally by 12% of R_p, distances that scale with implantation energy, so the effect has usually been ignored. Layers above the substrate have not been scaled vertically with decreasing line width—indeed, with the development of multilayer metallization, the tendency has been in the opposite direction. Self-aligned processes use the gate as an implantation mask, so a tall gate stack can cast a significant shadow at the edge of the source or drain. VLSI processing is designed to minimize diffusion in order to obtain fine line-widths, so gaps from shadowing may not be covered over. For example, if the gate stack is 0.5 μm high, with vertical sidewalls, then the shadow can be 600 Å long; see Fig. 9b. This will introduce an unexpected series resistance into the device. The length and side of the shadow for a particular case depends on the rotation angle of the wafer relative to the beam tilt. The practical situation is often complicated by the fact that automatic wafer loading may leave the rotation angle of the wafer unknown and variable from wafer to wafer.

The distribution of implanted ions can be calculated from the equations of ion scattering given in Section 8.2.1. Chapter 10 describes two sophisticated methods using these equations that allow the ion distribution and associated effects to be calculated accurately. Here we will summarize briefly a simpler approach that allows the moments of ion distributions to be calculated. The moments method was developed by Lindhard, Scharff, and Schiøtt[20] as an alternative to direct integration of the scattering integral along the ion path, and was refined by Sanders[21] to give moments of the projected range.

Using spherical coordinates, we define a probability $p(r, E, \theta)$ that an ion with energy E will come to rest at a distance r and angle θ to its current position and direction. If we allow the ion to move by a small vector distance $d\mathbf{r}$, the differential scattering cross section from Eq. 6 describes the probability that a collision will occur resulting in an energy loss T. We can therefore write

$$\frac{dp}{d\mathbf{r}} = N \int_0^{T_{max}} \left[p(r, E - T, \theta) - p(r, E, \theta) \right] d\sigma(T) \tag{19}$$

Recurrence relations for the moments of p can be found by expanding p in terms of Legendre polynomials, equating polynomial coefficients, and taking moments. Frequently, reduced variables are introduced to simplify the problem and scale the ion curves into a universal form independent of ion and target species—energy ϵ, distance ρ, and scattering parameter t.

$$\epsilon \equiv \left[\frac{M_2/(M_1 + M_2)}{Z_1 Z_2 q^2 / 4\pi\epsilon_0 a} \right] E \tag{20}$$

$$\rho \equiv N\pi a^2 \gamma x \tag{21}$$

$$t \equiv \frac{\epsilon^2 T}{\gamma E} \tag{22}$$

where a is the screening length, $a \equiv 0.88a_0(Z_1^{2/3} + Z_2^{2/3})^{-1/2}$, γ is the maximum energy transfer factor $\gamma \equiv 4M_1 M_2/(M_1 + M_2)^2$, and N is the number density of target atoms. The scattering parameter t can also be written $t^{1/2} = \epsilon \sin(\theta/2)$, combining together ion energy and ion energy loss and so simplifying the equation for $p(r, E, \theta)$. The results of moments calculations are available in the form of tables for various ion/target combinations. [13,15]

8.2.3 Damage

As each ion travels through the target, it undergoes a series of nuclear collisions. Every time the ion is scattered, a fraction of its energy is transferred to a target atom, which is displaced from its original position. The binding energy of a lattice site is only 10 to 20 eV, so it is easy to transfer enough energy to free an atom from its position and make it travel through the target as a second projectile. Now, both the ion and the displaced target atom travel and cause further displacements, so the energy is spread over many moving particles. Eventually the energy per particle becomes too small and the cascade stops. The result of one incident ion has been the displacement of many target atoms through nuclear scattering. After many ions have been implanted, an initially crystalline target will be so disrupted that it will have changed to a highly disordered state. If the target temperature is high enough, the competing process of self-annealing will occur, repairing some or all of the damage as it is generated. The result is therefore very temperature-dependent.

Because only nuclear scattering damages the target, the amount of damage done depends on the nuclear stopping power, which is a function of energy (Fig.2). For light ions, such as boron in silicon at 100 keV, stopping is mainly electronic. As the boron ions gradually lose energy, nuclear stopping increases and so more damage is done to the silicon lattice. For a medium-mass ion such as silicon in silicon, the same is true at slightly higher energies.

Figure 10 shows cross sections of a silicon target after various doses of silicon have been implanted at 300 keV, imaged by transmission electron microscopy. As the dose increases, a band of damaged silicon appears near the peak of the implanted silicon distribution and expands until, at the highest dose, the entire implanted region has been made amorphous. Figure 11 shows the energy deposited in nuclear collisions obtained from low temperature implantation experiments.[22] The transition from crystalline to amorphous occurs at a threshold of about 10^{24} eV/cm^3. For heavier ions such as arsenic, the results are similar except that nuclear stopping is already significant even at 100 keV, so the damaged region usually extends to the surface.

Once target atoms have been displaced, several things are possible. At low temperatures, the atoms are unable to move far through the lattice so local relaxations take place to accommodate them. Above room temperature, displaced atoms can diffuse to empty lattice sites and so repair damage, or can coalesce with other displaced atoms to form an extended defect. Vacancies that remain behind when an atom is displaced can also diffuse and combine to form extended defects. If the target temperature during implantation is high, perhaps due to the heating effect of the ion beam itself, self-annealing can be so effective that even heavy ions leave little damage. We can define a critical dose as the minimum necessary to amorphize the target.[23] The temperature dependence of the critical amorphization dose is shown in Fig. 12. Heavier ions displace a greater volume of target atoms per ion, so a higher temperature is necessary to allow complete recovery and so prevent amorphization.

Damage can affect the results of subsequent processing steps. Point defects are known to influence strongly diffusion in silicon, but the effects of implant-generated defects on diffusion is still the subject of controversy.[24,25] Damaged oxide etches faster than undamaged oxide because some of the bonds are already

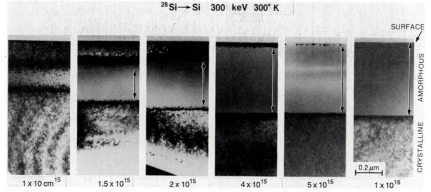

FIGURE 10

Transmission electron micrographs showing damage accumulation for silicon self-implantation as a function of dose. (*After Maszara and Rozgonyi, Ref. 22.*)

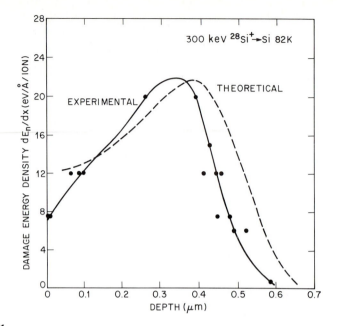

FIGURE 11
Deposited energy for silicon self-implantation. The experimental curve was based on the width of
the amorphous region as a function of dose, scaled to match the theoretical curve. (*After Biersack
and Haggmark, Ref. 7, Maszara and Rozgonyi, Ref. 22.*)

broken, and this can be used to modify the shape of oxidized regions. Dangling
oxide bonds can also act as electron traps in the finished device. Damage to
photoresist used as a mask can break down the organic chains to a carbon-rich
film, which is much harder to remove, making it unsuitable as a mask for high-
dose implants.[26] These process rate changes act in addition to the chemical and
doping effects caused by the implanted atoms themselves.

   One simple measure of the number of displaced atoms, neglecting any
annealing effects, is given by dividing the total energy deposited into nuclear
scattering, E_n, among the possible displaced atoms. If the displacement energy
(the energy needed to displace an atom from its lattice site) is E_d, then a moving
ion can increase the number of moving particles[27] only if it has an energy greater
than $2E_d$. If the moving ion energy is less than this, it can only create a recoil by
losing at least E_d and so being stopped itself. We therefore estimate the number
of displaced atoms as

$$N(E) \cong \frac{E_n}{2E_d} \tag{23}$$

We can find E_n by integrating the area under the curve for $(dE/dx)_{\text{nuclear}}$ (Fig.
11). The shape of the deposited energy curve can also be found using a moments
method in a similar way to the implanted ion profile.[28]

   When the energy deposited along an ion track is high, a simple cascade

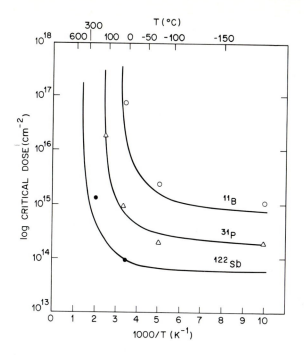

FIGURE 12

A plot of the critical dose necessary to make a continuous amorphous layer as a function of temperature. (*After Morehead and Crowder, Ref. 23.*)

theory is not adequate to explain the number of displaced atoms observed. The energy-per-lattice atom is much greater than the heat of melting, and the ion track causes what is known as a thermal spike.[29] This superheated region expands as it cools, resulting in an amorphous volume much larger than the original collision cascade. The number of displacements can be 10 times greater than predicted by Eq. 23.

8.2.4 Channeling

So far we have considered each collision that scatters the incident ions to be independent. Many targets are crystalline with regular arrangements of atoms. For ions moving in certain directions, the atom rows or planes line up so that there are long-range open spaces through which the ions can travel without significant scattering. Ions are steered down these channels by glancing collisions with the atom rows or planes, extending the final ion distribution deeper into the target. Figure 13a shows three ions entering a simple cubic lattice. Ion A is well aligned with a channel and so suffers only glancing collisions with the walls as it travels far into the lattice. Ion B is scattered into a channel after a short distance, perhaps because of a lattice imperfection. Ion C is not channeled and has random collisions with lattice atoms. Experimentally, the angular width of a channel can be found by measuring the number of ions back-scattered from the lattice as a function of angle for a well collimated beam (Fig. 13b). The result of channeling is to add

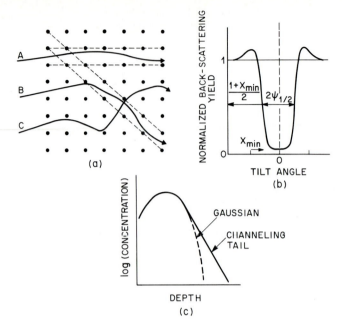

FIGURE 13
Schematic views of channeling. (a) Ion paths through a cubic lattice, showing channeled and nonchanneled cases. (b) Back-scattering yield around a channeling direction. Yield is a minimum when beam is well aligned with a channel. (c) The effect of channeling is to add a tail to the atom distribution.

a tail to the implanted distribution which, for silicon, can often be approximated by an exponential with decay length around 0.1 μm (Fig. 13c).

Channeling is characterized by a critical angle Ψ_1, which is the maximum angle between ion and channel for a glancing collision to occur. If we neglect thermal vibrations then for ions in the keV energy range this is given by[30]

$$\Psi_1 = 9.73\sqrt{\frac{Z_1 Z_2}{Ed}} \text{ (degrees)} \qquad (24)$$

where d is the atomic spacing along the ion direction (Å) and E is the ion energy (keV). This shows that channeling is more likely for heavier ions and lower energies. A stable channeled trajectory requires multiple glancing collisions, so the angular half-width of the channeled ion distribution, $\Psi_{1/2}$ in Fig. 13(b), is less than the critical angle. For axial channels, where a is the screening length defined in Section 8.2.2.

$$\Psi_{1/2} = 7.57\sqrt{\frac{a}{d}\Psi_1} \text{ (degrees)} \qquad (25)$$

The half-widths for planar channels are generally smaller than for axial

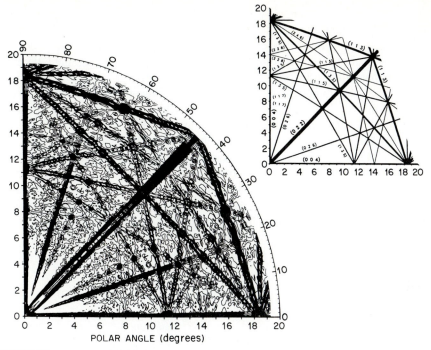

FIGURE 14

Back-scattering yield contour map for 1 MeV helium near the silicon [100] axis. Contours represent 2% intervals of normalized back-scattering yield. The inset shows indices of major planes. (*After Ziegler, Ref. 31.*)

channels, but planes span a greater range of angles than axes so the total channel acceptance area for planes can be greater. Figure 14 shows a channeling contour map for 1 MeV helium in silicon, where the contours represent 2% intervals of channeling probability.[31] This is a two-dimensional form of Fig. 13b. Planar and axial channels can clearly be seen for all low-order crystallographic directions. Notice that many of the channeling planes seem to be better described as dense strings of axial channels. To avoid channeling, most implantation machines tilt the wafer by 7°. Figure 14 shows that it is also important to choose an appropriate rotation angle, avoiding major planes. For heavier ions at lower energies, Eq. 24 indicates that channel acceptance angles will be greater and more of the back-scattering map will be occupied by channels.

For an ion inside a channel, the major energy loss is due to electronic scattering. There is little nuclear stopping because the ion is never scattered by a large angle. The electronic stopping power for a well-channeled ion is less than the value for an ion traveling in a random path because the electron density at the center of a channel is low. The reduction in stopping power is greatest for heavy ions, where nuclear scattering dominates random motion, and can increase the maximum range by factors of 2 to 4. Ions can be scattered out of channels by thermal vibrations, lattice defects, and foreign atoms. In practice, channeling is often limited by damage to the crystal caused by nuclear stopping of unchanneled ions. A preceding implant of silicon to amorphize the lattice will eliminate channeling effects.

At first, it might seem that the channeling effect provides an easy means for placing ions deep inside the target while minimizing lattice damage. Unfortunately, there are too many practical problems to make this a useful technique. The channeled ion profile is sensitive to changes on the order of 1° in wafer tilt and beam divergence, and ions are scattered by amorphous surface films and residual damage from previous processing steps. This sensitivity is illustrated in Fig. 15, which shows that measured ion profiles are a function of tilt angle for a well collimated ion beam.[32] Modeling of channeled ion distributions from first principles is difficult; the simplest solution is to add an exponential tail to the amorphous profile based on experimental measurements. Alternatively, the moments of a Pearson IV distribution can be adjusted to include channeling in some cases.[33]

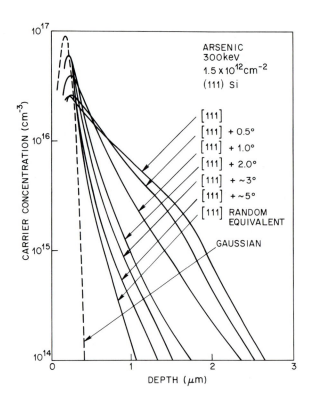

FIGURE 15

Electrically active arsenic distribution as a function of beam angle, tilted at 18° to the (110) plane. The "random equivalent" case is the usual 7° tilt used to avoid channeling, and still shows significant differences from a Gaussian. (*After Wilson et al., Ref. 32.*)

8.2.5 Recoils

In considering damage, we were concerned with the number of ions displaced from lattice sites. If the target is composed of two layers, atoms will be displaced from one layer into the next. Depending on the materials involved, this may be a useful technique or a harmful side effect. The most common case, implantation through a thin oxide layer into silicon, is shown in Fig. 16. Oxygen atoms are displaced from the oxide into silicon, giving a profile composed of two roughly exponential regions.[34,35] Close to the interface, the displacement cascades push many very-low-energy oxygen atoms just across the interface. Beyond this, primary oxygen recoils form a deeper exponential tail, which is similar in appearance to some channeling tails. The oxygen concentration near the interface exceeds the implanted arsenic concentration. Such high doses of oxygen can degrade carrier mobility and introduce deep-level traps.

Recoil mixing can be used positively for introducing dopant atoms that cannot readily be made into a source for implantation machines but can be deposited as a thin film onto a silicon substrate. Implanting silicon through the film will push a dopant tail into silicon. Self-implantation can also be used to

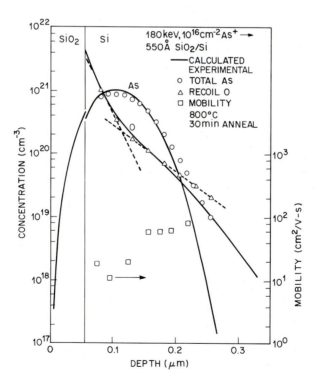

FIGURE 16

Oxygen recoils from the implantation of arsenic through an oxide layer. Solid lines are from Boltzmann calculations. Dashed lines indicate the two exponential regions of the recoil distribution. (*After Hirao et al., Ref. 34, and Christel, Gibbons, and Mylroie, Ref. 35.*)

break up thin contaminant layers under deposited films, prior to further processing such as the formation of a silicide.

8.3   IMPLANTATION EQUIPMENT

This section reviews the practical aspects of ion implantation. For a detailed discussion, see the Proceedings of the International Conferences on Ion Implantation Equipment and Techniques.[36]

8.3.1   Basic Layout

The basic requirement for an ion-implantation system is to deliver a beam of ions of a particular type and energy to the surface of a silicon wafer. Figure 17 shows a schematic view of a medium-energy ion implanter. Following the ion path, we begin on the left-hand side with the high-voltage enclosure containing many of the system components. A gas source feeds a small quantity of source gas such as BF_3 into the ion source where a heated filament causes the molecules to break up into charged fragments. This ion plasma contains the desired ion together with many other species from other fragments and contamination. An extraction voltage, around 20 kV, causes the charged ions to move out of the ion source into the analyzer. The pressure in the remainder of the machine is kept below 10^{-6} Torr to minimize ion scattering by gas molecules. The magnetic field of the analyzer is chosen such that only ions with the desired charge to mass ratio can travel through without being blocked by the analyzer walls. Surviving ions continue to the acceleration tube, where they are accelerated to the implantation energy as they move from high voltage to ground. Apertures ensure that the beam is well collimated. The beam is then scanned over the surface of the wafer using electrostatic deflection plates. The wafer is offset slightly from the axis of the acceleration tube so that ions neutralized during their travel will not be deflected

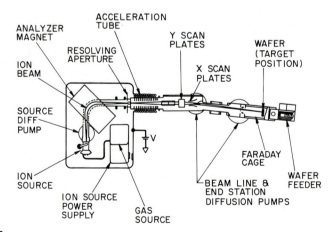

FIGURE 17
Schematic diagram of a typical commercial ion-implantation machine. (*After Varian-Extrion DF-3000 brochure.*)

onto the wafer. A commercial ion implanter is typically 6 m long, 3 m wide, and 2 m high, consumes 45 kW of power, and can process 200 wafers per hour (dose 10^{15} ions/cm^2, 100 mm wafers).

Three quantities define an ion implantation step: the ion type, energy, and dose. Given an appropriate source gas, the ion type is determined by the magnetic field of the analyzing magnet. In a magnetic field of strength B, ions of charge Q move in a circle of radius R, where

$$RB = \frac{M_1 v}{Q} = \sqrt{\frac{2M_1 V}{Q}} \tag{26}$$

where v is the ion velocity and V is the source extraction voltage. The magnetic field is adjusted so that R corresponds to the physical radius of the magnet for the desired ion. It is possible for other ions to be accepted if they have a similar value for M_1/Q, but since the source provides ions with some spread of energy, increasing the resolution of the analyzer too far will decrease the beam current. The selected ions are accelerated to the implantation energy by the voltage applied to the acceleration tube.

The total number of ions entering the target is called the dose. If the current in the ion beam is I, then for a beam swept over an area A, the dose Φ is given by

$$\Phi = \frac{1}{QA} \int I \, dt \tag{27}$$

where the integral is over time t. Completing the circuit between target and ion source allows the current to be measured. For an accurate current reading, care must be taken to recapture secondary electrons emitted from the target by the incident ions. A Faraday cage around the target, at a small positive bias voltage, collects this charge so that it can be included. Silicon wafers frequently have patterned surface layers of silicon dioxide, which is a good insulator. Implantation can charge up insulated regions of the surface high enough for dielectric breakdown to occur, which damages the materials. If the wafer is not well grounded, charging of the whole wafer can distort the ion beam. To avoid these effects, a low-energy electron beam can be directed onto the target surface during implantation. The electrons are drawn to charging regions where they neutralize the charge buildup.

Problem

Suppose you are using an ion implanter with a 30 kV extraction voltage, a 0.5 m magnet radius, and a source current of 100 μA for P^+ ions. Calculate values for the analyzer magnetic field, acceleration tube voltage, and exposure time so that the implanted profile has a peak concentration of 10^{20} cm^{-3} at a depth of 1000 Å for a 100 mm wafer. (Use Fig. 7 and a Gaussian range distribution.)

Solution

Using Eq. 26 for the magnetic field,

$$B = \frac{\sqrt{2M_1 V/Q}}{R} = 0.23 \ T$$

For a Gaussian, the peak coincides with the projected range. From Fig. 7, $R_p = 1000 \ \text{Å}$ when $E = 70 \ \text{keV}$, and this also gives $\sigma_p = 380 \ \text{Å}$. Using Eq. 9 for the peak concentration gives

$$\Phi = \frac{(\text{peak}) \ \sigma_p}{0.4} = 9.5 \times 10^{14} \ \text{cm}^{-2}$$

This would require an exposure time of 120 s using a 100 μA beam. The source extraction voltage is 30 keV, so the acceleration tube should be set at 40 keV.

Any implantation machine has design limits to its energy range. The minimum implantation energy is usually set by the extraction voltage, which cannot be reduced too far without drastically reducing beam current. Some special machines can operate in a deceleration mode, in which ions are extracted with a normal voltage but then slowed down in a reverse-biased "acceleration" tube resulting in energies as low as 5 keV. A more common technique uses implants of molecular ions containing the required dopant. For example, BF_2^+ ions could be implanted at 30 keV. When the molecule hits the target surface it immediately breaks up into its components, and the energy is divided according to the relative masses. In this case, we would have one 7 keV boron atom and two 11 keV fluorine atoms. The resulting profiles are the same as if 7 keV boron and 11 keV fluorine had been directly implanted, because the binding energy of the molecule is negligible compared to the implantation energy. If channeling is significant, the profiles will be different from that of 7 keV boron alone because the addition of fluorine will increase the lattice damage and reduce channeling. The fluorine distribution is shallower than the boron distribution, so ions can still enter channels when they have traveled past the fluorine peak. Fluorine has only a small effect on annealing and on electrical mobility in the final device.

The maximum implantation energy is set by the design of the high-voltage equipment. The only way to circumvent this is to implant a multiply-charged ion such as B^{++}. This ion would receive twice the energy of B^+ from the same accelerating potential, effectively doubling the energy of the machine. The price paid is a reduced beam current since the number of B^{++} ions in the source plasma is much smaller.

8.3.2   High-current and High-energy Machines

Any increase in beam current directly improves implanter throughput by reducing the time for each implantation step. High-current machines can deliver at least 5 mA ion-beam currents, so that 150 mm wafers can be given a dose of 10^{15} cm^{-2} in only 6 seconds per wafer. Multiple extraction electrodes and higher voltages are used to achieve this, combining acceleration and deceleration modes of operation as necessary to get a final beam of suitable energy. Much effort is also spent in the design of wafer-handling equipment so that as little time as possible is wasted in loading and unloading wafers.

A further consequence of high-beam currents is the need to use a different scanning system. Electrostatic scanning of high-current beams disrupts space charge neutrality and leads to beam "blow-up,"[11] so a mechanical scanning system is usually used. In this case, the wafer is scanned past a stationary beam. This method has the added advantage of keeping the same beam angle across the whole wafer, whereas an electrostatic system can vary by $\pm 2°$ for 100 mm wafers, as shown in Fig. 18. This advantage is usually canceled out by new requirements for the wafer holder. In order to conduct away the heat generated by the high-beam current, wafer stages are cooled, for example with liquid nitrogen. The stage flexes the wafer by $2°$ at the center in order to achieve good thermal contact.

The development of the idea of buried conductors and insulators (see Section 8.6) has led to a demand for implanters that can operate around 1 MeV. This introduces a new set of problems that must be overcome. At about 400 keV, the conventional design described above can no longer be used because of electrical breakdown of the air around the high-voltage equipment. One solution is to enclose the entire high-voltage section in a pressurized vessel containing sulphur hexafluoride at about 7 atmospheres, which increases the breakdown voltage by a factor of 20. The basic layout is unchanged, but routine tasks such as changing source gas or replacing a filament become more difficult. An alternative solution is to use a tandem accelerator arrangement,[37] as shown in Fig. 19. Here, negatively charged ions are injected into the accelerator. They are accelerated from ground to half of the desired voltage, then passed through a stripper canal where collisions with gas molecules change them to a positive charge. They are accelerated again as they move from high voltage to ground in the second acceleration section. High-pressure insulating gas is still required in the high-voltage sections, but the maximum voltage is only half of the conventional value and the pressurized section contains only simple components. Most of the equipment remains at ground potential. One disadvantage is that the stripper canal changes only 50% of the B^- into B^+. The remainder emerges as B^-, B^0, and B^{++}. At a different stripper gas pressure, B^{++} ions can be generated with an efficiency up to 50%, further increasing the effective energy.

High-energy implants frequently require wafer stages heated up to 600°C,

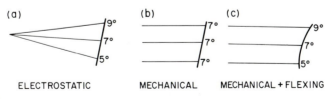

(a) ELECTROSTATIC (b) MECHANICAL (c) MECHANICAL + FLEXING

FIGURE 18
Schematic diagram of wafer scanning systems, with angles exaggerated for emphasis. (a) Electrostatic scanning causes a $\pm 2°$ variation across the wafer. (b) Mechanical scanning moves the wafer past a stationary beam, and causes no variation across the wafer. (c) Heat-sinking wafer holders flex the wafer to improve thermal contact.

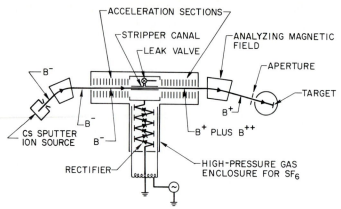

FIGURE 19
Schematic diagram of a tandem ion-implantation machine. (*After Purser, et al., Ref. 37.*)

so that self-annealing during implantation minimizes damage in the surface layer. Mechanical scanning is used because of the difficulty of electrostatically scanning a high-energy beam.

8.3.3 Uniformity

Progress in VLSI processing has resulted in a continuing trend to larger wafers, from 75 mm to 150 mm and beyond. This makes the task of uniformly implanting a wafer increasingly difficult. Nonuniformity can easily be seen from sheet resistivity maps across the wafer (see Chapter 7). Figures 20–22 show maps for 100 mm wafers made under various conditions to illustrate some common effects.[38,39] Contour lines mark 1% changes in sheet resistance. In Fig. 20, the 2° flex caused by the wafer holder leads to a variation in planar channeling across the wafer as the beam angle changes. As the dose is increased, lattice damage prevents channeling and the sheet resistance becomes much more uniform. Figure 21 shows the dependence of planar channeling on wafer rotation angle for 50 keV boron.

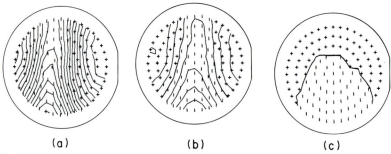

(a) (b) (c)

FIGURE 20
Sheet resistance maps: 50 keV phosphorus as a function of dose. (a) 5×10^{13} cm^{-2} (b) 2×10^{14} cm^{-2} (c) 2×10^{15} cm^{-2}. (*After Current, Ref. 38.*)

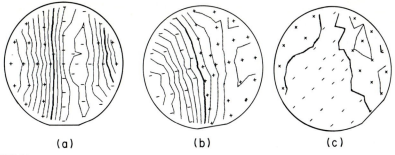

FIGURE 21
Sheet resistance maps: 50 keV boron as a function of rotation angle (dose $7 \times 10^{13}\,\mathrm{cm}^{-2}$, mechanical scan, $7°$ tilt, $2°$ flex). (a) $0°$ twist (b) $10°$ twist (c) $45°$ twist. (*After Current, Ref. 38.*)

The rotation angle is defined as the angle between the wafer flat and the direction of beam tilt. A $45°$ rotation is necessary to minimize channeling by the (220) planes. If the energy is lowered to 25 keV, channel acceptance angles are so large that even this is not effective. Figure 22a shows the case of implantation into an untilted wafer. Axial channeling is now dominant and the 2% wafer flex leads to a 15% variation in sheet resistance. The importance of a low-energy electron flood gun during high current implantation is dramatically shown in Fig. 22b and 22c. These wafers have no surface oxide, but the electrical connection to ground through the wafer holder is poor. Without a flood gun, the wafer charges up rapidly to a positive potential. Electrons are attracted out of the ion beam, disturbing the space-charge neutrality and causing the ion beam to expand so that the center of the wafer receives a lower dose than intended. The use of a flood gun effectively eliminates this effect.

8.3.4 Contamination

Ion implantation is basically a clean process because beam analysis separates contaminant ions from the beam before they hit the target. There are still several

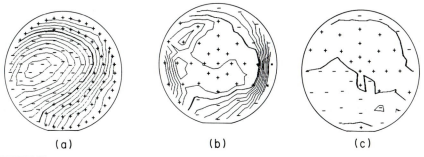

FIGURE 22
Sheet resistance maps: (a) 80 keV boron implanted at $0°$ tilt to show axial channeling. (b) 10 mA phosphorus implantation without an electron flood gun. (c) Same implant with an electron flood gun. (*After Current, Ref. 39.*)

sources of contamination possible near the end of the beam line, which can result in contaminant doses up to 10% of the intended ion dose. The primary source is metal atoms knocked from chamber walls, wafer holder and clips, masking apertures, and so on. Various hydrocarbons from pump oils may deposit onto the surface of the wafer and be recoiled deeper by the ion beam. After many implants, the internal surfaces of the machine become saturated with dopant. If the source gas is changed, subsequent implants can be cross-contaminated with old source atoms sputtered from the walls. The only solution to these problems is careful design to minimize beam contact with surfaces after the analyzer, particularly in the area of the wafer holder. Finally, particulates can drop onto the wafer during loading, preventing some areas from being properly implanted.

8.3.5   Focused Ion Beams

The implantation systems described so far have used a relatively broad beam scanned uniformly over the whole wafer surface. An alternative approach is to focus the beam to a small spot and directly write the desired dopant patterns onto the wafer surface without the use of a masking layer. This approach has been investigated for many years as a lithographic tool, where the heavy mass of an ion reduces scattering and improves resolution as compared to a focused electron beam (see Chapter 4). The development of high-brightness liquid metal sources over the past several years has opened the way for many new applications such as deposition, etching, and localized dopant implantation. As with all direct-write approaches, its usefulness is limited by the relatively long time it would take to pattern a whole wafer.

The first liquid metal sources used gallium, and since then alloy sources have been used to produce beams for most of the common dopant ions. For low-beam currents of around 10 pA, a 200 keV ion beam can be focused down to a 100 Å spot. This size increases with increasing current and with decreasing beam energy to about 1 μm for a 200 pA beam at 50 keV. As the beam hits the target, secondary electrons are emitted, which can be imaged to monitor the effects of the beam in real time. For etching, the target area contains a reactive gas such as chlorine. Beam damage enhances the local etch rate and volatile reaction products diffuse away into the chamber. For metal deposition, the target area contains a metallic compound such as trimethyl aluminum. In this case, the reaction product is not volatile and a layer of aluminum mixed with carbon accumulates. Opaque films can also be formed by ion damage of photoresist. These last two processes, together with the imaging capabilities of the method, have made possible the repair of both opaque and transparent defects on masks for lithography (see Chapter 4 for more detail). Maskless dopant implantation generally has the same results as conventional implantation. One difference is that the much higher beam current density increases local damage for light ions because there is not enough time between incident ions for point defects to diffuse and annihilate.

8.4 ANNEALING

Introducing dopant atoms into a semiconductor is only the first step in changing its electrical properties. As discussed in Section 8.2.3, implantation damages the target and displaces many atoms for each implanted ion. The electrical behavior after implantation is dominated by deep-level electron and hole traps, which capture carriers and make the resistivity high. Annealing is required to repair lattice damage and put dopant atoms on substitutional sites where they will be electrically active. The success of annealing is often measured in terms of the fraction of the dopant that is electrically active, as found experimentally using a Hall effect technique. The Hall effect measures an average effective doping, which is an integral over local doping densities and local mobilities evaluated per unit of surface area.

$$N_{\text{Hall}} = \frac{\left(\int_0^{x_j} \mu n \, dx\right)^2}{\int_0^{x_j} \mu^2 n dx} \tag{28}$$

where μ is the mobility, n is the number of carriers, and x_j is the junction depth. If the mobility is not a strong function of depth, N_{Hall} measures the total number of electrically active dopant atoms. If annealing activates all of the implanted atoms, this will be equal to the dose Φ.

For VLSI, the challenge in annealing is not simply to repair damage and activate dopant, which any long, high-temperature anneal will achieve, but to do so while minimizing diffusion so that shallow implants remain shallow. This has motivated much recent work in rapid thermal annealing (RTA), where annealing times are on the order of seconds. First, we will describe the properties of more conventional furnace annealing, where times are on the order of minutes.

8.4.1 Furnace Annealing

Annealing characteristics depend on the dopant type and the dose involved. There is a clear division between cases where the silicon has been made amorphous and where it has been only partially disordered. For amorphous silicon, regrowth is by solid phase epitaxy (SPE). The amorphous/crystalline interface moves towards the surface at a fixed velocity that depends on temperature, doping, and crystal orientation.[40] Rates for undoped silicon are shown in Fig. 23. Note that the activation energy for SPE is 2.3 eV, indicating that the process involves bond breaking at the interface. The presence of impurities such as O, C, N, or Ar slows down or disorganizes the regrowth, presumably because the impurities bind to broken silicon bonds. Impurities such as B, P, or As increase the growth rate (by a factor of 10 for concentrations of 10^{20} cm^{-3}), presumably because substitutional impurities weaken bonds and increase the likelihood of broken bonds.[41]

If the implantation conditions were not sufficient to create an amorphous layer then lattice repair occurs by the generation and diffusion of point defects. This process has an activaation energy of about 5 eV and requires temperatures

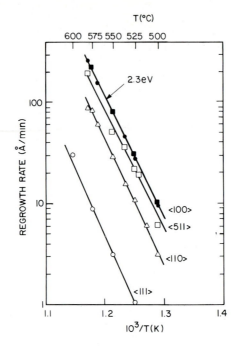

FIGURE 23
The solid-phase epitaxial growth rate of amorphous silicon as a function of temperature for various crystal orientations. (*After Csepregi et al., Ref. 40.*)

on the order of 900°C to remove all the defects.[42] It is therefore easier in many cases to repair a fully amorphized layer than a partially damaged one.

Annealing an amorphous layer at high temperatures causes competition between SPE and local diffusive rearrangement that can lead to polysilicon formation, so it is best to precede the high-temperature step by low-temperature regrowth. High-temperature defect diffusion can then repair extended defects remaining after SPE. Annealing a partially damaged layer at low temperatures can actually impair the process of lattice reconstruction because stable extended defects, such as dislocation loops, can be formed, which then require temperatures up to 1000°C to be removed. Even when implantation does form an amorphous layer, there must be a transition region of partially damaged material at the tail of the implanted distribution. This region is therefore susceptible to stable defect formation although, since there is little dopant in the tail, the activated dopant fraction is not greatly affected. Finally, if the surface layer remains crystalline, SPE can occur from both sides toward the middle and leave a string of misfit dislocations at the center due to mismatch of the layers.

The result of incomplete annealing is a reduction in the fraction of active dopant. This is most apparent around the peak of the dopant distribution, where damage is greatest. Figure 24 shows the result of annealing a boron implant at two different temperatures for a limited time.[11] The implant was not sufficient to amorphize the silicon, and annealing at 800°C leaves much of the dopant around

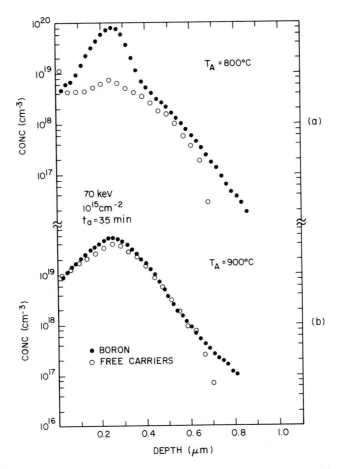

FIGURE 24
Concentration profiles of boron atoms (SIMS—solid circles) and corresponding free carrier concentrations (Hall data—open circles). (*After Hofker et al., Ref. 11.*)

the peak electrically inactive. A 900°C anneal is sufficient to achieve almost complete activation of the boron.

Annealing characteristics can be summarized by isochronal annealing plots, showing the fraction of activated dopant as a function of temperature for a fixed annealing time. As we would expect, these divide into two groups according to whether or not an amorphous layer is formed. Figure 25 shows the case of a partially damaged layer—room temperature implantation of boron into silicon. At low temperatures (region I), point defects dominate the free carrier concentration. As the temperature is increased, these defects can diffuse and recombine. Substitutional boron is pushed off lattice sites by this, but the net carrier concentration increases because so many traps anneal out. Above 500°C (region II), extended defects are formed that further reduce the number of substitutional boron atoms and cause a net decrease in free carriers. This is known as reverse annealing. Above 600°C (region III), the fraction of substitutional boron

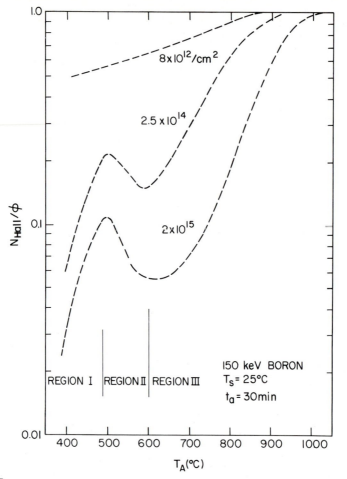

FIGURE 25
Isochronal annealing of boron. The fraction of activated dopant is plotted against anneal temperature for different implant doses. (*After Seidel and MacRae, Ref. 42.*)

atoms gradually increases as point defect generation and migration allows precipitates and dislocations to dissolve. The principal disadvantage of increasing the annealing temperature is that the implanted profile begins to broaden due to diffusion. For sufficiently low doses, (8×10^{12} ions/cm^2 in the figure), so little damage is created that dislocations never form and complete annealing occurs rapidly.

At sufficiently low doses, the annealing behavior of phosphorus is similar to that of boron because no amorphous layer is formed. This is shown by the dashed lines in Fig. 26, where increasing dose moves from left to right.[43] If the substrate is heated to 200°C during implantation, phosphorus will even show reverse annealing, just as for boron.[44] As soon as the silicon is made amorphous, a dramatic change is seen. The solid lines in Fig. 26 show high electrical activity

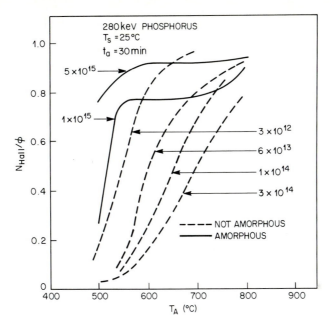

FIGURE 26
Isochronal annealing of phosphorus. The fraction of activated dopant is plotted against anneal temperature for different implant doses. Dashed lines are for partially damaging implants, solid lines for amorphizing implants. (*After Crowder and Morehead, Ref. 43.*)

at temperatures as low as 600°C, corresponding to the temperature for SPE. Substitutional dopants are incorporated into the lattice as it regrows, giving high electrical activity at low annealing temperatures. As would be expected, similar annealing characteristics are seen for boron if the substrate is made amorphous.

8.4.2 Rapid Thermal Annealing

Our goal during annealing is to repair lattice damage, a process with an activation energy of 5 eV, while minimizing diffusion, which has an activation energy in the range 3 to 4 eV. Because of these energy differences, at sufficiently high temperature, repair is faster than diffusion. Furnace annealing is capable of supplying high temperatures but the practical steps required to insert and remove wafers without stressing them lead to a minimum furnace anneal time of about 15 minutes. This is much longer than required to remove damage at high temperatures and so allows unnecessary diffusion. Rapid thermal annealing (RTA) is a term that covers various methods of heating wafers for periods from 100 seconds down to nanoseconds, allowing repair of damage with minimal diffusion.

RTA can be divided into three classes: adiabatic, thermal flux, and isothermal annealing.[45] In adiabatic annealing, the heating time is so short (less than 10^{-7} s) that only a thin surface film is affected. A high-energy laser pulse is used to melt the surface to a depth of less than 1μm, and the surface recrystallizes by liquid phase epitaxy with no memory of the previous damage. Dopant diffusion in the liquid state is very fast so the final profile is roughly rectangular, extending from the surface to the melt depth. By adjusting the pulse time and energy, shallow junctions can be obtained, but it is not possible to preserve other doping profiles or surface films, so the technique is not generally applicable to VLSI.

Thermal flux annealing occurs on time scales between 10^{-7} and 1 s, where heating from one side of the wafer with a laser, electron beam, or flash lamp gives a temperature gradient across the wafer thickness. Generally, the surface is not melted but surface damage can still be repaired by SPE before any diffusion has time to occur. Figure 27 shows an arsenic profile annealed with a scanned laser beam.[46] Almost complete electrical activation is obtained without diffusion. Unfortunately, the rapid quenching from high temperatures leaves many point defects which may condense to form dislocations, degrading the minority carrier lifetime of the material.

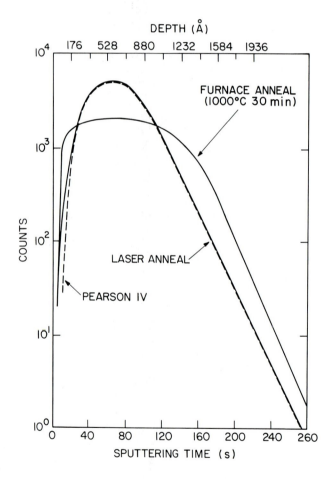

FIGURE 27
Comparison of annealed profiles using furnace and scanned-laser RTA methods. Laser annealing activates the dopant without significant diffusion. (*After Gat et al., Ref. 46.*)

Isothermal annealing covers heating processes longer than 1 s, including furnace steps. Rapid isothermal annealing uses tungsten-halogen lamps or graphite resistive strips to heat the wafer from one or both sides, as shown in Fig. 28. This offers significant advantages for VLSI processing because good activation can be obtained with less diffusion than with furnace annealing. The process is also cleaner because the walls of the chamber remain cold. Figure 29 shows annealing of very-high-concentration boron under various conditions. [47] Furnace annealing for 30 minutes at 1000°C leaves some dopant trapped by defects near the profile peak. A similar state occurs with rapid thermal annealing for 10 seconds at 1100°C, but with much less diffusion. Increasing the RTA time to 30 seconds removes the damage trapping but gives a final profile no deeper than the furnace step, which did not repair the damage.

RTA heating occurs because photons are absorbed by free carriers in the silicon, which then transfer their energy to the lattice. The heating rate depends on the number of carriers, which is a function of temperature and doping, and on surface emissivity. This makes temperature measurement difficult because the probe must have absorption characteristics similar to the wafers being processed. Small thermocouples and optical pyrometers are often used. Uncertainty about wafer temperature has led to controversy about the existence of a transient enhanced diffusion effect (TED), where the effective diffusion rate for some dopants over the first few seconds would be much greater than usual. In a recent review, Celler and Seidel conclude that TED of boron is well established but for arsenic the effect is small or nonexistent. [48] The cause of boron TED is also disputed. Fair attributes it to the diffusion of point defects away from the damage peak left by ion implantation, so enhancing defect-assisted diffusion in the profile tail. [24] Seidel and others attribute it to boron atoms that end up in interstitial positions in the tail after channeling, and so diffuse more rapidly until they fall into substitutional

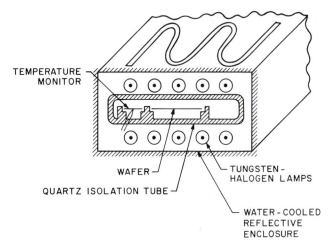

TEMPERATURE MONITOR

WAFER

QUARTZ ISOLATION TUBE

TUNGSTEN-HALOGEN LAMPS

WATER-COOLED REFLECTIVE ENCLOSURE

FIGURE 28
Schematic diagram of an isothermal rapid annealing system. (*After AG associates operation manual for 210M.*)

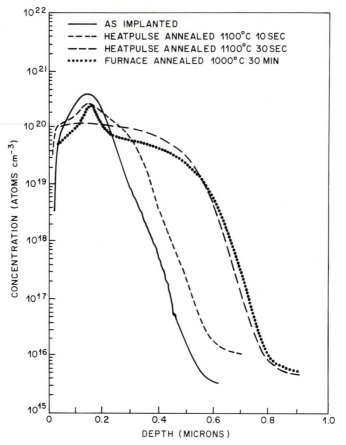

FIGURE 29
Boron profiles implanted at 35 keV showing the removal of precipitation effects in the highly damaged region using RTA. (*After Sadana, Shatas, and Gat, Ref. 47.*)

sites.[25] Enhancement factors for boron can be as great as 100, depending on the conditions involved.

8.5 SHALLOW JUNCTIONS

In scaling horizontally to micron and submicron dimensions, it is important also to scale dopant profiles vertically. We need to produce junction depths less than 1000 Å, including diffusion during dopant activation and subsequent processing steps. Modern device structures, such as the lightly-doped drain (LDD), require precise control of dopant distributions vertically and laterally on a very fine scale. This section considers the techniques that are available to achieve these goals.

Low energy implantation. For VLSI CMOS, we need to form shallow n^+ and p^+ layers. Arsenic is heavy enough to form shallow n^+ layers with implantation energies within the reach of standard machines. Figure 7 shows that, even at 75 keV, the range of arsenic in silicon is only 500 Å. At the same energy, boron has a range of 2500 Å. As described in Section 8.3.1, the effective boron energy can be reduced by implanting the molecular ion BF_2^+, which is equivalent to a boron implant of 11/49 times the energy. The extra energy carried by the fluorine increases the lattice damage, reducing channeling and improving annealing characteristics.

A profile can be moved closer to the surface by implanting through a surface film such as silicon dioxide. This shifts the profile by roughly the oxide thickness, but introduces the problem of recoiled oxygen atoms. If the film thickness is close to the ion range, it is hard to control accurately the dose penetrating through to the silicon. The recoil effect can be used directly if we dope by knocking dopant out from a deposited surface film using silicon or inert gas implantation. [49]

Tilted ion beams. It is the ion velocity perpendicular to the surface that determines the projected range of an implanted ion distribution. If the wafer is tilted at a large angle to the ion beam then the effective ion energy is greatly reduced. [50] Figure 30 illustrates this for 60 keV arsenic as a function of tilt angle, showing that it is possible to achieve extremely shallow distributions using comparatively high implantation energies. For large tilt angles, a significant fraction of the implanted ions is scattered out of the surface, so the total dose is reduced. As a practical technique, this is only useful when the wafer surface is unpatterned because large tilt angles cause long shadows (see Section 8.2.2) and asymmetries

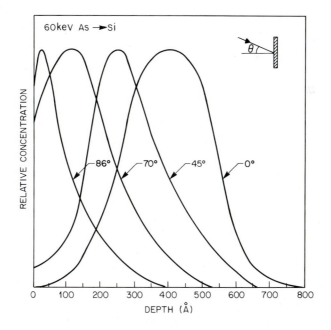

FIGURE 30
60 keV arsenic implanted into silicon, as a function of beam tilt angle. (*After De Cata, Williams, and Harrison, Ref. 50.*)

at mask edges. The same principles apply to doping of trench sidewalls, which are effectively silicon surfaces tilted at 83° to the ion beam.[51] In this case, dopant that scatters out of the surface of one side will probably enter the other sidewall lower down.

Implanted silicides and polysilicon. We can circumvent the problem of implanting a shallow layer in silicon completely if instead we implant entirely into a surface layer and then diffuse the dopant into the substrate. This is most often done when the surface film is to be used as a conductor making contact to the substrate. Diffusion results in steep dopant profiles without damage to the silicon lattice. Although this might seem to be a return to pre-implantation days when deposited glasses were often used, here much greater control is possible by choosing an appropriate dose and surface film thickness. Dopant diffusion in silicides and polysilicon is generally much faster than in single-crystal silicon, so the implanted atoms soon become uniformly distributed in the film.[52] This doping level drops slowly as atoms diffuse down into the substrate, giving a profile like the one shown in Fig. 31. The small peak at the interface may be due to grain boundary segregation or trapping by impurities at the interface. The presence of

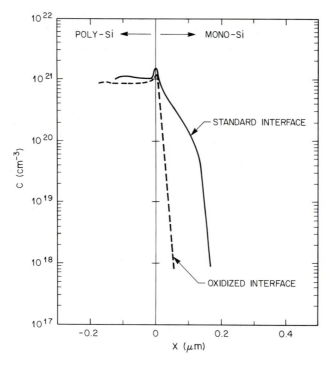

FIGURE 31
Arsenic diffusion into silicon from a polysilicon source, 950°C for 30 minutes. The presence of a 25 Å oxide is sufficient to block most of the diffusion. (*After Schaber, Criegern, and Weitzel, Ref. 52.*)

a 25 Å oxide between the polysilicon and the substrate is sufficient to block most of the diffusion.

For silicides, we also have the option of implanting before the heat treatment that forms the silicide. If the implant is beneath the metal, it will be "snow ploughed" forward as the silicide forms, resulting in a steep dopant gradient near the interface. If the implant is inside the metal, it will segregate out at the moving silicide–silicon interface giving a very sharply peaked dopant distribution.[53]

8.6 HIGH-ENERGY IMPLANTATION

Implantation at MeV energies makes possible several new processing techniques with application to VLSI. This section discusses these applications in two parts: implantation of dopant atoms to form conducting layers and very-high-dose implantation of chemically reactive atoms followed by annealing to form buried insulating layers.

8.6.1 Buried Conductors

VLSI CMOS circuits require high substrate doping to lower voltage drops from substrate currents and so prevent latchup, and at the same time low substrate doping to minimize junction capacitance. The traditional solution has been to use lightly doped epitaxial layers on heavily doped substrates so that the heavy doping does not enter the area where the transistor operates. A schematic device cross section is shown in Fig. 32. We would like to keep the epi-layer thin so that the good conducting substrate is close to the transistors, but we cannot allow heavy doping to reach the surface. This limit is set by the amount of outdiffusion from the substrate during the long tub drive-in diffusion step.

High-energy ion implantation offers three ways in which the traditional epitaxial CMOS process can be improved.[54] First, we can avoid a long diffusion step by directly implanting the tubs rather than diffusing them in from the surface. This will require a series of implants at different energies to form a roughly uniformly doped layer, followed by a short annealing step. Now the epi-layer thickness can be reduced and latchup resistance improved. This structure can be

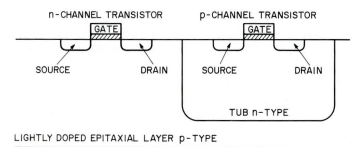

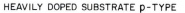

FIGURE 32
Cross section of transistors in an epi-substrate CMOS process.

improved further by retrograde doping of the tub—varying implant doses so that the tub surface is less doped than the tub bottom. This has the same advantages for the tub transistors as the use of epi-substrates has for the transistors outside the tub. Finally, we could dispense with the epi-substrate altogether and use a blanket high-energy boron implant as the first process step. This would form a buried, heavily doped layer serving the same function as a heavily doped substrate but at substantially reduced cost.

One problem that has not been fully resolved is the choice of masking material. A 2 MeV phosphorus implant requires a 2.5 μm oxide layer for a mask, or an even greater thickness of photoresist. Such thick masks would be very susceptible to shadowing effects. An alternative is to use thinner layers of a heavy metal such as tungsten.[55]

8.6.2   Buried Insulators

The idea of building devices in a thin silicon layer on an insulating substrate (SOI) has been around for many years. Initially, the main advantage was increased radiation-hardness due to reduced collection volume for charge generated by ionizing radiation. More recently, for VLSI technology, it offers a compact way to isolate devices from each other and to reduce parasitic capacitance. The recent availability of high-energy, high-current implanters has prompted much investigation into the formation of buried insulators by implantation of oxygen or nitrogen as an alternative to the epitaxial growth of silicon on sapphire or the recrystallization of deposited amorphous silicon films. This is a very active area, and the account given below is still somewhat provisional. A description of the latest work and its applications can be found in the proceedings of a recent symposium.[56] We will concentrate on buried oxide formation, and then mention the differences for buried nitride.

The principle behind formation of a buried oxide is very simple. If we implant oxygen at a dose above 10^{18} cm^{-2} then at the peak of the implant there should be more than twice as many oxygen atoms as silicon atoms in the substrate. During implantation and upon annealing, silicon dioxide should be formed. The details differ from more traditional implantation steps in several ways.

1. Our goal is to maintain a surface layer of high-quality, single-crystal silicon for device fabrication. If the surface becomes amorphous, we cannot regrow crystalline structure from the substrate because of the intervening buried oxide layer. The solution is to keep the substrate near 500°C during implantation so that self-annealing maintains the crystal structure (see Fig. 12).
2. The high doses involved substantially alter the target during implantation. Each incident ion sputters on the order of 0.1 silicon atoms, so a high dose will remove many layers of silicon atoms. Formation of oxide causes a 44% expansion of the buried layer. The net result is a slight swelling of the silicon surface.

3. The implanted oxygen profile will differ from that predicted by a simple linear theory because of sputtering, swelling, and oxide formation.[57] Detailed modeling of the oxide is complicated by varying diffusion rates as the material is altered, and by interaction of chemical and physical effects. Experiments show that the implanted profile saturates when the oxygen concentration reaches that of an oxide, and additional oxygen diffuses to the moving silicon–oxide boundaries, as shown in Fig. 33. The profile changes from Gaussian-like to a flat-topped distribution, limited by the concentration of oxygen in silicon dioxide.[58]

4. As implanted, the surface layer still contains a substantial amount of oxygen and much damage, although it is still single crystal. Annealing is usually performed under a deposited oxide cap, and can be divided into two regimes,[59,60] as shown in Fig. 34. Below 1300°C, annealing removes much of the damage near the surface but causes only local diffusion of oxygen to the buried layer. Many oxide precipitates are formed in a layer between the surface and the buried oxide, and much oxygen remains in the surface silicon where it will degrade device operation. There may also be a polysilicon region next to the oxide interface. Problems with surface film quality can be overcome

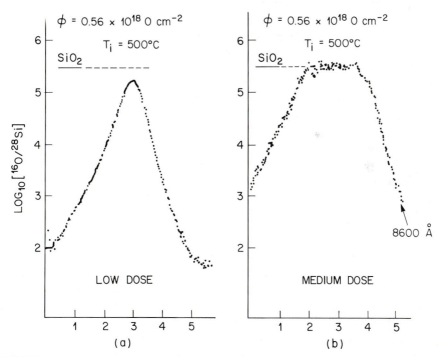

FIGURE 33

Unannealed profiles for 200 keV oxygen implanted into silicon. (a) For low doses, the profile has the usual shape. (b) For heavier doses, oxygen diffuses from the center region to the edges such that the maximum remains at the stoichiometric value for silicon dioxide. (*After Hemment et al., Ref. 58.*)

FIGURE 34
TEM cross sections of annealed high-dose oxygen implants. The upper figure is for an anneal of
1150°C for 2 hours. The lower figure is for an anneal at 1405°C for 30 minutes. The resulting oxide
layer is substantially thicker than for the first case, despite the fact that the dose was 10% less, and
the quality of the surface silicon is greatly improved. (*After Tuppen et al., Ref. 59, and Celler et
al., Ref. 60.*)

by epitaxially depositing a new silicon layer on the annealed substrate which
will be free of oxygen but will contain extended defects propagated up from
the flawed initial surface. Annealing at 1405°C causes a strong segregation of
oxygen into the buried layer from both sides, depleting the surface of almost
all of the implanted atoms. This leaves a high-quality surface film containing
little oxygen and less than 10^9 threading dislocations per square centimeter.
The segregation effect is strong enough to form a sharp buried oxide even
when the initial implant peak is below the level necessary to form an oxide
directly, allowing thiner oxides to be formed than by other methods. One
annealing technique uses bright lamps to superficially melt the backside of
the wafer (1412°C) such that there is a 7°C temperature drop across the wafer
from back to front.[60]

The case of nitrogen implantation, while similar in principle, has several
significant differences from oxygen. The diffusivity of nitrogen in nitride is ten
orders of magnitude less than for nitrogen or oxygen in silicon, or for oxygen

in oxide. For this reason, nitrogen implanted above the stoichiometric value for Si_3N_4 does not diffuse away during implantation or annealing. Instead, bubbles may form and the surface film may flake and peel off. Continuous buried nitride layers can be formed by annealing substoichiometric implants of nitrogen, where diffusion of nitrogen is mainly through silicon.[61] Segregation into the growing nitride at 1200°C is stronger than in the formation of oxide, so the surface silicon is depleted of nitrogen and is of high quality (Fig. 35). Remaining defects are at a similar or lower level than is the case for buried oxide. The nitride layer is usually polycrystalline, and there is no band of precipitates left in the silicon surface layer nor any polysilicon region. Current leakage, which occurs along the buried interface in a fabricated device, is more of a problem than with buried oxides.

8.7   SUMMARY AND FUTURE TRENDS

Over the past 25 years, ion implantation has been developed into a very powerful tool for integrated circuit fabrication. Our understanding of the underlying physics of atomic collisions allows us to predict implanted ion distributions, although channeling is still not accounted for satisfactorily. Implantation equipment is found in every fabrication line, with manufacturers emphasizing rapid wafer handling and high-beam currents to maximize throughput. Using suitable annealing cycles, lattice damage can be repaired to give high-quality silicon with good dopant activation. Development of this technique has been so successful that, in a modern VLSI process, it is used to introduce almost all of the dopant.

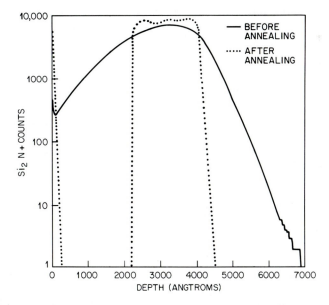

FIGURE 35
160 keV nitrogen implanted into silicon before and after annealing at 1200°C for 2 hours. (*After Nesbit, et al., Ref. 61.*)

Ion implantation continues to find new applications in VLSI technologies. Its attributes of controllability and reproducibility make it a very versatile tool, able to follow the trends to finer-scale devices. The lattice damage it causes can be removed using rapid thermal processing with minimal diffusion, and this technique should become commonplace in production environments in the next few years. For source/drain contacts, where concentrations need to be high but not too precise, diffusion from a surface layer such as a silicide forms a very shallow junction without shadowing problems, and this technique may displace direct implantation in some cases. At the same time, new applications such as buried conductors and insulators will be adopted as equipment suitable for production becomes available.

Compound semiconductors have not yet reached a VLSI level of integration, but much effort is being spent in developing the technology. Ion implantation plays an even more crucial role here because many compound semiconductors decompose at elevated temperatures so dopant cannot be diffused in from the surface. In addition to other damage, implantation of compounds can generate antisite defects—for example, gallium on an arsenic site in gallium arsenide—or net displacement of one component deeper into the target than the other. This can effectively dope the material enough to overwhelm the effect of the implanted atoms themselves. The annealing problem in compound semiconductors is, therefore, much more complicated than for silicon, and is an area of continuing research.

The statistical limits of ion implantation will begin to become apparent only when less than 100 atoms are required in each doping region, for example, in threshold control of a MOSFET. This corresponds to a dose of 10^{12} cm^{-2} for a 0.1 μm square transistor, giving doping levels around 10^{18} cm^{-3}. Many other obstacles will need to be overcome before this problem arises, so ion implantation will continue to be essential in VLSI device fabrication for the foreseeable future.

PROBLEMS

1 A 10 μA ion beam has a 10° half-angle divergence through a square aperture (8 cm $\times$ 8 cm) placed 6 cm away from the target. Using a current meter, how much time is needed to implant 10^{13} atoms/cm^2 for (a) a singly ionized, monatomic species (b) a triply ionized, diatomic species? (c) Using a charge integrator calibrated for singly charged monatomic species, what dose should be "set" to obtain a dose of 10^{13} atoms/cm^2 for the triply charged, diatomic species?

2 The drift-space vacuum between a mass-separation magnet and the target is approximately 10^{-6} Torr. Consider the possibility of a neutralizing charge exchange reaction

$$I^+ + N_2 \rightarrow I^0 + N_2^+$$

with a cross section of 10^{-6} cm^2/atom. What percentage of the ions are exchanged in a distance of 1 m? Take the fraction of unreacted particles to be $\exp(-x/\lambda)$, where λ is the mean free path.

3 From Fig. 23, estimate the solid phase epitaxial regrowth rate at 1100°C. How long does it take to regrow a 0.1 μm amorphous layer for a <100> silicon wafer? Suppose

that the minimum effective anneal time for your RTA system is 2 seconds. What will be the standard deviation of a low-dose 120 keV arsenic implant after minimum annealing?

4 Compare calculations of range profiles for implantation of 100 keV boron through 1500 Å of titanium silicide into silicon using the methods of R_p scaling and dose matching. Neglect skewness and assume a dose of 10^{15} cm^{-2}.

5 Suppose that a channeled ion experiences only electronic stopping. Integrate the electronic stopping to find the maximum range of a channeled 70 keV boron ion in silicon. What is the angular half-width of the [100] axial channel at this energy?

6 Add a third curve to Fig. 11 for the energy necessary to amorphize silicon based on the TEM cross sections of Fig. 10 for implantations at 300 K. Scale the curve to have roughly the same peak value as the theoretical calculations, just as has already been done for implantation at 82 K. Comment on the differences between the two experimental curves, and use your curve to calculate the energy density necessary to amorphize silicon under these conditions.

7 Suggest three experiments that would help to distinguish between the two proposed mechanisms for transient enhanced diffusion, discussed in Section 8.4.2. How would the results of your experiments distinguish between the mechanisms?

8 We would like to form 0.1 μm deep, heavily doped junctions for the source and drain regions of a submicron MOSFET. Compare the options that are available to introduce and activate dopant for this application. Which option would you recommend, and why?

9 Choose conditions for a set of four implants to form a CMOS buried layer and well structure to replace a conventional epitaxial process. The specifications are: buried layer depth 3.5 μm, buried layer doping 10^{19} cm^{-3}, well depth 2.5 μm, well doping 10^{18} cm^{-3}. The well should be formed using multiple implants to form a retrograde profile. Use the following expressions for ion moments (E in keV, distance in Å):

$$\text{(boron)} \quad R_p = 85.6E^{0.777} \qquad \sigma_p = 1760E^{0.106} - 1990E^{0.0192}$$

$$\text{(phosphorus)} \quad R_p = 15.6E^{0.958} \qquad \sigma_p = 22.4E^{0.659}$$

REFERENCES

1 W. Shockley, "Forming Semiconductive Devices by Ionic Bombardment," US Patent 2787564.

2 (a) J. F. Gibbons, "Ion Implantation in Semiconductors—Part I: Range Distribution Theory and Experiments," *Proc. IEEE*, **56**, 295 (1968).

(b) J. F. Gibbons, "Ion Implantation in Semiconductors—Part II: Damage Production and Annealing," *Proc. IEEE*, **60**, 1062 (1960).

3 (a) F. Eisen and L. Chadderton, Eds., *First International Conference on Ion Implantation*, Gordon and Breach, New York, 1971.

(b) I. Ruge and J. Graul, Eds., *Second International Conference on Ion Implantation*, Springer-Verlag, Berlin, 1972.

(c) B. L. Crowder, Ed., *Third International Conference on Ion Implantation*, Plenum, New York, 1973.

(d) S. Namba, Ed., *Fourth International Conference on Ion Implantation*, Plenum, New York, 1975.

(e) F. Chernow, J. Borders, and D. Brice, Eds., *Fifth International Conference on Ion Implantation,* Plenum, New York, 1976.

4 (a) J. Gyulai, Ed., *First International Conference on Ion Beam Modification of Materials*, Central Res. Inst. for Phys., H-1525 Budapest 114, POB49, Hungary, 1978.

(b) R. E. Benenson, E. N. Kaufman, G. L. Miller, and W. W. Scholz, Eds., *Second International Conference on Ion Beam Modification of Materials,* published in *Nucl. Inst. and Meth.*, **182/183**, North-Holland, New York, 1981.

(c) B. Biasse, G. Destefanis, and J. Pl. Gailliard, Eds., *Third International Conference on Ion Beam Modification of Materials*, published in *Nucl. Inst. and Meth.*, **209/210**, North-Holland, New York, 1983.

(d) *Fourth International Conference on Ion Beam Modification of Materials*, published in *Nucl. Inst. and Meth.*, **B7/8**, North-Holland, New York, 1985.

5 (a) J. S. Williams and J. M. Poate, Eds., *Ion Implantation and Beam Processing*, Academic Press, New York, 1984.

(b) F. Y. Wang. Ed., *Impurity Doping Processes in Silicon*, North-Holland, Amsterdam, 1981.

6 S. A. Cruz, "On the Energy Loss of Heavy Ions in Amorphous Materials," *Rad. Effects*, **88**, 159 (1986).

7 J. P. Biersack and L. G. Haggmark, "A Monte Carlo Computer Program for the Transport of Energetic Ions in Amorphous Targets," *Nucl. Inst. and Meth.*, **174**, 257 (1980).

8 J. Lindhard and M. Scharff, "Energy Dissipation by Ions in the keV Region," *Phys. Rev.*, **24**, 128 (1961).

9 Y. Xia and C. Tan, "Four-parameter Formulae for the Electronic Stopping Cross-section of Low Energy Ions in Solids," *Nucl. Inst. and Meth.*, **B13**, 100 (1986).

10 S. M. Sze, *Semiconductor Devices*, Wiley, New York, 1985, p.406.

11 (a) W. K. Hofker, D. P. Oosthoek, N. J. Koeman, and H. A. M. De Grefte, "Concentration Profiles of Boron Implantations in Amorphous and Polycrystalline Silicon," *Rad. Eff.*, **24**, 223 (1975).

(b) W. K. Hofker, H. W. Werner, D. P. Oosthoek, and N. J. Koeman, "Boron Implantations in Silicon: a Comparison of Charge Carrier and Boron Concentration Profiles," *Appl. Phys.*, **4**, 125 (1974).

12 W. P. Petersen, W. Fichtner, and E. H. Grosse, "Vectorized Monte Carlo Calculations for the Transport of Ions in Amorphous Targets," *IEEE Trans. Elec. Dev.*, **30**, 1011 (1983).

13 J. F. Gibbons, W. S. Johnson, and S. M. Mylroie, *Projected Range Statistics: Semiconductors and Related Materials*, 2nd Edition, Dowden, Hutchinson, & Ross, Stroudsburg, PA, 1975.

14 L. J. Parkes and J. P. Lavine, "Calculated Moments for the Implantation of Boron into Silicides," *J. Appl. Phys.*, **60**, 14 (1986).

15 A. F. Burenkov, F. F. Komarov, M. A. Kumakhov, and M. M. Tempkin, *Tables of Implanted Spatial Distributions*, Gordon and Breach, New York, 1986.

16 H. Ryssel, J. Lorenz, and K. Hoffmann, "Models of Implantation into Multilayer Targets," *Appl. Phys. A.*, **41**, 201 (1986).

17 S. Furukawa, H. Matsumura, and H. Ishiwara, "Theoretical Considerations on the Lateral Spread of Implanted Ions," *Jap. J. Appl. Phys.*, **11**, 134 (1972).

18 M. D. Giles, "Ion Implantation Calculations in Two Dimensions Using the Boltzmann Transport Equation," *IEEE Trans. CAD*, **5**, 679 (1986).

19 G. Hobler, E. Langer, and S. Selberherr, "Two-dimensional Modeling of Ion Implantation," in *Second International Conference on Simulation of Semiconductor Devices and Processes*, K. Board and R. Owen, Eds., Pineridge Press, Swansea, 1986.

20 J. Lindhard, M. Scharff, and H. Schiøtt, "Range Concepts and Heavy Ion Ranges," *Mat. Fys. Med. Dan. Vidensk. Selsk*, **33**, No. 14, (1963).

21 J. B. Sanders, "Ranges of Projectiles in Amorphous Materials," *Can. J. Phys.*, **46**, 455 (1968).

22 W. P. Maszara and G. A. Rozgonyi, "Kinetics of Damage Production in Silicon during Self-implantation," *J. Appl. Phys.*, **60**, 2310 (1986).

23 F. F. Morehead and B. L. Crowder, "A Model for the Formation of Amorphous Silicon by Ion Implantation" in Ref. 3(a).

24 R. B. Fair, J. J. Wortman, and J. Liu, "Modeling Rapid Thermal Diffusion of Arsenic and Boron in Silicon," *J. Electrochem. Soc.*, **131**, 2387 (1984).

25 L. C. Hopkins, T. E. Seidel, J.S. Williams, and J. C. Bean, "Enhanced Diffusion in Boron Implanted Silicon," *J. Electrochem. Soc.*, **132**, 2035 (1985).

26 T. C. Smith, "Wafer Cooling and Photoresist Masking Problems in Ion Implantation," in Ref. 36a.

27 G. H. Kinchin and R. S. Pease, "The Displacement of Atoms in Solids by Radiation," *Rep. Prog. Phys.*, **18**, 1 (1955).

28 (*a*) K. B. Winterbon, "Heavy-ion Range Profiles and Associated Damage Distributions," *Rad. Effects*, **13**, 215 (1972).
(*b*) D. K. Brice, "Ion Implantation Depth Distributions: Energy Deposition into Atomic Processes and Ion Locations," *Appl. Phys. Lett.*, **16**, 3 (1970).

29 W. L. Brown, "Collision Cascades, Ionization, Spikes and Energy Transfer," *Mat. Res. Soc. Symp. Proc.*, **51**, 53 (1985).

30 D. S. Gemmell, "Channeling and Related Effects in the Motion of Charged Particles through Crystals," *Rev. Mod. Phys.*, **46**, 129 (1974).

31 J. F. Ziegler, "Channeling of Ions near the Silicon <100> Axis," *Appl. Phys. Lett.*, **46**, 358 (1985).

32 R. G. Wilson, H. L. Dunlap, D. M. Jamba, and D. R. Myers, "Angular Sensitivity of Controlled Implanted Doping Profiles," *National Bureau of Standards Report SP 400-49*, November 1978.

33 M. Simard-Normandin and C. Slaby, "Empirical Modeling of Low Energy Boron Implants in Silicon," *J. Electrochem. Soc.*, **132**, 2218 (1985).

34 T. Hirao, G. Fuse, K. Inoue, S. Takayanagi, Y. Yaegashi, and S. Ichikawa, "The Effects of the Recoil-implanted Oxygen in Silicon on the Electrical Activation of Arsenic after Through-oxide Implantation," *J. Appl. Phys.*, **50**, 5251 (1979).

35 L. A. Christel, J. F. Gibbons, and S. Mylroie, "Recoil Range Distributions in Multilayer Targets," *Nucl. Inst. and Meth.*, **182/183**, 187 (1981).

36 (*a*) H. Ryssel and H. Glawischnig, Eds., *Proceedings of the Fourth International Conference on Ion Implantation Equipment and Techniques*, Springer Series in Electrophysics, vol. 11, Springer-Verlag, New York, 1983.
(*b*) J. F. Ziegler and R. L. Brown, Eds., *Proceedings of the Fifth International Conference on Ion Implantation Equipment and Techniques*, North-Holland, Amsterdam, 1985.

37 K. H. Purser, M. Cleland, H. Taylor, and T. H. Smick, "The IONEX MeV Implantation System," *SPIE*, **530**, 14 (1985).

38 M. I. Current, N. L. Turner, T. C. Smith, and D. Crane, "Planar Channeling in Si (100)," *Nucl. Inst. and Meth.*, **B6**, 336 (1985).

39 M. I. Current, "Current Status of Ion Implantation Equipment and Techniques for Semiconductor Fabrication," *Nucl. Inst. and Meth.*, **B6**, 9 (1985).

40 L. Csepregi, E. F. Kennedy, J. W. Mayer, and T. W. Sigmon, "Substrate-orientation Dependence of the Epitaxial Regrowth Rate from Si-implanted Amorphous Si," *J. Appl. Phys.*, **49**, 3906 (1978).

41 L. Csepregi, E. F. Kennedy, T. J. Gallagher, J. W. Mayer, and T. W. Sigmon, "Reordering of Amorphous Layers of Si Implanted with P, As, and B Ions," *J. Appl. Phys.*, **48**, 4234 (1977).

42 T. E. Seidel and A. U. MacRae, "The Isothermal Annealing of Boron Implanted Silicon," in Ref. 3(*a*).

43 B. L. Crowder and F. F. Morehead, "Annealing Characteristics of n-Type Dopants in Ion Implanted Silicon," *Appl. Phys. Lett.*, **14**, 313 (1969).

44 M. Tamura, T. Ikeda, and T. Takuyama, "Crystal Defects and Electrical Properties of Ion-implanted Silicon," in Ref. 3(*b*).

45 C. Hill, "Beam Processing in Silicon Device Technology," *Laser and Electron Beam Solid Interaction and Material Processing*, J. F. Gibbons, L. D. Hess, and T. W. Sigmon, Eds., North-Holland, New York, 1981.

46 A. Gat, J. F. Gibbons, T. J. Magee, J. Peng, V. R. Deline, P. Williams, and C. A. Evans, Jr., "Physical and Electrical Properties on Laser-annealed Ion-implanted Silicon," *Appl. Phys. Lett.*, **32**, 276 (1978).

47 D. K. Sadana, S. C. Shatas, and A. Gat, "Heatpulse Annealing of Ion-implanted Silicon: Structural Characterization by Transmission Electron Microscopy," *Inst. Phys. Conf. Ser.*, **67**, 143 (1983).

48 G. K. Celler and T. E. Seidel, "Transient Thermal Processing of Silicon," in *Silicon Integrated Circuits*, D. Kahng, Ed., Suppl. 2C to *Applied Solid State Science*, Academic Press, New York, 1985.

49 H. Ishiwara and S. Furukawa, "Formation of Highly-doped Thin Layers Using Knock-on Effect," in *Ion Implantation in Semiconductors*, F. Chernow, J. A. Borders, and D. K. Brice, Eds., Plenum, New York, 1976.

50 L. De Cata, J. S. Williams, and H. B. Harrison, "The Generation of Shallow Implanted Dopant Layers by Low-angle Implantation," *Nucl. Inst. and Meth.*, **B4**, 368 (1984).

51 G. Fuse, H. Umimoto, S. Odanaka, M. Wakabayashi, M. Fukumoto, and T. Ohzone, "Depth Profiles of Boron Atoms with Large Tilt Angle Implantations," *J. Electrochem. Soc.*, **133**, 996 (1986).

52 H. Schaber, R. v. Criegern, and I. Weitzel, "Analysis of Polycrystalline Silicon Diffusion Sources by Secondary Ion Mass Spectrometry," *J. Appl. Phys.*, **58**, 4036 (1985).

53 M. Horiuchi and K. Yamaguchi, "SOLID II: High Voltage High-gain Kilo-Ångstrom Channel Length CMOSFET Using Silicide with Self-aligned Ultra Shallow (3S) Junction," *IEEE Trans. Elec. Dev.*, **33**, 260 (1986).

54 M. I. Current and R. A. Martin, "MeV Implantation for CMOS Applications," *SPIE*, **530**, 23 (1985).

55 T. I. Kamins, P. J. Marcoux, J. L. Moll, and L. M. Roylance, "Patterned Implanted Buried-oxide Transistor Structures," *J. Appl. Phys.*, **60**, 423 (1986).

56 A. Chiang, M. W. Geis, and L. Pfeiffer, Eds., "Semiconductor-on-Insulator and Thin Film Transistor Technology," *Mat. Res. Soc. Symp. Proc.*, **53**, Boston, 1986.

57 E. A. Maydell-Ondrusz and I. H. Wilson, "A Model for the Evolution of Implanted-oxygen Profiles in Silicon," *Thin Solid Films*, **114**, 337 (1984).

58 P. L. F. Hemment, E. Maydell-Ondrusz, K. G. Stevens, J. A. Kilner, and J. Butcher, "Oxygen Distributions in Synthesized SiO_2 Layers Formed by High Dose O^+ Implantation into Silicon," *Vacuum*, **34**, 203 (1984).

59 C. G. Tuppen, M. R. Taylor, P. L. F. Hemment, and R. P. Arrowsmith, "Effects of Implantation Temperature on the Properties of Buried Oxide Layers in Silicon Formed by Oxygen Ion Implantation," *Appl. Phys. Lett.*, **45**, 57 (1984).

60 G. K. Celler, P. L. F. Hemment, K. W. West, and J. M. Gibson, "Improved SOI Films by High Dose Oxygen Implantation and Lamp Annealing," p. 227 in Ref. 56.

61 L. Nesbit, S. Stiffler, G. Slusser, and H. Vinton, "Formation of Silicon-on-Insulator by Implanted Nitrogen," *J. Electrochem. Soc.*, **132**, 2713 (1985).

CHAPTER 9

METALLIZATION

S. P. MURARKA

9.1   INTRODUCTION

Conductive films are required to provide interconnection between contacts on devices and between devices and the outside world. Figure 1 shows a cross section, a simplistic scheme, and a designer's symbolic representation of a typical MOSFET (metal-oxide-semiconductor field effect transistor). The central region is called the gate. In this region the substrate silicon is isolated from the metal electrode by an insulating layer, which is, generally, a thin thermally grown SiO_2 layer (see Chapter 3). The metal electrode, generally a polysilicon layer (see Chapter 6), is called a gate electrode and it controls the working of the MOSFET as an off or on device. The work function of the metal plays an important role in determining the voltage required to activate the device. The two identical regions neighboring the gate region, are called the source and the drain. The metal contacting these regions is generally aluminum. The metal work function again plays an important role in determining the current flow characteristics in these regions.

Metallization use in bipolar devices is similar to that in MOSFETs. In a bipolar transistor, the central region is the semiconductor (base), and is directly in contact with the metal. The two neighboring regions are now called emitter and collector, and are also directly in contact with the metal. Thus a bipolar transistor, in cross section, looks very similar to a MOSFET with no gate oxide. The metal work function plays an important role in all three regions in determining the current flow characteristics, in a manner similar to that in the source and drain regions of the MOSFET.

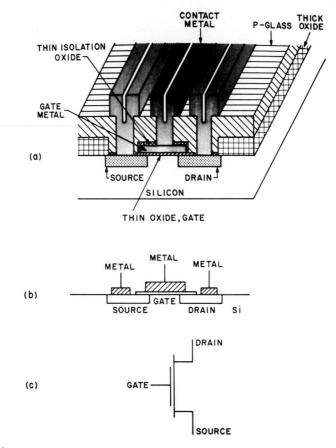

FIGURE 1
A typical MOSFET in (a) cross section, (b) a simple schematic representation, and (c) a designer's symbolic representation.

Primary metallization application can, therefore, be divided into three groups: (1) gate, (2) contact, and (3) interconnects. Interconnection metallization, which interconnects thousands of MOSFETs or bipolar devices using fine-line metal patterns, is generally the same as gate metallization. All metallization directly in contact with semiconductor is called contact metallization. As mentioned earlier, polysilicon film has been the form of metallization used for gate and interconnection MOS devices.[1] Aluminum has been used as the contact metal on devices and as the second-level interconnection to the outside world.[2] Several other metallization schemes have been proposed to produce ohmic contacts to a semiconductor.[3] In several cases a multiple-layer structure involving a diffusion barrier has been recommended.[4] Platinum silicide (PtSi) has been used as a Schottky barrier contact and also simply as an ohmic contact for deep junctions.[5] Titanium/platinum/gold or titanium/palladium/gold beam lead technology has been successful in providing high-reliability connection to the outside world.[6,7]

Very-large-scale integration (VLSI) and the continued evolution of smaller and smaller devices has aroused concern about these existing metallization schemes and interest in the development of new metallization schemes for gates, interconnections, and ohmic contacts, and has motivated continued study of the reliability of aluminum and its alloys as current carriers.

The applicability of any metallization scheme in VLSI depends on several requirements, which are listed[8] in Table 1. Most important of these requirements are the stability of the metallization throughout the IC fabrication process and its reliability during the actual use of the devices. Thus, both the mechanical (physical) and electrical characteristics of the metallization, and the devices they cover, must be preserved. In addition, the metallization schemes should be easy and economical to implement, that is, easy to form and pattern. A considerable amount of research and development, therefore, goes into looking for a metallization scheme. Such efforts have led to a large volume of work involving thin metallic films: the deposition, characterization, and interaction between films and between films and substrates.

In Section 9.2 we shall briefly discuss gate and interconnection metallization, contact metallization, and other applications. In the following sections, metallization choices, methods of deposition, patterning, metallization problems, and a brief introduction to the new role of metallization will be presented.

9.2 METALLIZATION APPLICATIONS

9.2.1 Gates and Interconnection

Gate and interconnection metallization plays two important roles. First, it controls the speed of the circuit by virtue of the resistance of the interconnection runners.

TABLE 1
Desired properties of the metallization for integrated circuits

1. Low resistivity.
2. Easy to form.
3. Easy to etch for pattern generation.
4. Should be stable in oxidizing ambients; oxidizable.
5. Mechanical stability; good adherence, low stress.
6. Surface smoothness.
7. Stability throughout processing, including high temperature sinter, dry or wet oxidation, gettering, phosphorus glass (or any other material) passivation, metallization.
8. No reaction with final metal, aluminum.
9. Should not contaminate devices, wafers, or working apparatus.
10. Good device characteristics and lifetimes.
11. For window contacts — low contact resistance, minimal junction penetration, low electromigration.

The RC time constant of these runners varies as $R_s L^2/d_{ox}$, where R_s is the ohm-per-square (sheet) resistance, L is the length of the runner, and d_{ox} is the oxide thickness. For faster circuits, a reduction in R_s is essential.

Problem

Calculate the RC time constant for a 1 cm long doped polysilicon interconnection runner on 1 μm thick SiO_2. The polysilicon has a thickness of 5000 Å and a resistivity ρ of 1000 $\mu\Omega$-cm.

Solution

$$RC = \frac{R_s L^2 \epsilon_{ox}}{d_{ox}} = \frac{\rho}{d_{Poly-Si}} \frac{L^2 \epsilon_{ox}}{d_{ox}}$$

where

$$\epsilon_{ox} = \text{permittivity of the } SiO_2$$

$$\epsilon_{ox}/\epsilon_0 = \text{dielectric constant of } SiO_2 = 3.9$$

$$\epsilon_0 = \text{permittivity of free space} = 8.86 \times 10^{-14} \text{ farads/cm}$$

Substituting the given values and changing farads into coulombs per volt, we get

$$RC = \frac{1000 \times 10^{-6}}{5000 \times 10^{-8}} \frac{1^2 \times 3.5 \times 10^{-13}}{10^{-4}} \text{ s}$$

$$= 7 \times 10^{-8} \text{ s}$$

$$= 70 \text{ ns}$$

Besides its role in circuit speed, metallization also controls the so-called flat-band voltage V_{FB}. V_{FB} is the voltage required to counter balance the work function difference between metal and semiconductor so that a flat-band condition is maintained in the semiconductor.[9] In absence of any other charges in the oxide or at the oxide-semiconductor interface

$$V_{FB} = \phi_m - \phi_s = \phi_{ms} \tag{1}$$

where ϕ_m and ϕ_s are the work functions of the metallization (at the gate) and the semiconductor, respectively. Generally, there are other charges which will change Eq. 1. See Reference 9 for details.

Flat-band voltage contributes to threshold voltage V_T. V_T is the voltage required at the gate metal (measured with respect to source voltage) to achieve conduction between source and drain regions. Thus V_T determines the gate-to-source voltage that will switch the MOSFET to an on condition. It is important, in addition to the requirements listed in Table 1, that the gate metallization be chosen judiciously for the desired MOS behavior and speed of the circuits.

9.2.2   Ohmic Contacts

A good ohmic contact, usually formed by depositing a metal on the semiconductor, does not perturb device characteristics and is stable both electrically and mechanically. It has a resistance, called contact resistance, that should be neg-ligible compared to the device resistance. The specific contact resistance R_c (Ω-cm^2) is defined as[9]

$$R_c = \left(\frac{dV}{dJ}\right)_{v=0} \qquad (2)$$

which can be obtained from current density (J) and voltage (V) characteristics.

R_c is related to ϕ_B, the so-called Schottky barrier height of the metal, and doping density N_D in the semiconductor.[9] Figure 2 shows the relationship between R_c, ϕ_B, and N_D for contacts on a n-type silicon substrate. It is clear that R_c decreases with increasing dopant concentration and decreasing barrier height. The occurrence of the so-called Schottky barrier between the metal and semiconductor is due to the requirement that Fermi levels in two materials be the same or matched. In a simple case, unmatched Fermi levels exist because the work functions of the metal and the semiconductor are different. Ideally, we can calculate the difference and predict the behavior of the metal on a semiconductor. For example, for an n-type semiconductor, if the work function ϕ_m of the metal is greater than the work function ϕ_s of the semiconductor, the contact between the two is rectifying. On the other hand, if ϕ_m is less than ϕ_s, the contact is ohmic. Thus, as in the case of gate metallization, the metal work function plays an important role in determining contact properties. However, as seen in Fig. 2, if the dopant concentration is very high (greater than 10^{19} per cm^3) and the contact resistance is very low, then, theoretically, good contacts can be made with any metal.

9.2.3 Other Applications

There are several other applications of metallization in integrated circuits, the most important one being the top-level metal that provides a connection to the outside world. In general, this metal film is deposited and patterned during the last stages of processing, whereas gate and interconnection and contact metallization is formed during earlier stages. Top-level metal is simply a current carrier and is therefore very thick. This metal film also acts as a corrosion resistance coating for protecting active devices. To reduce interconnection resistance and save area on a chip, multilevel metallization, as discussed later in this chapter, may be used in the future.

Metallization is also used to produce rectifying (Schottky barrier) contacts, guard rings, redundancy in memories, and diffusion barriers between reacting metallic films.

9.3 METALLIZATION CHOICES

Table 1 listed the desired properties of the metallization for integrated circuits. None of the metals satisfies all the desired characteristics. Even aluminum, which has most of the desired properties, suffers from a low melting point limitation.

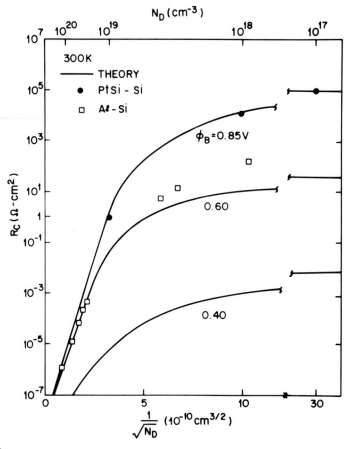

FIGURE 2

Theoretical and experimental values of specific contact resistance R_c as a function of the dopant concentration N_D and barrier height ϕ_B (*from Sze, Ref. 9.*)

Aluminum also exhibits undesirable electromigration behavior, as discussed later. For present and future integrated circuit applications, different metallizations for different applications may be desirable. The changing world of silicon integrated circuit (SIC) fabrication has led to fast shrinkage of device dimensions—laterally and vertically—requiring processing temperatures to be 900°C or lower. Also, wet chemical etching has been replaced by dry etching, which permits more selectivity in etching various materials on a wafer. These changes have opened doors to re-examination of some metals and alloys for VLSI application in SICs.

9.3.1 Metals or Alloys

Table 2 lists the possible metallization choices. Of these, polysilicon has been the usual gate metallization for MOS devices. Only recently, polysilicon/refractory

TABLE 2
Possible metallization choices for integrated circuits

Application	Choices
Gates and interconnection and contacts	Polysilicon, silicides, nitrides, carbides, borides, refractory metals, aluminum, and combinations of two or more of above
Diffusion barrier	Nitrides, carbides, borides, Ti-W alloy, silicides
Top level	Aluminum
Selectively formed metallization on silicon only	Some silicides, tungsten, aluminum

metal silicide bilayers have replaced polysilicon so that lower resistance can be achieved at the gate and interconnection level. By preserving the use of polysilicon as the "metal" in contact with the gate oxide, well-known device characteristics and processes have been unaltered.

Refractory silicides, formed on top of polysilicon, have provided the highest compatibility. Disilicides of molybdenum ($MoSi_2$),[10] tantalum ($TaSi_2$),[11] and tungsten (WSi_2)[12] have been developed and have found their way into the production of microprocessors and random-access memories. More recently, $TiSi_2$[13,14] and $CoSi_2$[15] have been suggested to replace $MoSi_2$, $TaSi_2$, and WSi_2. Aluminum and refractory metals tungsten and molybdenum are also being considered for the gate metal.[16-18]

For contacts, aluminum has been the preferred metal because of (1) the ease of processing, (2) aluminum's ability to reduce native SiO_2, which is always present on silicon wafers exposed to atmosphere, and (3) its low resistivity. However, the lower melting point of aluminum, limits its use to processing steps after which no high-temperature (normally greater than 450°C) operation is permitted. In spite of this limitation, aluminum has satisfied the required metallization criteria. For VLSI applications, several other factors such as shallower junctions, step coverage, electromigration (at higher current densities), and contact resistance can no longer be ignored. Each of these factors is briefly discussed below.

As device dimensions shrink, junction depths of 2500 Å or less become desirable. Such shallow junctions require metal contacts that do not lead to spiking shorts (shorts related to deep penetration of metal into the semiconductor) commonly observed with aluminum contacts, or to contact depths that are a significant fraction of the junction depth.

The decreasing area of contacts also causes step-coverage problems. Windows in the passivating layer are deep, and most metals are deposited using techniques (evaporation, sputtering) that do not produce the good step coverage required for these deep holes.

The increased current density in reduced-size conductors and increased chip temperature promote electromigration-induced problems in aluminum runners and contacts. Thus, the contact metallization must also be resistant to electromigration.

Finally, a contact, although stable both with respect to electromigration resistance and junction spiking, could still be poor due to high contact resistance that could result from the presence of an interfacial layer between the silicon and the contact metallization. Thus, elimination of this interfacial layer is essential for good contacts.

Contact failure can be a result of junction spiking, incomplete step coverage, electromigration-induced open circuits, and high contact resistance. Numerous investigations to provide an understanding of these phenomena have been carried out. Simultaneously, several possible solutions to the contact problems have been considered. These include use of: (1) dilute Si–Al alloy; (2) polysilicon layers between source, drain, or gate and top-level aluminum; (3) selectively deposited tungsten, that is, deposited by CVD methods so that the metal is deposited only on silicon and not on oxide; and (4) a diffusion barrier layer between silicon and aluminum, using a silicide, nitride, carbide, or combination thereof. Use of a self-aligned silicide, such as PtSi, guarantees extremely good metallurgical contact between silicon and silicide.[5] Silicides have been recommended in processes where shallow junctions and contacts are formed at the same time.

Even the use of a silicide as contact metallization will require a diffusion barrier to protect the silicide from interaction with aluminum used as the top metal. Aluminum interacts with most silicides in the temperature range of 200–500°C. Because of their high chemical and thermodynamical stabilities, transition metal nitrides, carbides, and borides offer potential applicability as a diffusion barrier between silicide (or silicon) and aluminum. Among other suggested barriers, Ti–W alloy has been used in various metallization schemes.[19]

The most important requirement of an effective metallization scheme in VLSI is that metal must adhere to the silicon in the windows (or to the polysilicon or polysilicon/silicide on gates) and to the oxide that defines the window. In this respect, metals (e.g., Al, Ti, Ta, etc.), that form oxides with a heat of formation higher than that of SiO_2 are the best. This is the reason that titanium is the most commonly used adhesion promoter. Titanium reduces SiO_2 at relatively low temperatures and forms a strong bond with it, and thus acts as the desired "glue" layer. Metals like tungsten or molybdenum, on the other hand do not reduce SiO_2, and, therefore, it is unlikely they will be used directly on SiO_2.

9.3.2 Properties

Properties of various metallizations of interest are compared in Table 3. Electrical resistivity values given in the second column are typical for polycrystalline films deposited in the thickness range of 100 to 10,000 Å. Lower resistivity values belong to purer, thicker, and large-grained (or single-crystalline) films. This is because impurities, grain boundaries, and surfaces all scatter electrons and thus

TABLE 3
Properties of various metallizations

Metal or alloy	ρ^a ($\mu\Omega$–cm)	$T_m{}^b$ (°C)	α^c (ppm/°C)	Reaction with Si at (°C)	Stable on Si up to (°C)
Al	2.7–3.0	660		~250	~ 250
Mo	6–15	2620	5	400–700	~ 400
W	6–15	3410	4.5	600–700	~ 600
MoSi$_2$	40–100	1980	8.25	—	>1000
TaSi$_2$	38–50	~2200	8.8–10.7	—	≥1000
TiSi$_2$	13–16	1540	12.5	—	≥ 950
WSi$_2$	30–70	2165	6.25, 7.9	—	≥1000
CoSi$_2$	10–18	1326	10.14	—	≤ 950
NiSi$_2$	~ 50	993	12.06	—	≤ 850
PtSi	28–35	1229	—	—	≤ 750
Pt$_2$Si	30–35	1398	—	—	≤ 700
HfN	30–100	~3000		450–500	450
ZrN	20–100	2980		450–500	450
TiN	40–150	2950		450–500	450
TaN	~200	3087		450–500	450
NbN	~ 50	2300		450–500	450
TiC	~100	3257		450–500	450
TaC	~100	3985		—	—
TiB$_2$	6-10			> 600	> 600

[a] ρ = resistivity, typical thin film value.
[b] T_m = melting point.
[c] α = linear thermal expansion coefficient from Refs. 20 and 21.

increase resistivity. For silicides, nitrides, carbides, and borides, variation in composition will also affect resistivity. Deviation from the stoichiometry of the compound leads to defects such as vacancies or interstitials that behave like impurities and increase resistivity. Thus, for metallization applications, the resistivity of a given film may vary from one deposition to another unless all deposition parameters are controlled to yield identical films. Finally, the resistivity of the deposited film may change when subjected to a high-temperature anneal. Such anneals may lead to impurity contamination, interaction with local environment, grain growth, or simply annealing of point defects. Most as-deposited films will exhibit an initial reduction in defects and an increase in grain size during low-temperature anneals, which lead to lower resistivities. This occurs prior to any chemical or metallurgical interaction. Interaction may cause resistivity to increase or decrease, depending on the resistivity of the product. Titanium's resistivity increases when it absorbs oxygen or reacts with it to form an oxide. On the other hand, when titanium reacts with silicon to form $TiSi_2$, the resistivity is lowered.

Grain growth, annealing of defects, and interactions in solid state are all diffusion controlled (see Chapter 7). Solid-state diffusion is appreciable only at temperatures that are larger than one-third the melting point (in Kelvin) of the solid in which diffusion is occurring. The melting points of various metallizations are given in the third column of Table 3. Reaction temperatures with silicon and the highest temperature at which these materials are considered stable on silicon are given in the fifth and sixth columns, respectively. Note that the reaction temperatures appear to be very low compared to the melting points of the intermetallics, Mo, or W. This can be understood if one considers the melting point of the silicon, 1412°C, and the fact that the temperatures cited in this table refer to thin polycrystalline metal film on silicon. Diffusion, leading to reaction, in polycrystalline films can occur, even at low temperatures, via grain boundaries and dislocations.

Thin films on any substrate are in a condition of stress determined by two factors. The first one, called intrinsic stress, is related to lattice mismatch between the substrate and the film, film structure and purity, and defects in films. Some of these stresses can be changed by annealing. The second factor is the difference between thermal expansion coefficients of the film and the substrate. The larger the difference, the greater the stress. Thus silicides, in general, end up with large stress conditions. The higher the temperature of forming the silicide, the larger will be the stress in the resulting film. The stress in the film must be balanced by an opposite stress induced in the substrate. Therefore, the stress in the film will not only determine the mechanical stability of the film, but may also affect the electrical properties of the substrate and the devices formed on such substrates. Thus knowledge of film stresses is essential in properly utilizing a particular film in integrated circuits.

In addition to having desirable properties, the chosen metallization scheme must be one that can be implemented in a processing sequence. Table 4 compares the processing capabilities of various metallizations. Sections 9.4, 9.5, and 9.6 discuss these in detail, and 9.3.3 discusses stability on semiconductors and insulators.

9.3.3 Stability on Semiconductors and Insulators

The last column in Table 3 gives the highest temperature of applicability of various metallizations on silicon. These temperatures are found by experience and experimentation. At higher temperatures, either the materials react with silicon or they loose their character and properties due to decomposition, agglomeration, or mechanical failure. For example, a Pd_2Si film will agglomerate at temperatures above 700°C, a tungsten film will react with silicon, and all nitrides will react with silicon leading to the decomposition of the metal-nitrogen bonds. Thus, the use of these metallizations on silicon (or polysilicon) above the cited temperatures is not recommended.

The stability of various popular metallizations on SiO_2 is summarized in

TABLE 4
Processing capabilities of various metallizations

Metal or alloy	Deposition[a] method	Patterning[b]	Oxidation for films on		Stability of films	
			Si	SiO$_2$	on SiO$_2$	with Al as top metal (°C)
Al	E,S	DE	Yes	Yes	Good	—
Mo	E,S,CVD	DE	No	No	Poor	500
W	E,S,CVD	DE	No	No	Poor	500
MoSi$_2$	CD(E,S,CVD)	DE	Yes	No	Ok[d]	500
TaSi$_2$	CD(E,S,CVD)	DE	Yes	No[c]	Ok[d]	500
TiSi$_2$	CD(E,S,CVD),R	DE,SA	Yes	No[c]	Ok[d]	500
WSi$_2$	CD(E,S,CVD)	DE	Yes	No	Ok[d]	500
CoSi$_2$	(E or S) + R	SA	Yes	No	Ok[d]	400
PtSi	(E or S) + R	SA	Yes	No	Ok[d]	250
TiN	S,R,S.	DE	No	No	Good	450

[a]E = evaporation, S = Sputtering, CVD = chemical vapor deposition, CD = codeposition,
 R = reaction with Si, R.S. = reactive sputtering.
[b]DE = dry etching, SA = self-aligned.
[c]Silicide oxidizes to form solid oxides (on surface) that may prevent further oxidation.
[d]In inert ambients.

Table 4. Only Al and TiN are unconditionally good on SiO$_2$, because both Al and Ti reduce SiO$_2$ to form interfacial metal-oxide bonds that promote adhesion and stability. W and Mo do not reduce SiO$_2$ and are found to have poor adhesion on SiO$_2$ surfaces. WSi$_2$, MoSi$_2$, CoSi$_2$, and PtSi adhere well to SiO$_2$ surfaces, but have poor stability when exposed to oxidizing ambients. These silicides readily decompose, in such ambients, to form SiO$_2$ and metal, and, on continued oxidation, to form SiO$_2$ and metal oxide. TiSi$_2$ and TaSi$_2$ on Si, on the other hand, are oxidized to form silicon and metal oxide and on continued oxidation to form SiO$_2$ and metal oxide. Silicides on silicon are stable in all ambients up to the temperatures given in Table 3.

The last column in Table 4 gives the temperatures to which various metallization/aluminum composites can be heated without noticeable change in properties. Aluminum is a natural choice for top-level metal. Thus, it is used on top of all metallization schemes. After deposition of Al film, the structure is generally subjected to annealing at 400–450°C to promote adhesion and eliminate electrical damages associated with the metal deposition process (see Section 9.4). As can be seen, except for CoSi$_2$ and PtSi, all metallizations listed will survive such annealing.

CoSi$_2$ and PtSi react with aluminum at 400 and 250°, respectively. Thus, when using aluminum on top of these silicides, a diffusion barrier of a nitride, a carbide, or a refractory metal will be necessary to stabilize the structure to a temperature in the range of 400–450°C. We shall discuss diffusion barriers again in Sections 9.6 and 9.7.

9.4 PHYSICAL VAPOR DEPOSITION

There are two types of deposition processes that are useful. The first one, called chemical vapor deposition (CVD), was presented in Chapter 6. CVD offers several advantages of which three are very important: (1) excellent step coverage, (2) large throughput, and (3) low-temperature processing. Tungsten and molybdenum have been very successfully deposited using LPCVD (low-pressure chemical vapor deposition). Other metallizations, such as WSi_2, $MoSi_2$, $TiSi_2$, and $TaSi_2$ have also been deposited by a LPCVD process, although the general applicability of such processes has not yet been established. Selective CVD deposition, in which the metallization is selectively deposited only on top of the silicon surface and not on exposed SiO_2 surfaces, is a very promising technique. Laser CVD of metals has promise for both direct writing and area deposition of metals.

In this chapter, we shall consider only physical vapor deposition methods, namely evaporation and sputtering. In both cases, the formation of a deposit on a substrate away from the source consists of three steps: (1) converting the condensed phase (generally a solid) into a gaseous or vapor phase, (2) transporting the gaseous phase from the source to the substrate, and (3) condensing the gaseous source on the substrate, followed by the nucleation and growth of the film. In cases where a compound, such as a silicide, nitride, or carbide, is deposited, one of the components is a gas and the deposition process is termed reactive evaporation or sputtering. In both types of deposition, the deposition material is transformed into a gas phase, and interactions with the residual gases and sputtering gases become important in determining the property of the deposited film. More detailed information on deposition processes can be found in Reference 21.

9.4.1 Deposition Methods

Deposition of films by thermal evaporation is the simplest method. In the evaporation method, a film is deposited by the condensation of the vapor on a substrate, which is maintained at a lower temperature than that of the vapor. All metals vaporize when heated to sufficiently high temperatures. Several methods, such as resistive, inductive (or radio frequency, rf), electron bombardment (e-gun), or laser heating, can be used to attain these temperatures. For transition metals, especially the refractory metals, evaporation using an electron gun is very common. For low-melting metals like aluminum, any of the these heating methods can be utilized. Resistive heating is, however, limited in application because of lower throughput associated with the smaller metal charge generally used in such systems. On the other hand, e-gun evaporations cause radiation damage, but high-temperature postdeposition heat treatments can anneal this out. This method is advantageous because the evaporations take place at pressures considerably lower than sputtering pressures and, therefore, the gas entrapment in the film is negligible or nonexistent. Radio frequency heating of the evaporating source in an appropriate crucible could prove to be the best compromise in providing large throughput, clean environment, and minimal levels of radiation damage.

The evaporation method can also be used to deposit an alloy or a mixture of two or more materials by the use of two or more independently controlled evaporation sources. The individual component evaporation rates are determined independently under experimental conditions. Then, conditions are adjusted to deposit an alloy or a mixture in the desired composition. However, this method of coevaporation, which has been used to deposit refractory silicides, requires good calibration that must be maintained during a run and from run to run. Coevaporation from a single source containing the alloy constituents is generally not possible, because fractionation occurs as the evaporation proceeds.

In sputter deposition, the target material is bombarded by energetic ions to release some atoms. These atoms are then condensed on the substrate to form a film. Sputtering processes, unlike evaporation, are very well controlled and are generally applicable to all materials—metals, alloys, semiconductors, and insulators. Radio frequency, dc, and dc-magnetron sputtering can be used for metal deposition. Alloy-film deposition by sputtering from an alloy target is possible because the composition of the film is locked to the composition of the target. This is true even when there is considerable difference between the sputtering rates of the alloy components. At the early stages of sputtering the component with high sputtering rate is sputtered off the target preferentially, leaving the target surface deficient in this component. The deficient region soon becomes deficient enough to compensate for the higher sputtering rate, leading to deposits of a composition similar to that of the target. The difference between the target and film compositions depends on the type of the equipment, the sputtering parameters, and the alloy constituent. By proper choice of the equipment and sputtering parameters, several complex alloys have been deposited with identical composition in the target and the film. Alloys can also be deposited with excellent control of composition by use of individual component targets. In certain cases, the compounds can be deposited by sputtering the metal in a reactive environment. Thus, gases such as methane, ammonia or nitrogen, and diborane can be used in the sputtering chamber to deposit carbide, nitride, and boride, respectively. This technique is called reactive sputtering.

Sputtering is carried out at relatively high pressures (0.1 to 1 pascal, or Pa). Because gas ions are the bombarding species, the films usually end up including small amounts of the gas. These gases in the film are nonreactive except when the sputtering gas is contaminated with chemically active gases. However, trapped gases cause stress changes, which are a function of the sputtering conditions.[21]

Sputtering is a physical process in which the deposited film is also exposed to ion bombardment. Such ion bombardment causes sputtering damage, which leads to unwanted charges and internal electric fields that affect device properties. However, such damages can generally be removed by annealing at relatively low temperatures ($<500°C$), unless the damage is so severe as to cause an irreversible breakdown of the gate dielectric.

9.4.2 Deposition Equipment and Process

VLSI metallization is currently done in vacuum chambers.[22] The decision to select or build a given vacuum system depends on specific deposition needs.

A mechanical pump, usually called a roughing pump, brings the system from atmospheric pressure to about 10 to 0.1 Pa (7.5 $\times 10^{-2}$ to 7.5 $\times$ 10^{-4} Torr). Such pressures may be sufficient for low-pressure chemical vapor depositions. An oil-diffusion pump, backed up by the roughing pump, can bring the pressure down to 10^{-5} Pa and with the help of a liquid nitrogen trap to as low as 10^{-7} Pa. A liquid nitrogen trap is essential for minimizing the oil contamination streaming into the main working chamber. For most sputtering and evaporation systems a combination of a liquid nitrogen trap, an oil-diffusion pump, and a mechanical pump should be adequate. For faster pumping, an oil-diffusion pump can be replaced with (1) a turbomolecular pump, which could bring the pressure down to 10^{-8}–10^{-9} Pa or (2) a cryopump or sputter-ion pump, which is capable of bringing the pressure down to 10^{-9} Pa. Pumps of the latter group are oil-free and are recommended where oil contamination must be avoided, such as in an MBE (molecular-beam epitaxial) system.

In choosing a vacuum system, the pumping speed of the pumps and resistance offered by the plumbing between the main chamber and the pump must be carefully evaluated. The pumping speed is defined as the ratio Q_a/P in units of liters per second at 20°C. Q_a is the quantity of the gas that flows through the intake cross section of the pump in a unit time and is pressure dependent. P is the partial pressure of the gas at a point near the intake port of the pump. Most pumps operate at constant speed in a large pressure range, above and below which the speed drops very quickly. Thus, during early pumpdown period (from atmospheric pressure) the pumping speed is low. The pumpdown time $t(s)$, in seconds, can be calculated using the equation[21,22]

$$t(s) = \frac{276V}{S_o + S} \ln \frac{P_o}{P} \tag{3}$$

where V is the volume of the vacuum chamber, S_o and S are the pumping speeds in liters per second at starting and final pressures P_o and P respectively. Values for S_o and S can be obtained from the pump manufacturer.

Since pumping speed is determined at the intake cross section of the pump, the pump intake should be as close to the main vacuum chamber as possible. Some distance will be involved because of the need to isolate the chamber from the pumping station. To optimize pumping, therefore, interconnecting tubes and valves should be as large as possible. In addition the outgassing from the interconnections and the main chamber will increase pumpdown time and ultimate achievable pressure. Outgassing is temperature dependent, and this effect is utilized in minimizing its effect. This is done by heating or baking the chamber to temperatures higher than operating temperatures while pumping is continued. After an ultimate pressure at high temperature is achieved, the baking is discontinued, leading to significant improvement in the final ultimate pressure. More on outgassing is presented later.

Besides the pumping system, the need for pressure gauges and controls, residual gas analyzers, temperature sensors, ability to clean the surface of the wafers by backsputtering, contamination control, and gas manifolds, and the use of automation, should be carefully evaluated. These are discussed in detail in References 21 and 22.

Figure 3 shows a schematic view of a deposition system. The chamber shown is a bell jar, a stainless-steel cylindrical vessel closed at the top and sealed at the base by a gasket. Beginning at atmospheric pressure the chamber is evacuated by a roughing pump, such as a mechanical rotary-vane pump or a combination mechanical pump and liquid-nitrogen-cooled molecular sieve system. The rotary-vane pump can reduce the system pressure to about 20 Pa, and the combination pump system can achieve about 0.5 Pa. At the appropriate pressure, the chamber is opened to a high-vacuum pumping system that continues to reduce the pressure of the process chamber. The high-vacuum pumping system may consist of a liquid-nitrogen-cooled trap and an oil-diffusion pump, a trap and a turbomolecular pump, or a trap and a closed-cycle helium refrigerator cryopump. In a low-throughput system, a trap, a titanium sublimation pump, and an ion pump could also be used. The choice of pumping system depends on required pumping speed, ultimate pressure attainable (in a reasonable time), desired film quality, method of film deposition, and expense. Traditional systems have used oil-diffusion pumps, but fear of contaminating the films with oil has led to the use of turbomolecular pumps and cryopumps. The cryopump acts as a trap and must be regenerated periodically; the turbomolecular and diffusion pumps act as transfer pumps, expelling their gas to a forepump.

The high-vacuum pumping system brings the chamber to a low pressure that is tolerable for the deposition process. This low pressure is considered the "working" or base pressure. As an example, the desired base pressure may be 6.6×10^{-5} Pa (5×10^{-7} Torr) for an aluminum evaporation system, but when auxiliary heaters are turned on the chamber pressure may rise by an order of magnitude or more. The pressure may rise still further when the evaporation source is heated. To reduce the time required for the deposition process, including the pumpdown period, system cleanliness is an absolute necessity on several levels. All components in the chamber are chemically cleaned and dried. Generally,

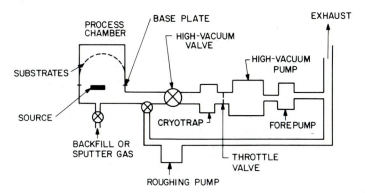

FIGURE 3
Schematic view of a high-vacuum chamber with substrates mounted in a planetary substrate support above the source.

warm water circulates in the coolant channels of the vacuum chamber to reduce the adsorption of water vapor on the freshly coated interior when it opened to the atmosphere. Any interior film buildup is removed frequently to avoid a major source of trapped atmospheric gas. Freedom from sodium contamination is vital when coating MOS devices.[23] This requirement involves cleaning the substrate to be coated in HF solutions, avoiding skin contact with any interior portion of the coating system, and using pure metal sources.

Sputter deposition demands similar precautions. The system operates with about 1 Pa of argon pressure during film deposition. Despite the relatively high system pressure, sputtering is as demanding a process as evaporation, because other gases, such as water vapor and oxygen, may be detrimental to film quality at background pressures of about 10^{-2} Pa. The purity of the argon sputtering gas is also a factor. Thus, to maintain purity, the lines connecting the gas source to the sputter chamber should be clean and vacuum-tight. For sputtering, a throttle valve should be placed between the trap and the high-vacuum pumping system. The argon gas pressure can then be maintained by reducing the effective pumping speed of the high-vacuum pump, while the full pumping speed of the trap for water vapor is utilized.

Assuming that a vacuum station of volume V has no leaks and is equipped with pumps of adequate capacity, and that the ultimate limitation is set by outgassing of water vapor, the chamber pressure P at any time t after pumpdown has been initiated is given by the approximate relation[22]

$$P = P_o \exp\left(\frac{-St}{V}\right) + \frac{Q}{S} \tag{3a}$$

This equation is a modified form of Eq. 3. Here again P_o is the initial pressure, S is the pumping speed, and Q is the rate of outgassing within the system. After the first hour of pumping the second term dominates and $P = Q_1/S$, where Q_1 is the outgassing rate after approximately 1 hour. Note that Q is a slowing varying function of time, since the source of outgassing, in principle, will eventually be depleted. This characteristic of vacuum systems has led to the introduction of "load-lock" systems, where the substrates are introduced into the process chamber through a lock chamber that cycles between atmospheric pressure and some reduced pressure. At the reduced pressure the substrates are transferred from the lock into the process chamber, and only the substrates have to be outgassed rather than the whole chamber interior. After completion of the process the substrates are transferred through the same or another lock and removed from the system. The use of such systems in production facilities is increasing, because with them, the number of wafers processed per day can exceed that which can be processed in a simple chamber (when silicon wafers 100 mm in diameter and larger are used). We used the term "process" in the above description because, in addition to film deposition, reactive sputter etching and plasma etching are also performed in load-lock systems.

Control of conductive film thickness is essential, because a film thinner than desired can cause excess current density and failure during operation. Conversely, excessive thickness can lead to difficulties in etching. The use of thickness monitors is common in evaporation deposition, and in magnetron sputter[22] deposition where planetary systems support the substrates. In some magnetron deposition processes, the film is deposited without monitoring during the deposition, but is checked after the deposition.

The most common thickness monitor is a resonator plate made from a quartz crystal. The plate is oriented relative to the major crystal axes, so that its resonance frequency is relatively insensitive to small temperature changes.[24] The acoustic impedance and the additional mass of any film deposited on the resonator cause a frequency change that can be measured accurately. After calibrating the monitor in the deposition system, it may be used to control the deposition rate as well as the final thickness of the deposited film. The resonator crystal has a finite useful life and must be replaced; however, no recalibration is necessary if the deposition system has not been modified. The resonator has a finite useful life because $\Delta f \; \alpha \; \Delta m$ holds true only for $\Delta f / f_o \leq 0.05$, where Δf is the resonator frequency change, Δm is the additional deposited mass, and f_o is the initial resonator frequency.

We can calibrate such systems and measure film thickness in at least two different ways. The simplest is to use a microbalance and weigh the substrate before and after film deposition. The film is assumed to have bulk density ρ_D, so that the increase in mass Δm is related to the film thickness x by

$$\text{Volume} = \frac{\Delta m}{\rho_D} = Ax \qquad (4)$$

and

$$x = \frac{\Delta m}{\rho_D A} \qquad (5)$$

where A is the area of the film.

Another technique uses surface profile measuring equipment. A fine stylus, usually diamond, is drawn over the surface of the substrate and encounters a step where the film has been etched or masked during deposition. The entire height of the step is detected by differential capacitance or inductance measurements. Calibration is maintained by using standard film samples which can be checked periodically. Films as thin as 100 Å or less can be measured by such equipment. Other techniques for measuring conductor film thickness include optical interference techniques and eddy current measurements.

9.4.3 Fundamentals of Physical Vapor Depositions

Although evaporation and sputtering processes are physically very different, certain behaviors of the species in the gas phase are governed by the same principles. One is the phenomenon of atomic or molecular scattering and randomization during travel from the source to the substrate. Scattering occurs due to collisions with atoms or molecules of all kinds—vapor species and residual gas molecules in the chamber. Thus, scattering is related to the density or pressure of atoms or molecules in the gas phase and is defined in terms of a quantity called mean free path (MFP). MFP is defined as the average distance of travel between subsequent collisions. From the kinetic theory of gases MFP (λ) is calculated as[21]

$$\lambda = \frac{kT}{P\pi\sigma^2 \sqrt{2}} \tag{6}$$

where k is Boltzmann's gas constant, T the temperature in degrees Kelvin, P the pressure, and σ the molecular diameter. Thus the mean free path is directly proportional to the temperature of gas and inversely proportional to the pressure and the square of the molecular diameter. At room temperature and for a typical diameter of 3 Å, one obtains

$$\lambda = \frac{1.455}{p(\text{Pa})} \text{ cm} \tag{7}$$

For a typical evaporation chamber with a pressure of 10^{-4} Pa, evaporated species can travel very long distances before they can get scattered, and most molecules get from the source to the substrate unscattered. On the other hand, during sputtering in a typical pressure of 0.5 Pa, the MFP is small and atoms or molecules could get scattered several times prior to deposition on the substrate. The scattering probability can then be defined as the fraction n/n_0 of the molecules that are scattered in a distance d during their travel through the gas. This fraction is given as

$$\frac{n}{n_0} = 1 - e^{-d/\lambda} \tag{8}$$

Here n_0 is the total number of molecules and n is the number that suffered collision.

Problem
Calculate the percent of molecules that suffer collisions during travel from a source to the substrate in a deposition system at 0.5 Pa and 10^{-4} Pa. The source-to-substrate distance is 50 cm. Assume a typical molecular diameter of 3 Å.

Solution
Using Eq. 7 for a 3 Å diameter molecule in the evaporation chamber with pressure 10^{-4} Pa,

$$\lambda = \frac{1.455}{10^{-4}} \text{ cm}$$

$$= 1.455 \times 10^4 \text{ cm}$$

whereas in the sputtering chamber at 0.5 Pa

$$\lambda = \frac{1.455}{0.5} \text{ cm}$$

$$= 2.91 \text{ cm}$$

To obtain the percent of molecules scattered, Eq. 8 is used with $d = 50$ cm.

$$\text{Then for } 10^{-4} \text{ Pa} \quad \frac{n}{n_o} = 0.003 \text{ or } 0.3\%$$

$$\text{and for } 0.5 \text{ Pa} \quad \frac{n}{n_o} \cong 1 \text{ or } \cong 100\%$$

In the above example, if the source-to-substrate distance is 50 cm (a typical distance in equipment with a planetary substrate holder), during evaporation in 10^{-4} Pa pressure only about 0.3% molecules will suffer collision, whereas during sputtering at 0.5 Pa pressure practically all molecules will suffer collision. This would then mean that during evaporation molecular motion is more or less nonrandomized and there is line-of-sight deposition. During sputtering, however, there is considerable randomization of direction of travel (unless a bias is applied to provide directionality to charged species) leading to better uniformity of deposition on stepped surfaces.

The other important behaviors to be considered relate to film growth. Film growth can be considered as condensation, nucleation, and growth in sequence. Molecules from the gas phase interact with the substrate. When the kinetic energy is small, the impinging molecule is adsorbed on a surface with a certain probability of getting desorbed. The adsorption is favored when the binding energy between the adsorbed molecule and the surface is large as when some type of chemical reaction occurs or when the surface mobility is low, as in the case of heavy atoms. If the binding energy is small, nucleation of a thin film may never occur, depending on the temperature and vibrational energy on surface. Nucleation of the film occurs when adsorbed atoms diffuse on the surface and form aggregates. Growth of the film occurs when the nuclei exceed a certain critical size that is determined by the condensate-species interfacial free energy and the free energy per unit volume of the growing phase.[24]

The cosine law of deposition. The evaporation rate R from a clean surface is related to the equilibrium vapor pressure P_e of the evaporating species by the Langmuir expression.[21]

$$R = 4.43 \times 10^{-4} \left(\frac{M}{T} \right)^{1/2} P_e \text{ g/cm}^2\text{-s} \tag{9}$$

where P_e is in pascals, M is the molecular weight of the evaporating species in grams, and T is the temperature in degrees Kelvin. The directionality and spatial distribution of the evaporated species in the evaporating chamber can be calculated following Holland's discussion of evaporation from various types of sources.[25] For evaporation from a small-area source and deposition on a plane receiver, the geometric relationships are shown in Fig. 4. In such a case, the mass deposited per unit area is given as

$$R_D = \frac{M_e}{\pi r^2} \cos\phi \cos\theta \tag{10}$$

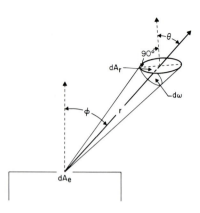

FIGURE 4
Vapor deposition from a source of a small area dA_e onto a substrate surface element of area dA_r. Geometric relationships are shown that lead to Eq. 10. (*From Maissel and Glang, Ref. 21.*)

where M_e is the total mass of the evaporated material, and r, ϕ, and θ are as shown in Fig. 4.

Equation 10, known as the cosine law of deposition, clearly shows that emission from finite-area evaporation sources is not spherically symmetrical as it would be from a point source. Maximum deposition occurs in directions normal to the emitting surface, where $\cos\phi$ is maximum, that is, ϕ is zero. In practice, small-area evaporation sources and therefore Eq. 10 are realistic. However, if the deposition surface and the evaporation source are both mounted on a spherical surface of radius r_0, it can be shown that $\cos\phi = \cos\theta = r/2r_0$, so that the mass deposited per unit area is now given by

$$R_D = \frac{M_e}{4\pi r_0^2} \tag{11}$$

Thus, the amount of deposit should be the same at all points on the spherical surface. This is the reason that most modern deposition systems use planetary wafer holders as shown in Fig. 5. Three substrate holders are mounted on a circular rail and are rotated along this rail in a way that is similar to planetary motion. There is a circular motion about the center axis normal, and each holder moves around in space above the target, as shown by the arrows in Fig. 5. Such motion causes multiple rotation of the substrates and continuously varying (1) angle of incidence, (2) instantaneous deposition rate, and (3) source-substrate distance. This leads to improved thickness uniformity of the deposit over the steps, that is, step coverage. However, because of the increased volume needed to accommodate the planetary fixture, the source-to-substrate distance is increased, leading to reduced deposition rates.

Impurity trapping during deposition. There are several sources of impurities: the deposition source and the gaseous environment being the most important ones. The source will release impurities in the gas phase. These impurities and those

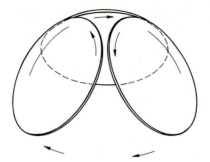

FIGURE 5
A schematic view of a planetary wafer holder.

in the environment will then bombard the surface of the growing film and get trapped during continued deposition. The fraction f_i of the species i trapped in the film is given by[21]

$$f_i = \frac{\alpha_i N_i}{\alpha_i N_i + R} \tag{12}$$

where N_i is the number of atoms of species i that bombard the film per unit area per unit time, α_i is the effective sticking coefficient of species i during deposition, and R is the deposition rate of the film.

In the case of sputtering, where a bias may be applied to the substrate, Eq. 12 is modified.[21]

$$f_i = \frac{\alpha_i N_i - (j/q)(As - \beta)}{\alpha_i N_i - (j/q)(As - \beta) + R} \tag{13}$$

where

$$A = \frac{\alpha_i N_i + \beta j q^{-1}}{\alpha_i N_i + j(s + \beta)q^{-1}}$$

and β is the fraction of the bias current due to the impurity ions, j is the bias current density, q is the electronic charge, and s is the sputtering yield for the impurities.

It is clear from Eqs. 12 and 13 that to reduce contamination, the deposition rate should be increased and α_i and N_i should be reduced. Application of bias during sputtering will increase or decrease the impurity contamination depending on the contribution of impurities to the bias current. For negative bias, positively ionized impurities will be trapped, whereas negatively ionized impurities such as $O_2^=$ or O_2^- will be repelled.

9.4.4 Optimization of the Deposition Process

Deposition processes for metallization need to be tailored for specific schemes. For example, for gate metallization polysilicon, a polysilicon/silicide sandwich, silicide alone, or refractory metal alone may be appropriate. Similarly, aluminum, aluminum on a silicide, aluminum on a silicide with a diffusion barrier in between, or aluminum on tungsten could be used for contacts. Top metal, generally, is aluminum. Individual or multiple deposition processes must be optimized to yield the best product. Engineering considerations, in such applications, are related to the mechanical stability of the films, uniformity of thickness and other properties within a run and from run to run, coverage of the stepped surfaces, equipment cost, and cost per chip. Each of these considerations vary from metal to metal and from substrate to substrate and therefore must be evaluated carefully for each metallization scheme. Sputtering from a single target or multiple targets offers flexibility and versatility and is easy to adopt. Evaporation, on the other hand, provides purer films. Use of rotating and planetary substrate holders enhances uniformity and step coverage. In certain cases, in situ cleaning (by back-sputtering) and heating of the substrates help adhesion and produce better properties in the deposit and the devices. Back-sputter cleaning, also known as back-sputter etching, is generally carried out by reversing the sputtering process. The substrates and the shutter, which isolates the target from the substrates, are biased in such a way that a controlled sputtering of the substrate surface occurs, leading to removal of undesirable surface layers and impurities. However, the back-sputter cleaning must be properly characterized and calibrated for each system before it is employed because of the possibility of cross-contamination.[26] Several deposition problems are discussed in Section 9.6.

9.5 PATTERNING

VLSI applications require very demanding controls on the metallization dimensions. This necessitates the use of anisotropic etching techniques as discussed in Chapter 5. Most chemical etchings are isotropic and therefore generally not suitable for fine-line patterning, except for the self-aligned processes discussed later. In this section, we shall briefly look at dry etching processes used to etch various metallizations. We also give some chemical recipes for etching that are useful for large-area etching such as removal of metal for reworking on the wafers.

9.5.1 Dry and Chemical Etching

Reactive ion etching (RIE) processes are carried out in a manner similar to back-sputtering. Accelerated ions hit the substrate and cause sputtering of the material from the surface, effectively etching the material on the surface. The difference lies in the nature of the gases employed. For back-sputtering or sputter etching or ion milling, inert gases such as argon and neon are used. For RIE, reactive gases (such as Cl_2 and CCl_3F) are used, hence the name reactive ion etching. Such gases play a dual role; they sputter-etch and react with the material, producing reaction product that is volatile so that it can be pumped out. Because the etching medium is a flux of ions directed towards the substrate, etching is anisotropic, that is, preferential etching in the direction normal to the surface. Table 5 lists popular gases used for reactive ion etching of various metals.

The etching of single-layer metals is less complicated than that of multilayer structures. Etching of a silicide-polysilicon bilayer sandwich is not trivial. This etch is more demanding because the substrate in this case is a thin (<250 Å) layer of gate oxide that must be preserved. The etching of silicide-polysilicon bilayers on very thin oxides requires extremely good process control, a good etch selectivity between silicide and polysilicon, and a high degree of selectivity between polysilicon and the oxide. The requirement of vertical walls without undercutting the polysilicon and attacking the gate oxide is extremely severe because of the over-etching required to clear both the silicide-polysilicon from the sloped surfaces and also the deposits on the oxide walls.

Table 6 lists the popular chemical etches that can be used to etch metals. All silicide etches contain HF and HNO_3 acid in varying ratios. They are not listed here because such etches attack everything else on the substrate and are to be used cautiously for specific experimental purposes only.

9.5.2   Self-Aligned Silicides

The initial thrust of the use of silicide in silicon integrated circuits concentrated on PtSi as the contact metallization.[5] This application was concerned with the reliability of contacts. Formation of the PtSi (and of Pd_2Si) by a self-aligned process had two important advantages: (1) the process did not require any additional lithography, and etching and alignment were predetermined and (2) it resulted in very clean silicon-silicide interfaces and, thus, highly reproducible contacts. The process was easily adaptable in fabrication lines that had sputtering equipment with back-sputter etch capability. Annealing was easy because both platinum and palladium were rather insensitive to small amounts of oxygen. In fact, small amounts of oxygen were added during silicide formation in order to form a thin SiO_2 layer on the silicide surface. The SiO_2 layer improved the selectivity of the chemical etch for the metal over the silicide. PtSi was adopted more readily because of better reproducibility. Contact to outside world was made using a

TABLE 5
Gases for reactive ion etching of metallizations

Metal	Gases
Polysilicon	Flourine- or chlorine-containing gases (CF_4, SF_6, Cl_2, CCl_3F, etc.) with/without oxygen
Al and Al containing small amounts of Si, Cu, Ti	CCl_4, $CCl_4 + Cl_2$, BCl_3, $BCl_3 + Cl_2$
Tungsten	Fluorinated gases
Refractory silicides	Fluorinated plus chlorinated gas with or without oxygen
Thin layers of TiN, TiC	Same as Al etch

TABLE 6
Chemical etches for metals

Metal	Chemical etch and conditions
Polysilicon	HF + an oxidizing acid, usually HNO_3 or CrO_3. Various compositions to achieve desired etch behavior. Acetic acid is added to reduce the etch rate and enhance uniformity across the surface.
	23.4% KOH (by weight), 13.3% n-propanol, and 63.3% water will remove polysilicon without attacking oxide.
Al and Al containing small amounts of Si, Cu, Ti.	CH_3COOH: HNO_3:H_3PO_4:H_2O in 5 : 5 : 85 : 5 mixture at 40 – 45°C. (Does not attack SiO_2 or silicon, will attack Si_3N_4).
W	0.25 molar KH_2PO_4/0.24M KOH/0.1M $K_3Fe(CN)_6$. (Will attack polysilicon or silicon.)
	HNO_3: HF in 1 : 1 or diluted solution. (Will attack Si, polysilicon, SiO_2, Si_3N_4.)
	Boiling water.
Mo	CH_3COOH: HNO_3:H_3PO_4:H_2O in 5 : 1 : 85 : 5 mixture followed by HCl : H_2O in 1 : 3 mixture.
Ti	H_2O_2: EDTA in 1 : 2 mixture at 65°C, H_2O_2–HCl, or H_2O_2–H_2SO_4.

Ti/Pt or Ti/Pd diffusion barrier between the silicide and gold or aluminum as the outermost layer.

More recently, other, lower-resistivity group VIII metal silicides—namely those of cobalt, nickel, and titanium—have been evaluated. The resistivities of these silicides were included in Table 3. $CoSi_2$ offers the lowest resistivity. A comparison of these self-aligned silicides and their processes is given in Reference 27.

The conventional self-aligned silicide process for contacts is described here. The metal film is deposited over the patterned oxide and silicon in the window. In situ cleaning, such as by back-sputtering in the sputtering chamber, is essential for reproducibility. After the metal deposition, the silicide is formed by annealing. Finally, unreacted metal over SiO_2 is removed by a selective wet etch that etches metal without attacking silicide or SiO_2.

To use the same approach at the gate level, a few changes are necessary. In Fig. 6, a refined sequence of forming the silicide at gate, source, and drain simultaneously is shown schematically.[27] In this sequence, polysilicon is deposited on the gate oxide and doped. Then, the oxidation mask Si_3N_4 is deposited by LPCVD process. The polysilicon–Si_3N_4 sandwich is defined to form the gate and interconnection pattern. Source and drain are formed by ion implantation and the photoresist is removed. Following this, oxidation is carried out to form oxide walls on the polysilicon. This heat treatment activates and diffuses the dopants.

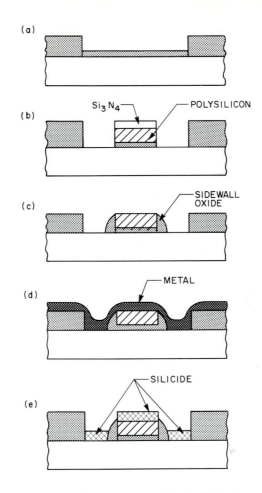

FIGURE 6
Showing process steps for self-aligned gate, source, and drain-silicide formation.

The oxide is now removed anisotropically from the source and drain region, leaving oxide sidewalls on the polysilicon. The remaining nitride is removed by a selective chemical or dry etch, leaving exposed polysilicon and source and drain surfaces. Metal is then deposited following an in-situ surface cleaning. A silicide is formed by annealing, and unreacted metal is removed by selective chemical etch, leaving silicide at the gate, source, and drain regions and on the interconnection lines. Note that the use of Si_3N_4 on top of the polysilicon film serves two important roles. First, it serves as the oxidation mask for the polysilicon film, allowing the oxidation to occur only on the sidewalls of the polysilicon and at the source and drain regions. Secondly, it allows the formation of the silicide with excellent step coverage that matches the polysilicon step coverage.

A successful application of the self-aligned silicide technology will, however, occur only after a clear understanding is obtained of the metal-silicon reaction kinetics and the role of various factors (such as interfaces, dopants, and impurities) on the reaction kinetics.

9.6   METALLIZATION PROBLEMS

In this section, various metallization problems associated with (1) deposition, (2) processing, (3) metallurgical and chemical interactions, (4) electromigration, and (5) device performance are considered. Each is examined separately. Solutions, in some cases, are related to careful experimentation and control. In other cases, there are no simple solutions and research continues to find solutions.

9.6.1   Deposition

Impurities in the film, adhesion, stress, cracks, grain size, stoichiometry in alloys, step coverage, and thickness nonuniformity are possible problems associated with deposition. With the exception of the step coverage problems, all can be solved by tailoring the physical vapor deposition process. Impurities in the films can be minimized by using pure evaporation and sputtering sources, high vacuum and pure gases, and clean surfaces. The use of oil-less vacuum systems will eliminate contamination resulting from hydrocarbons. Surface treatments that reduce sticking coefficients may be formulated to minimize impurity trapping in the film. Adhesion of the film is promoted by (1) inducing strong atom-to-atom bonding (between the film and the substrate) within the interfacial region, (2) reducing the stresses, (3) the absence of easy deformation or fracture modes such as those caused by surface unevenness and particulates, and (4) the absence of long-term degradation modes such as those due to exposure to moisture and air. Surface cleanliness and localized deposit-surface interaction will improve localized atom-to-atom bonding. Deposition parameters are mainly responsible for stresses—although stresses caused by thermal expansion coefficient differences and intrinsic factors (see below) could play an important role.

A thin film deposited on a substrate is in a state of stress that can be tensile (i.e., the film wants to contract) or compressive (i.e., the film wants to expand parallel to the surface). Such stresses arise from several factors and are generally grouped into two components, the thermal stress σ_t and the intrinsic stress σ_i.

$$\sigma = \sigma_t + \sigma_i \tag{14}$$

The thermal stress σ_t results from a difference between the coefficient of thermal expansion of the film, α_f, and that of the substrate, α_s. One material expands with temperature at a different rate than the other. Quantitatively, σ_t is given by

$$\sigma_t = \frac{E_f}{1 - \nu_f} \int_{T_1}^{T_2} (\alpha_s - \alpha_f) dT \tag{15}$$

where E_f and ν_f are the Young's modulus and the Poisson's ratio of the film, respectively, and T is temperature.

Equation 15 is strictly true only if the coefficients of thermal expansion do not change with temperature (i.e., $d\alpha/dT = 0$), which is not the case for silicon. The expansion coefficient α_{Si} changes from $\sim 2.6 \times 10^{-6}/°C$ at 25°C to $\sim 3.3 \times 10^{-6}/°C$ at 900°C.[28]

Intrinsic stress, also referred to as growth stress, is less understood quantitatively. There are a variety of sources of intrinsic stress in polycrystalline films. Among these are volume changes, grain growth, microstructure, impurities, defects, and film composition. These stresses are low when the temperature is greater than $T_m/3$ (T_m = melting temperature in Kelvin). At this point, atom mobility is high and atoms readily diffuse along grain boundary paths. Holding the film above this temperature will minimize growth stresses. In sputtered films, the entrapment of the sputtering gases (e.g., argon) causes intrinsic stresses. Thus the sputter pressure–stress and sputter rate–stress relationships must be carefully established for each sputtering system to find the sputtering conditions that yield the lowest stress condition in the deposits.

Cracks in the deposits are the result of poor adhesion, the presence of particulates or unevenness, and/or the relief of stresses in the film. Control of these effects will eliminate cracks in practically all cases. Cracks may, however, result following a heat treatment of the film—usually due to a volume change or a chemical interaction.

Grain size variation in the deposits is of the least concern in most cases. However, in cases where the film is subjected to postdeposition anneal or electromigration, grain size variation and grain growth may cause failure as discussed below for aluminum films. Larger grains minimize grain boundary area, and therefore, diffusion and electromigration are reduced. Thermally, larger grains provide more stability.

When depositing alloy films, the stoichiometry of the alloy in the film should be controlled. As discussed earlier in this chapter, sputtering from a given alloy target provides the best control for a given sputtering condition. Varying the sputtering rates and the chamber pressure has small effect on the composition of the deposit. However, variability in the composition of the alloy target with depth could lead to compositional variations in the films deposited in different runs.

Step coverage is a serious problem for physical-vapor-deposited films. Poor step coverage results because of (1) the directionality of the deposition from the evaporating or sputtering sources, (2) low mobility of the deposited atoms, molecules, or ions, and (3) the enhanced topography (deeper steps) resulting due to processing prior to metallization.

Solutions to the step coverage problem have been approached in several ways. First, raising the temperature of the substrate during film deposition to about 300°C creates greater surface mobility of the deposited material, thus reducing the severity of cracks that exist in corner regions. Next, orientation of the substrate relative to the source can be optimized.[29,30] This optimization is sometimes especially important, because shadowing occurs in the deposition process when a point source such as an e-beam or an inductively heated melt is

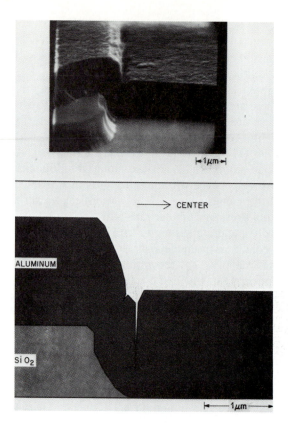

FIGURE 7

(a) The photograph on top shows what is actually obtained. The bottom sketch shows the prediction by a computer model. The step in SiO_2 is perpendicular to the planet radius and the arrow points to the planet center.

used. Computer simulation has been useful in modifying the supporting planetary system.

Since most planetary systems do not use rotation of the individual substrate about its own axis, orientation within the planet is significant in reducing step coverage problems.[29] Step edges that are parallel to the planet radius are coated symmetrically. Steps with edges placed perpendicular to the planet radius tend to be coated asymmetrically, and also tend more to exhibit cracks (Fig. 7).

If small contact windows are to be coated, the course of action may be different than outlined above. For VLSI, a plane surface may be approximated by depositing the interlevel dielectric by bias-sputter deposition or by using planarization.[31] Planarization is a low-temperature process that reduces surface features (see Section 9.7.1). A thick resist layer is applied to the dielectric, and a plasma-etch process is used that attacks the dielectric and the resist at equal rates. To accommodate this process, a thicker than normal (usually by a factor of 2) intermediate dielectric layer is needed. The extensive heat treatment that normally would be used to make the dielectric flow, thus reducing the severity

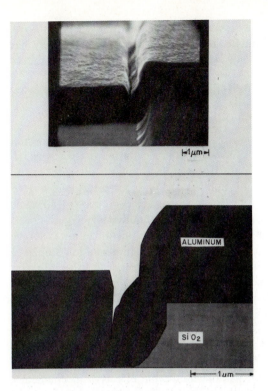

FIGURE 7

(b) Actual (top) and computer model prediction (bottom) are shown for the step parallel to the planet radius. (*From Blech, Fraser, and Haszko, Ref. 29.*)

of the step contours, cannot be tolerated in VLSI where implanted dopants are not permitted to diffuse extensively. Contact window step coverage remains a problem even on planar surfaces, because extensive taper-etching of the window edge would consume excessive wafer area.

The use of sources that have larger areas than point sources, such as magnetrons, relieves many of the step coverage problems. If the substrates are relatively distant (20 to 30 cm) from the source, such as planetary-mounted wafers, the directionality of the sputtered metal vapor becomes more random. Randomness occurs because at pressures of about 0.5 Pa the mean free path of the argon atoms is of the order of 1 cm. Thus the vapor incident on the planetary-mounted substrate during sputtering is more random in direction than the evaporated vapor, but it is also "colder" because it transfers energy to the argon gas. The vapor's lower energy, which is characteristic of the incident vapor, leads to less movement of the deposited species on the substrate surface. Decreased movement can limit the grain growth and the development of ordered (fiber texture) structures. The substrates can be relatively close and stationary, or they can move slowly in front of a large-area magnetron. This proximity to the source permits high deposition rates with material that has undergone an order of magnitude less travel through the argon. Significantly more heating of the substrate can be achieved, resulting in improved step coverage. Sidewall and flat-surface film-thickness ratios ranging from 50 to 100% have been obtained on steps. In windows this ratio is dependent on the aspect ratio (depth/width) of the window.

9.6.2   Processing

Metallization is applied at different stages of integrated-circuit processing. The gate metal is deposited and patterned during early stages. The contact metal is formed toward the end. If the contact metal is different from the top-level metal, the latter is generally applied after the former. Thus, a gate metal is subjected to considerable processing, whereas the contact metal experiences very little. Postmetal processing steps are listed in Table 7.

It is evident from Table 7 that gate metallization is subjected to several high-temperature steps and at least one ion-implantation step. It is also subjected to oxidizing environments. All of these steps lead to stability problems. Polysilicon alone has been used as metal at this level to solve practically all of the problems. In recent years a polysilicon–silicide sandwich has replaced polysilicon in VLSI devices. The stability of this sandwich during processing and during device operation is related to the tight control required on the silicide deposition parameters. Each silicide has its limitation.[32] Impurities, especially carbon and oxygen, play varying roles in different silicides. Silicides of titanium and tantalum tolerate small amounts of oxygen without noticeable effect on mechanical stability. Other silicides behave differently. Silicides of tungsten and molybdenum need special precautions, especially during oxidation. This is due to the inability of these metals to

TABLE 7
Postmetal processing steps

Gate and interconnection metal	Contact metal	Top metal
1. Patterning by use of RIE	1. P-glass deposition flow or planarization	1. Metal pattering
2. Silicide anneal	2. Diffusion barrier	2. Si_3N_4 deposition if necessary
3. Sidewall oxide formation by thermal oxidation or CVD followed by selective anisotropic etch	3. Aluminum (top metal) deposition	3. Si_3N_4 patterning
4. Ion implantation for source and drain	4. $450°C/H_2$	4. $450°C/H_2$ anneal for 30 – 60 min
5. Source and drain diffusion		
6. Contact metallization		

reduce native interfacial-oxide layers on polysilicon. The stability of the refractory silicides can be enhanced by the use of silicon-rich silicides. Excess silicon provides bonding at the interface and an improved oxidation characteristic. Silicides of titanium and cobalt, recently recommended for gate applications because of their self-aligned features and low resistivities, need special attention because of oxygen affinity and low-temperature (<900°C) stability respectively.[27]

Contact metallization stability requirements are less severe. Shallow junctions (100–2000 Å deep) require contacts that will not penetrate and ruin such junctions. Also, dopant redistribution during silicide formation must be controlled or compensated for reliable junction/contact formation.

Aluminum is the most popular top-level metal. Because its melting point is low, the 350–450°C postmetal anneal leads to various metallurgical and chemical interactions as discussed in Section 9.6.3. Except for these interactions, aluminum is a very stable material.

9.6.3 Metallurgical and Chemical Interactions

An introduction to the metallization problems associated with the stability of various metallizations was given in Section 9.3.3. Metallizations can be completely destroyed by reactions induced by thermally activated processes with the substrates or the layers on top. Pure silicide films on polysilicon or silicon are fairly stable in their metallurgical interactions with the substrate. Most of them are affected only slightly by the various chemical solutions used in IC fabrication. $TiSi_2$, however, dissolves at a very fast rate in HF-containing solutions. Thus, exposure of $TiSi_2$ to such solutions should be avoided.

Aluminum on silicon or silicide, however, leads to metallurgical interactions causing serious instabilities. Annealing of aluminum on silicon (typically at 450°C) causes dissolution of silicon by diffusion into the metal and leads to pit formation (Fig. 8).[33] The dissolution process is highly nonuniform due to the presence of a nonuniform interfacial layer that leads to isolated crystallographic etch pits, bounded by {111} surfaces.[34] The pits in silicon (Fig. 8) become visible after selectively removing aluminum from the surface. In actual devices, diffused aluminum is present in silicon below the silicon-aluminum interface. If the penetration is deep, it leads to contact and junction failure. This interaction between aluminum and silicon is a thermally activated process and, therefore, depends on the temperature and duration of the anneal. The interaction has been followed by etching the aluminum off and measuring the pit size as a function of time and temperature. Pit growth is found to be a thermally activated process with 0.8 eV activation energy.[33] The median pit size $\bar{\ell}$ is directly related to the square root of the anneal time (Fig. 9) and to $w^{-2/5}$, where w is the window area (Fig. 10). The (time)$^{1/2}$ dependence indicates a diffusion-controlled process. The window-area dependence can be understood by noting that the amount of aluminum is infinite as long as runners go over the windows. With increasing window size the Si-to-Al contact area increases, allowing pits with different sizes to form and therefore reducing the pit size, since the amount of silicon dissolved in aluminum

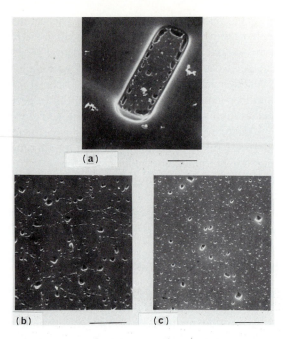

FIGURE 8
Aluminum/silicon interpenetration structures observed after a 450°C 30 min H$_2$ anneal and aluminum etch. (a) Typical contact window that had been covered with In-source aluminum. Recrystallized silicon particles can be seen dispersed randomly on the wafer surface. (b), (c) Pit structure and silicon-precipitate distribution associated with e-gun and In-source aluminum, respectively. The micrographs correspond to the central region of a large contact pad. The marker in each case represents 5 μm. (*From Vaidya, Ref. 33.*)

is independent of the window size and only depends on aluminum film thickness, length of the runners, and anneal temperature and time.

Contact failures, in Al-Si systems, also occur due to precipitation of dissolved silicon (from aluminum) on cooling. The silicon precipitates (see Fig. 11) may cause an undesirable increase in the contact resistance, especially for Al-on-n-Si contacts.[35]

To prevent junction shorts caused by the preferential dissolution of silicon into aluminum, silicon is added to the aluminum during the deposition of the metal film. The amount of silicon required in the aluminum is determined by the maximum process temperature and the solid solubility, which can be obtained from the Al-Si phase diagram. Normally, slightly more than 1 wt. % silicon is added. If the contacts are all to p silicon, this method of solving the junction spiking problem is acceptable. Once again, there is a problem if the contacts are to n silicon. Because of excess silicon present in aluminum, on cooling, some precipitation of silicon occurs in the contact window. This leads to a nonohmic contact to n silicon, because the recrystallized silicon precipitate contains aluminum which is

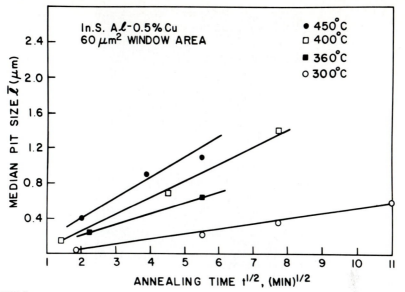

FIGURE 9

Median pit size ($\bar{x}$) as a function of the square root of the annealing time at 300, 360, 400, and 450°C; the window size was 60μm². (*From Vaidya, Ref. 33.*)

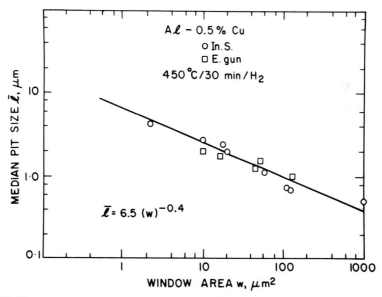

FIGURE 10

Log-log plot of median pit size ($\bar{\ell}$) vs. window area (w) for In-source and E-gun aluminum after 450°C 30 min H₂ anneal. (*From Vaidya, Ref. 33.*)

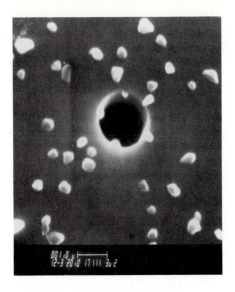

FIGURE 11
Silicon precipitates in and around a window in SiO_2 on silicon; aluminum has been etched off.

a p-type dopant. Figure 12 shows the resistance changes in resistors of various widths as a function of the anneal time at 200°C.[36] Thus, even at 200°C, Si-Al interaction occurs, leading to changes in the properties of the metal.

To solve this problem of Al-Si interaction a diffusion barrier between silicon and aluminum is recommended. Of the barriers suggested in Section 9.3.1, self-aligned silicides of Pt, Pd, Co, Ni, and Ti have been considered most often. They have been used both as Schottky and as ohmic contacts to silicon. The formation of silicides by metallurgical interaction between pure metal film and silicon leads to the most reliable and reproducible Schottky barriers. The properties are reliable and reproducible because silicide formation by metal-silicon interaction frees the silicides-silicon interface of surface imperfections and contamination. Similarly,

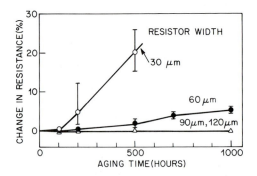

FIGURE 12
Change in the resistance of diffused resistors with silicon-doped aluminum metallization during a 125 mA and 200°C ambient stress test. (*From Mori, Kanamori, and Ueki, Ref. 36.*)

the formation of the silicide-silicon contacts atomically cleans the interface, thus avoiding the variability in contact properties that may otherwise occur when the surface is contaminated or imperfect.

The formation of self-aligned silicides, however, will consume silicon from the shallow junction regions—although the process does not require development of a patterning technique. The possibility of using silicides deposited directly into contact windows offers the advantage of preserving shallow junctions, which may be penetrated by a conventional silicide formed by reacting metals with the silicon. For deposited silicides, one must develop a patterning process. Among several other ways that have been suggested to form shallow silicide contacts are: (1) use of a snowploughing effect (the dopant diffuses ahead of the silicide) during silicide formation, (2) diffusion of dopant from the deposited silicide, (3) the process of solid-phase epitaxy, (4) use of metal-rich codeposited silicide, and (5) use of two-metal deposits. Of these ways, the first three form the shallow junction at the same time the silicide is formed. The last technique forms the contact silicide (such as platinum or palladium silicide) together with the diffusion barrier (such as tungsten metal) on top. All these techniques have advantages and disadvantages that need further investigation.

Aluminum interacts with silicides in a similar manner. The reaction leads to the decomposition of silicide, the dissolution of silicon in aluminum, and the formation of binary aluminum-metal or ternary aluminum-metal-silicon intermetallics depending on the heat treatments involved. The result is aluminum penetration of the silicide into the underlying silicon substrate. Various silicides react with aluminum at different temperatures. Refractory silicides of Mo, Ta, Ti, and W react with aluminum at temperatures higher than 450°C. Pd_2Si and $PtSi$ react at about 250°C. $CoSi_2$ reacts at about 400°C. Use of the last three silicides will therefore require a diffusion barrier between them and the aluminum. Diffusion barriers like TiN, TiC, and Ti–W are found to be stable at 450°C, and no metallurgical or electrical degradation of the silicide-barrier-aluminum contact metallization has been reported at 450°C as long as these barrier films are pinhole free. Generally a thickness in the range of 500–1000 Å is appropriate. Note that TiN and TiC are the compounds of nitrogen and carbon with titanium, respectively, whereas Ti–W is a deposited mixture of the two elements, which may form intermetallic compound or compounds depending on the heat treatment following the deposition.

9.6.4 Electromigration

Electromigration can cause considerable material transport in metals. It occurs because of the enhanced and directional mobility of atoms caused by (1) the direct influence of the electric field and (2) the collision of electrons with atoms, which leads to momentum transfer. In thin-film conductors that carry sufficient current density during device operations, the mode of material transport can occur at much lower temperatures (compared to bulk metals) because of the presence of grain boundaries, dislocations, and point defects that aid the material transport. Thus, in devices using aluminum, failures associated with discontinuities in the conductor caused by the electromigration have been observed.[37-40] Figure 13 shows electromigration failures in several Al–Si and Al–polysilicon structures. Aluminum is gathered in the direction of electron flow, leading to a discontinuity in the current-carrying lines. This type of device failure seriously affects the reliability of silicon integrated circuits.

The drift velocity v of migrating ions is given by

$$v = j\rho \frac{qZ^*}{kT} D_o e^{-Q/kT} \tag{16}$$

This equation, derived theoretically from Einstein's drift velocity equation,[41] has been modified[42] for use in determining the relationship between the median time to failure (MTF) and j and Q. MTF is the time, for a given testing condition, at which 50% of the testing sites will have failed. A general expression for MTF is

$$\text{MTF} \ \alpha \ j^{-n} e^{Q/kT} \tag{17}$$

In Eqs. 16 and 17, j is the current density, n an exponent between 1 and 3, ρ the film resistivity, qZ^* the effective ion charge,[41] k Boltzmann's constant, T the absolute temperature, D_o the preexponential diffusivity, and Q the diffusion activation energy.

Electromigration-induced failure is the most important mode of failure in aluminum lines.[39] Most MTF measurements, for a given current, are for MTF as a function of the temperature and the linewidth of the conductor. The results are found to be strongly influenced by the method of deposition of the metal, the temperature of the substrate during deposition, the alloying element, if any, the substrate type, the film thickness, and the length of the conductor.

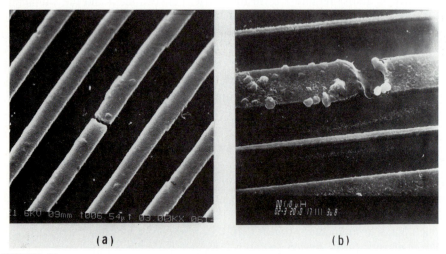

(a)   (b)

FIGURE 13
SEM micrographs of electromigration failure in aluminum runners, for (a) S-gun magnetron–deposited Al–0.5% Cu alloy and (b) In-source-evaporated Al–0.5% Cu alloy. (*From Vaidya, Fraser, and Sinha, Ref. 38.*)

Activation energy Q, measured for electromigration in aluminum and its alloys, ranges from 0.4 to 0.8 eV. This can be compared with the diffusion activation energies of 1.4 eV for bulk, 0.28 eV for surface, 0.4–0.5 eV for grain boundary, 0.62 eV for grain boundary plus bulk, and a value of $0.62 < Q < 1.4$ eV for defect-assisted bulk diffusion.[43] The measured activation energy lies in the range of those for pure grain boundary diffusion and grain boundary (and defect) plus bulk diffusion. This is very significant, because the relative contributions of the grain boundary and bulk diffusions depend on the grain size and the grain structure of the film. It explains the observed spread in the Q values obtained for a variety of films prepared by different methods. Within experimental error and allowing for the grain size variations, the activation energy Q is found to be independent of the film-deposition conditions and the alloying element (or elements).[39,43]

Detailed investigations[38] of the electromigration lead to the conclusion that MTF increases with increasing grain size S, degree of $\{111\}$ preferred orientation, and decreasing spread in the grain size distribution σ. MTF is proportional to an empirical microstructural quantity η, given by[38]

$$\eta = \frac{S}{\sigma^2} \log(I_{111}/I_{200})^3 \qquad (18)$$

where I represents the intensity of corresponding X-ray diffraction peak. In these findings, for the first time, a rationale has been presented to take into account the various structural factors that influence electromigration resistance.

Figure 14 shows MTF as a function of linewidth.[37] MTF decreases with decreasing linewidth; however, at linewidths of less than 2μm MTF increases, thus reversing the trend of the wider lines. This effect has also been rationalized on the basis of the grain structure and size distribution. TEM studies[37] have revealed that in narrow lines, the grain structure takes on a "bamboo" appearance with grain boundaries generally running perpendicular to the direction of the current flow. In contrast, wider, less stable lines of the same material contained much more heterogeneous grain structure with several triple points across the width (Fig. 15).

Addition of copper to aluminum or Al–Si alloys considerably increases the electromigration resistance, which is also found to depend strongly on the method of the deposition of the metal film, in addition to the linewidth (see Fig. 14). The apparent superiority of the Al-Cu alloys[44] and e-gun evaporated metal or alloys is associated with the resulting preferred $\{111\}$ texture and improved grain size distribution such as in the bamboo structure.[37] At present one can speculate that in addition to changing the texture of the aluminum films, the addition of copper increases the electromigration resistance by enhancing the activation energy of the self-diffusion of aluminum. Other elements that have shown similar effect are Ni,[45] Cr,[46] Mg,[47] and O.[48]

It has been shown that MTF first decreases with the length of the aluminum runner, and then becomes independent of the length for all practical purposes.[49] A current density threshold, which is found to be inversely proportional to the

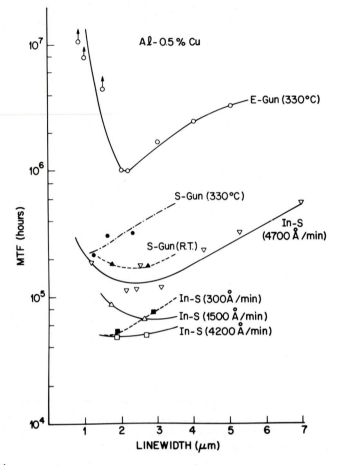

FIGURE 14
Extrapolated median time to failure (MTF) as a function of line width for E-gun, S-gun, and In-source (In-S) aluminum films at 80°C and 1×10^5 A-cm^{-2}. The film deposition conditions are shown on the plot. The in-source films were evaporated onto heated (300°C) substrates at different rates. R.T. = room temperature. (*From Vaidya, Sheng, and Sinha, Ref. 37.*)

aluminum runner length, exists, below which the aluminum mass transport due to electromigration stops.[50,51] Thus, for a given current through the runner, the electromigration-induced mass transport is higher in the small runners. It has been suggested the stress gradients in the film are the prime cause for this reverse mass transport and current-density threshold.[50,51]

The use of polysilicon as the sacrificial layer between the source, drain, or gate and the top-level aluminum offers two very important advantages.[52] First, polysilicon is deposited by LPCVD techniques that produce excellent step coverage, and eliminate the possibility of contact break at the window edges. Second, it provides enough silicon to satisfy the solubility requirements of aluminum. Doping of polysilicon can be achieved by autodoping from the source and drain

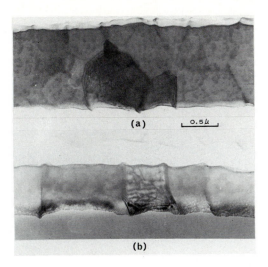

FIGURE 15
Transmission electron micrographs of the grain structure of e-gun aluminum–0.5% copper in (a) 2.2 μm, and (b) 1.0 μm wide lines. (*From Vaidya, Sheng, and Sinha, Ref. 37.*)

areas, in situ doping during deposition, or by ion implantation. This process, called the "poly plug" process, is used in fabrication of NMOS devices.[52] For CMOS devices requiring both n- and p-type polysilicon, extra masking and lithographic steps are necessary to implant n and p dopants in the desired windows.

Junction leakage and electromigration studies of the aluminum over n^+-polysilicon composite have been carried out.[53-56] It was found that, while such a composite can provide excellent step coverage and prevent thermally induced Al–Si interpenetration, the devices are still susceptible to failure by junction leakage at positively biased windows, resulting from the reduced electromigration resistance of the composite. Figure 16 shows the reduced lifetime for e-gun or in-source aluminum over polysilicon compared to that for e-gun or in-source aluminum alone. The process involved a localized diffusion of silicon (from polysilicon) into aluminum, followed by the transport of silicon across the metal-semiconductor interface. The lifetimes are found to exhibit a strong dependence on current, varying as I^{-n}, where n = 10, indicative of steep temperature gradients in the vicinity of the contacts.[48] These studies[53-56] indicated that the thermal activation energy associated with this failure mode is 0.9 eV, corresponding to the diffusion of silicon in aluminum alloy that contains silicon in excess of 0.4%.[57]

9.7 NEW ROLE OF METALLIZATION

New metallizations will play increasingly important roles in electronic devices and integrated circuits. Metallizations will not only be current carriers, but play an active role in determining device properties, as in the case of the gate electrode or Schottky barrier diode.

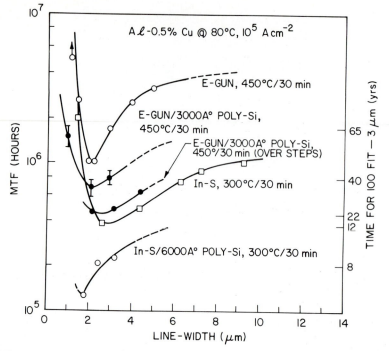

FIGURE 16

MTF as a function of the runner width for plain Al and Al/poly-Si composites at 80°C and 1×10^5 A-cm^{-2}. The respective heat treatments are shown on the plot. The time taken to react a 100-FIT failure rate for the 3 μm lines of each structure is listed on the right-hand side. (*From Vaidya and Sinha, Ref. 53.*)

9.7.1 Multilevel Structures

To minimize interconnection resistance and to save valuable chip area, multilevel metallization schemes have been proposed. In this concept, as demonstrated in Fig. 17, which shows a three-level metallization scheme, metal interconnects run in the third dimension as well. Theoretically, several levels of metals can be employed. Dielectric films are used as the insulators between various metal layers. The interconnections are made by opening holes in the dielectric layer.

There are several problems associated with making a multilevel metallization scheme work. These are all related to deposition, etching, and metal characteristics. Aluminum is the most favored metal. Its use limits the maximum temperature for dielectric film deposition to 400–450°C. The most serious processing problem is associated with the step coverage and filling of the vias. Figure 17 shows the ideal step coverage and via filling such that each level is flat and parallel to the original substrate surface. It has been very difficult to produce such deposits—either the metal or the dielectric. CVD methods, known for their ideal step coverage, have not yet been successful for aluminum. CVD aluminum deposits contain large amounts of impurities (carbon and oxygen) that increase the

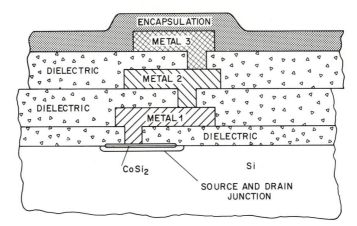

FIGURE 17
A schematic drawing of a multilevel metallization structure.

resistivity.[58] Also, the deposits are very rough deposits and make patterning very difficult. Low-temperature CVD depositions of SiO_2 produced reasonably good coverage, but have been found to be inadequate due to poor dielectric strength. Several other dielectrics, like polyimides and spin-on glasses, have been tried with success in very limited cases.

New developments that may lead to realization of structures of the type shown in Fig. 17 are deposition by bias-sputtering and the planarization process. In bias-sputtering, the substrate is biased so that back-sputtering is facilitated on the surface. Thus sputter deposition and back-sputter etching occur simultaneously. Back-sputtering is asymmetric and is favored at sharp corners and edges in such a manner that, when optimized, it gives excellent step coverage of the deposit, ending up as flat surface. Bias-sputter deposition rates are, however, low. Also, ion damage occurs during these depositions. More work is necessary to prove this process.

In the planarization process, the surface is planarized parallel to the substrate surface. After thick CVD deposits that fill the via and cover around it have been made, a thick photoresist is deposited. Photoresist is usually so thick that it covers all the steps well and the top surface is flat. A plasma or reactive-ion-etching process is chosen to etch the resist and the CVD deposit at the same rate. The surface is etched until all resist and some glass is etched off, leaving the surface flat. The process steps are shown schematically in Chapter 6, Fig. 19. The planarization process is becoming popular in VLSI fabrication.

A mention of selective metal deposition, a technology yet to be proven, must be made. In this type of deposition tungsten metal is selectively deposited on silicon only; no deposit occurs on the SiO_2 layers.[59-61] Such a method of deposition, when proven reliable and reproducible, is probably the best for multilevel structures as well as for contact metallization. It provides excellent contact to silicon and the deposit could fill the via. In addition, the deposit of tungsten will not require a diffusion barrier when contacted with an aluminum top layer. Finally, one can see the selective tungsten technology's applicability to forming the gate and interconnection metallization on polysilicon. In fact, selective tungsten could be deposited simultaneously on polysilicon gates and interconnections, sources, and drains, in a manner similar to that described for self-aligned silicide technology. The advantage over silicide technology lies in the minimal consumption (100–200 Å) of the silicon in the source and drain regions. However, several problems remain: (1) reproducibility of the selectivity of the deposition; (2) occurrence of microstructural defects at the W–Si interface (such defects are associated with the deposition process); (3) preferential erosion of n^+ silicon compared to p^+ silicon; and (4) sporadically observed contact instability.[61]

9.7.2 Epitaxial Metals, Three-Dimensional Devices, and Heterostructures

Recent developments in forming epitaxial silicides[62-64] and aluminum[65-66] open the door to new methods for contact metallization and to the possibility of forming three-dimensional devices and heterostructures. Epitaxial schemes will provide grain-boundary-free and possibly dislocation-free films. Such films practically eliminate metallurgical interaction problems associated with grain boundary (and dislocation) diffusions. Metallizations will be stable at higher temperatures and have longer electromigration lifetimes. Epitaxial growth of metallization on silicon also suggests the possible epitaxial growth of silicon on top of the metal. Thus, heteroepitaxial structures of silicon-metal-silicon can be created. This scheme opens a way to make three-dimensional circuits and new devices. Metal-base transistors and permeable-base transistors are two examples of new devices.[10] Silicon–$CoSi_2$–silicon heteroepitaxial structures and metal-base transistors have recently been fabricated.[64]

Epitaxy of aluminum on silicon has recently been demonstrated.[65,66] In this case, it was demonstrated that Al–Si interaction did not occur, at least up to 450°C, a remarkable improvement on the stability of Al–Si contacts. Since diffusion activation energy in single-crystal aluminum films is nearly 1.4 eV, the electromigration will be significantly reduced in epitaxial aluminum lines on silicon.

9.7.3 Diffusion Barriers

In the absence of epitaxial metal films, diffusion barriers will be required between silicon or silicide and aluminum for devices that will be subjected to poly-aluminum anneals at 400°C or higher temperatures. As discussed earlier, transition metal borides, carbides, and nitrides have been suggested for this application with reasonably good success. A new approach employs a codeposited two-metal mixture. A low-temperature reaction between a substrate silicon with a refractory metal–silicon alloy mixture always leads to the formation of the noble metal silicide on silicon and to the accumulation of the refractory metal (or the silicide formed at sufficiently high temperatures) at the top. The amount of silicon consumed can be controlled by varying the composition of the alloy and limiting the reaction temperature so that only noble metal silicides are formed. The advantage lies in the observed layered phase separation,[67] so that the refractory metal top layer can act as the diffusion barrier between the silicide and subsequently deposited aluminum. Such an approach, however, will require the development of a pattern generation process for the two-metal deposit.

An extension of the two-metal approach was made by the use of metal-rich titanium carbide films.[68] The titanium-rich carbide films can be reactively sputtered, and can react with silicon leading to a phase separation with $TiSi_2$ near the silicon substrate plus an outer carbide layer. Such a heat-treated structure with an aluminum top layer maintains its integrity even after 550°C 30 min heat treatment. Thus, a heat treatment of titanium-rich carbide ($Ti_{3.1}C$) film on silicon at 650–750°C results in a good $TiSi_2$ contact and a built-in diffusion barrier for aluminum penetration.

Recently titanium boride (TiB_2) films of excellent quality have been deposited.[69] They were found to provide excellent stability to silicon-TiB_2-aluminum contact systems. One can envision using TiB_2 on silicon and heating to temperatures high enough to initiate a reaction between the boride and silicon at the interface. This reaction leads to the release of boron to dope silicon and forms silicide at the interface. Thus, in one anneal, a shallow (pn) junction, a silicide contact, and a boride diffusion barrier can be made.

9.7.4   Redundant Metal Links

Many large MOS memory chips are designed, of necessity, with redundant metal (fusible) links to significantly improve the yield. These spare metal rows and columns of bits, which can be exchanged for the faulty ones, are part of the circuit design. They can be opened by laser or electrical means, thus disconnecting the faulty ones. After the faulty links are disconnected, their memory in the memory array is transferred to the spare row or column by opening additional fusible links in the memory-decoding circuitry. Thus, use of metallization in the form of these fusible links allows considerably enhanced yields and reduces the cost of large memory chips.

9.8   SUMMARY AND FUTURE TRENDS

Metallization for VLSI requires considerable understanding of the application, metallization choices, properties, stability, and selection of optimum deposition processes. This chapter presented an explanation and discussed the limitations of various metallization schemes. It also discussed some of the significant new trends such as the use of metallization to create three-dimensional device structures and devices.

Future applications will demand lower resistance at all levels and sophisticated deposition techniques. It is apparent that presently used or conceived metal-lization schemes are inadequate for future device concepts. Selective tungsten or aluminum deposition, epitaxial silicides or aluminum, and self-aligned silicides show promise and could pave the way for better and faster devices and circuits. Multilevel or three-dimensional metal networks will reduce the interconnection resistance that could result as chips get more complex, and will also reduce the chip area significantly.

PROBLEMS

1 Show that scaling down to smaller linewidths does not affect the RC time constant except for the effects of fringing fields.

2 Assuming there are no charges in the oxide except that Q_f (the fixed oxide charge) at the oxide (1000 Å thick)–silicon interface is $2 \times 10^{11}q$ per cm^2: calculate the ϕ_m that will be necessary to produce a condition of $V_{FB} = 0$. The ϕ_s for silicon is 4.8 V.

3 High-temperature stability of metal films on lightly doped silicon can be investigated by measuring the resistivity changes of the metal film as a function of the annealing temperature. Design the experiment and state the assumptions you need to make for this study.

4 Metal interconnection lines of varying widths are formed. Prior to patterning, the sheet resistance was measured to be 2.5 $\Omega/\square$. Calculate the total resistance of 0.5, 1, 2, and 5μm wide and 1 cm long lines. What other information will you need to guess the type of metal used?

5 Calculate the mean free paths of Al, Si, Co, and Pt atoms sputtered in pressures of 0.5 and 1 Pa at 300 K. Estimate the number of collisions an aluminum atom will encounter in a path length of (a) 2.5 cm and (b) 40 cm. Which of the two deposits will have larger grains?

6 Schottky barrier height is generally determined by measuring I-V characteristics of a Schottky diode. Invariably, the error in defining the contact area leads to erroneous results. Derive an expression for the error in the Schottky barrier height as a function of the error in the area.

7 A metal film ($\alpha_v = 27 \times 10^{-6}$ per degree K) is deposited on silicon ($\alpha_v = 9 \times 10^{-6}$ per degree K) and heated to 1000°C. Calculate the stress generated in the film when cooled to 25°C. What will be the stress in the substrate (given α_v is the volume expansion coefficient and that $E = 1 \times 10^{12}$ dyn/cm^2). If a metallurgical reaction occurs between the metallic film and the substrate, how will the room-temperature stress be affected?

8 The activation energy Q of the electromigration in aluminum has been reported to be 0.3, 0.4, 0.6, 0.8, 1.1, and 1.4 eV by different authors. At room temperature, plot an MTF vs. Q curve for three current densities through the conductor—10^3, 10^5, and 10^7 A/cm^2. Compare the current density and lifetime of these with a typical, tungsten-filament 100 watt bulb. (Use a constant of proportionality $= 2.78 \times 10^{-6}$ h-cm^4/A^2, and $n = 2$.)

9 In an experiment, several aluminum 1μm thick runners of 0.5, 1, 2, 4, and 6μm widths are tested for electromigration failure. If the diffusion activation energies in these runners are 1.3, 1.0, 0.7, 0.6, and 0.5 eV, respectively, plot MTF as a function

of width. A fixed current of 5 mA is forced through the conductors at temperature 25°C. (Use factors given in Problem 8.)

10 Do Problem 8 at 100°C and −100°C. Explain the significance of these results.

11 A circuit's design requires a maximum permissible current density of 5×10^5 A/cm^2 through a conductor 1 mm long, 1 μm wide, and nominally 0.5 μm thick. Assume that 10% of the conductor length passes over steps and is 50% of the nominal metal film thickness. What maximum voltage may be used across the conductor if the sheet resistance is 5.6×10^{-2} Ω/☐? (Neglecting the thinner cross sections at steps can lead to reliability problems.)

REFERENCES

1 For earlier works on polycrystalline silicon see papers in *Semiconductor Silicon 1969, 1973*, and *1977*, The Electrochemical Society, Princeton, N.J., 1969, 1973, 1977.

2 T. Chung, "Study of Aluminum Fusion into Silicon," *J. Electrochem. Soc.*, **109**, 229 (1962). Also see references listed in Ref. 3.

3 A. G. Milnes and D. L. Feucht, *Heterojunctions and Metal-Semiconductor Junctions*, Academic, New York, 1972, pp. 293–305.

4 M-A. Nicolet, "Diffusion Barrier in Thin Films," *Thin Solid Films*, **52**, 415 (1978).

5 M. P. Lepselter and J. M. Andrews, in *Ohmic Contacts to Semiconductors*, ed. by B. Schwartz, The Electrochemical Society, Princeton, N.J., 1969, p. 159.

6 M. P. Lepselter, "Beam-Lead Technology," *Bell Syst. Tech. J.*, **45**, 233 (1966).

7 S. P. Murarka, H. J. Levinstein, I. Blech, T. T. Sheng, and M. H. Read, "Investigation of the Ti-Pt Diffusion Barrier for Gold Beam Leads on Aluminum," *J. Electrochem. Soc.*, **125**, 156 (1978).

8 S. P. Murarka, *Silicides for VLSI Applications*, Academic, New York, 1983.

9 S. M. Sze, *Physics of Semiconductor Devices*, 2nd ed., Wiley, New York, 1981.

10 T. Mochizuki, K. Shibata, T. Inoue, and K. Ohuchi, "A New MOS Process Using MoSi$_2$ as a Gate Material," *Jpn. J. Appl. Phys.*, suppl., **17-1**, 37 (1978).

11 S. P. Murarka and D. B. Fraser, "Silicide Formation in Thin Cosputtered (Tantalum + Silicon) Films on Polycrystalline Silicon and SiO$_2$," *J. Appl. Phys.*, **51**, 1593 (1980).

12 B. L. Crowder and S. Zirensky, "1-μm MOSFET VLSI Technology: Part VII - Metal Silicide Interconnection Technology - A Future Perspective," *IEEE J. Solid State Circuits*, **SC-14**, 291 (1979).

13 S. P. Murarka, D. B. Fraser, A. K. Sinha, and H. J. Levinstein, "Refractory Silicides of Titanium and Tantalum for Low-Resistivity Gates and Interconnections," *IEEE J. Solid State Circuits*, **SC-15**, 474 (1980).

14 R. Haken, "Applications of the Self-Aligned Titanium Silicide Process to very Large Scale Integrated n-metal-oxide-Semiconductor and Complementary Metal-oxide-Semiconductor Technologies," *J. Vac. Sci. Technol.*, **B3**, 1657 (1985).

15 H. J. Levinstein, S. P. Murarka, and A. K. Sinha, "Cobalt Silicide Metallization for Semiconductor Integrated Circuits," US Patent 4,378,628, dated April 5, 1983.

16 S. P. Murarka, "Self-Aligned Silicides or Metals for Very Large Scale Integrated Circuit Applications," *J. Vac. Sci. Techn.*, **B4**, 1325 (1986).

17 D. M. Brown, W. E. Engler, N. Garfinkel, and P. V. Gray, "Self-Registered Molybdenum-Gate MOSFET," *J. Electrochem. Soc.*, **115**, 874 (1968).

18 P. L. Shah, "Refractory Metal Gate Processes for VLSI Applications," *IEEE Trans. Electron Devices*, **ED-26**, 631 (1979).

19 P. B. Ghate, J. C. Blair, C. R. Fuller, and G. E. McGuire, "Applications of TiW Barrier Metallization for Integrated Circuits," *Thin Solid Films*, **53**, 117 (1977).

20 G. V. Samsonov and I. M. Vinitskii, *Handbook of Refractory Compounds*, IFI/Plenum, New York, 1980.

21 L. I. Maissel and R. Glang, Eds. *Handbook of Thin Film Technology*, McGraw-Hill, New York, 1970.

22 J. F. O'Hanlon, *A User's Guide to Vacuum Technology*, Wiley, New York, 1980.

23 A. S. Grove, *Physics and Technology of Semiconductor Devices*, Wiley, New York, 1967.

24 J. H. Hollomon and D. Turnbull, "Nucleation," *Prog. Metal. Phys.*, **4**, 333 (1957).

25 L. Holland, *Vacuum Deposition of Thin Films*, Wiley, New York, 1956.

26 J. Vossan, J. H. Thomas, III, J.-S. Maa, and J. J. O'Neill, "Preparation of Surfaces for High Quality Interface Formation," *J. Vac. Sci. Technol.*, **2A**, 212 (1984).

27 S. P. Murarka, "Self-Aligned Silicides or Metals for Very Large Scale Integrated Circuit Applications," *J. Vac. Sci. Technol.*, Nov. 1986.

28 H. F. Wolf, *Silicon Semiconductor Data*, Pergamon, New York, 1969.

29 I. A. Blech, D. B. Fraser, and S. E. Haszko, "Optimization of Al Step Coverage through Computer Simulation and Scanning Electron Microscopy," *J. Vac. Sci. Technol.*, **15**, 13 (1978); errata, *J. Vac. Sci. Technol.*, **15**, 1856 (1978).

30 W. Fichtner, Chapter 10, this volume.

31 A. C. Adams, "Plasma Planarization," *Solid State Technol.*, **24**, 178 (1981).

32 S. P. Murarka, "Refractory Silicides for Integrated Circuits," *J. Vac. Sci. Technol.*, **17**, 775 (1980).

33 S. Vaidya, "Pit Formations at Al/(100) Si Contact Windows," *J. Electron. Mater.*, **10**, 337 (1981).

34 J. Black, "Reliability Study of Doped Aluminum Conductor Films," US NTIS, AD Report #AD-A050677 (1977).

35 S. Vaidya, unpublished.

36 M. Mori, S. Kanamori, and T. Ueki, "Degradation Mechanism in Si-doped Al/Si Contacts and an Extremely Stable Metallization System," *IEEE Trans. Components, Hybrids, Manufacturing Technol.*, **CHMT-6**, 159 (1983).

37 S. Vaidya, T. T. Sheng, and A. K. Sinha, "Linewidth Dependence of Electromigration in Evaporated Al-0.5% Cu," *Appl. Phys. Lett.*, **36**, 464 (1980).

38 S. Vaidya, D. B. Fraser, and A. K. Sinha, "Electromigration Resistance of Fine Line Al," *Proceedings of the 18th Annual Reliability Physics Symposium*, IEEE, New York, 1980, p. 165.

39 P. B. Ghate, "Electromigration-Induced Failures in VLSI Interconnects," *Solid State Technol.*, **26**, 113 (1983). See other references on electromigration, listed in this paper.

40 D. Pramanik and A. N. Saxena, "VLSI Metallization Using Aluminum and its Alloys," *Solid State Technol.*, **26**, 131 (1983).

41 D. A. Blackburn, in *Electro- and Thermo-transport in Metals and Alloys*, ed. by R. E. Hummel and H. B. Huntington, AIME, New York, 1977, p. 20.

42 J. R. Black, "Electromigration: Brief Summary and Some Recent Results," *IEEE Trans. Electron Devices*, **16**, 338 (1969).

43 H. U. Schrieber and B. Grabe, "Electromigration Measuring Techniques for Grain Boundary Diffusion Activation Energy in Aluminum," *Solid State Electron.*, **24**, 1135 (1981).

44 F. M. d'Heurle, "The Effect of Copper Additions on Electromigration in Aluminum Thin Films," *Metal. Trans.*, **2**, 683 (1971).

45 F. M. d'Heurle and A. Gangulee, in *Nature and Behavior of Grain Boundaries*, ed. by H. Hu, Plenum, New York, 1972, p. 339.

46 F. M. d'Heurle, A. Gangulee, C. F. Aliotta, and V. Ranieri, "Electromigration of Ni in Al Thin-Film Conductors," *J. Appl. Phys.*, **46**, 4845 (1975).

47 F. M. d'Heurle, A. Gangulee, C. F. Aliotta, and V. A. Ranieri, "Effects of Mg Additions on the Electromigration Behavior of Al Thin Film Conductors," *J. Electron. Mater.*, **4**, 497 (1975).

48 H. J. Blatt, "Superior Aluminum for Interconnections of Integrated Circuits," *Appl. Phys. Lett.*, **19**, 30 (1971).

49 B. N. Agarwala, M. J. Attardo, and A. P. Ingraham, "Dependence of Electromigration-Induced Failure Time on Length and Width of Aluminum Thin-Film Conductors," *J. Appl. Phys.*, **41**, 3954 (1970).

50 I. A. Blech and C. Herring, "Stress Generation by Electromigration," *Appl. Phys. Lett.*, **29**, 131 (1976).

51 I. A. Blech, "Electromigration in Thin Aluminum Films on Titanium Nitride," *J. Appl. Phys.*, **47**, 1203 (1976).

52 J. T. Clemens, US Patent No. 4,291,322, 1981.

53 S. Vaidya and A. K. Sinha, "Electromigration-Induced Leakage at Shallow Junction Contacts Metallized with Aluminum/Polysilicon," *Proceedings of the 20th Annual Reliability Physics Symposium*, IEEE, New York, 1982, p. 50.

54 H. M. Naguib and L. H. Hobbs, "The Reduction of Poly-Si Dissolution and Contact Resistance of Al/n-Poly-Si Interfaces in Integrated Circuits," *J. Electrochem. Soc.*, **125**, 169 (1978).

55 J. R. Lloyd, M. R. Polcari, and G. A. MacKenzie, "Observation of Electromigration in Heavily Doped Polycrystalline Silicon Thin Films," *Appl. Phys. Lett.*, **36**, 428 (1980).

56 Y. Fukuda and S. Kohda, "Solid Phase Reactions of a Polycrystalline Silicon Film with an Overlapping Aluminum Film," *Appl. Phys. Lett.*, **42**, 68 (1983).

57 G. J. van Gurp, "Diffusion-Limited Si Precipitation in Evaporated Al/Si Films," *J. Appl. Phys.*, **44**, 2040 (1973).

58 R. A. Levy, P. K. Gallagher, R. Contolini, and F. Schrey, "Properties of LPCVD Aluminum Films Produced by Disproportionation of Aluminum Monochloride," *J. Electrochem. Soc.*, **132**, 457 (1985).

59 R. S. Blewer and V. A. Wells, *Proc. 1st IEEE VLSI Multilevel Interconnection Conf.*, New Orleans, June 1984, IEEE Cat. #84CH1999-2, p. 153.

60 M. L. Green and R. A. Levy, "Structure of Selective Low Pressure Chemically Vapor Deposited Films of Tungsten," *J. Electrochem. Soc.*, **132**, 1243 (1985).

61 R. H. Wilson, "The Use of Selective Tungsten Deposition. Achieving the Goals of the $1/4\mu$m CMOS Technology," presented at the SRC Topical Research Conference, $1/4\mu$m CMOS Technology, held at Cornell University, Ithaca, N.Y., Dec. 5–6, 1985.

62 R. T. Tung, J. M. Gibson, and J. M. Poate, "Formation of Ultrathin Single-Crystal Silicide Films on Si: Surface and Interfacial Stabilization of Si-NiSi$_2$ Epitaxial Structures," *Phys. Rev. Lett.*, **50**, 429 (1983).

63 M. Liehr, P. E. Schmidt, F. K. LeGoues, and P. S. Ho, "Schottky Barrier Heights of Epitaxial Ni-Silicides on Si (111)," *J. Vac. Sci. Technol.*, **A4**, 855 (1986).

64 J. C. Hensel, R. T. Tung, J. M. Poate, and F. C. Unterwald, "Electrical Transport Properties of CoSi$_2$ and NiSi$_2$ Thin Films," *Appl. Phys. Lett.*, **44**, 913 (1984).

65 A. S. Ignatier, V. G. Mokerov, A. G. Petrova, A. V. Rybin, and N. M. Manzha, "Auger Electron Spectroscopy of an Aluminum-Silicon System," *Sov. Tech. Phys. Lett.*, **8**, 174 (1982).

66 T-M. Lu, Physics Department, RPI, Troy, N.Y., private communication.

67 K. N-Tu, "Shallow and Parallel Silicide Contacts," *J. Vac. Sci. Technol.*, **19**, 766 (1981).

68 M. Eizenberg and S. P. Murarka, "Reactively Sputtered Titanium Carbide Thin Films: Preparation and Properties," *J. Appl. Phys.*, **54**, 3190 (1983).

69 J. R. Shappirio, J. J. Finnegan, and R. A. Lux, "Diboride Diffusion Barriers in Silicon and GaAs Technology," *J. Vac. Sci. Technol.*, **B4**, 1409 (1986).

CHAPTER
10

PROCESS
SIMULATION

W. FICHTNER

10.1 INTRODUCTION

The reduction in active device dimensions to micron and submicron geometries has resulted in an intimate coupling of process conditions, device behavior, and circuit performance to a degree unknown a few years ago. It becomes more and more difficult to develop new processes due to the inherent complexity of IC fabrication.

The use of computer-aided design tools has proven to be invaluable in the development of new technologies and in IC design. A modern IC process contains several hundred individual steps. Computer simulations have emerged as a very elegant way to aid process and device engineers in their task of finding an optimum process.

Traditionally, new technology development has been guided by an experimental "trial-and-error" approach. Starting with an existing process, certain steps in the process are changed, together with structural dimensions. The modified process is then used to fabricate several lots of ICs (1 lot = 20–25 wafers). Completed test structures are subsequently evaluated to investigate whether the original goals have been met. In Figure 1, route A schematically describes this approach, which might require many iterations to optimize the new process. [1] Based on selected parameters of the new process, new circuits can be designed.

The application of software tools in the development of new processes and novel device structures has become a worthwhile, albeit challenging, alternative to the experimental route (see path B in Figure 1). Fabricating one lot in a modern process can cost considerably more than 10,000 dollars, and consume

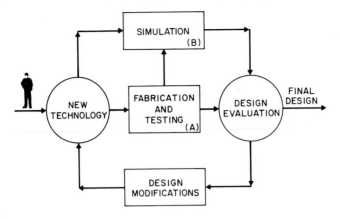

FIGURE 1
Simulation and experiment are alternative routes to technology development and CAD. Route A schematically represents the experimental approach. Route B is the software approach; simulation stands for process, device, and circuit simulation. (*From Fichtner et al., Ref. 1.*)

weeks or even months of effort. The use of accurate simulation tools in the proper computing environment allows for comparatively inexpensive "computer experiments."

Figure 2 shows schematically the various software aids for IC technology development, and their coupling. Process simulation deals with all aspects of IC fabrication. With the proper input parameters (processing steps, layout geometry), process simulation determines the details of the resulting device structure, including the boundaries of the different materials of the structure and the distribution of impurity ions within the structure.

The output of the process simulation, together with applied terminal voltages and currents, is the input to a device simulation program, which determines the electrical characteristics of the device. Furthermore, important insight can be gained by analyzing the behavior of internal variables such as the electrostatic potential and the carrier densities.

The current-voltage characteristics of the device are then used as the essential data to construct a compact model of the device. Such a model can be used as input to a circuit simulation program to determine the electrical characteristics of a circuit comprised of multiple interconnected devices.

In Chapter 11, NMOS, CMOS, and bipolar fabrication sequences are

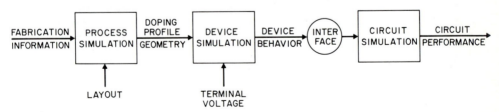

FIGURE 2
Coupling between the various types of simulation. (*From Fichtner et al., Ref. 1.*)

discussed. Figure 5 in Chapter 11 shows, as a typical example of a modern technology, the major process steps of an NMOS process. Modern processes like this one allow the classification of IC fabrication steps into three areas: (1) thermal processing and doping; (2) pattern definition (lithography); and (3) pattern transfer (etching and deposition). Each of these areas contains several subcategories, as shown in Table 1.

The material of this chapter focuses on process simulation. In the following sections, the fabrication steps of major importance—ion implantation, diffusion, oxidation, epitaxy, etching, and deposition—are discussed.

10.2 ION IMPLANTATION

Successful application of ion implantation depends strongly on the ability to predict and control electrical and mechanical effects for given implant conditions. In the past, the basic theory of ion stopping in solids has been the LSS theory, named after its developers, Lindhard, Scharff, and Schiott (see Chapter 8). This theory has been used widely to predict primary ion range and damage distributions in amorphous, semi-infinite substrates. According to the LSS theory, the ion distribution has a Gaussian shape with a projected range R_p and a standard deviation ΔR_p. Range data for different ion-target combinations have been derived on the basis of LSS and are available in the literature.[2,3]

In VLSI processing, however, it is quite common to implant into a non-planar substrate that is covered by one or more thin layers of different materials. Typical examples are threshold-adjust implants, chanstop, and source/drain implants into gate- and field-oxide regions, which may be covered by Si_3N_4. Furthermore, implantations may be performed through thin layers of heavy metals (e.g., Ta or $TaSi_2$). The existence of multilayered structures results in implant profile discontinuities at the interfaces between layers. Additionally, atoms from surface layers may be knocked into deeper layers by impinging ions. This *recoil* effect might degrade the electrical performance of the finished device.

The basic assumptions of the LSS theory do not allow its application to multilayered structures. In the following sections, we apply results from Chapter 8 that are pertinent to the theory of ion collisions in solids.

TABLE 1
A classification of IC process steps

Thermal processing and doping	Pattern definition	Pattern transfer
Epitaxy	Optical lithography	Wet chemical etching
Ion implantation	Electron beam lithography	Ion milling
Predeposition	Ion beam lithography	Reactive ion etching
Annealing	X-ray lithography	CVD
Drive-in		Evaporation
Oxidation		Sputtering

Let us consider the case of a 100 keV ^{75}As$^+$ implant through a double layer of Si_3N_4 and SiO_2 into silicon[4] (Figure 3). An arsenic atom entering the system may be scattered not only by silicon atoms, but also by nitrogen atoms in Si_3N_4 and oxygen atoms in SiO_2. If the transferred energy T is high enough, the target atoms or "recoils" are set into motion, possibly creating recoils themselves until they come to rest. Particles at each position z are described by their energy E and the direction θ in which they are traveling with respect to the z axis.

The Monte Carlo (MC) and Boltzmann transport equation (BTE) methods are widely used to simulate ion implantation phenomena in solids. In the following, the major features of both methods are described, together with representative simulation results.

10.2.1 Monte Carlo Methods

Monte Carlo (MC) methods are based on the simulation of individual particles through their successive collisions with target atoms. The final result is based on the summation of the nuclear and the electronic scattering events (see Chapter 8, Eq. 1) occurring in a large number N ($N > 1000$) of simulated trajectories within the target. By following N histories, distributions for the range parameters of the primary and recoiled ions and the associated damage can be obtained.

Each history begins with a given energy, position, and direction. Figure 4 shows schematically the history of one particle. The ion is assumed to change direction with each elastic binary nuclear collision and to move in a straight, free-flight path between collisions. Within a given layer, the average path length of this particle is given by the inverse cube root of the atomic density. Between elastic collisions, the incoming energetic ion loses energy via inelastic electronic stopping independent of elastic contributions.

With the availability of high-speed digital computers, ion-transport calculations based on the MC method have been used by a variety of authors. Major

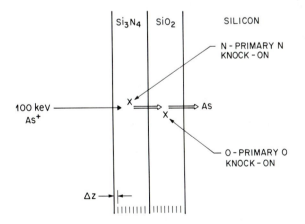

FIGURE 3
Arsenic implantation through a Si_3N_4–SiO_2 double layer. (*After Smith and Gibbons, Ref. 4.*)

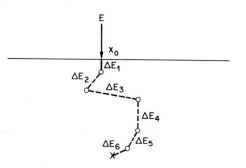

FIGURE 4

Movement of ions in MC calculations. History of one particle implanted into an amorphous target. The particle experiences five elastic collisions before it stops.

differences between the various approaches have been in the treatment of elastic nuclear scattering.[5]

In order to solve the scattering problem numerically, the target is divided into a number of discrete elements (see Δz in Figure 3). A typical example showing the potential of MC calculations is shown[6] in Figure 5. Common elements (Sb, As, B, P) have been simulated for implant energies of (a) 50 keV and (b) 100 keV. For all elements, 10^4 ion trajectories have been simulated. The curves were obtained from full three-dimensional MC calculations by summing up the x and y components. The jagged character of the curves is due to the discretization of the target.

10.2.2   The Boltzmann Transport Equation Method

In the Boltzmann transport equation (BTE) method,[7] the scattering processes of the ions in the target are described by changes in the statistical momentum distribution function $F(\mathbf{p}, z)$. The number of particles with energies and angles in the momentum interval $d\mathbf{p}$ that go through a unit-area element at depth z normal to the surface is given by $F(\mathbf{p}, z)\, d\mathbf{p}\, dz$.

For each species s, the spatial evolution of this momentum distribution function is determined by the Boltzmann equation

$$\frac{\partial F_s(\mathbf{p}, z)}{\partial z} = \int \left[\frac{F_s(\mathbf{p}', z)d\sigma(\mathbf{p}' \to \mathbf{p})}{\cos\theta_{p'}} - \frac{F_s(\mathbf{p}, z)d\sigma(\mathbf{p} \to \mathbf{p}')}{\cos\theta_p} \right] + Q_s(\mathbf{p}, z) \quad (1)$$

where θ denotes the angle between the ion motion and the z axis. Ions can be scattered from a state $\mathbf{p}$ into a final state $\mathbf{p}'$ or they can be scattered out of $\mathbf{p}'$ into $\mathbf{p}$. For recoil distributions, Q_s describes the creation of recoil particles from rest. In the initial state, the distribution functions in the surface plane $z = 0$ are given by the implant dose and the implant angle.

In the numerical solution of Eq. 1, the distribution function F is represented as a two-dimensional matrix $[F_{ij}]$ with entries corresponding to particles of energy E_i ($0 \leq E_i \leq E_o$) moving in direction $\theta_j \leq \pi/2$ (Figure 6). Reasonable computation times restrict the number of elements for $[F_{ij}]$. Fifteen equally spaced

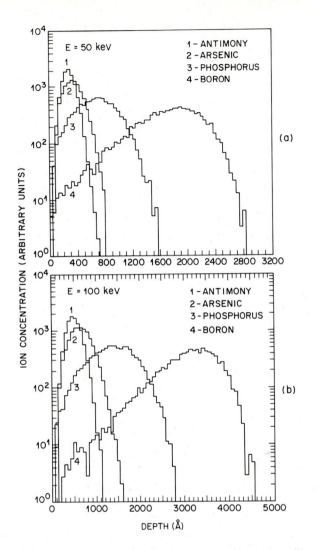

FIGURE 5
Ion concentration as a function of depth for common elements implanted into silicon at 50 and 100 keV. All results are obtained simulating 10^4 ion trajectories. (*After Fichtner, Ref. 6.*)

energy intervals and ten angular intervals have been found sufficient for 5 to 10% accuracy in the range distributions.[8] No azimuthal angle is necessary due to the circular symmetry of the problem. At each step in the calculation, the scattering of ions in a depth segment dz is calculated by Eq. 1. While ions with an energy below some threshold E_{min} are stopped, the others continue until they have lost enough energy in collisions to be at rest. Figure 7 gives a schematic representation of the one-dimensional situation. Several passes of the stepwise integration over target depth may have to be performed in order to include back-scattering phenomena of low-mass ions.[9]

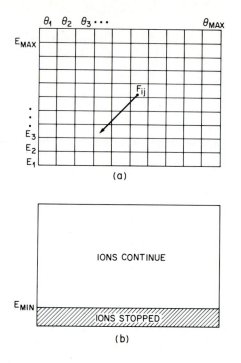

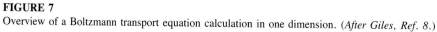

FIGURE 6
(a) Scattering in the distribution matrix. (b) Partitioning of the distribution after each integration step. (*After Giles, Ref. 8.*)

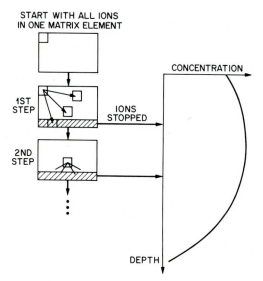

FIGURE 7
Overview of a Boltzmann transport equation calculation in one dimension. (*After Giles, Ref. 8.*)

Simulated profiles compare very well with experimental data and MC results. Figure 8 compares vertical profiles obtained from MC and BTE calculations with experimental data for a 70 keV, 10^{15} cm^{-2} ^{11}B implant into silicon.[10]

In the two-dimensional case, an azimuthal angle ϕ_k and a lateral index l are used to describe the lateral ion motion by the distribution function F_{ijkl}. For typical calculations, the following values are used: $i_{max} = 16$, $j_{max} = 10$, $k_{max} = 6$ and $l_{max} = 64$. Equation 1 guides the scattering of the ions from states F_{ijkl} to all other states $F_{i'j'k'l'}$.

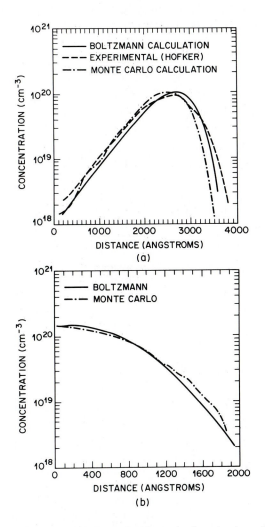

FIGURE 8

(a) Vertical projected profiles calculated for 70 keV boron implantation into silicon, dose 1×10^{15} cm^{-2}. (b) Lateral projected profiles calculated for 70 keV boron implantation into silicon, dose 1×10^{15} cm^{-2} (*After Giles, Ref. 8, and Hofker et al., Ref. 10.*)

Figure 8b shows a comparison of MC and BTE results for the lateral projected profile. The implant conditions are the same as for figure 8a.

The two-dimensional BTE method allows the calculation of very general target structures. This is especially important for the nonplanar surfaces found in VLSI structures. Ions scattered into free space travel without scattering until they re-enter the target.

Figure 9 illustrates the power of the BTE method with the simulation for a 50 keV, 10^{15} cm^{-2} ^{75}As implant at a 63° mask edge. Such a geometry is typical for a VLSI structure. In Figure 9a, the concentration contours for the As implant show some depletion of dopant near the mask edge. Apart from the ion-concentration profiles, the BTE method allows the calculation of damage and

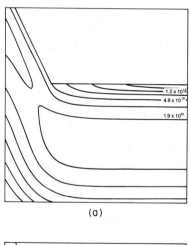

(a)

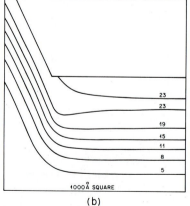

(b)

FIGURE 9

(a) Concentration contours for a 50 keV arsenic implant into silicon near a mask edge with a slope angle of 63°and dose 1×10^{15} cm^{-2}. (b) Damage profile for the same implant in units of eV/Å. (*After Giles, Ref. 8.*)

recoil distributions. For the same arsenic implant, Figure 9b illustrates simulated damage profiles near the tapered mask edge.

10.3   DIFFUSION AND OXIDATION

10.3.1   Basic Equations

In full generality, impurity diffusion in silicon is described by the following set of equations:

1. Flux equations for all charged particles:

$$\mathbf{J}_i = -D_i \nabla S_i + Z_i \mu_i S_i \mathbf{E} \tag{2}$$

   where $\mathbf{J}_i$ is the particle flux, S_i stands for the concentration of diffusing species (donors, acceptors, interstitials, vacancies), D_i and μ_i are the corresponding diffusivities and mobilities, respectively, Z_i is the charge state, and $\mathbf{E}$ is the electric field.

2. Continuity equations for all charged particles:

$$\frac{\partial S_i}{\partial t} + \nabla \cdot \mathbf{J}_i = G_i \tag{3}$$

   where G_i stands for the generation-recombination rate.

3. Poisson's equation:

$$\nabla \cdot (\epsilon \mathbf{E}) = q(p - n + N_D^+ - N_A^- + \text{other charged species}) \tag{4}$$

   where n and p are the electron and hole densities, respectively, ϵ is the dielectric permittivity, and N_D^+ and N_A^- are the ionized donor and acceptor concentrations, respectively.

   The diffusion of a substitutional impurity in silicon is governed by the local concentration of point defects,[11] with diffusivity

$$D = D_I^i \frac{C_I}{C_I^{eq}} + D_V^i \frac{C_V}{C_V^{eq}} \tag{5}$$

where D_I^i and D_V^i are the intrinsic diffusivities of the impurity atoms, C_I and C_V are the concentrations of interstitials and vacancies, and C_I^{eq} and C_V^{eq} are the corresponding equilibrium values.

   This equation can be rewritten

$$\frac{D}{D^i} = f_I \frac{C_I}{C_I^{eq}} + (1 - f_I) \frac{C_V}{C_V^{eq}} \tag{6}$$

where $f_I = D_I/(D_I + D_V) = D_I^i/D^i$. Equation 6 assumes that the diffusivity of an impurity in silicon is proportional to the concentrations of vacancies and interstitials weighted by the interstitialcy component f_I.

In most situations of interest, Eq. 4 can be eliminated by assuming local charge neutrality, $\nabla \cdot \mathbf{E} = 0$. For a two-dimensional problem, Eqs. 3 have to fulfill a set of initial and boundary conditions:[12]
Condition 1:

$$C_i(x, z, 0) = f(x, z) \tag{7}$$

Condition 2:

$$C_i(x, \infty, t) = 0 \quad \text{or} \tag{8}$$

$$C_i(x, \infty, t) = C_B \quad (= \text{bulk concentration})$$

Condition 3: No impurity flux is allowed along the lines of symmetry ($x = x_R$ and $x = x_L$).

$$\frac{\partial C_i}{\partial x} = 0 \quad \text{for } x = x_R \text{ and } x = x_L \tag{9}$$

Condition 4: The boundary condition at the surface depends on whether the surface is being oxidized,

$$D_i \frac{\partial C_i}{\partial z}\bigg|_{z=0} = C_i \left(\frac{1}{m} - b \right) \dot{z} \tag{10}$$

or is exposed to an impurity gas source:

$$D_i \frac{\partial C_i}{\partial z}\bigg|_{z=0} = h(C_i - C_i^*) \tag{11}$$

In Eq. 10, m is the segregation coefficient, given by the ratio of the dopant concentrations in silicon and SiO_2

$$m = \frac{C_i^{Si}}{C_i^{SiO_2}} \tag{12}$$

and b accounts for the volume change associated with the formation of SiO_2 (1 unit of SiO_2 consumes 0.44 units of Si). Equation 10 is valid under the assumption that the diffusion coefficient in the oxide is much smaller than in the silicon. If this is not true, Eq. 10 must be modified, and Eq. 4 has to be solved also in the oxide.

In Eq. 11, h is the mass transfer (or evaporation) coefficient and C^* is the dopant concentration in the gas phase.

The diffusion coefficient D_i in Eq. 10 is, in general, a function of the concentrations, of the impurities for high dopant concentrations (see Chapter 7). All process simulation programs reported include the concentration dependence of D_i obtained from the vacancy-diffusion model:

$$D_i = D_i^x + D_i^- f + D_i^= f^2 + \frac{D_i^+}{f} \tag{13}$$

with $f = N/n_i$. The variables D_i^x, D_i^-, $D_i^=$, and D_i^+ are neutral, single and double negative, and positive vacancy states in silicon, N is the electron concentration that depends on all C_i, and n_i is the intrinsic concentration at the diffusion temperature.

At low impurity concentrations, N is approximately equal to n_i and the diffusion coefficient reduces simply to the sum of the various vacancy states, independent of concentration:

$$D_i = D_i^x + D_i^- + D_i^= + D_i^+ \tag{14}$$

The individual diffusivities in Eq. 13 or Eq. 14 are given in Arrhenius form

$$D_i^* = D_{io}^* \exp\left(-\frac{Q_i^*}{kT}\right) \tag{15}$$

with the prefactor D_{io}^* and the activation energy Q_i^* (see Chapter 7).

The electron concentration N can be approximated by

$$N = \frac{C_{net} + \sqrt{C_{Net}^2 + 4n_i^2}}{2} \tag{16}$$

with

$$C_{net} = -\sum_{i=1}^{n} Z_i N_i \tag{17}$$

where Z_i and N_i stand for the valence and concentration, respectively, of species i.

Diffusion of all important group III (B) and group V (As, Sb) elements in silicon is described well by the above diffusion model.[13] The diffusion of phosphorus, however, is governed by a rather complex diffusion behavior and is modeled by the three-region model (see Chapter 7).

The following example shows the results of two-dimensional BICEPS simulation for an NMOS process with 1 μm design rules.[12] After growing the gate oxide of 250 Å we implant boron at high energy to adjust the threshold voltage and to prevent punch-through. Polysilicon is deposited and doped, and the source-drain regions are opened up by lithographic and etching steps. Source and drain are formed by a high-dose arsenic implant. Several drive-in steps follow until the device is finally processed. Figures 10a and 10b are surface plots of the total concentration and boron concentration, respectively, in the source-drain regions and under the gate. Note the redistribution of boron in the source and drain junction areas caused by the emitter-dip effect (Figure 10b).

10.3.2 Oxidation and Segregation

In thermal oxidation, oxidant from the gas phase diffuses, in the form of O_2 molecules, through holes in the SiO_2 network toward the interface to form new SiO_2 material. This "growth" is accompanied by a large volume increase. At

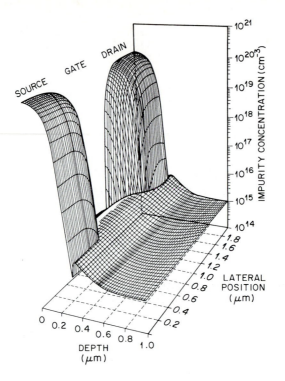

FIGURE 10
Final arsenic and boron doping profiles. (a) Surface plot for the total concentration in a 1 μm gate
length MOS device.

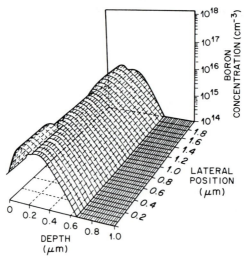

FIGURE 10
(b) Surface plot of the boron concentration. (*After Penumalli, Ref. 12.*)

sufficiently high temperatures, the reaction is aided by viscoelastic flow of the oxide film towards the surface.

For one-dimensional problems, the growth rate of SiO_2 on top of Si can be obtained by integrating the oxygen flux that diffuses through an SiO_2 layer of thickness d_{ox} to react at the surface:

$$d_{ox}^2(t) + Ad_{ox}(t) = B(t + \tau) \tag{18}$$

where A and B are the rate constants and τ accounts for the initial oxide thickness at time t_0 (e.g., a natural oxide, etc.).

Whereas oxidation simulation in one dimension has been successful, no well-established theory is available at this point that would allow a first-principles solution of two-dimensional oxidation phenomena. Typical examples would be the lateral oxidation near a Si_3N_4 mask edge, giving rise to the bird's beak phenomenon, or the oxidation of a silicon trench.

Several numerical models are available that allow approximate numerical solutions in two dimensions.

The coordinate transformation method. The bird's beak problem has been simulated by using a coordinate transformation method to obtain approximate numerical solutions.[12] The diffusion equation and the associated boundary conditions are transformed from the physical domain to a coordinate system where the moving boundary remains stationary in time. Figure 11 illustrates this transformation schematically. With this approach, the solution domain is simplified at the expense of complicating the underlying equations, which can be solved by straightforward numerical methods.

Figure 12 shows the region and the boron profile before (a) and after (b) local oxidation. In Figure 12a, the as-implanted boron profile is shown. Oxidizing this profile for several hours in wet and dry atmospheres not only redistributes the boron considerably, but also results in the bird's-beak geometry in Figure 12b.

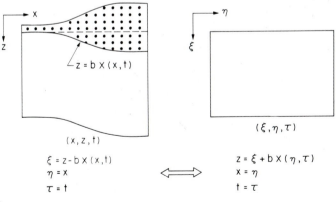

FIGURE 11
Transformation of the physical domain to a coordinate system.

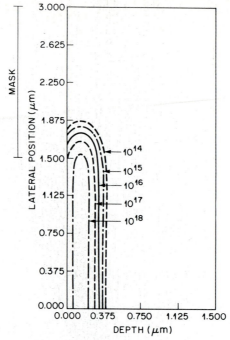

FIGURE 12
Effect of local oxidation. (a) Region and boron profile before oxidation.

The viscous flow model. A serious shortcoming of the coordinate transformation model is the neglect of oxide flow during oxidation. The growth of a new SiO_2 layer across the interface amounts to the insertion of oxygen atoms into the positions of silicon atoms, which, if unrelaxed, would lead to a linear expansion of about 30% in x, y, and z directions.[14] In the z direction, this linear expansion can happen directly by lifting the free SiO_2 surface. In the x-y plane, the required expansion cannot be realized that easily. Initially, the SiO_2 material experiences elastic plane strain with an associated compressive plane stress, which transports the excess material upward (viscoelastic flow).

For temperatures above 960°C , the viscous flow model is capable of simulating bird's beak geometries.[15] The oxidation process is split into two mechanisms: (1) the flow of oxidant through the SiO_2, and (2) the flow of oxide due to the volume expansion and associated stress buildup.

For temperatures above 960°C, the hydrodynamic equations reduce to a creeping-flow equation

$$\mu \nabla^2 \mathbf{v} = \nabla P \tag{19}$$

where μ and P are the density and pressure, respectively, and $\mathbf{v}$ is the velocity. The two-dimensional model has been applied to the problem of semirecessed oxidation.[16]

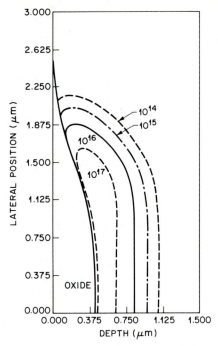

FIGURE 12
(b) Region and boron profile after oxidation. (*After Penumalli, Ref. 12.*)

Simulation results for the growth of two field oxides (0.7 and 0.53 μm) at 1025°C in a wet ambient are shown next. The initial pad oxide has a thickness of 480 Å, and the nitride layer thicknesses are 250 Å and 1700 Å, respectively.

Figures 13a and 13b present the flow pattern for both cases, revealing the complicated nature of two-dimensional oxide growth. The oxide begins to move in a direction normal to the interface. In the case of the thin nitride layer, this motion more or less continues during the later stages of oxide growth. The growth of the oxide with the thick Si_3N_4 layer is strongly affected by the nitride. The compressive stress due to the nitride reduces growth by forcing SiO_2 to flow towards the open surface.

The elastic model. For temperatures below 960°C, oxides no longer exhibit viscous behavior. In the elastic model, oxidant diffuses through the oxide and reacts at the interface.[17] The new oxide forces the SiO_2 layer above it to deform according to the equations of elasticity.

Under the assumption that both SiO_2 and Si_3N_4 are linear elastic bodies, the bird's beak effect has been numerically investigated. Figure 14 shows a typical result for a long oxidation (4 h, 1000°C). Good agreement with experimental data has been achieved.[18]

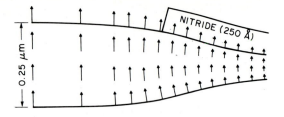

(a)

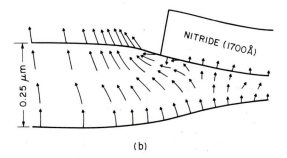

(b)

FIGURE 13
Velocity field distribution illustrating the oxide flow for (a) a thin and (b) a thick nitride mask. The arrows represent the direction as well as the magnitude of the velocity vector. (*After Chin et al., Ref. 15.*)

10.3.3   A Two-dimensional LOCOS Simulation Example[19]

In the following, an example of a two-dimensional LOCOS simulation using the SUPREM IV (Stanford University Process Engineering Models) program will be given. Figure 15 shows the SUPREM IV input file for a LOCOS simulation

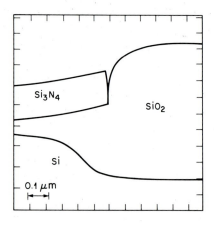

FIGURE 14
Final shape for a local oxide with an initial pad oxide thickness of 500 Å. (*After Matsumoto and Fukuma, Ref. 18.*)

```
# Grid Definition

line x loc=-0.75 spacing=0.1 tag=1

line x loc=0 spacing=0.05 tag=m

line x loc=0.75 spacing=0.1 tag=r


line y loc=0 spac=0.05 tag=s

line y loc=0.3 spac=0.05

line y loc=0.5 tag=b


#Structure Definition

region silicon xlo=1 xhi=r ylo=s yhi=b

bound exposed xlo=1 xhi=r yhi=s ylo=s

initial ori=100


#pad oxide and nitrid

deposit oxide thick=0.02

deposit nitride thick=0.03

etch nitr left pos=0

plot.2d grid


#Numerical Methods

symb lu min.fill symm=f


diffuse elas weto2 time=100 temp=1025 init=1

end
```

FIGURE 15
SUPREM IV input file for a LOCOS simulation. (*After Law, Rafferty, and Dutton, Ref. 19.*)

using the elastic model. The first nine statements define the structure and the initial grid for the simulation. The *line* statements build a rectangular grid from $x_l = -0.75$ μm to $x_r = 0.75$ μm and from $y_s = 0$ to $y_b = 0.5$ μm. For this case, the oxide will grow to about 0.65 μm. The mask edge will be at 0.0 μm. After oxide and nitride deposition, the nitride is etched away from $x = -0.75$ μm to $x = 0$. The *plot.2d* statement generates a plot of the initial grid (Figure 16a), and the *symb* line defines the numerical methods to be used. The local wet oxide is grown at 1025°C for 100 minutes. New oxide elements created at the interface force the existing SiO_2 layer in an upward direction. The Si_3N_4 mask has to bend upward to allow the volume expansion. Figure 16b shows the result of the simulation.

The generality of the numerical models provides the means to study oxide growth on top of nonplanar surfaces such as silicon trenches.[20] This fact is illustrated in Figs. 17a and 17b where the initial grid, and the final structure after a dry oxidation of 240 minutes at 950°C, can be seen.

10.4 EPITAXY

This section describes a model that simulates epitaxial doping profiles in a variety of growth conditions.[21-23]

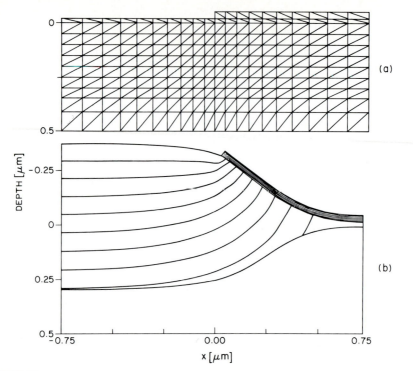

FIGURE 16
LOCOS simulation with SUPREM IV. (a) Initial grid, (b) Final oxide shape. (*After Law, Rafferty, and Dutton, Ref. 19.*)

10.4.1   Epitaxial Doping Model

In this model, silane (SiH_4) is used to grow silicon in a hydrogen ambient in an atmospheric-pressure reactor. A typical doping species could be arsine (AsH_3). Fick's second law is applied throughout the silicon to account for the thermal redistribution of impurities during epitaxial growth.[24] Thus,

$$\frac{\partial C(z,t)}{\partial t} = \frac{\partial}{\partial z}\left(D\frac{\partial C}{\partial z}\right) \qquad z_f < z < \infty \tag{20}$$

has to be solved in a region as shown in Figure 18. C is the dopant concentration in the silicon, D is the diffusion coefficient, and z and t are the space and time variables, respectively.

The solution of Eq. 20 must satisfy the following initial and boundary conditions:

$$C(z,0) = f_1(z) \tag{21}$$

$$D\frac{\partial C}{\partial z}\bigg|_{z\to\infty} = 0 \tag{22}$$

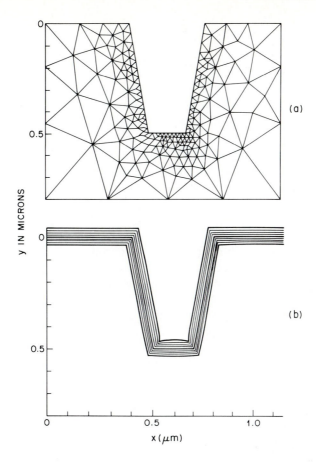

FIGURE 17
Oxidation of a trench structure. (a) Initial grid, (b) final oxide profile. Multiple lines indicate oxide growth.

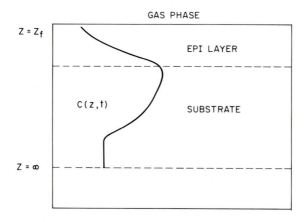

FIGURE 18
Schematic cross section of a silicon wafer for the purpose of solving Fick's second law. (*After Reif and Dutton, Ref. 22.*)

$$D\frac{\partial C}{\partial z}\bigg|_{z=z_f} = f_2(t) \tag{23}$$

where $f_1(z)$ represents the impurity diffusion just before the epitaxial deposition, and Eq. 22 states that the flux of impurities deep inside the silicon is zero. Equation 23 accounts for the fact that during epitaxial growth, the diffusion flux of impurities in the solid at the gas-solid interface is a function of time. An expression for $f_2(t)$ can be derived from a mathematical description of the mechanisms that control the incorporation of the impurities into the silicon-host lattice during the growth process.[22,25]

Based on the various mechanisms involved, Eq. 23 can be written as[22]

$$D\frac{\partial C}{\partial z}\bigg|_{z=z_f} = f_2(t) = -k_{mf}\left[P_D^o - \frac{C(z_f)}{K_p}\right] + gC(z_f) + K_A\frac{\partial C(z_f)}{\partial t} \tag{24}$$

The first term describes the flux of the dopant species leaving the boundary layer by adsorbing at the surface. The variable k_{mf} is a kinetic coefficient associated with the mechanism dominating the dopant-incorporation process. P_D^o is the input partial pressure, $C(z_f)$ is the dopant concentration at the interface, and K_p is a segregation coefficient relating the epitaxial dopant concentration to the concentration of dopant species in the gas phase. The second term, $gC(z_f)$ represents the rate at which the adsorbed layer decreases its concentration of dopant species due to the silicon-covering step. The last term represents diffusion of dopant atoms between the adsorbed layer and the bulk silicon. The variable K_A relates the epitaxial dopant concentration to the concentration of the dopant species in the adsorbed layer.

Fick's second law (Eq. 20) can now be solved subject to the boundary and initial conditions specified in Eqs. 21 to 24.

Figure 19 compares a doping profile simulated using this model with a profile measured by the spreading resistance technique. For the comparison shown in the figure, two consecutive independent arsenic-doped epitaxial films were deposited as indicated in the inset. The arsine flows corresponding to the first and second layers were adjusted to produce epitaxial doping levels of approximately 10^{17} and 10^{15} cm^{-3}, respectively.

Between the end of the first deposition cycle and the beginning of the second, the reactor was purged with H_2 for 8 min at 1050°C. The transition between the high and low doping levels is typical of cases where a lightly doped layer is grown epitaxially on top of a heavily doped substrate or buried layer. This transition is at first abrupt and then becomes gradual. This graded transition is a result of the autodoping phenomenon.

10.5   LITHOGRAPHY

Microelectronic fabrication technology is based on the use of lithography to create the very small features and patterns that make up an integrated circuit (see Chapter 4). In this section, we present the background and typical simulation results for optical and electron-beam (e-beam) lithography.

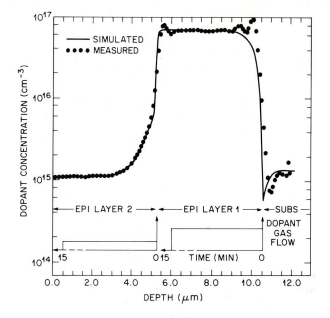

FIGURE 19

Measured and simulated doping profiles corresponding to two consecutive epitaxial depositions. The growth rate is 0.35 μm/min at $T = 1050°C$. (*After Reif and Dutton, Ref. 22.*)

10.5.1   Optical Projection Lithography

In projection printing, the image of an object (the mask) is projected onto the wafer through a complicated, diffraction-limited optical system with a demagnification between one and twenty times.[26]

A generalized optical system is shown in Fig. 20. The projection lens is located between the mask and the wafer, which means that mask-to-lens and lens-to-wafer distances have to be controlled, together with the orientations of wafer, mask, and lens. The information to be replicated is contained on the mask. The mask pattern is transferred by the exposure system to form an aerial image, consisting of spatially-dependent light-intensity patterns in the vicinity of the wafer. Exposure of the resist-coated wafer to the aerial image changes the composition of the resist. This process makes the resist sensitive to a chemical developer.

The simulation of the above processes consists of three parts:

1. Optical computations. The end product of the optical computations is the two-dimensional net (incident and reflected) intensity distribution I. The necessary input information for computing I relates to the optical system, the intensity distribution pattern of the light source, and the resist and substrate parameters.

2. Exposure computations. The interaction of the exposing radiation I with the resist reduces the local inhibitor concentration M. Calculation of the local

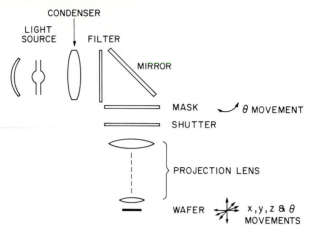

FIGURE 20
Elements of a typical refractive-type projection mask aligner. (*After Lin, Ref. 26.*)

instantaneous value of M requires a knowledge of specific exposure parameters that depend on the resist.

3. Development calculations. The development response of the resist to the developer requires a knowledge of empirical resist constants that permit computation of the development rates from M. The development rates then permit profile calculations for any particular development time.

The deviations from the ideal diffraction-limited system are measured by the modulation transfer function (MTF) of the lens. The MTF includes all aberrations such as spherical aberration, coma, astigmatism, and field curvature. For a mask consisting of lines and spaces of spatial frequency ξ_p, the MTF is defined as the ratio of the mask modulation to the image modulation

$$\text{MTF}(\xi_p) = \frac{M_{\text{image}}(\xi_p)}{M_{\text{mask}}(\xi_p)} \tag{25}$$

where

$$M_{\text{mask}} = \frac{I_{\text{max}} - I_{\text{min}}}{I_{\text{max}} + I_{\text{min}}} \tag{26}$$

and a similar expression for M_{image}. The factors I_{max} and I_{min} are the maximum and minimum intensities, respectively. For an idealized optical system as in Figure 21, the angle α_0 between the maximum diameter of the exit pupil and the image plane determines the resolution. This is described by the numerical aperture, NA, defined by

$$\text{NA} = \bar{n} \sin \alpha_0 \tag{27}$$

or the effective f/number,

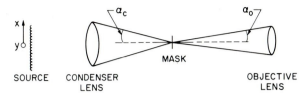

FIGURE 21
Definition of symbols in a partially coherent system. (*After O'Toole and Neureuther, Ref. 27.*)

$$\frac{f}{\text{number}} = \frac{1}{2\text{NA}} \qquad (28)$$

where $\bar{n}$ is the refractive index of the surrounding medium. The quality of the image depends critically on the illumination conditions. The illumination condition of most practical interest is partially coherent illumination. Partial coherence is quantified by the ratio

$$\sigma = \frac{\text{NA}_c}{\text{NA}_o} \qquad (29)$$

where NA_c and NA_o are the numerical apertures of the condenser and objective lens, respectively (Fig. 21). The parameter σ describes the degree of filling of the entrance pupil of the imaging lens by the source. It is the ratio of the imaged source at the entrance pupil to the pupil diameter. An incoherent source ($\sigma = \infty$) is a source of infinite dimension, whereas a coherent source ($\sigma = 0$) is a point source.

The concept of the MTF is useful to specify the performance of optical projection lenses. However, because images of interest in microfabrication are not sinusoidal, the concept of the MTF is difficult to generalize. In recent years, the direct calculation of intensity distributions has become possible. The basic theory of imaging with partially coherent light has been applied to projection printing.[28-30] Based on this theory, a comprehensive computer program called SAMPLE[31,32] has been developed, which can be used to explore the effects of spatially coherent illumination of a periodic mask pattern.

The basic effects of imaging with partially coherent light can be seen in Figure 22, which shows the calculated image intensity near the edge of a mask pattern consisting of 2 μm lines and 6 μm spaces.[30] Since the mask is periodic, it is symmetric around both the $x = -1\mu$m and the $x = 3$ μm axes. The partial coherence factor σ has been defined in Eq. 29. The numerical aperture of the lens is 0.28 and the wavelength is 0.436 μm. The focus error for the curves is taken in units of 0.4 Rayleigh units; one Rayleigh unit is 2.78 μm $= \lambda/(2\text{NA}_o)^2$.

Two-dimensional calculations. Two-dimensional calculations were used to study nonperiodic diffraction-limited images in and out of focus with different partially coherent illumination conditions, assuming circular pupils, nonreflective substrates, absorption-free photoresists, and quasi-monochromatic illumination.[33]

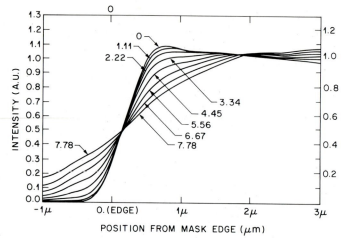

FIGURE 22

Effect of focus error in Rayleigh units for $\sigma = 0.6$ on the image of a mask pattern with 2 μm lines and 6 μm spaces. NA = 0.28, $\lambda = 0.436$ μm. (*After O'Toole and Neureuther, Ref. 30.*)

Figure 23 shows calculated diffraction images of a 1 μm $\times$ 1.2 μm contact hole and a 1 μm $\times$ 4 μm rectangular opening. An aberration-free lens with NA = 0.32 has been used for the calculations. The illumination is treated as monochromatic at $\lambda = 4047$ Å and a partial coherence factor of $\sigma = 0.78$ is used. The focal plane is specified by $z = 0$, which means that $z = 2$ μm is a 2 μm focus error from either side of the focal plane. Each point in the figure is assigned a number that indicates the intensity level. There are a total of twenty levels ranging from 0,1,2, ..., 9,1,2, ..., to 9. Each level is 1.5 dB from the next level. Again, the level 0 indicates all intensities larger or equal to unity, which is the intensity required to completely expose a large uniform transparent area at a predetermined development condition.

Calculations of photoresist exposure.[35] Calculations of resist exposure require knowledge of the optical constants of the substrate and any overlying layers, and of the thicknesses of all the corresponding layers. The key to describing the exposure-dependent optical properties of the photoresist are the exposure parameters A, B, and C. A and B describe the absorption constant α according to

$$\alpha = AM(z,t) + B \tag{30}$$

where M is the relative amount of photoactive inhibitor present at any position z and time t during exposure.

In the calculation, the complex index of refraction $\mathbf{n}$ of the photoresist is used:

$$\mathbf{n} = \bar{n} - ik \tag{31}$$

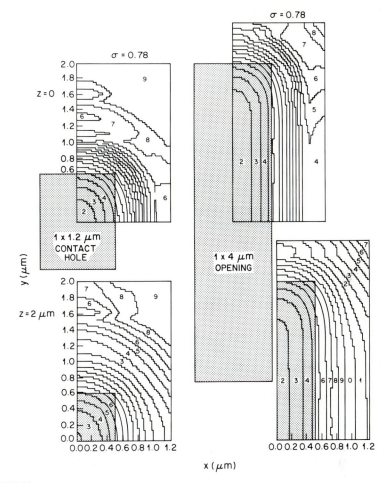

FIGURE 23
Images in focus ($z = 0$) and out of focus ($z = 2$ μm) of a 1 μm × 1.2 μm contact hole and a 1 μm × 4 μm rectangular opening. The imaging lens has a numerical aperture of 0.32. The wavelength of the light is 4045 Å. (*After Lin, Ref. 34.*)

where $\bar{n}$ is the real part of the index, and k is the extinction coefficient at the exposing wavelength λ:

$$k = \frac{\alpha\lambda}{4\pi} \tag{32}$$

The index **n** can be expressed with Eqs. 30 to 32 as

$$\mathbf{n} = \bar{n} - i\frac{\lambda[AM(z,t) + B]}{4\pi} \tag{33}$$

During exposure, **n** changes as the inhibitor is destroyed by the exposing light with intensity I. The optical sensitivity parameter C relates the destruction rate to the light intensity:

$$\frac{\partial M(z,t)}{\partial t} = -I(z,t)M(z,t)C \tag{34}$$

In order to calculate the exposure distribution within the resist, thin-film optical computation techniques are used.[35] The optical properties of the resist vary during exposure as a function of depth. Therefore, the resist is subdivided into sublayers thin enough to be treated as if they had isotropic properties. Interference effects (standing waves) have a periodicity of $\lambda/2\bar{n}$, so the sublayers must be thin compared to this value.[36] Another criterion for the sublayer thickness is that the change in inhibitor concentration between neighboring sublayers be small. The sublayer thickness, Δz_j, is usually in the range of 10 to 500 Å.

Figure 24 shows the intensity distribution $I(z)$ within a 0.584 μm photoresist film with a 6000 Å SiO$_2$ layer, at the beginning of exposure by uniform incident illumination at $\lambda = 4358$ Å.[37] Also included in the figure is the resulting inhibitor concentration after an exposure of 57 mJ/cm^2.

The final success of the calculation of resist exposure depends strongly on the choice of A, B, and C in the equations. The constants A, B, and C have been measured by a variety of workers for different resists under various exposure and development conditions.[38]

Photoresist development. The description of an image exposure in the photore-

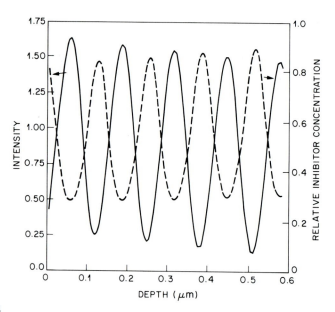

FIGURE 24
Simulated intensity and inhibitor profile within 0.584 μm AZ1350 photoresist on 600 Å of SiO$_2$ on Si. Solid line: intensity of exposing light; dashed line: inhibitor concentration after exposure to 57 mJ/cm^2 at $\lambda = 4358$ Å. (*After Dill et al., Ref. 37.*)

sist is given by a two-dimensional matrix of inhibitor concentration values $M(x, z)$. The development process is modeled as a surface-controlled etching reaction that is controlled by the local value of M. To model the development of exposed resist, an analytical relationship between inhibitor concentration and etch rate is used in SAMPLE, based on a least-squares fit to experimental data, of the form

$$R = \exp (E_1 + E_2M + E_3M^2) \tag{35}$$

A typical example of lithography simulation. Figure 25 contains the SAMPLE input file for a simple lithography simulation. A diffraction-limited optical system with NA $= 0.28$ is used to expose a periodic pattern with 1.25 μm lines and spaces. The wavelength of the source is *lambda* $= 0.4358$ μm. The *resmodel* line specifies the parameters of the positive resist for Eqs. 33 and 34 ($A = 0.551$ μm^{-1}, $B = 0.058$ μm^{-1}, and $C = 0.01$) and the refractive index **n**. The *layers* statement specifies that the wafer has a thick substrate with refractive index $(4.73, -0.136)$ and one additional layer of thickness 0.0741 μm and refractive index $(1.47, 0.0)$. *Run 3* executes the standard bleaching calculation to calculate the standing waves produced in the photoresist. The result of this calculation is the inhibitor matrix M. The *devrate* line sets the constants E_i in Eq. 35. The development time for the calculation is given by the *devtime* statement. *Run 4* executes the actual development calculation and determines the development contours of the photoresist layer.

The results of the SAMPLE run are summarized in Figure 26. Due to the periodicity of the mask-space pattern, only one half of a mask and space pattern is shown. The mask edge is located at 0.625 μm. In Figure 26a, the light intensity is plotted as a function of the horizontal distance. The five development contours

```
# Single wavelength projection lithography (default)    -- samop0

lambda 0.4358                        ; # lambda parameter

proj 0.28                            ; # numerical aperture

linespace 1.25 1.25                  ; # linespace parameters
         .
parcohdef 0 0.7 1.5                  ; # sigma and defocus

run 1                                ; # run image machine

resmodel ((0.4358))

        (0.551, 0.058, 0.010)

        (1.68, ((-0.02))) (0.7133)   ; # resist exposure parameters

layers (4.73,-0.136)

        (1.47,0.0,0.0741)            ; # layer parameters

dose 150                             ; # dose for exposure in mJcm⁻²

run 3                                ; # run exposure machine

devrate 1 (5.63, 7.43, -12.6)        ; # resist development parameters

devtime 15 75, 5                     ; # development times

run 4                                ; # run development machine

descumspec 0.02, 0.04, 3             ; # run descum
```

FIGURE 25
SAMPLE input file for lithography simulation. (*After Ref. 32.*)

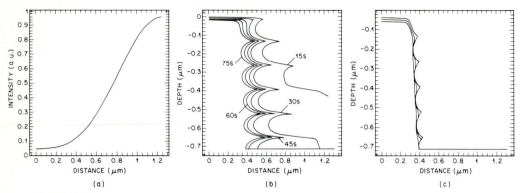

FIGURE 26
SAMPLE result for the input file in Figure 25. (a) Intensity distribution around the mask edge, (b) development contours, and (c) final profiles after descum. (*After Ref. 32.*)

at 15, 30, 45, 60 and 75 s provide an excellent illustration of the standing-wave exposure characteristics and their influence on the developed features (Figure 26b). After the *descum* operation, the final feature has grown about 0.4 μm (Figure 26c).

10.5.2   Electron-beam Lithography

In electron-beam (e-beam) lithography, finely focused electron beams are used to expose polymeric resist layers. Whereas optical lithography is based on diffraction phenomena, e-beam lithography utilizes the scattering of energetic electrons for structure definition. The interaction and scattering of electrons with the resist and the underlying substrate depend on the beam energy, the resist parameters and its thickness, and the substrate parameters.

For exposure simulations, the quantity of interest is the spatial distribution of energy dissipated by electrons in the resist due to interaction phenomena with the resist atoms. This stored energy forms an actual image of the desired pattern.

Fast electrons interact with electrically charged positive target nuclei and negative target valence electrons. As a result of these interactions, the incident electrons change both their direction and their velocity. Whereas electron-nuclear collisions are responsible only for a change in direction of the incident ions (elastic collisions), electron-electron collisions cause both an angular change and an energy loss. The complete scattering history of a particle is the accumulated result of small deflections produced by different nuclei, or it may be due to one single deflection produced by one target nucleus.

The commonly used approximation for elastic small-angle scattering of a charged particle by a nucleus of charge $Z_i q$

$$\left.\frac{d\sigma}{dE}\right|_{\text{elastic}} = \frac{Z_i(Z_i + 1)q^4}{4m^2v^4(\sin^2\frac{\theta}{2} + \alpha_i^2)^2} \tag{36}$$

where the subscript i indicates the atomic species, m and v are the mass and the velocity of the incoming electron, θ is the scattering angle, $\alpha_i = 2.33\, Z_i^{1/3}E^{-1/2}$

is the atomic screening parameter, and E is the energy. The probability for an electron with velocity v to collide with an atom of species i is given by

$$w_i = N_i v \sigma_i \tag{37}$$

$$\sum_i w_i = 1 \tag{38}$$

where N_i is the density of atoms i and σ_i is the total scattering cross section obtained by integrating Eq. 36 over the full solid angle:

$$\sigma_i = \frac{Z_i(Z_i + 1)\pi q^4}{m^2 v^4 \alpha_i^2 (\alpha_i^2 + 1)} \tag{39}$$

The inelastic collisions due to electrons in the target are modeled by the stopping power, which is the energy lost per unit path length. A quantum-mechanical treatment using the Mott electron-electron scattering formula gives

$$-\frac{dE}{ds} = \frac{4\pi N_{AVO} q^4 \rho Z}{m_0 v^2 A} \ln\left[\frac{m_0 v^2}{2I}\left(\frac{e}{2}\right)^{\frac{1}{2}}\right] \tag{40}$$

where N_{AVO} and A are Avogadro's number and the atomic weight of the solid, respectively, s is the path length, ρ is the density, m_0 is the electron mass, and I is the average ionization energy of an atom.[39] Equation 40 is used in most simulations of electron transport, and it is commonly referred to as the continuous-slowing-down-approximation (CSDA).

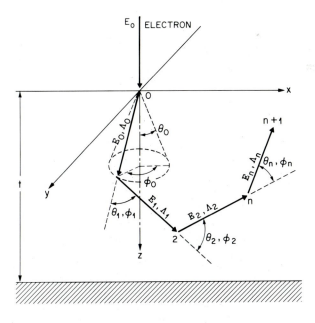

FIGURE 27
Schematic diagram showing the initial Monte Carlo step lengths for electron scattering in a thin resist film on a thick substrate. (*After Kyser and Murata, Ref. 40.*)

Calculation of energy deposition functions. Since high-speed digital computers have become available, Monte Carlo (MC) methods have been applied to study phenomena related to electron scattering in solids.

Figure 27 illustrates schematically the several steps in the history of an electron in a resist target. The distance along the particle trajectory between successive scattering events is given by the mean free path approximation[40]

$$\lambda = \frac{1}{\sum_i \sigma_i N_i} = \sum_i \frac{A_i}{N_{AVO}\rho\sigma_i C_i} \tag{41}$$

where A_i is the atomic weight of the scattering element i atom in grams per mole, and C_i is the weight fraction for element i.

In MC calculations, the free path is calculated from[41]

$$s = -\lambda \ln r \tag{42}$$

with a random number $0 \leq r \leq 1$. A second random number is used to determine the species i that acts as scattering center in the next elastic collision. Two further random numbers are used to calculate the angle of scattering and the azimuthal angle, respectively.

This model has been widely used to calculate energy deposition functions. The main concern in Monte Carlo calculations is to obtain a reasonably small standard error in the final result. The standard error is inversely proportional to the square root of the number of electrons. A reduction of the error by a factor k can be achieved only by increasing the number of samples by a factor k^2. Typical exposure calculations are performed using 20,000 to 50,000 trajectories for one particular target/exposure situation.

Figure 28 shows simulated particle trajectories of 50 electrons in 0.5 μm GMC resist on silicon for energies of 3, 10, 20 and 50 keV. These trajectories were obtained from an MC simulation assuming a free-path-length model (Eq. 42) using the CSDA for energy loss. The trajectories have been projected onto the x-z plane, and the y component is not shown. These figures provide qualitative results for the movement of electrons inside the target. Basically, one can distinguish between two different scattering components: a lateral forward scattering component, and a back-scattering component that can extend over large lateral distances.

MC calculations are usually performed for idealized point and line sources. The latent image, which is the absorbed energy density of the delta-function line source, allows the calculation of the spatial distribution of energy density for any arbitrary beam shape by Fourier transformation. If an exposure profile is to be written for a rectangular beam, as in Figure 29a, the profile for the absorbed energy density is obtained from the MC data by a convolution of a Gaussian distribution with itself over the square dimension.[42] The result of the convolution is

$$f(x) = K\left[\text{erf}\left(\frac{a - x}{\sqrt{2}\sigma} \right) + \text{erf}\left(\frac{a + x}{\sqrt{2}\sigma} \right) \right] \tag{43}$$

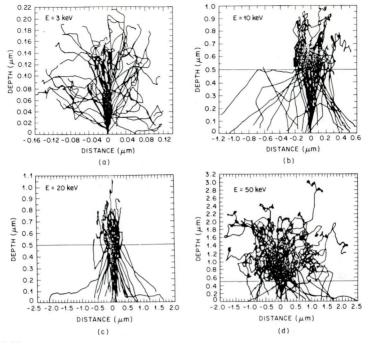

FIGURE 28

Simulated particle trajectories for 50 electrons in a target with 0.5 μm of resist for four different energies of (a) 3, (b) 10, (c) 20 and (d) 50 keV. Note the different depth scales in the four cases. (*After Fichtner et al., Ref. 1.*)

where the beam width (FWHM) $= 2a$, σ is the standard deviation, and K is a constant. For $a/\sigma \gg 1$, the edge slope is

$$\left.\frac{df}{dx}\right|_{x=\pm a} = \frac{2K}{\sqrt{2\pi}\sigma} \tag{44}$$

The edge width is given by $\sqrt{2\pi}\sigma/2$, and is defined by the tangent to $f(\pm a)$ intercepting $f(x) = 0$ and $f(x) = 2K \, \mathrm{erf}(a/\sqrt{2}\sigma)$. The edge of $f(x)$ is symmetric around its half-height.

If the exposure pattern is to be written with lines composed of one or more Gaussian shaped beams, possibly with different weights, as in Figure 29b, each beam is described by

$$f(x) = K'\exp\left(-\frac{x^2}{\sigma^2}\right) \tag{45}$$

Depending on the actual beam shape, either Eq. 43 or 45 is used as the envelope function for the digital convolution of the latent image from the ideal line source. This convolution assumes that superposition of electron exposure and subsequent energy deposition holds.

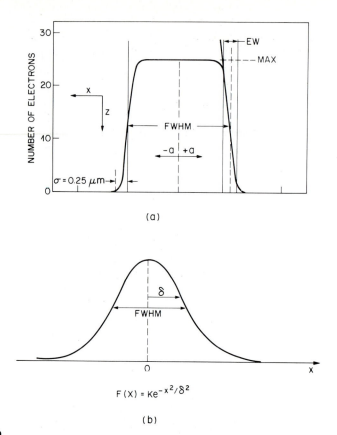

(a)

(b)

$$F(X) = Ke^{-x^2/\delta^2}$$

FIGURE 29
Exposure patterns for arbitrary beam shapes. (a) Schematic representation of a rectangular beam. The vertical axis is the number of electrons distributed over the incident line, normalized to 10^5 electrons. (*After Kyser and Pyle, Ref. 42.*) (b) Schematic representation of a Gaussian round beam.

Figure 30a shows the simulated MC result of the energy deposited within 1.8 μm of resist at three depths for a 25 keV beam. At the surface ($z = 0$), the distribution is very narrow, but for increasing depth, it becomes broader due to backscattering contributions from the substrate. To calculate the lateral distribution of deposited energy (see Figure 30a), the delta-function distribution is convoluted with the rectangular beam in Figure 29a. The energy deposited varies with z. The tails in the original line response deep in the resist are a significant part of the total distribution in Figure 30b.

As in optical lithography, we can calculate resist development. For positive resists, a general relationship between R and E is

$$R = (A + BE^n)[1 - \exp(-\alpha z)] + \epsilon(E) \qquad (46)$$

where R is the etch rate in Å/s, z is the distance below the surface, E is the

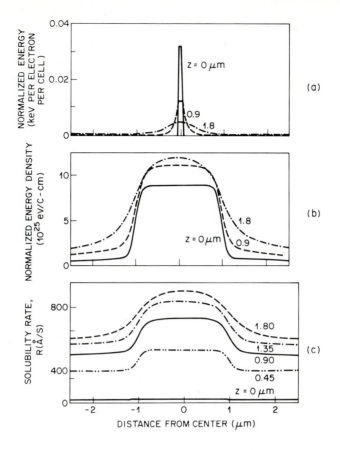

FIGURE 30
Energy distributions and etch rate. (a) Lateral distribution of energy deposited in the film of a 1.8 μm polymeric resist on silicon (25 keV, 2.0 μm written linewidth) using a Monte Carlo simulation for an ideal line source. (b) Lateral distribution of energy deposited within a 1.8 μm resist by 25 keV electrons. (c) Lateral distribution of etch rate for the same resist film and the latent image for a dose of 20 μC/cm^2. (*After Kyser and Pyle, Ref. 42.*)

local absorbed energy density in keV/cm^3, and A, B, α and n are appropriate constants. The dependence of ϵ is modeled as

$$\epsilon(E) = \epsilon_o + CE^k \tag{47}$$

where C and k are constants and E is evaluated at $z = 0$. The form of Eq. 46 implies that for $z \ll 1/\alpha$ and vanishing incident dose Q, $R = \epsilon_o$. For $z \gg 1/\alpha$ and $Q = 0$, $R = A + \epsilon_o$. This type of dissolution behavior has been observed experimentally for certain positive-resist materials under optical and e-beam exposure. If α becomes large and $C = 0$, Eq. 46 reduces to the solubility rate behavior of PMMA used in SAMPLE.[31,32] The parameter α can be interpreted to describe the distance the solvent must diffuse into the resist before any significant development reaction starts, and corresponds to a diffusion distance α^{-1}. The correction term $\epsilon(E)$ provides the proper surface rate.

Equation 46 transforms the latent image of Figure 30b into a solubility-rate image. Figure 30c gives the lateral etch rate distribution with the constants in Eqs. 46 and 47 set to $A = 50$ Å/s, $\alpha = 1.5\ \mu m^{-1}$, $B = 2.5 \times 10^{-18}$, $n = 1.05$, $C = 2.0 \times 10^{-30}$, $k = 1.5$, and $\epsilon_0 = 0.5$ Å/s. The development proceeds in the same manner as outlined before in the discussion of photoresist development.

Development modeling has been extended to three dimensions.[43] A polynomial like the one in Eq. 35 was fitted to experimental solubility results. As an example, Figure 31a shows the development simulation for a $2 \times 2\ \mu m$ square surrounded by 0.6 μm wide rectangles. The region of interest lies inside the dashed line. In Figure 31b, a top view of the developed resist pattern after a 10 s development time is shown. Vertical lines in the figures are 0.1 μm apart. Figure 31c illustrates the developed pattern after 70 s. The gap between the 0.6 μm wide rectangles has more or less vanished into a kink. After 200 s (Figure 31d), all

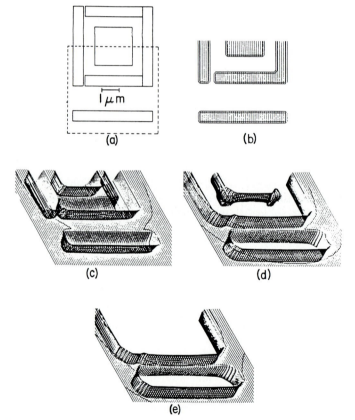

FIGURE 31
(a) Pattern layout of connected and separated rectangles. (b) Calculated profiles of exposed patterns after 10 s of development in a mixture of 1:1 MIBK:IPA. (c) Same after 70 s. (d) Same after 200 s. (e) Final developed profiles after 260 s development. (*After Jones and Parasczak, Ref. 43.*)

directly exposed regions have been fully developed, but a small bridge remains in the central portion of the pattern. Figure 31e gives the final result after 260 s.

10.6 ETCHING AND DEPOSITION

Etching ar´ deposition are integral process steps in any modern silicon technolo⸝y. In the schematic wafer process in Table 1, many etching and deposition steps occur; for example, the patterning of resists and the deposition of polysilicon, insulators, and metal layers. No basic physical theories exist that would allow first-principles solutions of etching and deposition processes.

To circumvent this problem and to obtain some answers through simulation, the modeling of etching and deposition steps can be viewed as a purely geometric problem. The resulting shape of the surface is determined by an initial profile which moves through the medium at a speed depending on the position and other variables such as etch rates.

In SAMPLE, the string model has been used in etching and development

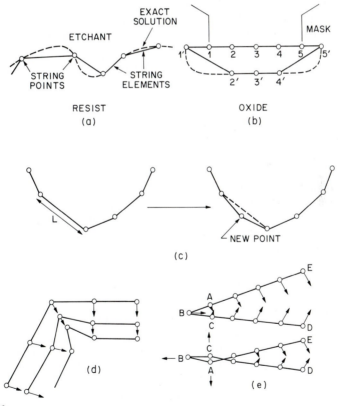

FIGURE 32
String algorithm. (a) String model approximation to an etch front. (b) Problems with uniform etching of an oxide. (c) Length reduction by bisection. (d) Strings during contraction. (e) Loop formation, budding and expansion. (*After Jewett et al., Ref. 44.*)

simulations.[44-46] Figure 32 summarizes the main features of the string model. A polygon consisting of straight line segments forms a string to approximate the etch front. In order to calculate the front position at the next time step, each point along the string advances in a direction perpendicular to the local etch front (Fig 32a). This procedure requires occasional reordering of the string point positions to keep the segments at approximately equal length. One of the major problems in the implementation of the algorithm is the proper choice of the segment length. Another problem is connected with the selection of proper time steps. Since the etch rate normally varies with position, errors can occur. A typical situation is shown in Figure 32b. During the first time step, points 1 and 5 move under the resist mask due to the etch undercut, and points 2, 3, and 4 move straight down. This has the effect of a large deviation from the accurate result as displayed by the broken line. The solution to this problem as adopted by SAMPLE is to calculate the proper position of points 2 and 4 by advancing them along the angle bisector of the two adjacent segments.

One advantage of the string model is its capability to allow the addition of further points along a string. This is achieved by breaking a large segment into two smaller ones (Figure 32c). For regions where the segments become shorter, contraction can occur. The string model allows the corner point to lag behind (Fig. 32d) till it can be deleted. In same cases, however, a sharp angle can occur as in Figure 32e. Segment AB is not allowed to contract before A has crossed the string between B and D. This process is called loop budding.

10.6.1 Deposition

Modern technologies with scaled design rules require excellent step coverage of evaporated films. Other situations might require complete coverage of conductors with insulating films.

Several different deposition sources are in use today. All simulations of deposition profiles are implemented using the string model algorithm previously described.[46] The deposition rate is an analytical function and the simulation proceeds through a number of time steps analogous (but in reverse direction) to etching. The following assumptions are taken:

1. The mean free path of atoms is larger than the distance R between deposition source and substrate.
2. The distance r is large compared to the step height.
3. The film grows at a rate proportional to $\cos \theta / r^2$, where θ is the angle between the surface normal and the vapor stream direction. This is called the *cosine growth* law. At the end of this section, new results are presented which challenge this assumption.
4. The film grows toward the direction of the vapor stream. Roughening effects due to large incidence angles are ignored.
5. For cold substrates, the sticking coefficient is one.
6. For elevated temperatures, a random walk model governs surface migration.[46]

Figure 33 shows several important source geometries. For a unidirectional source Fig. 33a, material arrives at the substrate from one direction only. The growth rate is given as

$$R(x, z) = 0 \qquad (48)$$

if the point at (x, z) is shadowed, and

$$R(x, z) = C(\sin \omega \hat{x} + \cos \omega \hat{z}) \qquad (49)$$

otherwise, where ω is the angle between the z axis and the vapor streams, $\hat{x}$ and $\hat{z}$ are unit vectors in the x and z directions, respectively, and C is the growth rate of an unshadowed surface in the direction of the vapor stream.

For a dual source evaporator, Fig. 33b, the vapor stream arrives from two different directions at angles ω_1 and ω_2, and the final result is a sum of two terms just like Eqs. 48 and 49.

For a hemispherical source, Fig. 33c, the flux F of the vapor stream is proportional to the area A:

$$F = KA \qquad (50)$$

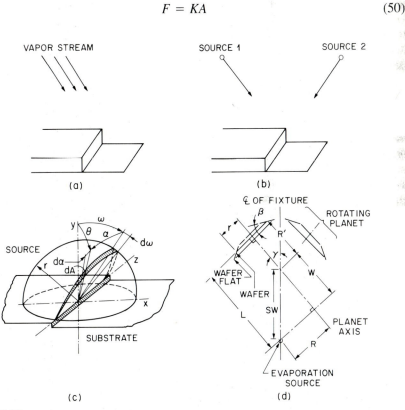

FIGURE 33
(a) Geometry for a unidirectional source. (b) Geometry for a dual evaporation source. (c) Geometrical relations between hemispherical vapor source and substrate. (d) Schematic planetary fixture for an e-beam evaporator. (*After Fichtner, Ref. 6.*)

where K is a proportionality constant. In Fig. 33c, a flux dF proportional to the area dA will pass through it, namely

$$dF = Kr^2 \cos \alpha \, d\alpha \, d\omega \qquad (51)$$

Assuming a cosine growth, the change in thickness in direction y due to the entire area enclosed by $d\omega$ is

$$dy = Kr^2 \int_{\alpha=-\frac{\pi}{2}}^{\frac{\pi}{2}} \cos \alpha \cos \theta \, d\omega \, d\alpha = 2K \cos \omega \, d\omega \qquad (52)$$

and, since assumption 4 holds,

$$dx = dy \tan \omega = 2K \sin \omega \, d\omega \qquad (53)$$

The final deposition rate is calculated by integrating Eqs. 52 and 53:

$$R(x, y) = 2K \left[(\cos \omega_1 - \cos \omega_2)\hat{x} + (\sin \omega_2 - \sin \omega_1)\hat{z} \right] \qquad (54)$$

Figure 33d shows a schematic of a planetary fixture.[47] The wafers are fixed on the planet, which rotates around its own planet axis and also around the planetary center line. The rotation around the planet axis does not alter the deposition equations. Figure 34 shows a more detailed view of the wafer geometry at position A, and the rotating moving source. It also contains a definition of "A" and "B" steps.

In case of the outer "B" step, the amount of material arriving per unit area and per unit rotation $d\omega$ is

$$I(\omega) = \frac{\dfrac{R}{L}\left[1 + \dfrac{WL}{R^2} - \left(\dfrac{r}{R}\right)^2 - \dfrac{rL}{R^2}\tan(\alpha - \beta) \right]}{\left[1 + \left(\dfrac{W}{R}\right)^2 \right]^{\frac{1}{2}} \left[1 + \left(\dfrac{R}{L}\right)^2 - \left(\dfrac{r}{L}\right)^2 - 2\left(\dfrac{r}{L}\right)^2 - \dfrac{2r}{L}\tan(\alpha - \beta) \right]^{\frac{3}{2}}} \qquad (55)$$

and the displacements are given by

$$\Delta x(\alpha)d\alpha = I(\omega)\cos \theta''\tan \alpha \, d\omega \qquad (56a)$$

$$\Delta z(\alpha)d\alpha = I(\omega)\cos \theta''d\omega \qquad (56b)$$

The total displacements due to deposition from all angles are obtained by integration of Eq. 56 over all $\alpha_{\min} \leq \alpha \leq \alpha_{\max}$.

Calculated metal deposition profiles for the inner and outer "B" steps are shown in Figure 35 for angles β from $-5°$ to $30°$. The metal thickness is 1.5 μm and the step height is 1.0 μm.

10.7 SUMMARY AND FUTURE TRENDS

Process simulation deals with all aspects of IC fabrication. This chapter summarizes the relevant results and provides examples of the state of the art of process simulation.

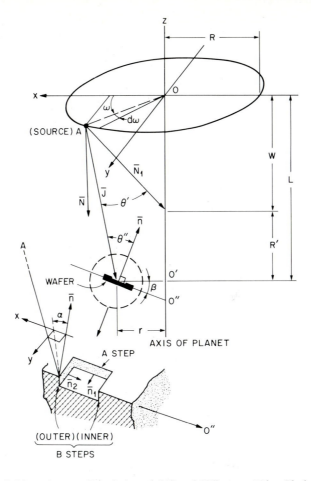

FIGURE 34
Geometry and definition of terms. Description of "A" and "B" steps. (*After Blech, Fraser, and Hazko, Ref. 47.*)

The use of computer simulations in the development of new processes has become a widely accepted technique to reduce the high cost and long turn-around times of real experiments. Programs such as SAMPLE and SUPREM are widely distributed in industrial and academic sites. Together with device and circuit simulation tools, these programs are already available on engineering workstations, enabling a convenient access for the process and device designer.

In the future, we can expect increased sophistication in computer programs (expert systems) combined with a better understanding of the physical processes to have an even stronger impact in the design of new technologies and device structures. Complete design systems will enable on-line process design to predict the desired device- and circuit-parameter sensitivities, and to facilitate circuit design and layout with given design rules.

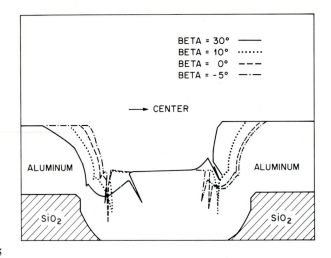

FIGURE 35
Dependence of step coverage profiles for "B" steps on the angle β. (*After Blech, Fraser, and Hazko, Ref. 47.*)

PROBLEMS

The following problems were chosen to illustrate that solutions for simple problems can be obtained by analytical solutions. Sometimes it is necessary to obtain quick estimates for problems when programs such as SUPREM are not available.

1 Consider the case of an undoped substrate with an epitaxial layer being put down with a doping concentration C_0. Solve for the doping concentration as a function of time.

2 The second problem is that of an undoped (or very low doped) layer being deposited on a heavily doped substrate. This case is more interesting, since it occurs quite often in standard MOS or bipolar processing. Some of the dopant will outdiffuse into the growing layer. If it diffuses all the way through, it can reach the growth surface and evaporate. A net loss of dopant will occur if the evaporation rate exceeds the adsorption rate of dopant species from the gas-boundary layer.

3 In device processing, it is quite common to implant into finite areas, which results in lateral doping distributions determined by the shape of the mask edge and the scattering of the ions during the slowing-down process. Calculate the two-dimensional distribution of ions in the vicinity of an arbitrarily shaped mask edge under the following conditions:

(a) Ions entering a target at the point $(0, 0, 0)$ will come to rest at (x, y, z) with a Gaussian probability function

$$f(x, y, z) = \frac{1}{(2\pi)^{2/3} \Delta R_p \, \Delta X \, \Delta Y} \exp\left[-\frac{x^2}{2\Delta X^2} - \frac{y^2}{2\Delta Y^2} - \frac{(z - R_p)^2}{\Delta R_p^2} \right]$$

ΔX and ΔY are the lateral standard deviations. In amorphous semiconductors, $\Delta X = \Delta Y$.

(b) The stopping power of the mask material is equal to the stopping power of the semiconductor. This assumption holds very well for the Si-SiO$_2$ system.

(c) Diffusion and channeling effects are ignored.

4 A thin, uniformly doped region at the silicon surface is used as a doping source with initial impurity concentration C_0. Under the condition that the total amount of diffusing substance, ϕ, remains constant and the boundary condition

$$\frac{\partial C}{\partial z} = 0 \text{ for } z = 0$$

the one-dimensional diffusion equation has the solution

$$C(z, t) = \frac{\phi}{\sqrt{\pi D t}} \exp\left(-\frac{z^2}{4Dt}\right)$$

Generalize this case to a three-dimensional structure in the vicinity of a diffusion mask.

5 Solve the diffusion problem of redistributing donor and acceptor impurities between Si and SiO_2 during thermal oxidation at high temperature. Assume that the initial distribution in the silicon is uniform.

REFERENCES

1 W. Fichtner, L. W. Nagel, B. R. Penumalli, W. P. Petersen, and J. L. D'Arcy, "The Impact of Supercomputers on IC Technology Development and Design," *Proc. IEEE,* **72**, 96 (1984).

2 J. Gibbons, W. S. Johnson, and S. Mylroie, *Projected Range Statistics*, 2nd edition, Wiley, New York, 1975.

3 B. Smith, *Ion Implantation Range Data for Silicon and Germanium Device Technologies*, Research Studies Press, Portland, Oregon, 1977.

4 D. H. Smith and J. F. Gibbons, "Application of the Boltzmann Transport Equation to the Calculation of Range Profiles and Recoil Implantation in Multilayered Media," in *Ion Implantation in Semiconductors 1976*, ed. by F. Chernow, J. A. Borders, and D. K. Brice, Plenum, New York, 1977.

5 J. P. Biersack and L. G. Haggmark, "A Monte Carlo Computer Program for the Transport of Energetic Ions in Amorphous Targets," *Nucl. Instrum. and Methods*, **174**, 257 (1980).

6 W. Fichtner, "Physics of VLSI Processing and Process Simulation," in *Silicon Integrated Circuits Part C*, ed. by Dawon Kahng, Academic Press, 1985.

7 L. A. Christel and J. F. Gibbons, "An Application of the Boltzmann Transport Equation to Ion Range and Damage Distributions in Multilayered Targets," *J. Appl. Phys.*, **51**, 6176 (1980).

8 M. D. Giles, "Ion Implant Calculations in Two Dimensions Using the Boltzmann Equation," *IEEE Trans. CAD*, **5**, 679 (1986).

9 M. D. Giles and J. F. Gibbons, "A Multipass Application of the Boltzmann Transport Equation for Calculating Ion Implantation Profiles at Low Energies," *Nucl. Inst. Meth*, **209/210**, 33 (1983).

10 W. K. Hofker, D. P. Oosthoek, N. J. Koeman, and H. A. M. de Grefte, "Concentration Profiles of Boron Implantations in Amorphous and Polycrystalline Silicon," *Radiat. Eff.*, **24**, 223 (1975).

11 P. M. Fahey, "Point Defects and Dopant Diffusion in Silicon," PhD Thesis, Integrated Circuits Laboratory, Department of Electrical Engineering, Stanford University, June 1985.

12 B. R. Penumalli, "A Comprehensive Two-Dimensional VLSI Process Simulation Model, BICEP-S," *IEEE Trans. ED*, **30**, 986 (1983).

13 S. E. Hansen, "SUPREM-III User's Manual Version 8520," PhD Thesis, Integrated Circuits Laboratory, Department of Electrical Engineering, Stanford University, May 7, 1985.

14 U. Gösele and T. Y. Tan, in *Defects in Semiconductors II*, ed. by S. Mahajan and J. W. Corbett, North-Holland, Amsterdam, 1983, p. 45.

15 D. Chin, S. Y. Oh, S. M. Hu, R. W. Dutton and J. L. Moll, "Stresses in Local Oxidation," *IEDM Technical Digest 1982*, 228 (1982).

16 D. Chin, S. Y. Oh and R. W. Dutton, "A General Solution Method for Two-Dimensional

Nonplanar Oxidation," *IEEE Trans. ED*, **30**, 993 (1983).

17 E. P. Eernisse, "Stress in Thermal Oxide During Growth," *Appl. Phys. Lett.*, **35**, 1 (1979).

18 H. Matsumoto and M. Fukuma, "A Two-Dimensional Si Oxidation Model Including Viscoelasticity," *IEDM Technical Digest 1983*, 39 (1983).

19 M. E. Law, C. Rafferty, and R. W. Dutton, "SUPREM IV Users Manual," Technical Report, Integrated Circuits Laboratory, Department of Electrical Engineering, Stanford University, July 1986.

20 C. S. Rafferty, W. E. Law, and R. W. Dutton, "Two-Dimensional Process Modeling and SUPREM IV" Second International Conference on *Simulation of Semiconductor Devices and Circuits*, Pineridge Press, Swansea, U.K., 1986.

21 R. Reif, T. I. Kamins, and K. C. Saraswat, "A Model for Dopant Incorporation into Growing Silicon Epitaxial Films, I. Theory," *J. Electrochem. Soc.*, **126**, 644 (1979); see also R. Reif, T. I. Kamins, and K. C. Saraswat, "A Model for Dopant Incorporation into Growing Silicon Epitaxial Film, II. Comparison of Theory and Experiment," *J. Electrochem. Soc.*, **126**, 653 (1979).

22 R. Reif and R. W. Dutton, "Computer Simulation in Silicon Epitaxy," *J. Electrochem. Soc.*, **128**, 909 (1981).

23 D. A. Antoniadis, S. E. Hansen, and R. W. Dutton, "SUPREM-II-A Program for IC Process Modeling and Simulation," Stanford Electronics Laboratories Technical Report No. 5019-2, June 1978.

24 P. H. Langer and J. I. Goldstein, "Impurity Redistribution during Silicon Epitaxial Growth and Semiconductor Device Processing," *J. Electrochem. Soc.*, **121**, 563 (1974).

25 M. M. Faktor and I. Garrett, *Growth of Crystals from the Vapor*, Chapman and Hall, New York, 1974.

26 B. J. Lin, "Optical Methods for Fine-Line Lithography," in *Fine-Line Lithography*, ed. by R. Newman, North-Holland, New York, 1980.

27 M. M. O'Toole and A. R. Neureuther, "Influence of Partial Coherence on Projection Printing," SPIE Vol. 174, *Developments in Semiconductor Microlithography IV*, 22 (1979).

28 M. Born and E. Wolf, *Principles of Optics-Electromagnetic Theory of Propagation, Interference and Diffraction of Light*, Pergamon Press, New York, 1959.

29 E. C. Kintner, "Method for the Calculation of Partially Coherent Imagery", *Appl. Opt.*, **17**, 2747 (1978).

30 M. M. O'Toole and A. R. Neureuther, "Influence of Partial Coherence on Projection Printing," SPIE Vol. 174, *Developments in Semiconductor Microlithography IV*, 22 (1979).

31 "SAMPLE User's Guide," Electronics Research Laboratory, Dept. of Electrical Engineering and Computer Science, UC Berkeley, 1982.

32 "SAMPLE Version 1.6a User's Guide," Electronics Research Laboratory, Dept. of Electrical Engineering and Computer Science, UC Berkeley, February 1, 1985.

33 B. J. Lin, "Partially Coherent Imaging in Two Dimensions and the Theoretical Limits of Projection Printing in Microfabrication," *IEEE Trans. ED*, **27**, 931 (1980).

34 B. J. Lin, "Theory and Practice of Deep-UV and Near-UV Lithography," in *Proceedings of the International Conference on Microlithography—Microcircuit Engineering 81*, compiled by A. Oosenbrug, September 28–30, 1981, Lausanne, Switzerland, p. 47.

35 P. H. Berning, "Theory and Calculations of Optical Thin Films," in *Physics of Thin Films*, Vol. 1, ed. by G. Hass, Academic, New York, 1963.

36 J. Cuthbert, "Optical Projection Printing," *Solid State Technology*, **20**, 59 (1977).

37 F. H. Dill, A. R. Neureuther, J. A. Tuttle, and E. J. Walley, "Modeling Projection Printing of Positive Photoresists," *IEEE Trans.*, **ED, 22**, 456 (1975).

38 A. R. Neureuther and W. G. Oldham, "Simulation of Lithography," in *Process and Device Modeling*, ed. by W. Engl, North-Holland, 1986.

39 M. J. Berger and S. M. Seltzer, *Tables of Energy Losses and Ranges of Electrons and Positrons*, US Department of Commerce, NTIS N65-12506.

40 D. F. Kyser and K. Murata, "Monte Carlo Simulation of Electron Beam Scattering and Energy Loss in Thin Films on Thick Substrates," in *Proceedings of the 6th International Conference on*

Electron and Ion Beam Science and Technology, ed. by R. Bakish, The Electrochemical Society, Princeton, 1974.

41 R. J. Hawryluk, A. M. Hawryluk, and H. I. Smith, "Energy Dissipation in a Thin Polymer Film by Electron Beam Scattering," *J. Appl. Phys.*, **45**, 2551 (1974).

42 D. F. Kyser and R. Pyle, "Computer Simulation of Electron-Beam Resist Profiles," *IBM J. Res. Develop.*, **24**, 426 (1980).

43 F. Jones and J. Parasczak, "RD3D (Computer Simulation of Resist Development in Three Dimensions)," *IEEE Trans. ED*, **28**, 1544 (1981).

44 R. E. Jewett, P. I. Hagouel, A. R. Neureuther, and T. van Duzer, "Line-Profile Resist Development Simulation Techniques," *Polym. Eng. Sci.*, **17**, 381 (1977).

45 J. L. Reynolds, A. R. Neureuther, and W. G. Oldham, "Simulation of Dry Etched Line Edge Profiles," *J. Vac. Sci. Technol.*, **16**, 1772 (1979).

46 W. G. Oldham, A. R. Neureuther, C. K. Sung, J. L. Reynolds, and S. N. Nandgaonkar, "A General Simulator for VLSI Lithography and Etching Processes: Part II—Application to Deposition and Etching," *IEEE Trans. ED*, **27**, 1455 (1980).

47 I. A. Blech, D. B. Fraser, and S. E. Hazko, "Optimization of Al Step Coverage through Computer Simulation and Scanning Electron Microscopy," *J. Vac. Sci. Technol.*, **15**, 13 (1978).

CHAPTER
11

VLSI
PROCESS
INTEGRATION

S. J. HILLENIUS

11.1 INTRODUCTION

It was first recognized by G. W. A. Dummar in 1952 that electronic equipment could be made from a single block of layers of insulating, conducting, rectifying, and amplifying materials with the electrical functions being connected by cutting out areas of the various layers.[1] This was the first description of an integrated circuit, but it was not until 1958 that J. S. Kilby invented and demonstrated an integrated circuit (IC). These first demonstration circuits were germanium transistors, resistors, and capacitors formed on a wafer by selectively etching the germanium by applying black wax to mask the active regions. The first circuits formed this way were phase shift oscillators and a flip-flop. From that primitive start, the mass production of semiconductor ICs developed in the early 1960s and has continued to grow exponentially. Today, ICs are complex devices with millions of individual components that are manufactured on a single silicon crystal and interconnected by tiny wires that are microns wide. The process of forming these devices out of a single piece of silicon crystal will be described in this chapter.

Integrated circuits consist of layers of insulating, semiconducting, and conducting regions. These layers are assembled in such a way as to form transistors that are interconnected to produce a certain desired electrical function. The individual transistors come in a variety of types. Each type of transistor has particular attributes that make it more or less desirable for particular types of electrical functions. Each type of transistor will also have different needs when it comes to processing. The major IC technologies that are used today are n-channel metal-oxide-semiconductor (NMOS), complementary metal-oxide-semiconductor (CMOS), bipolar, and integrated-injection-logic bipolar (I^2L). The order in which these technologies are presented is the order of increasing complexity of the processing. Table 1 shows the relative attributes of each of the IC technologies from the point of view of circuit characteristics.[2]

The purpose of this chapter is to present the fundamentals of integrating silicon processing steps to create silicon devices. The basic principles of operation of these devices will not be discussed, and it will be assumed that the reader is familiar with these fundamentals. These principles of operation will be referenced wherever it is necessary to include them to understand the processing constraints and particular features of each technology. Detailed explanations of the operation of these devices can be found in texts on device physics, such as those by Sze,[3,4] Grove,[5] and Muller and Kamins.[6]

11.2 FUNDAMENTAL CONSIDERATIONS FOR IC PROCESSING

The process of building devices in a single piece of silicon crystal is a process of building successive layers of insulating, conducting, and semiconducting

TABLE 1
Characteristics of integrated transistors[2]

	NMOS	CMOS	Bipolar n-p-n	Bipolar I^2L
1.General				
Supply voltage range	+	+ +	−	+
Power	+	+ +	−	+
Speed	+	+	+ +	+
Transconductance	−	−	+ +	+ +
Circuit density	+ +	+	−	+
Drive capability	−	+	+ +	+
2.Digital				
Switching speed	+ +	+ +	+ +	+
Power	+	+ +	−	+
Noise margins	+	+ +	−	−
Logic swing	−	+ +	−	−
3.Analog				
Gain per stage	−	+	+ +	
Bandwidth	−	−	+ +	
Input impedance	+ +	+ +	−	
Power	+ +	+ +	−	
Output swing	−	+	+ +	
Linearity	−	+	+ +	
Analog switches	+	+ +	−	
Precision elements	+ +	+ +	+	

The symbols represent moderate (−), good (+), and superior (+ +) behavior.

material. Each layer is patterned to give a distinct function and relationship with surrounding areas and subsequent layers. The layers are produced and patterned by using the techniques described in the first 10 chapters of this book.

11.2.1 Building Individual Layers

Figure 1 depicts some of the more important ways to create a layer in or on the silicon crystal. The layers are formed by oxidation, implantation, deposition, or epitaxial growth of silicon. Each one of these layers can be formed uniformly, as shown in the left side of Fig. 1, or selectively, as shown in the right side of the figure. The patterning of the layers for selective formation would be achieved by using the photolithography and etching procedures described in Chapters 4 and 5, as would any subsequent patterning of the layers.

The various techniques for forming individual layers have their own usefulness for particular applications. For example, if an insulating layer is needed, the oxidation of silicon described in Chapter 3 forms a layer of SiO_2, which is an excellent insulator and can be formed with a high degree of uniformity. However, the formation of the SiO_2 layer consumes silicon and may require relatively long thermal cycles for thicker layers. Therefore, if a thick insulating oxide is needed,

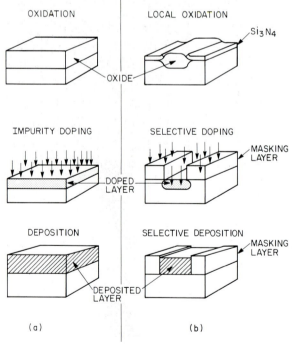

FIGURE 1
Schematic representations of forming layers in silicon: (a) uniform, (b) selective.

it might require an oxide deposition as described in Chapter 6. The deposited oxide in general has a poorer uniformity than the thermally grown oxide but can be deposited at much lower temperatures. This kind of analysis must be made at each step in the process development to determine which type of layer will give the desired results without adversely affecting previous layers. Oxidation and deposition are the most common ways of forming a dielectric layer. The oxidation can be made selective by depositing and patterning Si_3N_4 before oxidation, which allows the oxide to form only wherever the Si_3N_4 is removed, since the Si_3N_4 blocks the oxidation. This technique is sometimes called "local oxidation."

The junctions formed by n-type and p-type regions are the fundamental components of all transistors. The doping levels and depths of these doped regions will determine the characteristics of the devices. Ion implantation is the most important doping technique in modern ICs. The techniques and theory presented in Chapter 8, along with the diffusion of the dopants as described in Chapter 7, will determine the doping profiles. In designing a process sequence it must always be remembered that the doping profiles will change with every subsequent thermal cycle, and the total thermal cycle that impurities undergo must be taken into account. Selective doping, as shown in the middle of Fig. 1b, can be achieved using a variety of masking layers such as photoresist, polysilicon, or SiO_2. The deposition of materials can be achieved through the CVD methods described in Chapter 6 or the physical deposition techniques covered in Chapter 9. The choice of material will be determined by the particular application and function, the most common materials being silicon oxides, polysilicon, aluminum, and silicon nitride. Some deposited materials can be deposited selectively, as shown in the bottom of Fig. 1b, which may give an advantage for some applications. The epitaxial growth of single-crystal silicon can also be deposited uniformly or selectively. Selective epitaxial growth is not used commonly in ICs but has many advantages that may be useful for future devices.

11.2.2   Integrating the Process Steps

The process of building a succession of layers described in the previous section requires careful consideration of each layer's relative position to the others. A simple example of this process of building layers is shown in Fig. 2, which illustrates the sequence of forming a simple MOS capacitor out of four successive layers. The initial oxidation (Fig. 2a) forms an insulating layer between the final aluminum layer and the substrate. The oxide is then patterned and etched to form the active region of the capacitor. The active region of the capacitor consists of a boron-doped p-type region of the silicon, formed by ion-implanting boron into the silicon exposed by the etching of the oxide, as shown in Fig. 2b. The oxide acts as an implant mask in this step, effectively blocking the boron from doping the silicon under the oxide. The implant energy must be carefully chosen to ensure that the range of the ions in the oxide is not sufficient to penetrate to the silicon. An important feature of this particular step is that the implant is self-aligned to the oxide that serves as the isolation region of the final device. This self-aligned

MOS CAPACITOR

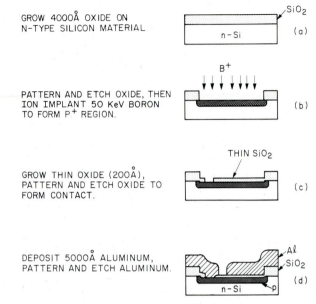

GROW 4000Å OXIDE ON
N-TYPE SILICON MATERIAL

SiO₂

n–Si

(a)

B⁺

PATTERN AND ETCH OXIDE, THEN
ION IMPLANT 50 KeV BORON
TO FORM P⁺ REGION.

(b)

GROW THIN OXIDE (200Å),
PATTERN AND ETCH OXIDE TO
FORM CONTACT.

THIN SiO₂

(c)

DEPOSIT 5000Å ALUMINUM,
PATTERN AND ETCH ALUMINUM.

Al
SiO₂

(d)

n–Si   p

FIGURE 2.
Steps to form an MOS capacitor.

feature is very important when device miniaturization is considered, and it will be covered in more detail with some of the VLSI technologies discussed later in this chapter.

The subsequent steps of growing and patterning a thin oxide, shown in Fig. 2c, and depositing and patterning aluminum (Fig. 2d) illustrates another important consideration of integrating layers—alignment. The capacitor that results from this sequence of steps will only work properly if the individual layers are isolated from each other except where an intentional contact is made. Therefore, if the aluminum over the thin oxide is misaligned so that it makes an electrical contact to the p region, the device will not function. This alignment consideration is one of the limiting factors in any device miniaturization and will be covered in detail in the next section.

11.2.3 Miniaturizing VLSI Circuits

In this section we will discuss some of the limitations to the efforts to reduce the size of ICs. The discussion will be restricted to some basic guides for reducing feature sizes and increasing density. Another aspect of miniaturization involves the way individual transistor characteristics change as the dimensions change. This topic will be covered in later sections, after the particular technologies are introduced.

Minimum Feature Size. The minimum feature size on each individual layer will be determined by the ability to reproduce and adequately resolve the feature routinely. This dimension is a function of both the minimum dimension that can be routinely resolved in a lithographic process and how accurately that feature can be transferred into the silicon during the pattern-transfer process. The minimum dimension the lithographic process will allow is a function of the limitations of the lithographic tools as discussed in Chapter 4.

The pattern-transfer process can take many forms and each will affect the minimum feature size differently. The anisotropic dry etching of a feature will cause very little change in the dimension of the feature etched in the patterned layer from the dimension of the feature formed by the lithographic mask. On the other hand, local oxidation (as shown at the top of Fig. 1b) will cause the feature to grow as the oxide becomes thicker, due to the lateral oxidation of the silicon. Also, ion implantation will involve some lateral scattering, increasing the patterned feature size from the mask dimension. Diffusion of dopants will also increase the feature size of a doped region of silicon. The final feature size will be determined by all of these factors and must be taken into account when mask dimensions are determined.

Nesting Tolerances. The distances required to "nest" the features of one level with respect to features on a previous level are called nesting tolerances. The nesting tolerance of an individual level is determined by (1) the overlay tolerance in the lithographic process, (2) the variation in finished feature sizes, and (3) the alignment sequence of the individual levels. The overlay tolerance is a characteristic dimensional variation associated with an alignment tool, and is a measure of the worst-case misregistration between two levels. If Δ is the worst-case misregistration distance between two adjacent aligned levels and there are n alignment steps between levels A and B, then the worst-case misregistration distance is $M_{A,B} = \sqrt{n}\Delta$, where Δ is assumed to be the same for all levels.[7]

The misregistration distance must be combined with the feature size variation to determine the minimum separation between levels. If $\pm\delta_A$ is the worst-case variation of an edge of a feature on level A, and $\pm\delta_B$ is the worst-case variation of an edge of a feature on level B, then the minimum separation S that may be allowed between them to ensure that the edges do not coincide is $S_{A,B} = \sqrt{n\Delta^2 + \delta_A^2 + \delta_B^2}$. Therefore, to minimize the separation between critical levels, the overlay error, variation in feature size, and number of alignment steps must be minimized. The alignment sequence will be optimized by considering the most critical alignments for each particular process sequence. Some alignments are critical while others are not. The optimum sequence will ensure that the critical alignments have the smallest separation distance while sacrificing separation between noncritical levels. This point will become clearer when the process sequence for particular technologies is considered.

Figure 3 shows the effect of both the misregistration between two levels and the worst-case variation in feature size. If the square feature in Fig. 3a must be placed entirely inside feature B, then the potential ill effects of both the feature

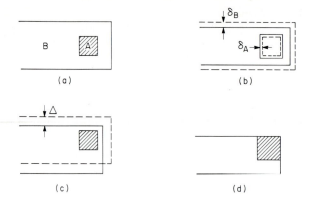

FIGURE 3
Effects of nesting tolerances on layout. (a) Ideal alignment, (b) worst-case feature size variation, (c) worst-case alignment, (d) both worst-case alignment and worst-case feature size variation.

size variation (Fig. 3b) and the misregistration (Fig. 3c) can be seen. Figure 3d shows the case where both the worst-case misregistration and the worst-case feature size variation are combined.

Problem

If the worst-case feature size variation is 0.5 μm and the worst-case misregistration is 0.25 μm for the features in Fig. 3, what is the minimum separation between levels if (*a*) feature A is aligned to feature B? (*b*) features A and B are aligned to a third feature C?

Solution

(*a*)
$$S = \sqrt{n\Delta^2 + \delta_A^2 + \delta_B^2} = 0.43 \ \mu\text{m}$$

$$n = 1, \ \Delta = 0.25 \ \mu\text{m}, \ \delta_A = \frac{0.5 \ \mu\text{m}}{2}, \ \delta_B = \frac{0.5 \ \mu\text{m}}{2}$$

(*b*)
$$S = \sqrt{n\Delta^2 + \delta_A^2 + \delta_B^2} = 0.5 \ \mu\text{m}$$

$$n = 2, \ \Delta = 0.25 \ \mu\text{m}, \ \delta_A = \frac{0.5 \ \mu\text{m}}{2}, \ \delta_B = \frac{0.5 \ \mu\text{m}}{2}$$

11.3   NMOS IC TECHNOLOGY

The first technology we will introduce is the n-channel metal-oxide-semiconductor (NMOS) technology. The process sequence for this technology is the least complex because it requires the least number of lithography levels, making it the logical starting point for an introduction to process integration. In addition, the fundamental aspects of the NMOS technology are also important for the complementary metal-oxide-semiconductor (CMOS) technology and the MOS memory technologies, which will be covered next.

The basic n-channel circuit consists of NMOS transistors, each consisting of a source, a drain, and a gate region. Each transistor is isolated from its neighbors by a thick field oxide. The basic device structure is shown in Fig. 4 with an insulating layer of phosphorus-doped SiO_2, called P-glass, covering the device. Also shown in this figure is a layer of dopant under the field oxide, called the chan-stop region, which serves to improve the isolation between transistors.

The NMOS transistor operates by causing a negative charge to move from the source to the drain in response to a positive charge on the gate. If the charge on the gate is sufficient to increase the source-to-gate voltage above a threshold voltage V_T, electrons will be attracted to the region under the gate to form a conducting path between drain and source.[8] In most applications, two kinds of NMOS transistors are used: enhancement mode devices (EMD) and depletion mode devices (DMD). The enhancement mode device has a V_T greater than 0, and for the depletion mode device V_T is less than 0.

11.3.1 Fabrication Process Sequence

Figure 5 shows a fabrication process sequence for a portion of an NMOS logic circuit.[9] The starting silicon wafer is a lightly doped p-type substrate. The first lithographic step, shown in Fig. 5a, is the isolation patterning, in which an Si_3N_4 layer is etched using anisotropic dry etching. This Si_3N_4 layer is deposited on a thin thermal oxide, called a pad oxide, which allows better adhesion between the nitride and the silicon. The etched regions of the SiO_2 expose the silicon that will subsequently be oxidized using the local oxidation technique shown in Fig. 1b. Before this oxide is formed, a boron chan-stop layer is selectively implanted into the isolation regions. The resist is then removed, and the field oxide is grown. This oxidation also serves to drive in the chan-stop implant. The nitride/pad oxide layer is now stripped (Fig. 5b) and a thin gate oxide is formed. A buried-contact window is then patterned in the gate oxide, followed by a boron threshold-adjust implant. The threshold-adjust implant is used to set the threshold voltage of the EMD. The threshold voltage of the DMD is set with a phosphorus or arsenic implant that is selectively implanted in the DMD regions. This is accomplished

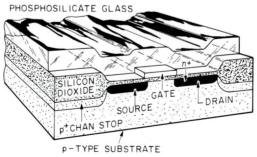

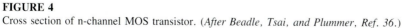

FIGURE 4
Cross section of n-channel MOS transistor. (*After Beadle, Tsai, and Plummer, Ref. 36.*)

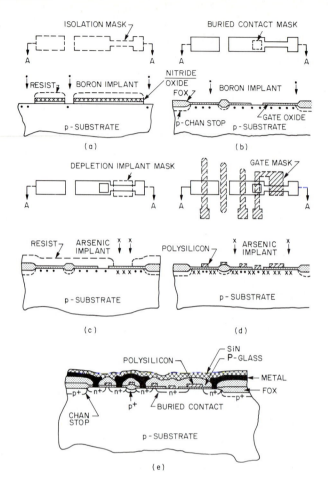

FIGURE 5

Top and cross section views of NMOS logic gate fabrication. (a) Isolation mask and cross section after nitride/oxide etch and boron chan-stop implant. (b) Buried-contact mask and cross section after field oxidation (FOX), gate oxidation, buried-contact window etch, and boron enhancement threshold-adjustment implant. (c) Depletion implant mask and cross section after resist-masked arsenic depletion implant. (d) Gate mask and cross-section after polysilicon gate definition and arsenic source/drain implant. (e) Final structure after P-glass flow, window etch, aluminum deposition, and aluminum etch. (*After Parrillo, Ref. 9.*)

by using a resist mask to protect the EMDs from the implant (Fig. 5c). At this point in the processing, the threshold voltages of both devices have been set by the two implants and the thickness of the gate oxide. The control of these implant doses and the quality and uniformity of the gate oxide are critical in determining the proper device characteristics.

The next series of steps will form the gate and source/drain region to complete the transistor structures. As shown in Fig. 5d, the polysilicon gate material is deposited and doped n-type with either an implant or vapor doping. After the polysilicon is patterned, the source/drain regions are implanted with arsenic or phosphorus. This implant is self-aligned to both the gate polysilicon and the field oxide. This is done by choosing the implant energy so that the dopant easily penetrates the thin gate oxide left in the source/drain regions but will not penetrate either the polysilicon gate or the field oxide. This self-aligned feature is an important aspect of MOS technology that allows the minimum possible overlap between the gate and source/drain regions. The polysilicon is in contact with the silicon substrate only where the opening was etched in the gate oxide using the buried contact mask. This contact from the gate to the source or drain of the transistor is formed when the phosphorus or arsenic used to dope the polysilicon diffuses into the silicon.

The final part of the processing serves to form metal connections between the devices and provide protection for the devices. A CVD oxide doped with phosphorus is deposited on the wafers. This P-glass serves a number of functions. The phosphorus is included in the glass to allow the viscosity to be reduced, so that at relatively low temperatures (950 to 1050°C) the glass will flow to smooth the surface topography. The phosphorus doping also helps to protect the devices from mobile ion (Na^+) contamination. The P-glass also serves as an intermediate dielectric to isolate the metal interconnect level from the polysilicon. The contact windows are etched into the P-glass to expose the source/drain regions or polysilicon wherever contacts are desired. Aluminum is usually the metal used to interconnect the devices. The contacts are usually sintered in a hydrogen ambient at temperatures up to 500°C to form a good ohmic contact with the silicon (Fig. 5e). The hydrogen also helps to anneal out any radiation damage that may have been introduced during metal deposition and etching. The last layer deposited on the wafer is a capping layer of plasma-deposited SiN or a deposited oxide that is used to protect the wafer from scratching or contamination. Windows are then etched in this capping layer to make external wire connections.

11.3.2   Special Considerations for NMOS ICs

The special considerations for NMOS ICs are all directed towards keeping the basic NMOS transistor characteristics as close to the ideal as possible. This means reducing parasitic conduction paths and series resistances as well as maintaining threshold voltage control and minimizing the leakage current of the device. All of these considerations will also apply to CMOS technology and the MOS memory technologies that will be discussed later.

Starting material. Virtually all MOS technologies use silicon wafers with the crystal surface in the <100> orientation. This <100> orientation is preferred over the <111> orientation because it introduces about 10 times less interface trap density, which can cause charging at the SiO_2-silicon interface.[4] This charging causes variations in the threshold voltage of the devices and is therefore undesirable. For most NMOS technologies, the silicon will be lightly doped p-type ($\sim 10^{15}$ atoms/cm^3).

Isolation. Isolation between transistors is accomplished by limiting the conduction of the parasitic transistor that occurs between two neighboring transistors. This parasitic NMOS transistor is shown in Fig. 6a. It consists of a polysilicon gate with the field oxide (FOX) acting as a gate oxide and the chan-stop region as the channel. The goal in any isolation scheme is to make this parasitic transistor have as high a threshold voltage as is practical without adversely affecting the transistor characteristics of the active device. This is accomplished by making the field oxide as thick as possible and the chan-stop as heavily doped as possible. Figures 6b through 6d show three possible approaches to growing the isolation field oxide. The local oxidation technique shown in Fig. 6b is the structure resulting from the process illustrated in Fig. 5. The structure in Fig. 6c uses a simpler technique, consisting of growing a thick oxide everywhere, then cutting windows where the active transistors will be formed. In the structure in Fig. 6d the silicon is partially etched before the local oxidation is performed to recess the field oxide.

The three structures shown in Fig. 6 illustrate the tradeoffs that must be considered in determining the field oxide structure. The simple approach of Fig. 6c allows the isolation regions to be most sharply defined because it is only limited by the oxide etch. However, the topography created by the field oxide step will cause difiiculty when the polysilicon gate is etched (see Chapter 5). The recessed oxide approach of Fig. 6d relaxes the topography, but increases the complexity of processing, and may introduce material defects in the silicon. [10]

The local oxidation process (Fig. 6b) is a compromise between a smoother topography and better isolation. The field oxide is usually several thousand angstroms thick to achieve adequate isolation and reduce capacitance to the substrate. The growth of the field oxide layer using the local oxidation technique causes the oxide to penetrate under the masking nitride layer. This causes the space between the transistors to grow during the oxidation. The oxide growth

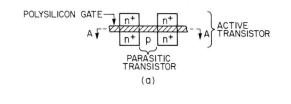

(a)

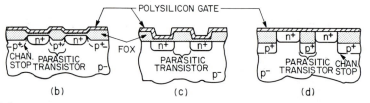

(b)   (c)   (d)

FIGURE 6
MOSFET isolation. (a) Top view of adjacent NMOS transistors with common polysilicon gate, illustrating active and parasitic transistor conduction paths. Cross section through A-A for (b) local oxidation isolation structures, (c) etched field-oxide isolation, (d) recessed oxide structure.

under the nitride layer is sometimes called "bird's-beak encroachment" and is one of the drawbacks of the local oxidation process. However, the topography of this process is less severe than simply etching windows in the field oxide, since some of the silicon is consumed during oxidation and the bird's-beak structure forms a slightly sloping edge.

One other consideration for the field oxidation step is how the chan-stop layer is affected by the formation of the field oxide. The field oxidation can cause significant lateral diffusion of the chan-stop layer. This diffusion raises the surface concentration of the substrate near the edge of the field oxide, causing the threshold voltage of that portion of the active device to increase. The edge of the device will not conduct as much as the interior portion and the transistor behaves as if it were a narrower device. This effect can be reduced by lowering the temperature of the oxidation.[11]

Channel doping. The doping level of the silicon in the channel region of the transistor determines the threshold voltage of the device, along with the gate oxide thickness and the work function of the gate material. For a given choice of gate material and oxide thickness, the channel doping can be chosen to give the desired threshold voltage. However, there are deleterious effects if the channel doping is either too high or too low. If the channel doping is too high, it can cause a reduction of carrier mobility at the surface.[12] If the channel doping is too low (depending on the drain junction depth and channel length), a situation can occur where the drain electric field will punch through to the source. This "punch-through" condition is illustrated in Fig. 7, showing the depletion regions of the source and drain nearly touching.

One technique for reducing the punch-through susceptibility for short-channel transistors is to implant boron sufficiently deep into the channel region to raise the substrate doping without changing the surface concentration. Figure 7 illustrates the subthreshold I-V curves for a device with a light substrate doping.[13] A short-channel-length device ($L = 1.2$ μm) punches through (curve A), while a long-channel device (curve B) shows the desired long-channel behavior. A boron implant dose of 8×10^{11} atoms/cm^2 was implanted into the short channel device at a projected range (R_p) of 0.3 μm. This implant was too shallow, and the threshold voltage (curve C) rose well above the desired value. By increasing the implant energy R_p (curves D to E), the boron implant was sufficiently deep to leave the surface concentration undisturbed and to prevent punch-through.

A shallow implant is sometimes used to adjust the threshold voltage. This can be a boron implant to raise the threshold voltage or a phosphorus or arsenic implant to reduce the threshold voltage. The phosphorus or arsenic implant will compensate the surface and, depending on the dose, can actually create a junction beneath the gate oxide. This would be the case if a depletion mode (normally on) transistor is fabricated, as in the process depicted in Fig. 5.

Gate material. Heavily doped n-type polysilicon is the most widely used gate material for modern NMOS technologies. The polysilicon can withstand the

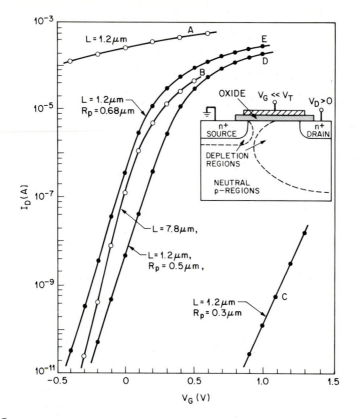

FIGURE 7
Drain current vs. gate voltage for n-channel devices with a substrate doping of 1.9×10^{15} atoms/cm^3, source/drain junctions 0.47 μm deep, 575 Å gate oxide, drain voltage of 5 V, and back-gate bias of 0 V. Devices A and B have no channel implant, and devices C, D, and E have a boron channel implant of 8×10^{11} atoms/cm^2 at various energies. The insert schematically illustrates the source and drain depletion regions for $V_G \ll V_T$ and $V_D > 0$. (*After Nihira et al., Ref. 13.*)

high temperature processing necessary to form the source/drain junctions. The polysilicon-SiO$_2$ interface is well understood, and the work function of n$^+$-doped polysilicon is also well suited for NMOS transistors with short channel lengths. [14] The major disadvantage to using the polysilicon gate is that the resistance of the gate may contribute to the RC delay of signals that are routed along it. Many approaches have been taken to reduce the resistance of the gate material from the 10 $\Omega/\square$ or more that is typical for the polysilicon gate. Certain refractory metals and the silicides of some refractory metals have been proposed as possible alternatives. This can reduce the sheet resistance of the gate interconnect to about 1 to 3 $\Omega/\square$, but may cause problems with adhesion or interface properties between the gate material and gate oxide.

One approach that combines the advantages of a polysilicon gate and reduces the sheet resistance is to use a refractory metal silicide on top of the polysilicon

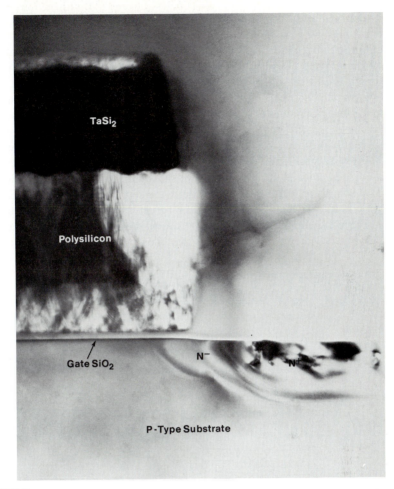

FIGURE 8
Transmission electron micrograph showing details of source/drain and gate regions. (*After Sheng, Ref. 17.*)

gate. This approach (called polycide)[15] preserves the polysilicon-SiO_2 interface while lowering the overall sheet resistance to about 1 to 3 $\Omega/\square$. The most common types of silicides used for this application are $MoSi_2$, WSi_2, $TaSi_2$ and $TiSi_2$.[16] Each of these materials has relative advantages or disadvantages for deposition and patterning (see Chapter 9), but the goal of reducing the sheet resistance is the same for all of them.

Source/drain formation. The n^+ source/drain region of the NMOS transistor should be as low in resistance as possible. Modern devices also require the junction to be as shallow as possible to allow the small dimensions needed for VLSI applications. Arsenic is used as the dopant for the source/drain since it has a high solubility and low diffusion rate. Source/drain implants are typically in the high-10^{15} to 10^{16} atoms/cm^2 dose range to produce low-resistance source/drain regions. Figure 8 shows the details of the source/drain region.[17] In this particular device a lightly doped drain structure is also shown; the purpose of the lightly doped drain will be discussed later.

Reducing the series resistance of the source/drain regions is an important factor in the design of smaller transistors. As channel conductance increases with shorter channel lengths, the resistance of the shallow source/drain region stays fixed or actually increases because of the need for shallower junctions. The result is that the resistance of the source/drain limits the current-delivering capability of short-channel devices. The resistance of the source/drain can be substantially reduced by using a refractory metal silicide in the source/drain region similarly to the way it was used to reduce the polysilicon resistivity in the previous section. In addition, the technique illustrated in Fig. 9 allows the simultaneous silicidation of source, drain, and gate. This process is sometimes called salicide for self-aligned

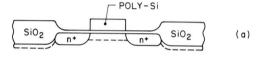

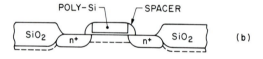

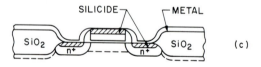

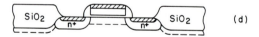

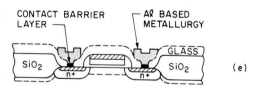

FIGURE 9

Process sequence for forming silicided source/drain and gates. (a) Cross section of transistor after source/drain implant, (b) oxide spacer formation, (c) metal deposition and reaction, (d) unreacted metal is removed, (e) P-glass is deposited and aluminum contacts are formed. (*After Ting, Ref 16.*)

silicide.[16] The key steps are shown in Fig. 9a – d, in which the oxide spacers are formed and the metal is selectively reacted with the exposed silicon. The oxide spacer is formed by depositing an oxide film and then reactive-ion-etching the oxide film in the source/drain region and on top of the polysilicon. A thin oxide spacer is left on the side of the polysilicon. A thin layer of metal is then deposited on the wafer and reacted to form the silicide wherever there is exposed silicon. There are many problems associated with this process (see Chapter 9) and a number of different silicides have been proposed, including PtSi, $MoSi_2$, $CoSi_2$, and $TiSi_2$. However, the advantages of only having to etch the polysilicon gate rather than a composite gate structure, in addition to substantially reducing the source/drain resistance, makes this structure one of the most promising approaches for scaled devices. The effect on the device characteristics can be seen in Fig. 10, in which the I-V characteristics are shown for device structures specially designed to emphasize the source/drain resistance.[18] The characteristics of two devices are shown, one with a silicided source/drain region and one without. The difference can clearly be seen in the current characteristics, showing a substantial increase in drive for the silicided device.

One last concern that has become very important for smaller devices is the "hot electron problem" that arises when device dimensions are reduced but the supply voltage is maintained constant (e.g., 5 V) and results in an increase in the electric field generated in the silicon. These intensified electric fields can make it possible for electrons in the channel of the transistor to gain sufficient energy to

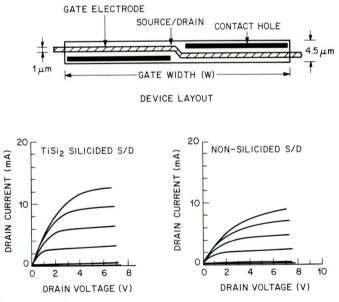

FIGURE 10
Reduction of source/drain resistance with $TiSi_2$. (a) Device layout of transistor used to emphasize the source/drain resistance. (b) I-V characteristics with $TiSi_2$ source/drain. (c) I-V characteristics without silicided S/D. (*After Kobayashi et al., Ref. 18.*)

be injected onto the gate oxide. The charging of the gate oxide causes a long-term device degradation, which raises the threshold voltage of the device and reduces the transconductance. One approach to minimize the degradation is to reduce the electric field at the drain region to prevent the electrons from achieving sufficient energy to be injected into the oxide. This is achieved by grading the junction of the drain by doing two implants into the source/drain region. One of the implants is designed to create a lighter doped region beyond the normal n^+ drain junction. This can be done by offsetting the heavier implant with a sidewall spacer forming the drain structure shown in Fig. 11b, sometimes called a lightly doped drain (LDD).[19] Another approach is to simply do two implants of phosphorus and arsenic into the same region to form the structure shown in Fig. 11a, which is sometimes called a double-doped drain (DDD).[20] The electric field

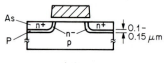

(a)

(b)

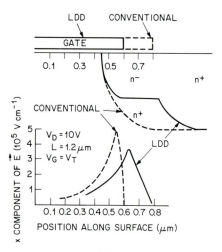

(c)

FIGURE 11

Cross section of MOSFET structure showing (a) double-doped and (b) lightly doped drain, used to reduce the peak electric field, (c) electric-field profile for a MOSFET, with and without the LDD structure. (*After Ogura et al., Ref. 19.*)

in the drain region is reduced for both of these structures. This is illustrated in Fig. 11c, where the magnitude of the electric field at the SiO_2 interface is plotted as a function of the position along the surface for both the LDD case and the conventional drain.

11.4 CMOS IC TECHNOLOGY

The next step in processing complexity is complementary metal-oxide semiconductor (CMOS) technology. CMOS technology employs both NMOS and PMOS transistors to form the logic elements. The advantage of CMOS technology is that the particular logic elements only draw significant current during the transition from one state to another but draw very little current between transitions, allowing power consumption to be minimized.

The circuit diagram of a CMOS inverter is illustrated in Fig. 12a. The cross section of the inverter structure (Fig. 12b) shows the n-channel transistor formed in a p-region called a tub or well. The p-channel transistor is formed in the n-substrate. The gates of the transistors are connected to form the input. If we define the threshold voltage of the n-channel and p-channel transistors as

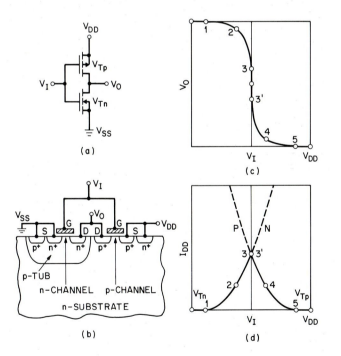

FIGURE 12
CMOS inverter. (a) Circuit schematic; V_{DD} and V_{SS} are the highest and lowest circuit potentials respectively. (b) Device cross section. (c) Output (V_O) versus input (V_I) voltage of inverter. (d) Current through inverter as a function of input voltage (solid curve); I-V characteristics of n- and p-channel transistors (dashed curves). The numbers correspond to different points on the inverter transfer characteristic. (*After Hoefflinger and Zimmer, Ref. 2.*)

V_{Tn} (e.g., 0.6 V) and V_{Tp} (e.g., −0.6 V) then we can track the output voltage V_O, shown in Fig. 12c as a function of the input voltage V_I. The transistor current for the individual transistors is shown in Fig. 12d as a function of the input voltage. At $V_I = 0$, the n-channel transistor is off, since $V_I < +0.6$ V, but the p-channel is on, since $V_I = 0 > V_{Tp}$ (= −0.6 V). As the gate voltage of the n-channel transistor is raised above V_{Tn}, it begins to conduct current which flows into the p-channel transistor. The p-channel transistor is on at this point since the magnitude of its gate voltage is well above the threshold voltage. Continuing to increase V_I will bring the gate voltage of the p-channel device closer to the p-channel threshold voltage and eventually turn the device off. The important point to remember is that in either logic state, where V_O is either V_{DD} or V_{SS}, one of the two transistors is off and since the transistors are in series, negligible current is conducted through the inverter. Significant current is conducted through the inverter only when both transistors are on during the transition, as shown in Fig. 12d. Therefore, in either logic state very little current is conducted from V_{SS} to V_{DD} and very little power is consumed. The low power consumption of CMOS is one of its most important attributes for high-density applications.

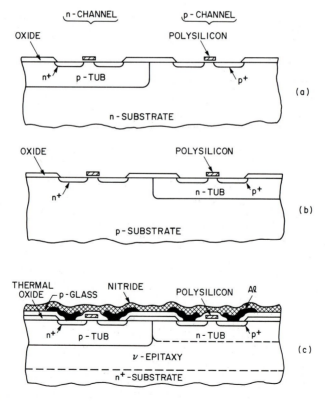

FIGURE 13
Various CMOS structures: (a) p-tub, (b) n-tub, (c) twin-tub. (*After Parrillo et al., Ref. 21.*)

11.4.1   Fabrication Process Sequence

The fabrication sequence for CMOS devices is similar to the sequence for NMOS devices. The differences occur where the doping must be individually adjusted for the n- and p-channel devices. The first consideration in the process is to determine how the substrates will be formed for the two types of transistors. The n-channel transistors need a p-substrate and the p-channel transistors require an n-substrate. There are three approaches to forming the two different substrates, which are referred to as p-tub, n-tub and twin-tub processes.

The p-tub (also called p-well) process involves implanting or diffusing p-type dopant into an n-substrate at a concentration that is high enough to over-compensate the n-substrate and to give good control over the desired p-type doping (Fig. 13a). The n-substrate doping must be high enough to ensure that the p-channel device characteristics are adequate. The p-tub doping typically must be five to ten times higher than that in the n-substrate to ensure this control. Excessive p-tub doping produces deleterious effects in the n-channel transistor, however, such as increased back-gate bias effects, reduction in mobility, and increased source/drain to p-tub capacitance.

The n-tub process is an alternative approach in which an n-tub is formed in a p-type substrate. The n-channel device is formed in the p-type substrate, so this approach is compatible with standard NMOS processing. In this case, the n-tub overcompensates the p-substrate, and the p-channel transistor suffers from excessive doping effects (Fig. 13b).

The twin-tub process allows two separate tubs to be implanted into very lightly doped silicon.[21] This allows the doping profiles in each tub region to be tailored independently so that neither type of device will suffer from excessive doping effects (Fig. 13c). The lightly doped silicon is usually an epitaxial layer grown on a heavily doped silicon substrate. The substrate can be either n-type or p-type.

The process sequence for a twin-tub CMOS process is outlined in Fig. 14. In this case the starting material is lightly doped ν-epitaxy over a heavily doped n^+ <100>-orientation substrate. Figures 14a to 14c show how the self-aligned twin tubs are formed using one lithographic mask step. A composite layer of SiO_2 and Si_3N_4 is defined and silicon is exposed over the n-tub region. Phosphorus is implanted as the n-tub dopant at low energy and enters the exposed silicon, but is masked from the adjacent region by the Si_3N_4 (Fig. 14a). The wafers are then selectively oxidized over the n-tub regions. The nitride is stripped and boron is implanted for the p-tub (Fig. 14b). The boron enters the silicon through the thin oxide but is masked from the n tub by the thicker SiO_2 layer. All oxides are then stripped and the two tubs are driven in (Fig. 14c).

After the tubs are formed, the formation of the field oxide and gates is identical to that discussed for the NMOS process. The threshold adjustment implants can be made into the channel regions of the devices to adjust the threshold voltages of the n- and p-channel transistors individually. The sources and drains are self-aligned to the gate as they were in the NMOS process. However, it is necessary to selectively implant the n-channel and p-channel

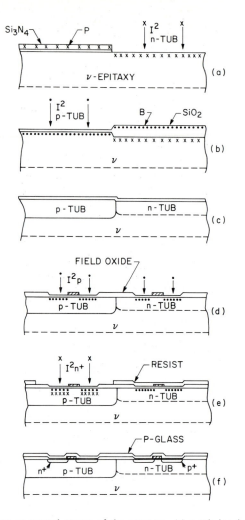

FIGURE 14
Twin-tub CMOS structure at several stages of the process: (a) n-tub ion implant (I^2); (b) p-tub implant; (c) twin-tub drive-in; (d) nonselective p^+ source/drain implant; (e) selective n^+ source/drain implant using photoresist mask; (f) P-glass deposition. (*After Parrillo et al., Ref. 21.*)

source/drains to form n^+ regions for the n-channels and p^+ regions for the p-channels. This can be done by using one mask. The one mask technique is illustrated in Figs. 14d and 14e. This technique consists of implanting boron nonselectively into all the sources and drains. This is followed with a selective implant of phosphorus or arsenic into the n- channel source/drain regions at a higher dose, so that it overcompensates the boron.[21] After the subsequent thermal cycles, the phosphorus or arsenic completely covers the boron vertically and laterally. A phosphorus glass layer is later deposited (Fig. 14f) and windows are etched. Aluminum metallization is defined using dry etching. The final layer is a plasma-deposited silicon-nitride layer that seals the devices and provides mechanical scratch protection. The final cross section of the device is shown in Fig. 13c.

11.4.2 Special Considerations for CMOS ICs

The considerations that were discussed for NMOS technologies (field isolation, source/drain structure, etc.) are also important factors for CMOS design. However, the fact that there are two separate kinds of transistors on the same substrate creates another set of parasitic effects that are a result of parasitic MOSFETs forming between the devices. In addition, there are unique problems involved with trying to optimize the characteristics of both types of transistors simultaneously, which will affect the choice of materials and dopings to form the gates and channel regions.

Isolation and latchup. The same principles involved with NMOS isolation apply to the isolation of CMOS devices of the same type within a given tub region. An added concern in CMOS circuits is isolating the two different types of transistors. Figure 15a shows the top view of n- and p-channel transistors straddling the common tub border. Figure 15b shows a cross section of the structure beneath the polysilicon. A parasitic n-channel transistor exists between the n^+ source and the adjacent n-tub. Similarly, a parasitic p-channel transistor exists between the p^+ source and the p-tub. The diffusion necessary to drive the tub in causes a reduction in the surface concentration of each tub near the border, and hence a reduction in the parasitic transistor's threshold voltage. Figure 15c shows the threshold voltage of each type of parasitic device as a function of the separation between the transistor edge and the tub border.[21] The upper curve on the left shows the parasitic n-channel threshold voltage reduction near the tub border, which occurs with no adjacent n tub and characterizes the effect of a long diffusion of a single-p-tub process. The lower curve on the left shows the further reduction of the threshold voltage due to the interdiffusion of the two types of tub impurities. This threshold voltage will determine the minimum separation necessary between the n- and p-channel transistors to avoid undesirable parasitic currents between the transistors.

Another problem that arises from diffused source/drain regions in close proximity to an adjacent tub is a condition called "latchup."[22] A latchup condition is caused by parasitic bipolar transistors formed by the source drain regions and the tubs, as illustrated in Fig. 16a. The bipolar transistors make up a thyristor (pnpn) device (Fig. 16b). The thyristor can be biased such that the collector current of the pnp device supplies a base current to the npn device in a positive feedback arrangement. This will produce a large sustained current between the positive and negative terminals of the thyristor (V_{SS} to V_{DD}) and will cause the CMOS circuit to cease functioning or even self-destruct. Although there are many aspects of circuit design and triggering events that are important to the subject of latchup, there are a few basic features that can be incorporated into the CMOS technology that will reduce the susceptibility to latchup. The most important of these are techniques that decouple the bipolar transistors. The decoupling of the bipolar

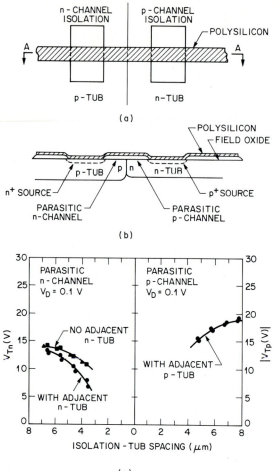

FIGURE 15
Isolation of n- and p-channel transistors. (a) Top view of adjacent n- and p-channel transistors sharing a common polysilicon gate. (b) Cross section under the polysilicon rail. (c) Parasitic n- and p-channel threshold voltages vs. transistor-edge-to-tub spacing. (*After Parrillo et al., Ref. 21.*)

transistors removes the thyristor pnpn arrangement and prevents the latchup condition.

The decoupling of the bipolar transistors can be understood by examining the four-terminal model illustrated in Fig. 16b. This figure shows the bipolar components and resistive components of the pnpn device. The two bipolar transistors can be decoupled by making the resistances R_{s1} and R_{w1} as small as possible or by physically separating the base from the collector of the two transistors. The R_{s1} resistance can be made small by locating the substrate (tub) contact as close as possible to the source of the transistor and by forming the tubs in a lightly doped epitaxial silicon region that is grown on a heavily doped substrate.[22] This is a

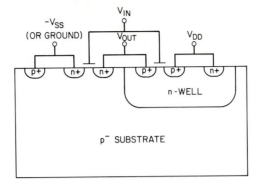

(a)

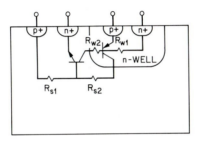

(b)

FIGURE 16
(a) Schematic cross section of a CMOS inverter. (b) Simplified pnpn thyristor model for the CMOS inverter. (*After Troutman, Ref. 22.*)

common practice for most modern CMOS devices. Another technique to reduce the resistance is to use high-dose, high-energy implants to form a heavily doped region in the p tub.

A technique that both decouples the bipolar transistors and improves the isolation of the MOS transistors is to form a trench between the two tubs. This "trench isolation" involves etching a narrow, deep groove in the silicon and then filling it with oxide or polysilicon.[23] Figure 17 illustrates a process sequence used to form a trench isolation structure. A layer of deposited Si_3N_4 is defined (Fig. 17a) and used as an etch mask during a reactive ion etch of the silicon (Fig. 17b). After the trench is formed a sidewall oxide is grown to form an insulating layer. The trench is then filled with polysilicon by depositing and etching back so that only the polysilicon in the trench is not removed (Fig. 17c). An oxide is then grown to cap the trench and the Si_3N_4 is removed (Fig. 17d). The processing can then proceed as in the standard CMOS process. This trench structure can be used along with the epitaxial silicon substrate. The trench is then made deep enough to penetrate through the epitaxial layer, which effectively decouples the bipolar transistors.

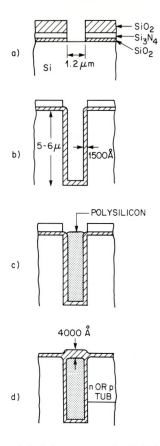

FIGURE 17
Process sequence for forming trench isolation structure for CMOS. (a) Trench mask patterning, (b) trench etching and oxide growth, (c) polysilicon deposition to fill trench, (d) oxide grown to cap trench. (*After Rung, Ref. 23.*)

Gate material and threshold adjustment. The choice of the gate material and the channel doping level will determine the threshold voltage for the transistors. To maximize the performance of the devices, the threshold voltages of the n- and p-channel transistors should be comparable, and, ideally, of equal magnitude. The threshold voltages should be as low as possible without introducing excessive off currents. The most common choice for gate material for modern CMOS technologies is n-type polysilicon that is doped heavily enough to make the polysilicon degenerate. This is usually combined with a silicide layer to lower the sheet resistance as discussed for the NMOS technology in Section 11.3. The work function of n^+ polysilicon is ideal for an n-channel device since it will yield a threshold voltage of less than 0.7 V for reasonable values of channel doping and oxide thicknesses. This point can be best understood by examining the equation for the threshold voltage of a MOSFET:

$$V_T = \phi_{ms} - \frac{Q_f}{C_{ox}} + 2\psi_B - \frac{Q_B}{C_{ox}}$$

where ϕ_{ms} is the metal-silicon work function difference of the gate material, Q_f is the charge in the gate oxide, ψ_B is the substrate Fermi potential, Q_B is the depletion region charge in the substrate, and C_{ox} is the capacitance of the gate. The gate capacitance is inversely proportional to the gate oxide thickness, and Q_B is proportional to the square root of the channel doping. These are the two parameters that can be used to set the threshold voltages of the transistor for a given choice of gate material.

Figure 18 shows the threshold voltage V_T of both n- and p-channel devices as a function of substrate doping and for three gate oxide thicknesses d. This plot assumes that Q_f is not a significant contribution to V_T. This is a reasonable assumption for any modern MOS process. The resulting plot for the n$^+$ polysilicon gate demonstrates the advantage of using n$^+$ polysilicon to obtain the proper threshold voltage for the n-channel transistors. The threshold voltage V_{Tn} is easily adjusted from 0 to $+1$ V with a substrate doping range of 10^{15} to 10^{17} cm^{-3}, which is the proper doping range to be compatible with other device constraints (such as short-channel effects). When the n$^+$ polysilicon gate is used, the p-channel threshold voltage is not as ideally adjusted with the substrate doping, as can be seen in the V_{Tp} curves in Fig. 18. The p-channel device cannot be adjusted for $|V_{Tp}|$ less than 0.7 V by simply reducing the p-channel substrate doping. For doping levels necessary to prevent short-channel effects, the threshold voltage magnitude is above 1 V.

A technique that allows the p-channel threshold voltage to be adjusted to the desired level is to implant a shallow boron layer into the channel region.[24]

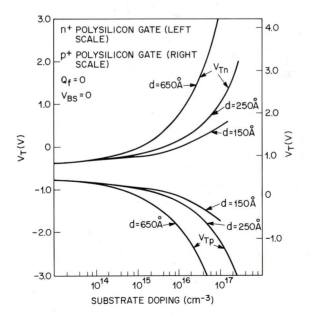

FIGURE 18

Calculated threshold voltages of n-channel (V_{Tn}) and p-channel (V_{Tp}) transistors as a function of their substrate's doping, assuming n$^+$ polysilicon gate (left scale) and p$^+$ polysilicon gate (right scale). Curves for gate oxide thicknesses d of 150 Å, 250 Å and 650 Å are shown.

The boron shifts the threshold voltages towards more positive values by forming a compensating layer. The boron threshold adjustment can also be used to raise the threshold voltage of the n-channel transistor. A single boron implant dose can be used to set the threshold voltages of both the n- and p-channel transistors if the background dopings are chosen correctly. An example of this technique is shown in Fig. 19, where the threshold voltages are shown for both the n- and p-type transistors as a function of this boron implant. The magnitude of V_{Tp} is reduced, because the boron implant adds negative change to the depletion region of the p-channel. The lower n-well doses allow the boron implant to have a greater effect on the value of V_{Tp}. One disadvantage to this approach is that the p-channel transistor is more susceptible to short-channel effects with the compensated surface channel.

Another choice for the gate material is p^+ polysilicon.[25] The work function of p^+ polysilicon is about 1.1 V greater than for n^+ polysilicon. This makes it ideal for the p-channel transistor in terms of having the threshold voltage easily adjusted to -0.7 V or less with channel doping of 10^{15} to 10^{17} cm^{-3}. However, the n-channel transistor must now be compensated to reduce the threshold voltage to reasonably low values. This is shown in Fig. 18 by using the scale on the

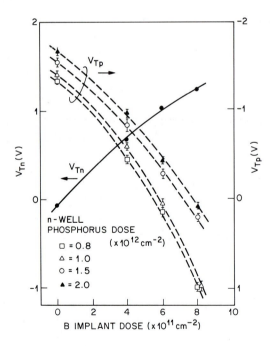

FIGURE 19

Threshold voltages of n-channel (V_{Tn}) and p-channel (V_{Tp}) transistors as a function of boron threshold-adjustment dose. The CMOS structure uses an n-well implanted into a p-type substrate whose doping level is 6×10^{14} atoms/cm^3. V_{Tp} results are shown for various implant doses of the n-well. (*After Ohzone et al., Ref. 24.*)

right side to determine the threshold voltages as a function of substrate doping. Other choices of gate material, such as $MoSi_2$, have metal work functions that are between those of n^+ and p^+ polysilicon, which allows both transistors to be slightly compensated to get the desired threshold voltages.[26] To optimize both devices simultaneously, both n^+- and p^+-type polysilicon gates can be used for the n- and p-channel transistors respectively. This approach allows both transistor threshold voltages to be easily adjusted to the desired threshold voltages without sacrificing short-channel effects, but adds to the processing complexity.

11.5   MOS MEMORY IC TECHNOLOGY

The MOS memory circuit has been the one IC application that accounts for most of the ICs produced. Simplicity in design and relative ease in taking advantage of new technology advances make this circuit a natural vehicle for technology development and a benchmark circuit for technology demonstrations. The memory circuits have unique needs and structures, which will be addressed in this section.

There are several different kinds of memory circuits. Random-access memory (RAM) allows any bit of information in a matrix of bits to be stored and accessed independently. RAM comes in two varieties: static and dynamic. The static RAMs (SRAMs) retain the data stored indefinitely, unless the power is removed from the circuit. Dynamic RAMs (DRAMs) require that the data in the cells be periodically refreshed to recharge the stored bit of information.[27]

Read-only memories (ROMs) are randomly addressable memories in which the data is stored permanently during the manufacturing process. The data can only be read from a ROM, and there is no way of modifying the data. Programmable read-only memories (PROMs) are read-only memories that can be programmed externally after the manufacturing process. If the data stored can also be erased through some means then the circuit is an erasable PROM or EPROM. All of the data stored in the ROMs, PROMs, and EPROMs are nonvolatile, meaning that the data is maintained even if the power is removed.

In this section we will discuss some of the processing constraints and unique structures that are associated with making high-density dynamic and static RAMs. The circuits are made using the NMOS or CMOS technologies discussed in the previous two sections. However, the need for high density results in structures that are not encountered in other CMOS or NMOS applications.

11.5.1   Dynamic Memory

The basic building block of any memory circuit is the cell in which a single bit of information is stored. The DRAM has the simplest cell structure, consisting of a single transistor and a capacitor. The simplicity of the cell allows the density of the circuit to be maximized even with the added overhead of having to design the circuitry needed to refresh the bits. Since the cell is the most important part of the circuit, the cell design will dictate the processing needs.

The simplest example of a DRAM cell layout is shown in Fig. 20. The cell consists of a single NMOS transistor (access transistor) and a capacitor. As shown in Fig. 20a, the gate of the transistor is connected to a "row select" or "word line," and the source of the transistor is connected to the "bit line." When both the word line and bit line are brought to a high value, the transistor is on, and charge can flow to the capacitor. If the capacitor initially had no charge (stored zero), then charge flows into the capacitor. If the capacitor initially is charged (stored one), then very little charge flows into the capacitor. To read an individual bit, the sensing circuitry measures the amount of charge that flows into the capacitor and determines whether it is a "zero" or a "one." The capacitor is then "refreshed" by either fully charging it or completely depleting it of charge, depending on which state it was in initially. The write cycle simply charges a capacitor to the state desired (0 or 1) and the amount of charge needed is not measured.

The cell shown in Figs. 20b and 20c uses a diffused region as the bit line and a polysilicon plate to form a storage capacitor. The capacitance of the

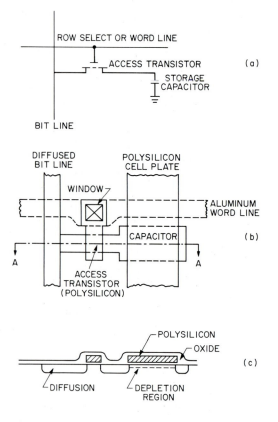

FIGURE 20
Single-transistor dynamic RAM cell with storage capacitor. (a) Circuit schematic. (b) Cell layout. (c) Cross section through A-A. (*After Hunt, Ref. 27.*)

capacitor needs to be as high as possible to give as high a signal as possible to the sensing circuitry, and the bit line capacitance and resistance needs to be made small to improve the performance of the cell. The efforts in DRAM processing and cell design are directed toward these two goals. The charge stored on a capacitor of area A, dielectric thickness d, and dielectric permittivity ϵ, with a voltage across the capacitor of V_s is given by

$$Q_s = \epsilon A \frac{V_s}{d} = \epsilon A \mathscr{E}$$

where $\mathscr{E}$ is the electric field across the dielectric. The above equation shows the three parameters that can be adjusted to increase the capacitance. However, the trend to higher packing density requires the cell size to shrink and the area of the capacitor to decrease. This makes it necessary to increase the charge stored by either increasing the electric field or increasing the dielectric constant.

Problem
What is the stored charge and the number of electrons on an MOS capacitor with an area of 4 μm^2, a dielectric of 200 Å thick SiO_2, and an applied voltage of 5 V?

Solution

$$Q_s = \epsilon \times A \times \frac{V_s}{d}$$

$$= 3.9 \times 8.85 \times 10^{-14} \text{ F/cm}$$

$$\times 4 \times 10^{-8} \text{ cm}^2$$

$$\times \frac{5 \text{ V}}{2 \times 10^{-6}} \text{cm}$$

$$= 3.45 \times 10^{-14} \text{ C or with } q = 1.6 \times 10^{-19} \text{ C/electron}$$

$$Q_s = 3.45 \times 10^{-14} \text{ C/}q = 2.15 \times 10^5 \text{ electrons}$$

The electric field can be increased by thinning the oxide. However, if the field strength is too high, there can be appreciable currents flowing in the oxide that will adversely affect the retention time and reliability of the capacitors. The practical limit to the field is somewhere around 2 to 4 MV/cm.[28] The dielectric constant can be increased by using other insulators for the capacitor such as Si_3N_4 or Ta_2O_5, which have dielectric constants of 8 and 22 respectively.[29,30] This is a significant increase over the dielectric constant of SiO_2 (which is 3.9), but requires sacrificing the processing simplicity of using SiO_2.

Another approach that effectively increases the field across the oxide is by selectively implanting both boron and shallow arsenic under the capacitor. This doping creates a positive fixed charge directly under the oxide, which increases the charge storage capacity. This is called a high-capacity (Hi-C) RAM cell.[31] A cross section of the Hi-C RAM cell is shown in Fig. 21a. A further improvement in the Hi-C RAM cell structure is shown in Fig. 21b. This improvement involves

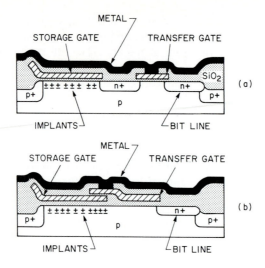

FIGURE 21

High-capacity (Hi-C) dynamic RAM cell structure with shallow arsenic ($+$) and deeper boron ($-$) implants. (a) One-transistor cell with single-level polysilicon. (b) Double-level polysilicon cell. (*After Tasch et al., Ref. 31.*)

using a second layer of polysilicon that serves to transfer the charge from the bit line to the storage capacitor, eliminating the drain region of the access transistor and allowing the cell size to be further reduced. The next stage in complexity to maintain the charge storage capacity and still decrease the cell size is to go to three-dimensional structures such as trench capacitors.[32] A trench capacitor is formed by etching a trench on the silicon so that the capacitor area is determined by the area of the walls of the trench. An example of a trenched DRAM cell is shown in Fig. 22a. An even more aggressive approach to the trenched capacitor structure is to use the same surface area for both the trench and the access transistor.[33] This structure is shown in Fig. 22b. The access transistor is formed along the trench sidewall by growing a thin oxide on the top portion of the trench sidewall and then depositing polysilicon to form a gate electrode. The access transistor then controls the charge stored in the polysilicon storage node in the trench.

The general trend in development of DRAM cells can be seen by indicating how each added level of complexity allows the overall cell size to decrease compared to the minimum feature size of the transistors. This trend is shown in Fig. 23. The various generations are shown in this figure along with the added features that add to the density and performance for each generation of memory. The structures described so far will allow DRAMs to be made with up to 4 million bits (4 Mb). However, even the trench cell is projected to be inadequate for a 16 Mb DRAM and a different approach is proposed that requires two transistors per cell, called a "gain cell."[34] This structure allows a smaller capacitance per cell to be tolerated by employing a second transistor on each cell to effectively amplify the signal charge on the capacitor.

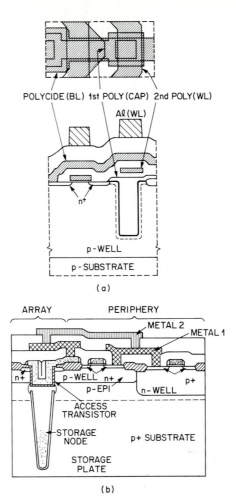

POLYCIDE(BL) 1st POLY(CAP) 2nd POLY(WL)

Aℓ(WL)

n⁺

p-WELL

p-SUBSTRATE

(a)

ARRAY   PERIPHERY

METAL 2

METAL 1

n⁺   p-WELL n⁺   p⁺

p-EPI   n-WELL

ACCESS
TRANSISTOR

STORAGE
NODE   p⁺ SUBSTRATE

STORAGE
PLATE

(b)

FIGURE 22

Trench capacitor structures for DRAM applications. (a) Top view and cross section of trench capacitor and access transistor. (b) Cross section of trench capacitor cell in which access transistor is formed along sidewall of trench. (*After Furuyama et al., Ref. 32, and Shah et al., Ref. 33.*)

11.5.2 Static Memory

The static random access memory (SRAM) has the advantage over the DRAM that there is no charge stored in a capacitor. The bit state information is stored in a pair of cross-coupled inverters. The inverter pair forms a "flip-flop," which forces one inverter to be at a high potential while the other is low and vice versa. The memory logic state is determined by which of the two inverters is high. The disadvantage of this cell arrangement is that it takes six transistors to make the memory cell, and therefore requires more area for each cell.

An example of an NMOS SRAM cell is shown in Fig. 24a. The cross-coupled inverter pair is represented by transistors T_1 to T_4.[27] Transistors T_5 and

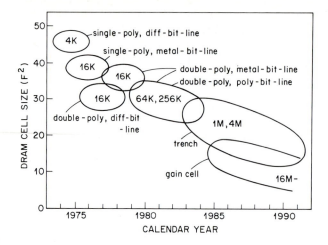

FIGURE 23
DRAM cell size normalized by minimum feature size squared, shown as a function of time of application. (*After Asai, Ref. 28.*)

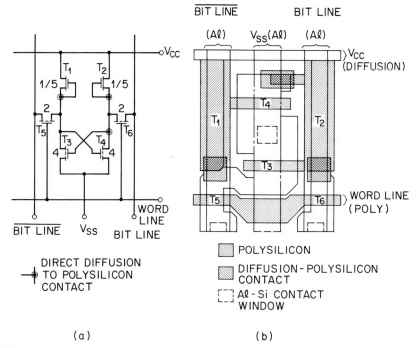

(a)   (b)

Fig. 24
Static RAM cell with transistor loads. (a) Circuit diagram of a six-transistor static RAM cell. V_{CC} is power supply and V_{SS} is ground potential. The numbers next to the transistors indicate the relative width-to-channel-length ratios. (b) Static RAM layout. (*After Hunt, Ref. 27.*)

T_6 are the access transistors that transmit the signal into and out of the cell when both the word lines and bit lines are simultaneously activated. The layout for this circuit is shown in Fig. 24b. The transistors labeled T_1 and T_2 are depletion transistors with a width-to-channel-length ratio of 1/5. This ratio is adjusted to provide enough current drive to meet the speed requirement of the cell without having excessive power dissipation. The depletion transistors simply act as loads to supply the current for the enhancement mode transistors T_3 and T_4. To minimize the cell area, buried contacts in which polysilicon is directly connected to the silicon diffusion region are required.

Since the depletion transistors are only acting as a load on this cell, they can be replaced by a high-value resistor. This can be a substantial savings in cell area if the resistor is made from polysilicon that has been ion implanted to provide the proper resistance.

Another approach to a SRAM cell is to use two coupled CMOS inverters. This is effectively the same as replacing the load transistors in the NMOS cell with p-channel transistors. The CMOS cell will require more area, but the power consumption when the circuit is quiescent will be virtually eliminated. This is a significant advantage when power is a major constraint, such as when battery backup systems are used.

One way to improve the packing density of CMOS static RAMs is to stack the transistors one on top of another. This arrangement allows the circuit design to extend into the third dimension. An example of a stacked CMOS inverter schematic is shown in Fig. 25a, with a cross section in Fig. 25b. [35] The second layer of transistors has to be fabricated on recrystallized silicon. The recrystallized silicon must be of sufficiently high quality to produce good transistor characteristics. This is achieved by depositing silicon on top of existing devices and crystalizing the silicon without significantly degrading the underlying structures. Since this must be done at temperatures that would cause excessive diffusion of the dopants in the underlying silicon (900–1100°C), laser annealing techniques are used to shorten the thermal cycle. These techniques are relatively complicated and difficult to control, and the area reduction is not as great as would be needed to justify the cost. However, as the techniques become more sophisticated and the need to form three-dimensional devices becomes greater, these devices will become more desirable.

11.6   BIPOLAR IC TECHNOLOGY

Bipolar IC technology has traditionally been the choice technology for high-speed applications. This is partly due to the fact that in this technology transit time was determined by the base width of the bipolar devices. The base width is not determined by lithography (as is the channel length of MOS devices) but by the difference between two impurity diffusion profiles. This distance can be precisely controlled and allows the base width to be made very thin, which allows high-speed operation of the bipolar transistors.

The processing of bipolar devices is more complicated than the MOS process

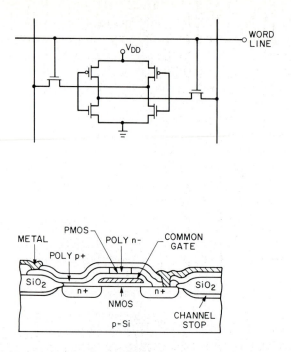

FIGURE 25
Stacked CMOS structure showing common gate of n- and p-channel transistors. (a) Schematic of SRAM cell for stacked structure. (b) Cross section of part of SRAM cell showing stacked structure. (*After Chen et al., Ref. 35.*)

sequences. The complication arises from the need to form a buried layer of dopant by growing an epitaxial layer of silicon on a patterned and doped silicon substrate.

In this section we will describe the fabrication process for bipolar ICs and the special process considerations that must be addressed.

11.6.1 Fabrication Process Sequence

Modern bipolar IC technology uses a thick silicon dioxide layer to electrically isolate the individual transistors. The resulting device cross section of an oxide-isolated bipolar transistor is shown in Fig. 26.[36] The process sequence used to make this device is illustrated in Fig. 27 at several key stages.

The starting material for these devices is a lightly doped p-type substrate with either a <111> or a <100> crystal orientation. A patterned buried layer is formed by implanting arsenic or antimony through windows etched in a thermal oxide, forming the heavily doped n^+ region of the collector. The implanted layer is driven into the substrate in an oxidizing ambient to form the "buried layer." Because of the differential rates of oxidation between the exposed buried layer and the surrounding area, a step is formed at the buried layer edge. All the oxide is then stripped and an n-epitaxial layer is grown on the substrate. The step in the buried layer propagates up through the epitaxial layer and serves as an alignment mark for the next lithographic level (Fig. 27a).

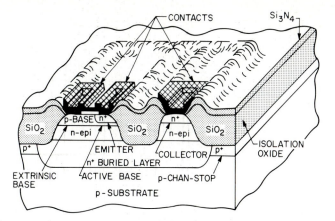

FIGURE 26
Three-dimensional view of oxide-isolated bipolar transistor. (*After Beadle, Tsai, and Plummer, Ref. 36.*)

The next step in the process is to form the oxide isolation between the devices. This is done in a similar manner to the field oxidation steps described in the MOS technologies (Fig. 27b). However, the oxide is thicker, and about half the epitaxial silicon layer is removed during the etching of the Si_3N_4 layer (Fig. 27c). A boron chan-stop layer is implanted into the silicon prior to oxidation. The oxidation is performed to convert the remaining epitaxial layer in the defined areas to SiO_2.

After the Si_3N_4 is removed, the silicon is oxidized and a base implant can be defined using resist (Fig. 27d). Contact holes to the intended base, emitter, and collector regions can then be made simultaneously with a single mask (Fig. 27e). A resist mask is then used to protect the base contact area while the emitter and collector contact regions are implanted using a high arsenic dose at low energy. An option here is to use a resist mask to cover the emitter and collector regions and implant a higher dose boron implant into the base contact region. This lowers the base contact resistance and decreases the resistance of the extrinsic-base region.

After the emitter and extrinsic-base regions are implanted, the emitter is driven into the desired depth in a nearly inert ambient. The emitter, base, and collector contacts can now be exposed by removing the thin oxide that covers them in an oxide etch. These regions can now be contacted without having to re-register a window inside the emitter area (Fig. 27f).

After the contact areas are exposed, a layer of Si_3N_4 can be deposited over the wafers to protect them from mobile ion contaminants such as sodium. Windows are then opened in the Si_3N_4, and the metallization layer is deposited and defined. The metallization layer can be a single layer of aluminum or composite structures such as PtSi for contact to the silicon, followed by successive layers of titanium, platinum, and gold.[37]

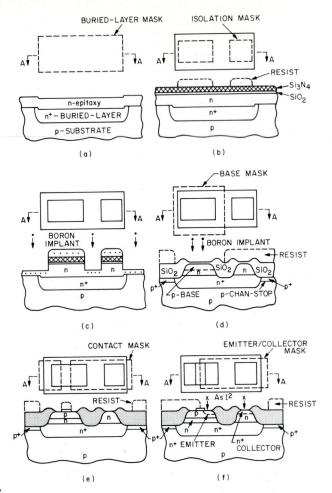

FIGURE 27

Top and cross-section view of bipolar transistor fabrication. (a) Buried-layer mask and cross section after n⁺-buried-layer and n-epitaxy growth. (b) Isolation mask and cross section with isolation resist mask on nitride/pad oxide (depression in epitaxy surface is omitted for clarity). (c) Cross section after nitride/pad oxide/silicon etch and chan-stop implant. (d) Base mask and cross section after isolation-oxide growth and boron base implant. (e) Contact mask and cross section after base, emitter, and collector contact opening. (f) Emitter/collector mask and cross section after arsenic emitter/collector ion implant (I^2). (*After Parrillo, Ref. 9.*)

11.6.2 Special Considerations for Bipolar ICs

The vertical structure of the bipolar transistor requires the doping and thickness of each layer to be carefully controlled. The effect of changing these parameters will be examined individually in this section.

Buried layer and epitaxial layer. The buried layer's sheet resistance should be as low as possible to reduce the collector resistance of the transistor. This requires the buried layer to be relatively heavily doped. However, doping the buried layer too heavily causes excessive outdiffusion into the more lightly doped n-epitaxial collector. For this reason, antimony or arsenic are commonly used as the buried-layer impurities rather than phosphorus, because of their smaller diffusion coefficients. The n-epitaxial layer should be as lightly doped as possible to minimize the base-collector capacitance of the transistor. The epitaxial layer should also be as thick as necessary to prevent the outdiffusion of the buried layer from reaching the base region and causing high collector-base capacitance. However, the epitaxial region cannot be too lightly doped or the resistivity can increase at high current, causing gain degradation.[3] Therefore, the typical buried layer sheet resistance values are about 5 to 15 $\Omega/\square$ and the n-epitaxial layer is typically doped with 10^{15} to 10^{16} atoms/cm^3, with a thickness of about 1–2 μm.

Base formation. The total integrated change in the active base (which is called the Gummel number) will determine the gain of the transistor for a given emitter structure. The lower the Gummel number, the higher the current gain.[3] However, if the base charge is too low it cannot support the reverse-bias voltage that is applied across the collector-base and/or emitter-base junctions, resulting in unwanted "punch-through" current. The narrower the base, the shorter the diffusion time of minority carriers across the base. This means that the base should be as narrow as possible for high performance, but the doping level must be increased as the width is decreased to ensure adequate "punch-through" protection. Typically the base Gummel number is from 10^{12} to 10^{13} atoms/cm^2 with the base width about 0.2 μm.

The extrinsic-base region (the p$^+$ region surrounding the emitter) must be doped high enough to provide a low-resistance path to the active-base region (the p region directly under the emitter). The active and extrinsic base regions can be formed with a single implant (as shown in Fig. 28a) or with a double implant (Fig. 28b). In the latter case, the higher-dose, low-energy implant is used to provide the extrinsic base, while the higher-energy, lower-dose implant establishes the active base properties. Figure 28c shows the actual doping profile of an npn transistor that uses a single base implant. Because of strong cooperative diffusion effects between the arsenic emitter (see Chapter 7) and the boron base, the base profile does not smoothly decrease into the silicon as shown schematically in Fig. 28a.

Emitter formation. The emitter must be as heavily doped as possible to maximize the current gain and minimize emitter resistance. The emitter profile should also be as shallow as possible to maintain sharp profiles and well-controlled junction depths. Abrupt and shallow arsenic emitter profiles can be obtained because of arsenic's concentration-dependent diffusion, making it an attractive choice for an emitter impurity. As emitters become shallower, the technique used to contact the emitter will have an increasing effect on the current gain of the device. This is illustrated in Fig. 29, which shows the common emitter current gain vs. collector current characteristics for three different shallow-emitter (0.2 μm) devices processed identically, except for the emitter contacts.[38] Contact to the emitter was made using n$^+$ polysilicon (poly), aluminum, or Pd$_2$Si on separate devices. The insert in Fig. 29 shows a schematic representation of the minority-carrier profile

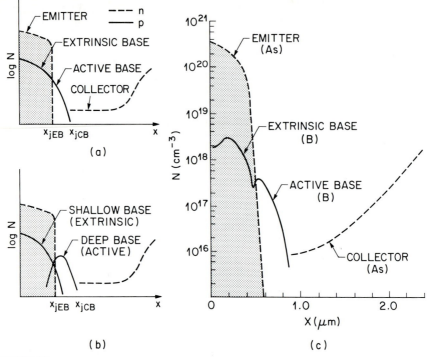

FIGURE 28
Doping profiles of npn transistor. (a) Single base implant (schematic). (b) Double base implant (schematic). (c) Single base implant (actual). (*After Parrillo, Ref. 9.*)

in the emitter for the three contact schemes at fixed base-emitter potential. The gain is determined by how steep the gradient of minority carriers is in the emitter.[4] The gradient of holes in the emitter establishes the base current. The hole gradient is made smaller (less base current) when 1000 Å of arsenic-doped polysilicon is placed between the emitter and aluminum contact. This is because the recombination velocity at the interface between the polysilicon and single crystal silicon is no longer infinite. The current gain can increase by three to seven times when polysilicon, rather than metal, is used to contact the emitter.

One severe yield-limiting problem that occurs in fabricating the emitter of bipolar devices is emitter-to-collector leakage or shorts. This effect occurs when the emitter dopant has locally enhanced diffusion through the base in the vicinity of material defects.[39] The problem becomes more severe as devices require narrower bases and shallower emitters to improve performance.

11.6.3 Self-aligned Bipolar Structures

The recent progress in bipolar devices has come from using self-alignment techniques. One example of how self-aligned structures can decrease the area

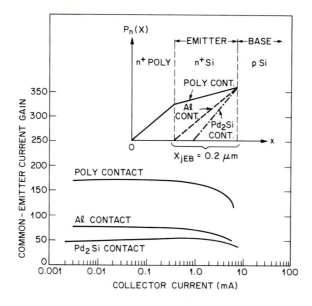

FIGURE 29

Common emitter current gain vs. collector current for shallow-emitter-junction (0.2 μm) npn transistors. (*After Ning, Tang, and Solomon., Ref. 38.*)

of the bipolar transistor is shown in Fig. 30. The base and emitter of this device are self-aligned to the base and emitter contacts.[40] This is achieved by using polysilicon to make contact to the heavily doped regions of the emitter and the extrinsic base, as well as to provide the doping source for these regions. A process sequence for this structure is shown in Fig. 30. The only lithographic step necessary to form the emitter is the patterning of the emitter stripe on the p^+ polysilicon shown in Fig. 30a. The poly is then oxidized. The p^+ polysilicon base contacts are formed by etching the Si_3N_4 and redepositing polysilicon (Fig. 30b). The emitter n^+ region and the base p^+ region are then formed by outdiffusing the dopant from the polysilicon (Fig. 30d). This results in the minimum possible spacing between the base and emitter contact, along with the advantage of having a polysilicon emitter.

11.6.4 Integrated Injection Logic

A particular circuit application of bipolar devices called integrated injection logic allows the bipolar transistors to be used in high-density applications with lower power consumption. Integrated injection logic (I^2L), also called merged transistor logic (MTL), allows the close packing of transistors, because the basic logic element consists of a pnp transistor connected to a series of npn transistors. The collector of the pnp transistor and the base of the npn transistors are electrically connected and share a common p-type diffusion region. The npn transistors also share a common emitter.[41] The circuit diagram is shown in Fig. 31b along with

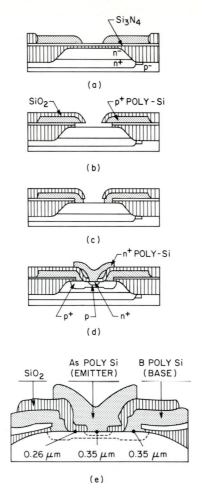

FIGURE 30
Process sequence for self-aligned transistor structure. (a) Patterning the emitter stripe, (b) etching the Si$_3$N$_4$, oxidizing the polysilicon, and etching the Si$_3$N$_4$, (c) depositing polysilicon and anisotropically etching to fill in under the oxide, (d) depositing and defining the polysilicon emitter, (e) the final structure showing the dimensions of the emitter and base contacts. (*After Konaka, Yamamoto, and Sakai, Ref. 40.*)

a cross section of the device in Fig. 31a. The npn transistors are operated in an inverse mode, with the buried layer acting as the emitter.

The I^{2}L structure shown in Fig. 31 also uses the technique of self-alignment described in the previous section to form the collector contacts for the npn transistors. The collector contacts, as well as the n$^+$ collector regions, are formed by using arsenic-doped polysilicon. The arsenic is outdiffused from the polysilicon to form the n$^+$ region. An oxide layer is formed and then etched away anisotropically to leave an oxide layer on the side of the polysilicon. This oxide layer allows the subsequent aluminum base contact to be isolated from the collectors.

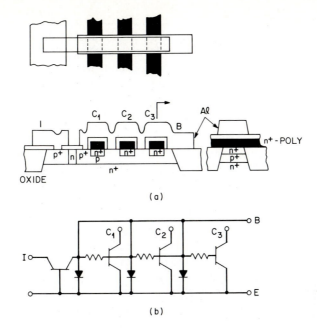

FIGURE 31
Self-aligned I^2L/MTL. (a) Structure. (b) Equivalent circuit. (*After Tang et al., Ref. 41.*)

11.7   IC FABRICATION

Throughout this chapter, attention was focused on various techniques to produce specific components of fine-line semiconductor structures. To successfully manufacture VLSI circuits, these process steps must be carried out in an environment that is meticulously controlled with respect to cleanliness, temperature, humidity, and orderliness. Process monitoring and control are other important areas of IC fabrication. These topics will be briefly considered in this section.

11.7.1   Fabrication Facilities

Pure water system. IC fabrication uses large quantities of water. Making ultrapure, electronic-grade water is a fundamental requirement in fabrication ICs. Particulates in the water, as well as contaminants such as sodium, copper, and iron, are deposited on the wafers and lead to device degradation. The water system must be engineered to deliver a continuous and large supply of ultraclean water with a low ionic content, as measured by monitoring the water's resistivity. Water with a specific resistivity of 18 $M\Omega$-cm is considered to have a low ionic content.

A pure water system consists of several sections.[42] In-flowing water is passed through charcoal filters and into electrodialysis units that filter and demineralize the water. It then passes through resin tanks, followed by mixed-bed ion-exchange resin units—to remove more minerals—and nuclear-grade resin cannisters to minimize sodium ion content. Undissolved solids, bacteria, and other organic matter are removed by a series of filters ranging in pore size from 10 μm down to 0.01 μm. Bacterial content is minimized by continuously circulating the water through an ultraviolet sterilization unit. The resistivity of the water is monitored continuously by electrical means, and particulates in the water are monitored continuously with an automatic light-blockage type of liquid-borne particle counter.

Clean room. In general, all process steps are implemented in a clean room. Here the air is maintained at a well-controlled temperature and humidity level and is continuously filtered and recirculated. Particulates must be avoided on wafers since they cause improperly defined features, undesirable surface topography, leakage through insulating layers, and other deleterious effects. Air in the clean room is monitored and classified with respect to particulates. A "class 100" environment has a maximum of 100 particles per cubic foot ($\sim$ 3500 particles per cubic meter) with particle size larger than 0.5 μm, and a maximum of 10 particles per cubic foot with particle size larger than 5.0 μm.[43] In the highly critical lithography area, particle densities are typically maintained below 10 particles per cubic foot. The filtered air flows from ceiling to floor at more than 85 linear feet per minute ($\sim$ 26 linear meters per minute), which is approximately the threshold for laminar flow (i.e., for uniform velocity of air following parallel flow lines without turbulence). Particulates can emanate from process equipment as well as from humans; personnel must wear proper clothing to protect the wafers. The size of the particulates that must be controlled will decrease along with the device dimensions. Figure 32 shows the correlation between the design rule of a device and the critical particle diameter that can affect the device yield.[44]

11.7.2 Process Monitoring

The electrical, mechanical, and visual tests used to evaluate and characterize the various process steps are an integral part of IC process development and manufacture. Nondevice wafers are used to monitor the calibration of equipment and to ensure that uniform results are obtained across wafers and lots of wafers. For example, sheet resistance measurements on monitor wafers are used to check ion-implantation machines, the thermal cycles of furnaces, and film deposition equipment. Oxide thicknesses are monitored using MOS and ellipsometric measurements. Capacitance-voltage measurements at elevated temperature are routinely used to monitor mobile-ion contamination in oxides, which may emanate from contaminated sources.

In addition to these in-process monitoring techniques, many measurements are performed on test chips that are located at several sites on completed IC wafers. Test chips are designed to measure the device characteristics of the particular structures of the technology being manufactured, and to monitor particular process variables such as line width uniformity, lithographic alignment quality, wafer distortion, and random fault densities. The device information is used to

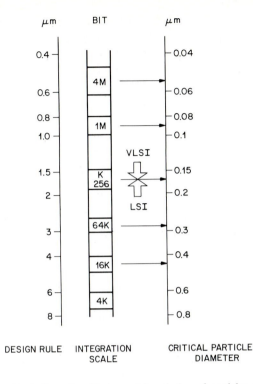

FIGURE 32

Correlation between critical dimension (design rule) and size of particles to be removed. (*After Tolliver, Ref. 44.*)

diagnose circuit performance. The process variable information is used to identify particular problems in the process.

An example of the kind of information that can be obtained by using the test structures is shown in Fig. 33. This figure is an automatically generated wafer map, showing the misalignment distances between levels in an NMOS process. The data is generated by measuring the resistance of electrical test structures with an automated computer-controlled data acquisition system. The resistance of these test structures is sensitive to the alignment between each level, and the measured values can be converted to misalignment distances. The wafer map then indicates the distance and direction of misalignment for each region of the wafer.[45] Figure 33a shows a wafer map of the misalignment of the polysilicon gate level with respect to the source/drain diffusion regions. The length of the arrows indicates the magnitude of the misalignment, and the direction of the arrows indicates the direction of the misalignment. The average vector magnitude and direction give the average translation and rotation of one level with respect to the other. Figures 33b and 33c show these same plots for the window level with respect to poly, and the metal level with respect to window.

Another important aspect of process monitoring requires careful monitoring of the electrical properties of the completed devices. Parameters such as sheet

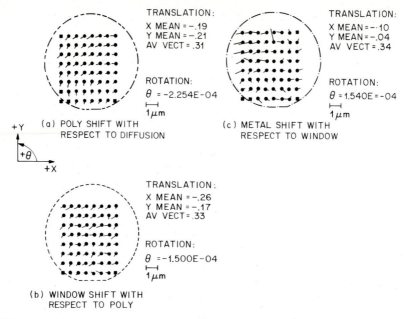

FIGURE 33
Electrical alignment maps for (a) polysilicide over diffusion, (b) window over polysilicide, (c) aluminum over window. (*After Orlowsky et al., Ref. 45.*)

resistances of diffused regions and threshold voltages of MOSFETs will give a valuable indication of the uniformity and reproducibility of the process steps. The measured parameters are sometimes displayed in distribution charts such as shown in Fig. 34. This chart shows the distribution of threshold voltages for MOSFETs on one wafer.[46]

11.8   SUMMARY AND FUTURE TRENDS

Throughout this chapter, technologies have been described that require hundreds of processing steps. The general trend for each individual technology has been towards smaller dimensions and more complicated processing. The shrinking of device dimension has allowed improvements of device performance as well as an increase in the number of devices. The basic trend for NMOS and CMOS development will continue to be driven by reducing the device dimensions. Guidelines have been established[20] for miniaturizing the dimensions of the MOS devices to maintain ideal transistor characteristics.[47] These guidelines were derived with the assumption that the internal electric fields would be maintained constant and the power supply voltages would be reduced. In practice, the supply voltages have not scaled with the device dimensions due to system constraints that require an IC to interface with many ICs that need to run at the same voltage. In addition, the simple scaling rules have not taken into account some of the parasitic effects that become dominant

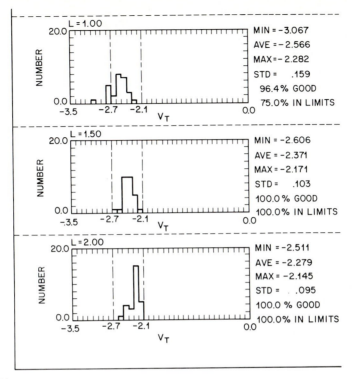

FIGURE 34
Distribution of threshold voltages for MOSFETs across a wafer. (*After Boulin and Bechtold, Ref. 46.*)

at the smaller dimensions, such as fringing capacitances and series resistances.[48] Therefore, the continuing trend in MOS devices will be to reduce the dimensions as much as possible and to continue to modify the structure to make devices that can sustain as high a supply voltage as possible. However, at some point, there will be a fundamental limit to how much device dimensions can be reduced. Some recent estimates have set this limit at 0.14 to 0.29 μm dimensions for the channel length of the transistors.[49] This may be the limit to simply scaling the devices in two-dimensional surface area, but the other trend to increase the density of the devices is to extend the structures into the third dimension. This means either changing the substrate configuration to improve isolation of the devices or making multiple layers of transistors so that the devices are stacked upon each other. In addition to multiple layers of devices, multiple layers of metallization will be needed to form high-density interconnects.

An example of the type of structure that can be obtained is shown in Fig. 35, in which three levels of active transistor are formed with successive silicon recrystallization.[49] The crystallization must be done at a sufficiently high temperature to form single-crystal silicon, but not high enough to adversely affect the devices already formed under the polysilicon film.

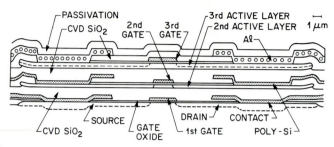

FIGURE 35
Three levels of active transistors fabricated on bulk silicon and two levels of recrystallized silicon. (*After Sugahara et al., Ref. 49.*)

The general trend for all the device technologies has been towards self-aligned structures and three-dimensional devices. The trench structures have enabled denser structures to be made for both CMOS and memory technologies. Recrystallized and epitaxial silicon has enabled both CMOS and bipolar devices to increase their density. As these trends continue, more layers of devices will be produced with a mixing of technologies applied to a single wafer. A mixture of CMOS, bipolar, NMOS, and memory applications will be used to optimize the performance of the IC.

PROBLEMS

1 Assuming $\Delta = \pm 1.0 \ \mu m$ and $\delta = 0.4 \ \mu m$ for all levels in the NMOS process, calculate the minimum worst-case separations for each level. Assume that all levels are aligned to the isolation mask, except the metal level (which is aligned to the window level) and the window level (which is aligned to the gate).

2 Consider the process steps necessary to make a simple diffusion resistor consisting of a long, rectangular p-type region in an n-type substrate with contacts formed on either end.

 (a) What are the minimum masking steps needed to form the resistor?

 (b) What would the dimensions (length L and width W) of a 500 Ω resistor have to be if the sheet resistance of the p region is 50 $\Omega/\square$ and the minimum feature size is 1 μm? (Ignore the resistance at the contacts.)

 (c) What would the resistor dimensions have to be for the case of (*b*) if the resistor must have $R = 500 \pm 50 \Omega$ and $\Delta W = 0.5 \ \mu m$?

3 A p-channel MOSFET is designed to have a threshold voltage of -0.8 V and a gate oxide thickness of 250 Å.

 (a) What channel doping is necessary to give the desired threshold voltage if p-type polysilicon is used as the gate material?

 (b) If the gate material is n-type polysilicon and the same channel doping is used as in part (*a*), what would the threshold voltage be?

 (c) If the structure of part (*b*) is threshold-adjusted to give a threshold voltage of -0.8 V, what is the necessary implant dose for the threshold adjustment? (Assume all the implanted charge is confined to the depletion region.)

4 A dynamic RAM must operate with a minimum refresh time of 4 ms. The storage capacitor in each cell is 10 μm square, has a capacitance of 50 fF (50×10^{-15} F), and is fully charged at 5 V.

(a) Calculate the number of electrons stored in each cell.

(b) Estimate the worst-case leakage current that the dynamic capacitor node can tolerate.

5 Calculate the area required to make a trenched capacitor cell with the same capacitance as the capacitor in Problem 4, where the trench depth is 1 or 5 μm and the trench width is 1 μm. Assume the capacitance per silicon surface area is the same as in Problem 4.

REFERENCES

1 J. S. Kilby, "The Invention of the Integrated Circuit," *IEEE Trans. Electron Devices*, **ED-23**, 648 (1976).

2 B. Hoefflinger and G. Zimmer, "New CMOS Technologies," in J. Carrol, ed., *Solid-State Devices 1980, Institute of Physics Conf. Ser. 57*, from 10th European Solid-state Device Research Conf., Sept. 1980.

3 S. M. Sze, *Semiconducting Devices: Physics and Technology*, Wiley, New York, 1985.

4 S. M. Sze, *Physics of Semiconductor Devices*, 2nd ed., Wiley, New York, 1981.

5 A. S. Grove, *Physics and Technology of Semiconductor Devices*, Wiley, New York, 1967.

6 R. S. Muller and T. I. Kamins, *Device Electronics for Integrated Circuits*, Wiley, New York, 1977.

7 R. J. Kopp and D. J. Stevens, "Overlay Considerations for the Selection of Integrated Circuit Pattern-level Sequence," *Solid State Technol.*, **22**, 79 (1980).

8 C. Mead and L. Conway, *Introduction to VLSI Systems*, Addison-Wesley, 1980.

9 L. C. Parrillo, "VLSI Process Integration," in S.M. Sze, ed. *VLSI Technology*, McGraw-Hill, New York, 1983.

10 K. Y. Chiu, J. L. Moll, K. M. Cham, J. Lin, C. Lage, S. Angelos, and R. L. Tillman, "The Sloped-Wall SWAMI-A Defect-Free Zero Bird's-Beak Local Oxidation Process for Scaled VLSI Technology," *IEEE Trans. Electron Devices*, **ED-30**, 1506 (1983).

11 A. N. Lin, R. W. Dutton, and D. A. Antoniadis, "The Lateral Effect of Oxidation on Boron Diffusion in <100> Silicon," *Appl. Phys. Lett.*, **35**, 799 (1979).

12 A. G. Sabnis and J. T. Clemens, "Characterization of the Electron Mobility in the Inverted <100> Si Surface," *IEDM Tech. Dig.*, 18 (1979).

13 H. Nihira, M. Konaka, H. Iwai, and Y. Nishi, "Anomalous Drain Current in n-MOS FETs and its Suppression by Deep Ion Implantation," *IEDM Tech. Dig.*, 487 (1978).

14 S. J. Hillenius and W. T. Lynch, "Gate Material Work Function Considerations for 0.5 Micron CMOS," *IEEE Proceedings of the Int. Conf. on Computer Design: VLSI in Computers, Port Chester, N.Y.*, 147 (1985).

15 H. J. Geipel, N. Hsieh, M. H. Ishaq, C. W. Koburger, and F. R. White, "Composite Silicide Gate Electrodes- Interconnections for VLSI Device Technologies," *IEEE Trans. Electron Devices*, **ED-27**, 1417 (1980).

16 C. Y. Ting, "Silicide for Contacts and Interconnects," *IEDM Tech. Dig.*, 110 (1984).

17 T. T. Sheng, private communication.

18 N. Kobayashi, N. Hashimoto, K. Ohyu, T. Kaga, and S. Iwata, "Comparison of $TiSi_2$ and WSi_2 Silicided Shallow Junctions for Sub-Micron CMOS," *IEEE 1986 Symposium on VLSI Technology*, San Diego, CA, 49 (1986).

19 S. Ogura, P. J. Tsang, W. W. Walker, P. L. Critchlow, and J. F. Shepard, "Design and Characteristics of the Lightly Doped Drain-Source (LDD) Insulated Gate Field-Effect Transistor," *IEEE Trans. Electron Devices*, **ED-27**, 1359 (1980).

20 R. H. Dennard, F. H. Gaensslen, H. Yu, V. L. Rideout, E. Bassous, and A. R. LeBlanc, "Design of Ion Implanted MOSFETs with Very Small Physical Dimensions," *IEEE J. Solid State Circuits*, **SC-9**, 256 (1974).

21 L. C. Parrillo, R. S. Payne, R. E. Davis, G. W. Reutlinger, and R. L. Field, "Twin-Tub CMOS–

A Technology for VLSI Circuits," *IEDM Tech. Dig.*, 752 (1980).

22 R. R. Troutman, *Latchup in CMOS Technology*, Kluwer, 1986.

23 R. D. Rung, "Trench Isolation Prospects for Application in CMOS VLSI," *IEDM Tech. Dig.*, 574 (1984).

24 T. Ohzone, H. Shimura, K. Tsuji, and T. Hirao, "Silicon-Gate n-Well CMOS Process by Full Ion-Implantation Technology," *IEEE Trans. Electron Devices*, **ED-27**, 1789 (1980).

25 L. C. Parrillo, S. J. Hillenius, R. L. Field, E. L. Hu, W. Fichtner, and M. L. Chen, "A Fine-line CMOS Technology that Uses P$^+$ Polysilicon/Silicide Gates for NMOS and PMOS Devices," *IEDM Tech. Dig.*, 418 (1984).

26 T. Noguichi, Y. Asahi, M. Nakahara, K. Maeguchi, and K. Kanski, "High Speed CMOS Structure with Optimized Gate Work Function," *IEEE 1986 Symposium on VLSI Technology*, San Diego, CA, 19 (1986).

27 R. Hunt, "Memory Design and Technology," in M. J. Howes and D. V. Morgan, eds., *Large Scale Integration*, Wiley, New York, 1981.

28 S. Asai, "Trends in Megabit DRAMs," *IEDM Tech. Dig.*, 6 (1984).

29 M. Taguchi, T. Ito, T. Fukano, T. Nakamura, and H. Ishikawa, "Thermal Nitride Capacitors for High Density RAMs," *IEDM Tech. Dig.*, 400 (1981).

30 K. Ohta, "VLSI Dynamic Memory Cell Using Stacked Ta_2O_5 Capacitor," in J. Nishizawa, ed., *Semiconductor Technology*, Ohmsha, 1982.

31 A. F. Tasch, Jr., P. K. Chattergee, H. S. Fu, and T. C. Holloway, "The Hi-C RAM Cell Concept," *IEEE Trans. Electron Devices*, **ED-25**, 33 (1978).

32 T. Furuyama, T. Ohsawa, Y. Watanabe, H. Ishiuchi, T. Watanabe, T. Tanaka, K. Natori, and O. Ozawa, "An Experimental 4-Mbit CMOS DRAM," *IEEE J. of Solid- State Circuits*, **SC-21**, 605 (1986).

33 A. H. Shah, C. P. Wang, R. H. Womack, J. D. Gallia, H. Shichijo, H. E. Davis, M. Elahy, S. K. Banerjee, G. P. Pollack, S. D. S. Malhi, C. J. Pilch, B. Tran, and P. K. Chattergee, "A 4-MBit DRAM with Trench-Transistor Cell," *IEEE J. of Solid-State Circuits*, **SC-21**, 618 (1986).

34 H. Schijo, "Tite RAM, A New SOI DRAM Gain Cell for MBit DRAMs," *Japan Society of Applied Materials 16th (1984) Conf. Solid-State Devices and Materials*, Kobe, 265–268 (Aug. 1984).

35 C. E. Chen, H. W. Lam, S. D. S. Malhi, and R. F. Pinizzotto, "Stacked CMOS SRAM Cell," *IEEE Electron Device Lett.*, **EDL-4**, 272 (1983).

36 W. E. Beadle, J. C. C. Tsai, and R. D. Plummer, *Quick Reference Manual for Silicon Integrated Circuit Technology*, Wiley, New York, 1985.

37 M. P. Lepselter, "Beam Lead Technology," *Bell System Technical Journal*, **45**, 233 (1966).

38 T. H. Ning, D. D. Tang, and P. M. Solomon, "Scaling Properties of Bipolar Devices," *IEDM Tech. Dig.*, 61 (1980).

39 F. Barson, "Emitter-Collector Shorts in Dipolar Devices," *IEEE J. Solid-State Circuits*, **SC-11**, 505 (1976).

40 S. Konaka, Y. Yamamoto, and T. Sakai, "30-ps Si Bipolar IC Using Super Self-Aligned Process Technology," *IEEE Trans. Electron Devices*, **ED-33**, 526 (1986).

41 D. D. Tang, T. H. Ning, R. D. Isaac, G. C. Feth, S. K. Wiedmann, and H. N. Yu, "Sub-nanosecond Self-Aligned I^{2}L/MTL Circuits," *IEEE Trans. Electron Devices*, **ED- 27**, 1379 (1980).

42 G. E. Helmke, "Anatomy of a Pure Water System," *Semiconductor Int.*, 119 (Aug. 1981).

43 E. C. Douglas, "Advanced Process Technology for VLSI Circuits," *Solid State Technology*, 65 (May 1981).

44 D. L. Tolliver, "Insights into Contamination Control for VLSI Processing" in W. M. Bullis and S. Broydo, eds., *VLSI Science and Technology/1985*, PV 85-6, 29, The Electrochemical Society, Pennington, NJ (1985).

45 K. J. Orlowsky, D. V. Speeny, E. L. Hu, J. V. Dalton, and A. K. Sinha, "Fabrication Demonstration of 1–1.5 μ NMOS Circuits Using Optical Tri-Level Processing Technology," *IEDM Tech. Dig.*, 538 (1983).

46 D. M. Boulin and P. F. Bechtold, private communication.

47 J. R. Pfiester, J. D. Shott, and J. D. Meindl, "Performance Limits of CMOS VLSI," *IEEE Trans. Electron Devices*, **ED-32**, 333 (1985).

48 K. K. Ng and W. T. Lynch, "Analysis of the Gate-Voltage-Dependent Series Resistance of MOSFETs," *IEEE Trans. Electron Devices*, **ED-33**, 965 (1986).

49 K. Sugahara, T. Nishimura, S. Kusunoki, Y. Akasaka, and H. Nakata "SOI/SOI/Bulk-Si Triple-Level Structure for Three- Dimensional Devices," *IEEE Electron Device Lett.*, **EDL-7**, 193 (1986).

CHAPTER
12

ANALYTICAL TECHNIQUES

J. B. BINDELL

12.1 INTRODUCTION

The analytical techniques used in the manufacture of semiconductor devices involve the use of many scientific specialties, primarily analytical chemistry, materials science, and physics. Most of the important physical methods allow ions, electrons, or electromagnetic radiation to interact with matter and then proceed to examine the secondary particles or radiations that are produced. The information obtained is then used to deduce the properties of the materials. For example, atoms exposed to energetic X-rays are elevated to excited states and decay via the emission of secondary X-rays, the energies of which are characteristic of the atomic identity. Measuring the intensity of these characteristic X-rays can permit the determination of the concentration of a specific element in the material being examined. When the volume of the material to be analyzed is large, one usually observes properties that relate to the bulk of the material; when the incident particles or radiation probe can be shaped into a very small (submicron) sized beam, microanalytical information can be obtained. These "analytical beam" methods have had enormous impact on VLSI technology.

The chemical methods used in VLSI technology include the traditional "wet chemical" approach as well as more modern instrumental methods such as mass spectroscopy and chromatography. These methods usually are applied to bulk materials but surface analyses are possible under special circumstances.

In a sense, VLSI and many of the modern analytical methods have "grown up" together. Semiconductor manufacturers have consistently exerted a strong pressure on the analytical instrument developers to keep pace with the demanding materials requirements of the industry. The development of smaller and smaller analytical-probe sizes can easily be shown to have been driven by the evolution

of smaller design rules during the past fifteen years. More recently, analytical instruments have become a part of the fabrication process itself. Scanning Electron Microscopes (SEMs) are now beginning to appear in cleanrooms as metrology tools and in testing areas as internal node voltage probes. Even Transmission Electron Microscopes, once found only in advanced materials (or biological) research laboratories are now being routinely used as process control tools in MOS memory chip manufacturing facilities.[1]

Analytical methods are usually applied to two somewhat overlapping areas: the evaluation (or characterization) of the materials to be used in VLSI circuits and the examination of the actual structures fabricated from these materials. The first of these uses the methods of traditional analytical chemistry while the second is more closely associated with the application of finely focused probes. Both of these areas are equally important because it is only through the use of well-characterized materials and processes that high-quality devices can be successfully fabricated.

There are five classes of materials or structures that we shall be concerned with in this chapter: bulk materials, external surfaces, thin films (or layered structures), interfaces, and actual devices. The characteristics associated with these classes are then further broken down into either distributed or localized properties. For example, a sample of bulk silicon may have a uniform level of iron contamination present at the parts per million (ppm) level, which may have no impact on device performance. The same number of iron atoms concentrated at defect sites might be catastrophic. It is always necessary to know how localized such properties are. Table 1 presents an overview of the properties of materials that we shall discuss. The table is not exhaustive.

The number of techniques used to examine all of these characteristics is very large, and to devote a paragraph or two to each method would be tedious and unproductive. In what follows we shall first examine the interactions that various particles and radiations have with matter in a number of energy regimes. We

TABLE 1
Characteristics of materials

Material class	Distributed property	Localized property
Bulk	composition, crystal structure, trace impurities	precipitates, inclusions, defects, grain size, crystal orientation
Surfaces	composition, contamination, structure	defects, cracks, morphology
Films (layers)	structure, composition, thickness, uniformity, orientation, adhesion, density	grain boundaries, composition versus depth, morphology, roughness, defects, continuity
Interfaces	contamination, chemical bonding, damage	contamination, localized defects
Devices	contamination	morphology, metrology, physical structure, localized stress, contamination

complement this discussion with an overview of the analytical instruments that measure these interactions, including brief examples of how these instruments provide analytical information. We will then discuss some of the more traditional (instrumental) analytical-chemistry techniques that have demonstrated their value in the VLSI arena.

12.2   ANALYTICAL BEAMS

Analytical beams, which are useful for VLSI, are formed with many types of charged (and sometimes neutral) particles. Electron beams used in such instruments as the Scanning Electron Microscope (SEM) are the most commonly encountered, but ion and neutron beams are also prevalent.[2] The energy of these beams ranges from a few eV to 2 MeV. Electromagnetic radiation ranging from (laser) light to UV, X-ray, and gamma rays also plays significant roles in the technology.

When an energetic particle (electron, ion, photon, neutron, etc.) strikes the surface of a material, several interactions take place simultaneously. The dominant process depends upon the energy and momentum of the particle. At very low energies, the particle may simply be reflected by the surface of the material but as the energy is increased, penetration also increases. Once it enters the material, the particle will begin to interact with the structure and lose energy. This energy is, of course, transferred to the material. After a number of atomic or nuclear collisions, the particle will either come to rest or be back-scattered and leave the material through the same surface that it initially encountered.

The atomic structure of the material will also have an effect on the interaction. If the structure is well ordered, cooperative effects, such as diffraction, will modify the result significantly. Finally, if the incident particle has sufficient momentum, it will change the structure of the material (radiation damage) and may, in fact, sputter some of it away.

The energy transferred to the target often leaves the target atoms in excited states and the decay of these atoms results in the release of secondary particles (electrons, X-rays). If the material was sputtered by the incident particle, the atoms or molecules leaving through the surface might either gain or lose electrons, thereby becoming secondary ions (+ or −) or they may leave as sputtered neutrals. In addition to all of this, electrons may be ejected from the surface by means of a number of mechanisms; the electron gas may be set into collective oscillation, or heat can be generated.

The beam-specimen interactions are, therefore, complex. In order to understand how most of these can lead to analytical information, we must examine each of these interactions. However, before doing so, it is convenient to introduce some of the generic features of the analytical instruments that are based upon these interactions.

12.2.1   Analytical Instruments

Most of the common analytical techniques depend on an analytical beam striking a material, after which one (or more) of the secondary particles (or radiations) that

result from the interaction can be monitored. For example, if a beam of electrons is used as a probe and the emitted X-rays are measured carefully, the resulting instrument is an electron microprobe (EMP).[3] If the energy distribution of the emitted electrons is determined, the technique of Auger Electron Spectroscopy (AES) is defined.[4] Table 2 presents a summary of the major incident–secondary combinations which have important VLSI applications. Since the primary energy of the probe is often important, that distinction is also included.

The instruments used for these techniques have a great deal in common and a generic instrument may conveniently be described (see Fig. 1). A high-voltage power supply provides the potential to extract and accelerate the primary particles from the source. For a primary electron beam, the source could be a tungsten filament, which is heated to the point of thermionic emission. The emitted electrons are then accelerated through an aperture, which forms a part of a lens system that focuses the beam to a cross-over point, forming an object. The successive (magnetic) lenses further focus the object into a small spot on the sample. Brighter sources can be focused into smaller spots, and the lower work function of LaB_6 makes it an excellent electron source for electron microscopes; sharply pointed sources which produce electrons through field emission do even better. Similarly, an ion beam can be produced by creating a confined plasma and then extracting the ions with a strong electrostatic field. In the case of an ion beam, a mass and/or energy spectrometer or filter is often added to the primary optics to produce a monoenergetic probe with a sharply defined mass range.

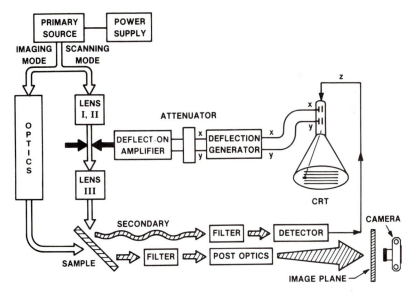

FIGURE 1

A schematic diagram of a generic analytical beam instrument. The instrument can be either a raster or a direct imaging construction. In either case, it is possible to filter the results in such a way that only the signals from one of the many components of the sample is actually imaged. This makes possible an elemental map. Direct measurement of the strength of the signal generated in either mode produces the information necessary for quantitative analysis.

TABLE 2
Analytical techniques used in VLSI technology

Primary beam	Energy range	Secondary signal	Acronym	Technique	Application
Electron	20–200 eV	Electron	LEED	Low energy diffraction	Surface structure
	300–30,000 eV	Electron	SEM	Scanning electron microscope	Surface morphology
	1 keV–30 keV	X-ray	EMP	Electron microprobe	Surface region composition
	500 eV–10 keV	Electron	AES	Auger spectroscopy	Surface layer composition
	100–400 keV	Electron	TEM	Transmission electron microscopy	High resolution structure
	100–400 keV	Electron, X-ray	STEM	Scanning TEM	Imaging, X-ray analysis
	100–400 keV	Electron	EELS	Electron energy loss spectroscopy	Local small area composition
Ion	0.5–2.0 keV	Ion	ISS	Ion scattering spectrometry	Surface composition
	1–15 keV	Ion	SIMS	Secondary ion mass spectrometry	Trace composition versus depth
	1–15 keV	Atoms	SNMS	Secondary neutral mass spectrometry	Trace composition versus depth
	1 keV and up	X-ray	PIXE	Particle induced X-ray emission	Trace composition
	5–20 keV	Electron	SIM	Scarning ion microscope	Surface characterization
	>1 MeV	Ion	RBS	Rutherford back-scattering	Composition versus depth
Photon	>1 keV	X-ray	XRF	X-ray fluorescence	Composition (μm depth)
	>1 keV	X-ray	XRD	X-ray diffraction	Crystal structure
	>1 keV	Electron	ESCA.XPS	X-ray photoelectron spectroscopy	Surface composition
	Laser	Ions	–	Laser microprobe	Composition of irradiated area
	Laser	Light	LEM	Laser emission microprobe	Trace elements (semiquantative)
Neutron	Reactor	Gamma	NAA	Neutron activation analysis	Bulk (trace) composition

When the probe strikes the surface, it may be moved from point to point by using a mechanical stage for coarse motion. Submicron movements can be obtained with electromagnetic (or electrostatic) deflection of the beam. Repeated raster deflections are accomplished by separate x and y deflection optics. By using the same x and y deflection signals to control the position of the beam on a cathode ray tube (CRT) while modulating the brightness (z) with the emitted analytical signal, a scanned image can be formed on the screen.

An example of such a scanned image is shown in Fig. 2. It was obtained with a primary electron beam being used as the probe and with low-energy secondary electrons producing the analytical signal. This example corresponds to the SEM. Figure 2a shows an exaggerated situation with only a few (SEM) raster lines being used; Fig. 2b shows the use of many more lines. By reducing the amplitude of the signal applied to the deflection coils by a factor of two (via a voltage attenuator) while keeping the CRT deflection signals the same, the magnification of the image would be doubled.

A second type of analytical instrument operates in an imaging mode. The same source (see Fig. 1) is focused by a primary series of lenses onto an extended region of the sample, and the secondary signals thus produced are imaged onto a fluorescent screen from which the image may be photographed. In some cases,

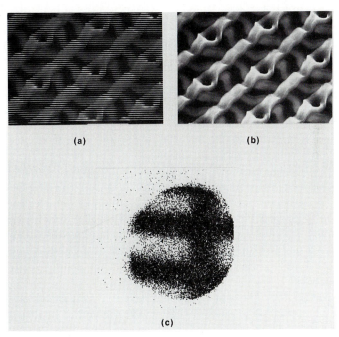

FIGURE 2
Examples of types of images. (a) shows a raster image with only a few lines to demonstrate the method of image formation. (b) The same image as (a) but with a sufficiently high density of lines that the individual scans are no longer visible. (c) A direct image of a boron distribution collected from a small area of an integrated circuit by the SIMS technique (discussed in Section 12.3.5). The diameter of the image field is 25 μm.

direct film or plate exposure is used. Such instruments often have provisions in the optical chain to filter the signal so that the image contains only restricted information. In the ion microprobe (Secondary-Ion Mass Spectroscopy–SIMS), the filter is a mass spectrometer and the image can display the response from a selected mass (elemental or molecular ion). In the Transmission Electron Microscope, a particular diffraction spot (see Section 12.3.3) can be isolated, thereby producing an image only of areas of the sample that have similar structure and crystallographic orientation. Figure 2c is an example of an ion image produced by an imaging SIMS instrument.

The switch from one type of instrument to the other often involves minor changes of the instrumental parameters; that is, the same instrument can be used in both modes. The Transmission Electron Microscope (TEM) is an example in which the direct TEM image mode can be switched to the scanning transmission mode (STEM) by replacing the screen with an electron detector and by scanning the primary beam as in an SEM.[5] An SEM mode is also possible in the TEM with a secondary-electron detector placed above the lens.

12.3 BEAM–SPECIMEN INTERACTIONS AND ASSOCIATED ANALYTICAL TECHNIQUES

We now turn to a detailed discussion of the interactions that the various analytical beams have with materials. We will review both electron–solid interactions and ion–solid interactions, but it should be recognized that there are many processes common to both. The case of radiation–solid interactions will also be discussed briefly.

12.3.1 Electrons

The inner (electron) potential of a material is very low, so that an incident electron striking a target with an energy in this range will not have sufficient energy to overcome this barrier and might simply be reflected by the surface as shown in Fig. 3a. As the energy is increased, the electron is able to penetrate the surface and will immediately begin to lose energy by interacting with the electrons contained in the solid. Three possible trajectories are shown in Fig. 3b. In one of them, a few large-angle elastic collisions with the nucleus have reversed the direction of the electron so that it actually leaves the material as a *back-scattered* electron. Other electrons are shown penetrating further into the solid. (We will return to this figure later.)

The energy loss per unit of path traveled by a particle is found to be a function of energy. The slowing down of an electron in the energy range commonly used in analytical instrumentation was first described by Bethe in a continuous energy loss model that expresses the energy loss per unit of distance traveled as

$$\frac{dE}{dX} = -2\pi e^4 N_{AVO} \frac{Z\rho}{AE_m} \ln(1.166\frac{E_m}{J}) \tag{1}$$

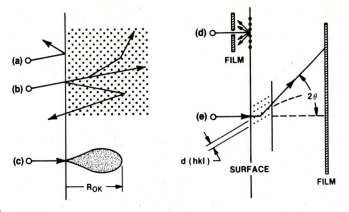

FIGURE 3

Electron–specimen interactions include (a) surface reflection, (b) penetration and scattering, (c) internal scattering interactions defining an interaction volume from which a particular analytical signal will be emitted, (d) surface diffraction, and (e) transmission diffraction. The last two cases are discussed in a later section.

where J, an average ionization potential, fits the empirical relationship

$$J = (9.76Z + 58.5Z^{-0.19}) \times 10^{-3} \text{ keV} \tag{2}$$

Here e is the electronic charge, N_{AVO} Avagodro's number, Z the atomic number of the target nucleus, A the atomic weight, ρ the density of the material, and E_m a mean electron energy along the path.[6,7,8] According to this equation, the amount of energy lost per unit of path traveled decreases with increasing electron energy and increases with the atomic number of the target. Thus, we would expect the ultimate distance that the electron travels (actually along the path) to increase with electron energy and to decrease with increasing Z.

As an electron travels through the solid, it will proceed approximately along straight lines, losing energy as it goes, until it is deflected by a nucleus. This deflection can be described in terms of the classical Rutherford scattering process. The same result is obtained from quantum mechanics. The result of such an interaction (for a mass M_1 of charge eZ_1 striking a second particle of mass M_2 and charge eZ_2 at rest with an energy E and a scattering geometry as sketched in Fig. 4a) can be described as follows with a formalism that applies to both positive and negative incident particles (electrons or ions), as suggested by Fig. 4b. The energy transferred to the target atom is[9]

$$T = \frac{4EM_1M_2}{(M_1 + M_2)^2} \sin^2 \frac{\theta}{2} \tag{3}$$

where θ is the scattering angle. For an electron, $M_1 \ll M_2$ and T is approximately $4E(M_1/M_2) \sin^2(\frac{\theta}{2})$, which is much less than E. Consequently, very little of an electron's energy is transferred to the nucleus. The probability that a scattering event produces a deflection greater than θ is expressed by the scattering cross section

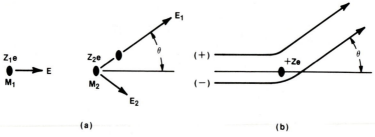

FIGURE 4
The scattering geometry for Rutherford scattering. In (a) the various masses and energies are defined. In (b) the geometry for both positive and negative incident particles is shown to produce similar trajectories under a central force so that the formalism applies directly to both cases.

$$Q(> \theta) = C\frac{(Z_1 Z_2)^2}{E^2}\mathrm{ctn}^2\left(\frac{\theta}{2}\right) \tag{4}$$

where C is a constant. The total path traveled by the electron (albeit in a zig-zag fashion) can be expressed as

$$R = \int_E^0 \frac{1}{(dE/dX)}dE \tag{5}$$

where R is the electron path length and the integration is along the actual path of the electron. From the expression for the cross section, Eq. 4, we see that as the energy is increased, the average scattering angle at each nuclear encounter decreases and the trajectory of the electron will tend to be stretched further into the solid. For the same target, increased electron energy will also increase the electron range because of the reduced scattering angle as well as the decrease in the energy loss per unit path.

A graphical example of this, taken from Monte Carlo calculations[7] is shown in Fig. 5. A particular electron's trajectory ends either when its energy is computed to be zero or when the path exits the surface (back-scatter). If the calculation was intended to provide information as to where some particular process (such as X-ray emission) was originating, the trajectories would be ended when the energy of the electron becomes less than the energy needed to excite the process, E_c. The right-hand portion of the figure thus gives a visual picture of the shape of the X-ray interaction volume of an electron in a solid.

Referring again to Fig. 3c, the distance from the surface to the "bottom" of the interaction volume is referred to as the maximum range, for which a useful expression (for normal incidence) is

$$R_{OK} = 0.0276\frac{AE^{1.67}}{Z^{0.889}\rho} \tag{6}$$

where if E is expressed in keV and ρ the density in g/cm^3, the result is in μm. A is the atomic mass in amu.[10] At the penetration range, the energy of the electron is assumed to have dropped to zero. Although the relationship is not really valid

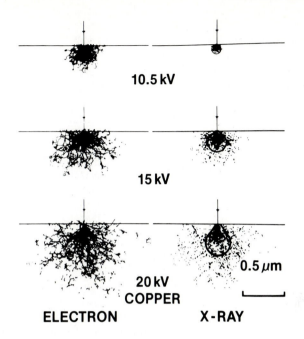

10.5 kV

15 kV

0.5 μm

20 kV
COPPER

ELECTRON   X-RAY

FIGURE 5
Electron and X-ray interaction volumes for 10.5, 15, and 20 keV electrons striking a copper sample. As the energy increases, the volume of the sample exposed to electrons also increases. The X-ray volume (shown in the right-hand side diagrams) corresponds to the volume in which the electrons have sufficient energy to produce the X-ray emission. These volumes are somewhat smaller. (*After Goldstein, et al., Ref. 7.*)

over an extended energy range, it can still provide a useful feel for the depth of electron–solid interactions. Table 3 shows Eq. 6 evaluated for three common materials.

The energies used in creating the table are (with the exception of the last row) comparable to the electron energies used in the SEM. Thus, in this particular instrument, the volume of interaction is usually comparable to typical VLSI dimensions. The interaction volume, in this case, is larger than the incident probe size. This suggests that the smallest volume that a technique can analyze is often determined by the underlying physics of the interaction and not by the size of the incident beam. In the case of the SEM, the dominant emitted secondary electrons (defined below) are produced in a region very close to the size of the primary beam. At the same time, some of the signal can arise from back-scattered electrons or X-rays exiting the surface a number of μm away from this site. Thus, the spatial resolution for analysis and for imaging is often not the same.

The electrons that are back-scattered through the surface also provide useful information. Many of these electrons have not traveled very far, and consequently they emerge with energies close to the primary beam energy. For any particular energy of the primary beam, the higher atomic numbered (heavier) material will produce more back-scattering than a lighter material. Back-scattering is

TABLE 3
Computed electron penetration range values in μm over an extended energy range[10]

Energy (keV)	Electron penetration range (μm)		
	Silicon	Aluminum	Gold
0.5	0.01	0.01	—
1.0	0.03	0.03	0.01
2.0	0.09	0.10	0.02
5.0	0.41	0.47	0.09
10.0	1.32	1.48	0.27
20.0	4.20	4.72	0.86
30.0	8.26	9.29	1.70
100.0	61.70	69.40	12.67

Note: These numbers are only intended to give the reader a feeling for the data; the formalism does not apply over the entire range of energies.

characterized by a back-scatter coefficient η which is defined as the ratio of the back-scattered current to the primary-beam current. The electron back-scattering coefficient has been found to be relatively insensitive to the primary beam energy over the range of 10–100 keV. In this interval, the back-scattering coefficient varies from 0.1 to 0.5, almost linearly with atomic number.

The back-scattering coefficient for a compound is proportional to the average density of the material. For a binary alloy, the back-scattered signal would thus vary linearly with the concentration of the heavier component. If the alloy is a mixture of two elements A and B, then the total back-scatter η_T produced by such a sample would be

$$\eta_T = W_A \eta_A + (1 - W_A)\eta_B \qquad (7)$$

where W_A is the weight fraction of A in the sample. Since the only variable here is W_A, a measurement of the back-scattered electron current can be used to obtain the sample composition directly.

While the electrons that are back-scattered have energies comparable to the energy of the primary electron beam, a large number of electrons is emitted with lower energies. Those electrons which leave the solid with energies less than about 50 eV are referred to as "secondary" electrons. Secondary electrons (SE) originate within 10–50 Å of the surface and constitute the signal used most often in the SEM.

Secondary-electron emission is more difficult to characterize than is back-scattering. The dependence of secondary-electron yields on Z cannot be generalized easily, and the chemical bonding of the surface component is often a controlling factor. Insulators can have higher SE yields than metals. Many VLSI samples tend to be insulators. The SE yield of insulators is also dependent upon the energy of the primary beam as shown in Fig. 6, where the total electron yield is

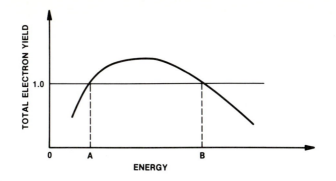

FIGURE 6
The total secondary and back-scattered electron yield as a function of the energy of the incident electron. When the yield is greater than 1.0, more electrons are produced than are incident. If the sample is an insulator, the surface will be depleted of electrons and will become positively charged. Point B is a stable operating point for insulators. Point A is not stable.

shown (SE + back-scattered) as a function of energy. At high energies ($E > B$), the total yield is less than unity and the sample will gradually charge (negatively) to the beam voltage unless some conduction mechanism interferes with this trend (usually the case). As this happens, the buildup of electrons repels the primary beam, reducing the effective primary-beam energy. At $E = B$, a stable situation will be reached. If the energy is slightly greater than B, the sample will charge more negatively, thereby reducing the effective beam energy until it reaches B, after which no charging can occur since the total yield is unity at that point. This suggests that lowering the beam energy to B will permit uncoated VLSI samples to be observed in an SEM. Although B varies with material, making it impossible for all of the materials in the sample to charge-balance at the same time, low-voltage imaging of semiconductor devices is nevertheless an important and usable technique. Low-voltage observation also reduces the chance that serious beam-induced damage will occur. B tends to be in the 1000–2000 eV range. Point A is not a stable operating point.

12.3.2   Excitation and Energy Loss

As the incident electron probe loses energy, that energy is transferred to the sample. Although the formalism suggests that the particles slow down in a continuous manner, the interaction is really a sequence of individual collisions that often leave a target atom in an excited state. The amount of energy transferred may be sufficient to remove an outer electron or a core electron from the atom, leaving a vacancy or hole in the electronic structure. It is also possible for the atomic electron simply to be excited into a higher energy state. In the latter case, the primary particle will have lost a discrete amount of energy that is characteristic of the atomic structure of the target atom. Since the atomic energy levels are well known, an examination of the energy spectrum of the primaries transmitted through the sample, revealing the losses, will identify the material through which

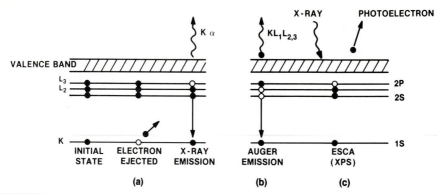

FIGURE 7
Electronic transitions involved in analytical signal production. In (a), initially full energy levels have
a vacancy created in the K shell. An electron dropping from a higher level into the K shell produces
an X-ray photon. In (b) the energy of the transition is carried off by a second electron in the Auger
effect. (c) represents the XPS case to be treated in a later section. The energy levels are the same
as those of (a) and (b).

the primaries traveled. This is usually done with very thin samples ($\sim$500 Å) in
the Transmission Electron Microscope and the technique is referred to as Electron
Energy Loss Spectroscopy (EELS).[11] The technique offers sensitivity to light
elements, making it a valuable auxiliary in the TEM.

The excited atoms eventually decay to the ground state with the emission of
either a photon or a so-called Auger electron.[4] This process is shown schematically in Fig. 7. The figure displays the electronic energy levels for an atom.
Specifically shown are the K and L shells, for which the principal quantum numbers are 1 and 2 respectively. The L shell actually consists of three closely spaced
energies corresponding to different values of the orbital angular momentum and
spin of the electron that normally occupies the level. The valence band is also
shown. An alternate notation often used for these same energy levels is also
indicated in the diagram.

Under normal circumstances, as depicted in Fig. 7a, all levels are occupied
as suggested by the solid circles. If an incident electron knocks out a K electron, as
suggested by the open circle, the atom is then excited and must decay. A transition
of an electron from the L to the K level will be associated with the emission of a
so-called K-alpha X-ray photon, whose energy will equal the difference between
the two energy levels involved. If, on the other hand, a second electron is emitted
to carry off this energy (Fig. 7b), no X-ray emission can occur and a radiationless,
or Auger, transition is said to have taken place. As suggested by Fig. 7b, this
transition is denoted by the energy levels involved. X-ray and Auger emission
are competing decay processes. The Auger yields are highest for light elements
while X-ray yields increase with increasing atomic number. Thus, X-ray methods
are not so sensitive to the lighter elements such as oxygen, nitrogen, or carbon
as is the Auger technique.

A third process (Fig. 7c), to be discussed later, occurs when an incident energetic X-ray simply knocks out an electron from a particular energy level, one of the two 2P levels in this case. The energy of this ejected photoelectron is the difference between the incident photon energy and the binding energy of the electron, corrected for the work function of the material and the difference between the Fermi levels of the sample and the electron energy detector. In the case of any of these processes, the energy of the emitted particle (electron or photon) remains characteristic of the atom from which it emerged and can be used for elemental identification as well as for compositional quantification.

Figure 8 shows the variation of the principal Auger (Fig. 8a) and characteristic X-ray (Fig. 8b) energies with atomic number. In general, overlaps are few in either technique, and most can be resolved because of other transitions that are excited at the same time. In terms of sensitivity (or detection limits), both X-ray emission and Auger spectroscopy are sensitive to the presence of about 0.1 weight percent; we shall see later, however, that Auger spectroscopy detects that concentration in the outer surface region only, and hence derives its signal from far fewer target atoms. Auger spectroscopy is also more sensitive to the lighter elements, permitting detection beginning with Li ($Z = 3$) and extending through the rest of the periodic table.

12.3.3 Electron-Beam Instruments

Three electron-beam instruments used extensively for VLSI are the Scanning Electron Microscope, the Transmission Electron Microscope, and the Auger Electron Spectrometer. Each has had a major impact on the technology, and it should be noted, as each instrument is described, that all have many features in

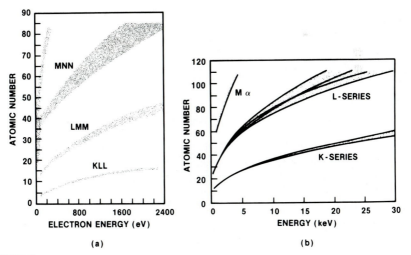

FIGURE 8

Transition energies for (a) Auger electron (*courtesy of the Perkin-Elmer Corporation*) and (b) X-ray emission (*courtesy of Princeton Gamma Tech., Inc.*). In each case the atomic number of the emitting species is plotted against the energy of the transition. By determining the energies, the elements can be identified. Note the energy scale differences between (a) and (b).

common. All three instruments, in their modern configurations, can be used to produce X-ray maps and secondary electron SEM images, but each is, of course, optimized for its own particular function.

The Scanning Electron Microscope. We have already encountered an SEM[12] image in Fig. 2b. That image was generated in the secondary-electron mode and is striking in both its three-dimensional presentation and in its "optical" quality. The finely focused beam has a relatively low angular divergence and consequently remains in focus over a relatively rough (on the micron scale) terrain. In addition to secondary-electron imaging, useful modes of the SEM also include back-scattered-electron imaging and X-ray mapping. The back-scattered image adds compositional information to the micrograph. This is because higher-density areas of the sample produce increased back-scattering signals and hence appear brighter in a micrograph. (Even a secondary image, in most instances, contains a significant contribution from back-scattered electrons.)

Figure 9 presents an example which illustrates these modes. As shown in Fig. 9a, an aluminum conductor is separated from an underlying tantalum silicide conductor (running at right angles to it) by a deposited oxide layer composed of borophosphosilicate glass (BPSG). The silicide is on a polysilicon layer. A secondary micrograph taken at 30 keV (Fig. 9b, top) shows the aluminum runners, as one would expect, but the silicide runners also appear bright, because of the

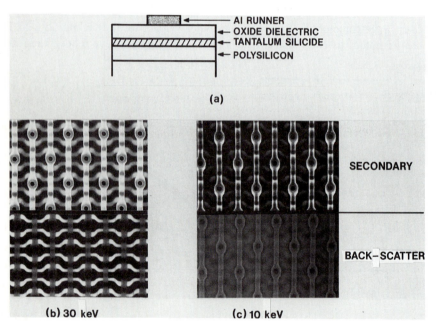

FIGURE 9

Scanning Electron Microscopy example. (a) shows the physical layer structure being examined. (b) represents 30 keV incident electrons while (c) represents 10 keV. The upper portions of each of these two photomicrographs correspond to secondary-electron images, and the lower portion of each is produced with back-scattered electrons. (*After B. J. Snyder, AT&T Bell Laboratories, unpublished.*)

high back-scattering of the electrons which penetrate the oxide (as well as the aluminum) and strike the (higher Z) silicide. Modulations of brightness in the aluminum result both from the back-scattering effect and from the way that the topography undulates, thereby changing the incident angle (tilt).

Secondary emission takes place only from an *escape depth* into the sample. As shown in Fig. 10, sloping surfaces will tend to produce increased emission, somewhat explaining the sensitivity of the SEM to surface topography. This also contributes to the "optical" appearance of an SEM micrograph. Edges and steps also tend to produce increased emission, due to bleeding of secondaries from neighboring surfaces. Back-scattered electrons produce additional secondary electrons as they exit the sample; this can be as far as 1 or 2 μm away from the entrance point.

The back-scatter images in Fig. 9 show this effect as well, but the aluminum (low Z) is less bright, in spite of the attenuated signal from the heavier (buried) silicide. As the energy of the primary electron beam is reduced, the silicide signal is reduced, as seen in the 10 keV micrograph (Fig. 9c). Here the lower-level conductor is barely visible while the Al and BPSG, similar in density, now show significantly reduced contrast.

An X-ray energy spectrum produced in the SEM can also reveal the effects of penetration of the primary beam. Figure 11a illustrates the X-ray spectrum generated when a nonrastered 10 keV electron beam is placed directly on the aluminum metal line of Fig. 9. Only the Al Kα X-ray line is observed on top of a slowly varying continuous X-ray spectrum generated by the deceleration of the electron when it strikes matter. Without moving the beam, increasing the energy to 20 keV produces a strong signal from the Si (oxide) underneath the Al. The reason that the Si peak is not observed in the first case is that by the time the electron has penetrated the aluminum and has reached the Si surface, it has lost energy. If its energy has dropped below the energy required to produce the silicon K-shell vacancy, Si Kα X-rays cannot be generated. Increasing the energy of the beam not only increases the penetration, but also increases the energy of

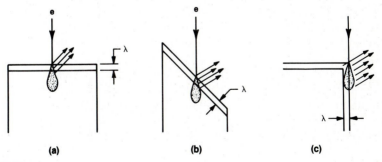

FIGURE 10
The effect of specimen tilt on secondary-electron emission. For the flat surface (a), only those electrons generated in the surface skin (escape depth) are emitted. For a tilted specimen (b), as well as an edge (c), there are more electrons produced in this region, so the signal is enhanced. λ is the escape depth.

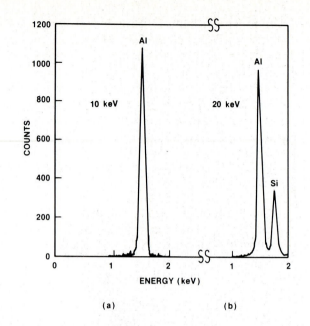

FIGURE 11
Energy-dispersive X-ray spectra created when an electron beam strikes the Al runner of Fig. 9. (a) 10 keV. (b) 20 keV, showing penetration to the underlying Si.

the electrons reaching the silicon, which allows Si X-ray production to occur.

The depth beyond which X-ray production is negligible can be estimated from a simple formula of the form[4]

$$R = 0.033(E^{1.7} - E_c^{1.7})\frac{A}{\rho Z} \tag{8}$$

where the result is expressed in μm. In this equation E is the primary beam energy and E_c is the critical energy required to excite the particular X-ray emission line of interest. A and Z have been previously defined. For most practical cases, this penetration is between 0.2 μm and a few μm. Thus X-ray spectra reveal very little about the outer surface layers of a sample. This is not the case for Auger spectra.

The control of VLSI device parameters requires a detailed knowledge of the cross-sectional geometry of individual devices. Figure 12a shows a cross section (taken in the SEM) of a memory device produced by directly cleaving a device wafer through the memory array. Since the circuit is highly repetitive, the chances of striking a particularly desired type of feature are good. A more difficult problem is to section a particular device in a circuit, as is often required for failure analysis. For this case, precision polishing of the cross section is required.[13] A chip is cut near the particular device of interest and then mounted in epoxy and placed in a special polishing fixture. Careful polishing (and junction staining when necessary) can produce micrographs similar to Fig. 12b.

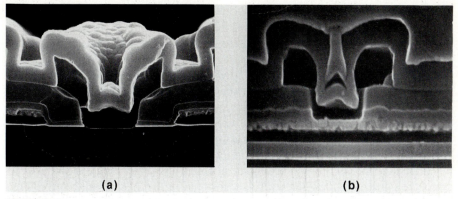

(a)   **(b)**

FIGURE 12
Comparison of (a) a cleaved and (b) a mechanically polished section. Both images were produced on the SEM. (*After B. J. Snyder and J. M. Szabo, AT&T Bell Laboratories, unpublished.*)

When the X-ray spectrum from a material is carefully determined, one can qualitatively identify most of the elements present. If a pure (or compound) elemental standard for every element in the material is available, one can compare the X-ray intensities generated by both the sample and the standard under the same experimental conditions. The ratio of the intensities of these two measurements, referred to as the "K-ratio," is a first approximation to the weight fraction of the element in the material.

The detailed physics of this process is now well understood and it is possible to correct these K-ratios for such factors as back-scattering differences between the sample and the standard, the actual standard's composition, X-ray absorption, and X-ray fluorescence effects. This is done in an iterative manner, and the quantitative results compare very favorably with wet chemical analysis. The Electron Microprobe (EMP) is an instrument capable of providing this quantitative analysis. It can be used to analyze small (10 μm diameter and smaller) areas and has been vital to the development of high-quality process control for VLSI products.

As mentioned in the Introduction, SEMs are now used as inspection and metrology tools in cleanrooms. For this VLSI application, instruments are required that have large (150–200 mm) wafer capacity and can observe the entire wafer and produce high-resolution images at relatively low (1 keV) beam energies. These new instruments are utilizing new lens geometries and are capable of making the submicron metrology determinations necessary for the next generation of device.

Auger Spectroscopy. The thin-film structures used in VLSI must often be examined for composition as a function of depth. Multilayer thin films often interact via diffusion or compound formation, and a detailed knowledge of these compositional variations is essential for the development of new structures and for the diagnosis of faulty processes. Further, the adhesion of films to substrates and to each other is dependent upon the cleanliness of the interfaces. Fractional monolayer of contamination often means loss of adhesion and ultimate fabrication failure. Information on layer composition is provided to the device engineer by the Auger technique.

Modern Auger instrumentation permits the incident electron beam to be focused to approximately a 500 Å diameter probe, so that it is possible to collect electron energy spectra from very small structures. As an example, we continue to examine the structure we previously used in our SEM discussion. Refer again to Fig. 9. Auger spectral lines are very weak and lie on top of strongly varying background signals. To enhance the Auger signals, derivative spectra, produced by modulating the energy selected by the analyzer and using a lock-in amplifier to detect the signal, are often displayed as shown in Fig. 13. In case (a) of this figure, two silicon transitions and an oxygen transition are visible in the spectrum, which is taken with the electron beam striking the SiO_2 surface. Using higher sensitivity instrument conditions, weak P and F signals also become evident, the first from the glass matrix and the second possibly from residual contamination produced by the etching or cleaning process. Figure 13b shows the Auger spectrum from the aluminum conductor. Two major Al lines are evident, but Ar, C, F, and Cu transitions are present as well, indicating the presence of surface contamination.

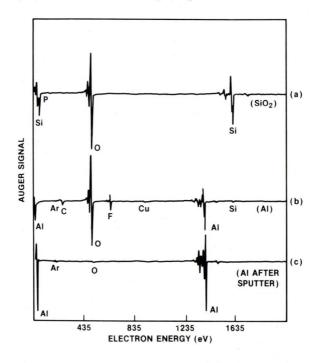

FIGURE 13

Auger spectra taken from (a) the SiO_2 surface, (b) the Al runner and (c) the Al runner, after a brief sputter etch. The very thin aluminum oxide in (b) is clearly evident. (*After J. A. Taylor, AT&T Bell Laboratories, unpublished.*)

By sputtering the surface of the sample for a short time (with a low keV argon ion gun located in the instrument), less than 50 Å of material is removed in situ. The resulting AES spectrum after such a treatment is shown in Fig. 13c. The contamination is essentially gone and only a hint of a surface oxide is evident. A more significant difference between Fig. 13b and Fig. 13c can be observed in the Al peaks; the shape and energy location of both transitions has changed in a measurable way. The Al and Al_2O_3 compositions can be easily differentiated in an Auger spectrum because the chemical bonding of Al in the metal and in the oxide is different. Further, the strong oxide peak of Fig. 13b arises almost entirely from a very thin layer. The impact that Auger Electron Spectroscopy (AES) has had on VLSI cleaning processes is easily inferred from this example; the small Cu transition in Fig. 13b represents contamination present as only a fraction of a monolayer.

It is useful to consider the reason for the high surface-specificity of the AES technique. If an Auger electron is released from a point in a material, it must travel to the surface before it can leave the surface and be detected. If, during this time, it suffers a collision with another electron, it will lose some energy and will no longer be distinguishable as an Auger electron; it will have become a part of the continuous background. The distance that an electron can travel in a solid before it has a high probability of collision is called the escape depth, and this distance is a function of energy. The escape depth is generally insensitive to the particular material, but oxides tend to have larger escape depths. At about 100 eV, the escape depth is close to 1 monolayer, and as the energy increases the escape depth increases with the square root of energy. Over the usual range of AES measurements, it rarely exceeds 10 atomic layers.

The formation of platinum silicide contacts in a bipolar transistor structure affords an example of the small-area-analysis capability of the Auger technique. Silicide contacts are formed by reacting sputtered platinum with silicon at moderate temperatures ($\sim$ 600°C). In the case covered by Fig. 14, the optical microscope revealed that some contacts had turned dark after the anneal, while others had not. Further, some localized structure seemed to develop in some of the contact windows. A secondary-electron micrograph of one of these bad contacts produced by the Auger instrument is shown in Fig. 14a. In the Auger technique, it is possible to follow the strength of a particular Auger emission line while sputtering the structure. In this way, a concentration versus depth profile can be produced. Figures 14b and c show the profiles taken in a bad and a good contact. The first, Fig. 14b, is taken over a bright (rough) area corresponding to what was typically observed in totally bad contacts. The surface film is only Pt while there is a small peak of oxygen at the Pt–Si interface. This oxide probably prevented the formation of silicide. The dark contact area, corresponding to a good region, shows (Fig. 14c) a uniform level of both Pt and Si until the Pt signal decreases and the Si signal increases. This corresponds to a uniform Pt–Si phase followed by the bulk silicon. An improvement in the precleaning step eliminated the interfering oxide layer and successfully resolved the problem.

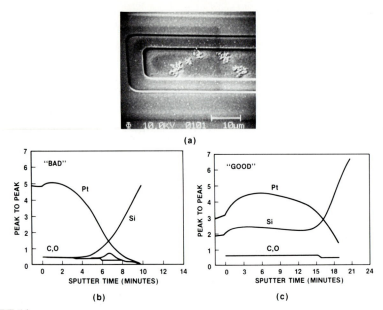

FIGURE 14
An example of a localized Auger analysis. (a) shows the contact window in which platinum silicide was to be formed. (b) A bad contact showing no silicide formation. (c) A good contact where the uniform Pt and Si signals indicate the presence of a silicide layer. The sputter time is proportional to the thickness. The original signal was taken in derivative mode, and the peak-to-peak response was taken as being proportional to the elemental concentration. Note the suggestion in (b) of a thin oxide layer which probably inhibited the silicide formation. (*After D. F. Lesher, AT&T Microelectronics, unpublished.*)

Diffraction Effects and the Transmission Electron Microscope. So far we have only discussed methods which provide compositional information via localized atomic properties such as energy level differences. Diffraction techniques can produce information concerning longer-range order. With the analytical techniques covered so far, amorphous silicon dioxide and crystalline quartz could not easily be distinguished.

Crystalline materials are ordered into periodic arrays that can scatter waves coherently via diffraction processes. DeBroglie's demonstration that particle beams can have wavelike properties has particular impact on electron beam methods.[14] The wavelength associated with a charged-particle beam which has been accelerated by a potential difference V is given by

$$\lambda = \frac{h}{\sqrt{2mVe}} \tag{9}$$

where h is Planck's constant and m is the particle mass. At very low electron energies corresponding to penetrations of only a few atomic layers, the wavelength is comparable to interatomic distances. Since only one or two layers contribute to the coherent scattering, there are many directions that correspond to constructive interference from these layers. Consequently, a photographic plate normal to the incident beam (Fig. 3d) will contain a periodic array of spots, the positions of which are related to the periodicities of the surface. The intensities of the spots can be related to the structure of the surface, the average position of impurity sites, or the terrace structures that are present (single-atom-layer steps). Although the method (LEED—Low-Energy Electron Diffraction) is far from a routine test, it is one of the few techniques that can provide detailed atomic scale information about semiconductor (as well as other) surfaces.[15]

At higher energies and for sufficiently thin sections (Fig. 3e), the atomic planes can diffract the beam if there is an atomic plane, with Miller indices (*hkl*), that is set for the usual Bragg condition

$$2d(hkl)\sin(\theta) = n\lambda \qquad (10)$$

where $d(hkl)$ is the interplanar spacing and n the order of diffraction. θ is the angle that the primary beam makes with the diffracting plane.[16]

For a single-crystal material, only a few spots will be evident, but for a polycrystalline film, there are enough crystallites present that all orientations of planes are represented. The result is the usual diffraction rings, as shown in Fig. 15a, produced by a 120 keV electron beam in a Transmission Electron Microscope. Each spot in a ring is produced by a different crystallite, which happens to be properly oriented for the particular Bragg condition, in the sample. By contrast, Fig. 15b shows the diffuse pattern from amorphous silicon; the rings in this case are produced by the presence of a continuous distribution of interatomic distances that are present in the material and not by the existence of

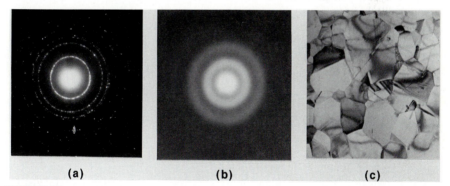

(a)   **(b)**   **(c)**

FIGURE 15

Three modes of Transmission Electron Microscope operation. (a) diffraction from a polycrystalline silicon film, (b) transmission through an amorphous silicon region showing broad, diffuse peaks corresponding to a distribution of interatomic distances, and (c) a transmission image in "bright field" showing the film's microstructure. (*After V. C. Kannan, AT&T Bell Laboratories, unpublished.*)

any planes of atoms. The peak of the distribution is related to the average inter-atomic distance, and this distance can be extracted from the diffraction pattern. Using Fourier transform methods, the distribution itself can be determined.

In its usual configuration, the Transmission Electron Microscope (TEM) operates as an imaging instrument, as opposed to the scanning instruments that we have discussed so far.[17] Electrons can be focused by magnetic lenses, similar to the way glass lenses focus light. Thus, as long as a sample can be thinned sufficiently for beam penetration, material density differences will yield images of high magnification (up to 300,000X is routine) that are relatively easy to interpret.

A microscope operates (simplistically) by a sequence of (electron) optical operations that produces a diffraction pattern of the specimen and then reconstructs this pattern into the enlarged image.[18] In the TEM, the diffraction pattern can be intercepted with an aperture, thereby permitting only one or a few of the diffraction spots to eventually recombine into a final image. This image would be sensitive to crystallographic features, including deviations from periodicity (defects). The TEM can image lattice strain as well as point defects and stacking faults. Since localized defects are of major importance to the operation of devices not much larger than the extent of these defects, the TEM has had great impact on guiding VLSI processing in directions that produce defect-free semiconductor structures.

The TEM in its imaging mode can produce highly magnified images; an example of the grain structure of polysilicon is shown in Fig. 15c. Recent developments in sample polishing have made it possible to cross-section VLSI devices and produce high-quality images from very thin sections ($\sim$ 100–500 Å). These sections scatter very little of the beam, compared to bulk sample cross sections that might be viewed in the SEM. Additional examples of device cross sections in the TEM are shown in Fig. 16. Corresponding SEM micrographs are also shown. The preparation of thin samples for TEM imaging requires slicing a wafer (through a periodic array because hitting a particular feature is almost impossible), mounting the slice in epoxy, mechanically polishing it, and then ion-milling it to the proper thickness.[1] The technique is very tedious and not always successful. Nevertheless, this method is gaining increased emphasis in VLSI diagnostics as the characterization of structural details smaller than 100 Å is becoming increasingly important.

12.3.4   Electromagnetic Radiation: ESCA and X-ray Methods

Incident electromagnetic radiation (light, X-rays, etc.) may also be used to examine solids. An important technique for VLSI is ESCA (Electron Spectrometry for Chemical Analysis), also known as X-Ray Photoelectron Spectroscopy (XPS).[4,19] This technique is very closely related to the Auger method. X-rays are used to excite the sample, and the emitted electrons are energy-analyzed. In the energy range under consideration here (a few keV or less), X-ray photon interactions with a solid differ from electron interactions in that the entire photon energy must be absorbed at once by the target atom. Thus, all of the photon's energy can

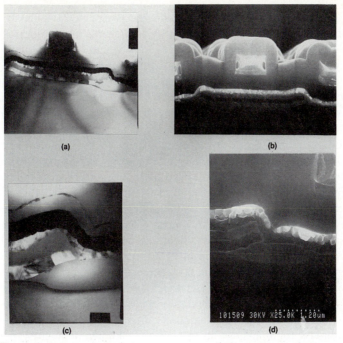

FIGURE 16

A comparison of TEM (a, c) and SEM (b, d) showing 256K DRAM device cross sections at two magnifications. Note the different contrasts. The TEM image contains more crystallographic information and also has higher resolution. The SEM image provides much of the same information but since the sample preparation involves only cleaving, it is obtained much more rapidly. (*After V. C. Kannan and B. J. Snyder, AT&T Bell Laboratories, unpublished.*)

go into ejecting an electron from a solid. The XPS measurement determines the actual kinetic energy of the ejected electron. This kinetic energy is equal to the incident energy of the X-ray less the binding energy of the electrons, after some corrections are made for the work functions of the specimen and the spectrometer. Superimposed on the XPS spectrum are Auger transitions.

X-rays produce less damage and less charging than do electron beams, thereby making the method particularly useful for examining insulators. ESCA, using a monochromatic X-ray line for excitation and a higher-resolution electron detector, has the ability to obtain more chemical information than does AES. For example, the binding energy of the carbon 1s shell of various carbon compounds varies by as much as 12 eV, and these differences are easily determined. Although the specific identification of a particular organic compound on the surface of a device is not usually possible, the method can distinguish between a few known possibilities. ESCA therefore has some advantages over AES for surface examination, but only when ESCA's principle drawback is not important: the large X-ray beam size ($\sim$ 1 mm in diameter). Newer instruments now offer $\sim$ 200 μm diameter beams.

ESCA is also a very surface-sensitive method. The Si 1s signals from Si and SiO_2 are easily separated, and an oxide less than 5 Å thick can be detected. ESCA is therefore very useful for evaluating surface-cleaning processes for VLSI wafers.

X-ray irradiation can also produce fluorescent X-rays (XRF) from the specimen. When measured with energy-dispersive X-ray detectors, these X-rays can be used for materials analysis as well.[20] An important recent application of this method is the determination of the phosphorus concentration in phosphosilicate glass (PSG).[21] After optical methods are used to measure the film thickness of the PSG, the X-ray fluorescence method permits the sample to be exposed to energetic X-rays, and the ratio of P X-ray signals to Si to be measured. This ratio can be calibrated with standards and used for rapid process monitoring and control. The X-ray signal can also be used to estimate the thickness of the film, but this procedure is not as reliable as the optical method.

X-ray fluorescence can be used more effectively for the determination of thin-film thickness when the X-ray signal can be distinguished from the X-rays produced by the substrate.[22] The primary X-rays penetrate through the film and into the substrate. As the film thickness is increased, the X-ray signal from the film will increase, while that from the substrate will decrease. Eventually, when the film is thick enough, the X-ray penetration will be contained within the film and the signal will saturate. Prior to that point, the method produces a signal which can be calibrated for thickness against a set of known standards. The method can either be set up in an absolute sense (using theoretical relationships), or be calibrated for thickness against some other method, such as direct microscopy or a sputter profile determination. Results can be obtained in just a few minutes.

X-rays are also diffracted by crystalline materials, and most crystal structures have been determined by X-ray diffraction methods. X-rays are used in VLSI for precision wafer orientation and for the examination of thin films. Diffraction studies of films yield information regarding preferential orientation with respect to the substrate, compound identification, and grain size. If a sufficient number of diffraction peaks can be utilized, precision lattice-constant measurements can be made. Since bulk values are available for most of the compounds of interest, deviations from these values can be attributed to the effects of stress, which can thereby be estimated.

12.3.5   Ion-Beam Techniques

The secondary processes discussed above also occur for the case of ion-beam bombardment of a material. A major difference is that ions carry sufficient momentum to cause structural damage to the material, ranging from local heating (phonon production) to defect production, compound formation, and sputtering. During these processes, crystalline materials are easily amorphized. All of these effects are energy-dependent for a constant incident-ion mass. The ion–solid interaction produces a broad spectrum of emitted products including sputtered and excited neutrals, positive and negative ions, and excited species. Both monomeric and polymeric sputtered ions are produced. Inelastic effects produce the emission of light, X-rays, and secondary electrons. Neutrals usually account for ~99% of the sputtered species. The secondary-ion yield for each incident ion is about 0.01% and the secondary-electron yield per incident ion is about 10% in the 0.5 keV to 1.5 keV primary-energy range.

At low ion energies (0.5 to about 1.5 keV), incident ions tend to be scattered from the outermost atomic layer. If the ion–surface-atom collision is assumed to be elastic (and classical), the ratio of the reflected to incident energy, K, may be calculated via the well-known Rutherford scattering model. The result is

$$K = \frac{E}{E_o} = \left\{ \frac{(M_2^2 - M_1^2\sin^2\theta)^{1/2} + M_1\cos\theta}{(M_1 + M_2)} \right\}^2 \qquad (11)$$

where θ is the scattering angle previously defined and E and E_o are, respectively, the scattered- and primary-beam energies.[4,23] M_1 and M_2 represent the primary- and surface-atom masses, respectively. The two common geometries are $\theta = 90°$ and $180°$. In the former case, with low energies, the technique is known as Ion Scattering Spectrometry (ISS) and noble-gas ions are usually used as the probe.[24]

Identification is made by measuring K, knowing M_1, and calculating M_2. The method is more sensitive to heavier target (M_2) nuclei and is very surface-sensitive. The case of back-scattering where $\theta \cong 180°$ will be discussed below in detail.

Secondary-Ion Mass Spectrometry (SIMS). As the primary energy is increased, penetration also increases and the onset of sputtering is reached. The emitted species may be either neutral (mostly) or charged ($+$ or $-$); the technique of Secondary-Ion Mass Spectrometry (SIMS) is based upon the detection of these ions.[4,25−29] SIMS has had an important impact on the development of VLSI technology because the detection limits are at or below the dopant and/or contaminant concentrations usually found in silicon devices. Some practical detection limits are listed in Table 4.

The sensitivity of the technique is illustrated by Fig. 17a, which shows the depth profile obtained for an iron implant; since the SIMS technique simultaneously sputters the surface, the time scale is directly convertible to a depth scale. The measurement, in this case, required an instrument with high mass resolution to separate ^{56}Fe from the $^{28}Si-^{28}Si$ dimer (a bonded and ionized pair of Si atoms). In spite of this resolution problem, a 5×10^{16} cm^{-3} detection limit was obtained.

Secondary-ion yields vary by as much as five orders of magnitude across the periodic table and are very sensitive to the chemical species making up the major components of the sample. These factors make the method quantitative, but it requires the use of standards produced for every material and matrix of interest. Ion implants are generally used for this purpose. Changes of ion yields near interfaces make interpretation particularly difficult, and the presence of thin oxides can easily be confused with concentration changes, since oxygen increases electropositive-ion yields. For this reason, an oxygen primary beam is usually

TABLE 4
Practical SIMS detection limits

Element	Primary species	Detected species	Detection limit atoms/cm^3	ppma*
B	O_2^+	$^{11}B^+$	1×10^{15}	0.02
P	Cs^+	$^{31}P^-$	1×10^{16}	0.2
As	Cs^+	$^{75}As^-$	1×10^{16}	0.2
Sb	Cs^+	$^{121}Sb^-$	1×10^{16}	0.2
Li	O_2^+	$^{7}Li^+$	1×10^{14}	0.002
Na	O_2^+	$^{23}Na^+$	1×10^{14}	0.002
K	O_2^+	$^{39}K^+$	1×10^{14}	0.002
Fe	O_2^+	$^{56}Fe^+$	1×10^{17}	2.0
Cu	O_2^+	$^{63}Cu^+$	1×10^{16}	0.2
Al	O_2^+	$^{27}Al^+$	1×10^{15}	0.02
F	O_2^+	$^{19}F^+$	1×10^{17}	2.0
Cl	Cs^+	$^{35}Cl^-$	1×10^{16}	0.2

*parts per million, atomic

used in SIMS. Cesium has similar yield-enhancing properties for electronegative species.

SIMS is an important technique for the study of impurity distributions in silicon and has been used extensively to characterize diffusion processes. [30,31] Implant standards are profiled, and the area under the SIMS intensity versus depth profile is integrated and equated to the dose in order to produce a relative sensitivity factor. As long as the peak impurity concentration remains low (below the one percent range) and does not affect the secondary-ion yields of the matrix, the conversion is linear and is valid over 5–6 orders of magnitude with an accuracy of about 5% to 10%.

SIMS profiles are often required over depths of a few microns. As the analysis continues, primary-beam ions can collide with impurities and modify the distributions via "knock-on" phenomena. The effect of "knock-on" is shown in Fig. 17b, which depicts profiles of a shallow 8 keV As implant obtained with an O_2^+ primary-ion beam at various energies. [32] The need, in this case, for using the lowest possible primary-ion energy was easily demonstrated.

Although SIMS has had its major impact on profiling low-level impurities in materials, it can also be useful for both major and minor compositional determinations. A particularly important advance has been the use of SIMS for profiling insulating glass structures. Normally SIMS has been ineffective on insulating materials because of the implantation of charge from the primary beam. Samples that are well insulated can develop electric fields during analysis, and these fields can both affect the beam–specimen interaction and cause motion of mobile alkali ions during the analysis. The recent advent of high-current-focused

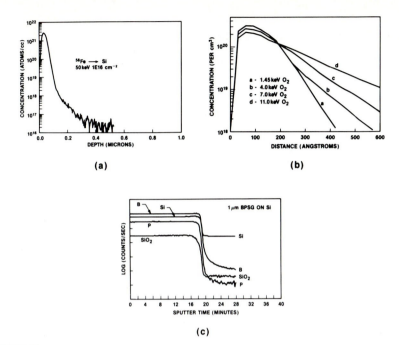

FIGURE 17

Three examples of SIMS concentration versus depth profiles. (a) An iron implant showing high mass resolution as well as a detection limit in the 5×10^{16} cm^{-3} range. (b) The effect of the incoming-ion energy on the shape of a shallow 8 keV arsenic implant. (*After Vandervorst, et al., Ref. 32. Replotted by M. Giles, AT&T Bell Laboratories, unpublished.*) (c) Charge neutralization effectively being used to allow a profile to be obtained from an insulating BPSG layer. Note the compositional uniformity of the film.

(and rastered) electron neutralizing beams has permitted the examination of such technologically important materials as BPSG, as shown in Fig. 17c. This particular profile demonstrates a compositional uniformity with depth into the film. The matrix signals are flat throughout the film, indicating effective neutralization. Similar results have been obtained for most of the insulating materials used in microelectronics processing. Quantitation of SIMS in the percent compositional range typically found for B and P in BPSG is very difficult because of strong matrix effects, but it can be accomplished.[33]

An important feature of SIMS is its mass discrimination, which often provides processing insight. Figure 18 shows a SIMS profile of a bipolar transistor structure. Both the emitter and the base contain boron doping, but the lack of ^{10}B in the base clearly suggests that the boron was implanted, whereas the emitter, which contains *both* naturally occurring isotopes, had to have been diffused. The difference in the emitter and collector Sb profiles suggests that they were implanted from two different machines.

SIMS is not usually considered a surface-sensitive technique, but surface analysis is possible under conditions of very clean vacuum and low beam current

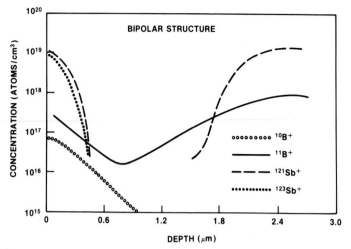

FIGURE 18
A SIMS profile through a bipolar transistor structure. The presence or lack of a normal isotopic ratio can be used to determine if the source of the dopant was a diffusion or a mass-selective ion implantation. (*After F. A. Stevie, AT&T Bell Laboratories, unpublished.*)

densities (large area). Although absolute quantitation is not yet available, "surface SIMS" can be very useful for the practical examination of semiconductor device surfaces. This has been demonstrated by carefully contaminating a silicon wafer with small amounts of Na, K, Al, Cr, and Cu and by then examining the wafers after various common VLSI cleaning techniques.[34] Definite differences in cleaning effectiveness were observed. A typical surface analysis of a silicon surface after an ammonium hydroxide–peroxide clean is shown in Fig. 19, where the presence of Al contamination is evident. Al was not detected by any other method; its source was found to be a contaminated reagent used in the cleaning process.

The sputtering process can also produce artifacts. Since sputtering yields are affected by the crystal planes, which are exposed, as well as by the slope of the surface, surfaces that have been sputtered for long periods of time begin to be roughened. This can cause a loss of depth resolution as the sample is etched, as well as cause changes in secondary-ion yield collection efficiencies.

A major new extension of the SIMS technique is SNMS, or Sputtered Neutral Mass Spectrometry.[35] This method uses an argon plasma to ionize the sputtered particles; the sputtering can be performed directly with the plasma or by means of an additional ion gun. Since the ionization takes place after ejection, matrix effects are minimal and the sensitivity for various species is more uniform. Detection limits in the few ppm range for all elements across the periodic table have been suggested.

The direct argon plasma bombardment approach produces excellent depth resolution. Figure 20 shows a film prepared with alternate ~ 20 Å depositions of Si and $WSi_{1.8}$. The figure clearly shows better than 40 Å depth resolution.[36] A

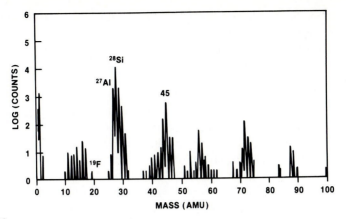

FIGURE 19
A surface SIMS mass scan taken on what was supposed to be a clean silicon surface. The high Al count at mass 27 was indicative of a contamination problem. The peak at mass 45 is from $SiOH^+$. (*After T. H. Briggs, AT&T Microelectronics and F. A. Stevie, AT&T Bell Laboratories, unpublished.*)

major drawback of this emerging technique is the large surface area (about 5 × 5 mm) currently required for the analysis.

Rutherford Back-scattering (RBS). As the primary energy is increased to the MeV region and if $^4He^+$ is used as the incident projectile, the ions can penetrate significant distances before being scattered by a nucleus. If a scattering angle close to 180° is used (170° is a practical value) and if the energy distribution

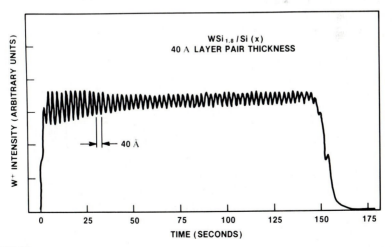

FIGURE 20
A SNMS depth profile of a repetitive tungsten silicide structure showing 40 Å resolution or better. The instrument used was an early prototype. (*After Evans, Ref. 36.*)

of these back-scattered ions is determined, information about the structure may be deduced. The analysis of these back-scattered ions forms the basis of the Rutherford Back-scattering Spectrometry method (RBS)[4,23] and it has seen extensive application to VLSI problems, particularly in metal-layer studies and damage evaluation of silicon structures.

The RBS technique is best illustrated with an example. Figure 21 shows the RBS spectrum produced when a 2.25 MeV He ion beam is scattered by a thin film of tantalum silicide which was deposited by sputtering from a compound target onto a bare silicon wafer. The physical structure of the specimen is shown in the inset of the figure. The spectrum shows the counts received by each channel of a multichannel analyzer plotted against the channel number. For this particular case, each channel is 4.3 keV wide so that the energy of the scattered particle at any point in the figure is $4.3 \times$ (channel #) and is expressed in keV.

If a series of elements were located only on the surface of the material, then each element would produce a sharp feature (Gaussian peak) at an energy equal to the kinematic factor for that element multiplied by the primary-beam energy. Thus, each surface contaminant could be identified by its energy in the RBS spectrum, providing that the substrate material did not produce strong scattering in the same scattering range. This usually means that the atomic mass of the substrate should be low compared to the atomic mass of the contaminants.

The most prominent feature in our example is the peak located approxi-

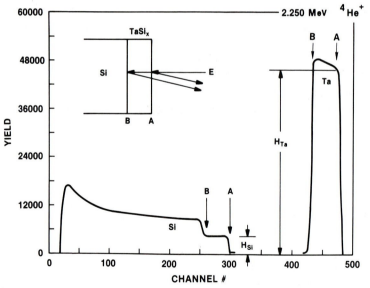

FIGURE 21

A Rutherford back-scattering spectrum taken from a tantalum silicide layer sputter-deposited onto a silicon substrate. The leading edges of both the Si and Ta are marked A. The inset shows the physical structure of the sample. The incident ^{4}He$^+$ ions had an energy of 2.25 MeV. (*After R. B. Irwin and A. J. Filo, AT&T Bell Laboratories, unpublished.*)

mately in channels 425–480. The high-energy "edge" of this feature occurs near channel 480 and corresponds to a scattered energy of 2.06 MeV. Table 5 contains the kinematic scattering factors for Si and Ta as calculated from Eq. 11. For Ta, the factor is 0.915, which would result in a scattered particle energy of $0.915 \times 2.25 = 2.06$ MeV. Thus the high-energy side of this feature (A) corresponds to scattering from the outer surface of the silicide film by a Ta atom. Since there are also silicon atoms at this surface, we would expect to find a feature corresponding to an energy of $0.56 \times 2.25 = 1.27$ MeV near channel 295. Such a feature is also evident in the lower energy range.

A film of finite thickness would produce a broadened spectral feature, as tantalum produces in the current example. This broadening results from the penetration of the He ions into the surface. As an ion travels through the material, it continuously loses energy. If it scatters at the silicide–silicon interface (B), it will lose energy due to the collision (kinematic factor) and then lose further energy as it returns through the film to the surface. Thus, the energy width of the feature increases with the film thickness. The "step" within the silicon spectrum corresponds to the change as the ion travels from the silicide into the silicon substrate.

The amount of energy lost by an ion traveling a distance x in a sample is suggested by

$$\Delta E = x \frac{dE}{dx} \tag{12}$$

where the stopping power (dE/dx) is available from tables as a function of energy.[23] Since the amount of energy lost by an ion traveling through the silicide is almost independent of whether it is scattered by the Ta or the Si, the width (A–B) in energy of both the Si and Ta portions of the spectrum arising from the film will be approximately the same, but shifted with respect to each other because of the differing kinematic factor values.

The actual composition at the surface of the structure can be determined by the ratio of the heights H_{Si} and H_{Ta} shown in the figure. A first approximation is given by

$$\frac{N_{Si}}{N_{Ta}} = \frac{H_{Si}}{H_{Ta}} \cdot \left(\frac{Z_{Ta}}{Z_{Si}}\right)^2 \tag{13}$$

where Z is the atomic number and where the squared atomic number ratios arise from the scattering cross sections (see Eq. 4). Some additional factors have been

TABLE 5
Kinematic factors for tantalum silicide samples

	He	Si	Ta
Atomic number	2	14	73
Atomic mass	4	28	181
K factor	0	0.5625	0.9154

neglected, including an energy-loss ratio factor which is usually between 0.9 and 1.1.

RBS is a method that is most useful for heavy elements on lower-density substrates. Since the composition of the sample is computed directly from the collected spectrum and known physical constants, it is an *absolute* method, in the sense that known standards are not required for the measurement. This makes the technique quite useful for the calibration of the implant doses of Sb or As into silicon.

If RBS is being used to examine single-crystal material, alignment along a major crystallographic direction will cause the back-scatter signal to drop due to channeling. This will permit the ions to penetrate to greater distances, with a corresponding loss of backscatter signal. If structural damage exists in the material, ions will be de-channeled at damage sites and will be subsequently or directly back-scattered. The RBS signal will therefore increase with the extent of this damage, demonstrating that the technique is also sensitive to atomic displacements. This approach has found great application in the study of ion-implant-induced structural damage (radiation damage) and the effects of subsequent annealing.

RBS has been increasingly applied to VLSI materials in recent years as relatively affordable instrumentation has begun to appear on the market. The major limitation for device studies has been the relatively large (mm size) beams. Recent microbeam developments have begun to attack this deficiency and it is anticipated that RBS will become an even more important technique as these developments continue.[37]

12.4   CHEMICAL METHODS

The analytical techniques that we have discussed so far have been nonchemical techniques, which have had particular impact on the development of VLSI technology. Many of the techniques of traditional analytical chemistry have had similar impact. We will discuss some of these in this section.

The field of analytical chemistry concerns itself with the accurate and precise determination of the elemental (or compound) composition of materials. The extreme sensitivity of semiconductor devices to the purity and stoichiometry of incorporated materials has caused the methods of analytical chemistry to play a central role in both the verification of materials' composition prior to use (quality control of incoming materials) and the solution to manufacturing problems.

The methods of analytical chemistry range from very sensitive and time consuming "wet chemical" procedures to the use of complex instrumental methods.[38] The semiconductor industry has pushed the development of the latter toward achievement of very high sensitivities; it is therefore appropriate that we focus our attention in this area.

12.4.1   Neutron Activation Analysis

Very-high-energy particle beams interact with materials in numerous ways, and some of them can be utilized for analytical purposes. Neutron beams have proven to be particularly useful in VLSI diagnostics. These particles have sufficient energy to penetrate the nuclear forces and affect the structure of the atomic nucleus. Large facilities are usually required for these measurements, and few semiconductor manufacturers have such equipment on site. The value of the results obtained compensates for the inconvenience of having to use remote (off-site) reactor or accelerator facilities.

In neutron activation analysis (NAA), a sample is placed in a nuclear reactor, where it is bombarded by neutrons.[39] When a neutron enters the nucleus of an element, its atomic mass is increased by one unit. The nucleus created in this process is usually in an unstable excited state (similar in concept to atomic excited states) and decays with a well-known half-life. The excited nucleus retains the neutron and emits one or more prompt gamma rays to return to the ground state. This final state may be radioactive, and if so will decay further, also with a well known half-life, to some excited state of a "daughter" (product) nuclide. This daughter then de-excites with the emission of a gamma ray whose energy uniquely identifies the element.

Consider a silicon wafer containing an impurity. The wafer is placed in the reactor and irradiated with neutrons. Both the silicon and the impurity capture neutrons during this exposure, and then both decay. The number of impurity atoms in an excited state is proportional to the time of irradiation, the neutron flux, the cross section for the reaction, and the concentration of the impurity. If the decay constants for the radioactive products are well known, then the activity level of the sample can be measured at a later time in the analytical laboratory and the concentration of the impurity can be calculated. We assume that the silicon activity is reduced to a level that will not saturate the counting system and that the impurity atoms have a longer half-life and retain a measurable activity. Sensitivities for impurities range from ppm to parts per trillion (ppt). A detection limit of 0.5 ppt for gold in silicon has been noted in the literature.[40]

NAA has found its greatest utility for VLSI in evaluating various furnace operations. Impurities from furnace tubes and liners have been shown to contaminate wafers during oxidation, and much understanding of the underlying mechanisms of furnace-tube degradation has been obtained through this sensitive technique.[40]

12.4.2 Chromatography

As semiconductor wafers are processed, they are continually being exposed to a variety of both inorganic and organic chemical reagents. A partial listing ranges from buffered HF, nitric or hydrochloric acids, solvents, and special cleaning solutions to polymers such as photoresist. Integrated circuit packages are also exposed to similar reagents and, of course, are molded from and contain a range of epoxies, polyimides, and silicone rubbers (RTV). These materials must be pure; contaminating compounds can leave residues, impede the curing of resins, or lead to reliability problems. Chromatography is a technique which permits the separation and identification of organic and inorganic mixtures at both major and trace levels.[41] Corresponding to the wide range of organic material properties is a wide range of chromatographic techniques. The various types of chromatography are differentiated by the nature of the *mobile phase* (gas, liquid, or ionic liquid) used to carry the material to be analyzed through the instrument.

The fundamental principles of chromatography may be illustrated by *gas* chromatography with the aid of Fig. 22. A long, hollow column is "packed," or coated, with a material with which the suspected contaminant as well as the major components will interact. This packing is referred to as the *stationary phase*. The *mobile phase*, the gas in this example, is the phase which contains the sample to be analyzed and which moves through the column with the samples. The strength of the interaction between the sample and the column packing will vary for different compounds, and this variation allows physical separation of the sample in the following manner. The column is maintained at an optimum temperature, and a carrier gas (usually helium) is forced through the column. To start the analysis a small volume of the sample to be analyzed is rapidly "injected" (point A) into the gas stream. The material must be either a liquid or a gas; if the material is a liquid but the injector temperature is sufficiently high, it will rapidly vaporize and also be swept into the column.

Assume that the material contains three components: "a," "b," and "c." As a molecule of "b" enters the column, it will interact with the stationary phase (coating or packing material) at a particular site. The molecule will remain on that site for some *residence* time and will then be swept on to the next site. The

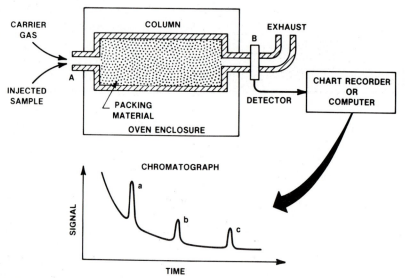

FIGURE 22

A simplified model of a gas chromatograph. The sample is injected into the carrier gas stream at A and its components elute from the other end of the column, B, one at a time. An example of a chromatogram is shown in the lower portion of the figure.

stronger the bond to the stationary phase, the longer the residence time will be and the longer the molecule will take to get to the end of the column and be detected (point B). A molecule that does not react with the material at all, "a," will arrive at point B very rapidly. Thus, the mixture of three materials will be "separated" by the column and each component will arrive at the detector after a different length of time. Once the *elution* or *retention* time for particular compounds is known (by calibration with known materials), the compound can be uniquely identified. Sensitivities of ppm are common with this technique. The graph of output from the detector as a function of time is referred to as a chromatogram.

The detector chosen is usually one which is sensitive to a particular property of the material or materials being analyzed. In gas chromatography, commonly used detection methods include thermal conductivity and flame ionization. For liquid and ion chromatography the refractive index of the *eluent* or its conductivity can be monitored. Alternatively, the absorption of radiation of a particular wavelength or wavelength range can be monitored. Since organic molecules have characteristic absorbances in the ultraviolet region (due to electronic transitions), UV absorption is often chosen.

If the detector is a mass spectrometer, each compound can be mass analyzed as it exits the column, and with rapid detection of the mass spectrum, each compound can be sequentially identified. The identification is usually done via a computer comparison of the mass spectrum of a compound with a large library of mass spectra which have already been identified. With the mass spectrometer as a detector and with the chromatograph being a gas chromatograph (GC), the technique is known as GC/MS.

An interesting GC study can be cited which was capable of examining surface contamination on a silicon wafer.[42] A contaminated wafer was cleaved into strips and placed in an oven, which in turn was connected to a GC column. The wafers were heated and the vapors collected onto a chilled column. The column was then heated, allowing the vapors to proceed through the column. The method was standardized by depositing known materials on a wafer and performing the same test. Figure 23 shows the resulting chromatograms for a wafer immersed in water after a Reverse Osmosis (RO) treatment and for a wafer which had been exposed only to a cleanroom environment for two hours. Each peak corresponds to a different compound, and most of the peaks could be identified. The RO-immersed wafer had an estimated total organic contamination of 3.0×10^{-11} g while the exposed wafer contained 1.5×10^{-9} g.

A useful application of GC/MS involves a technique known as headspace chromatography. The term *headspace* refers to the method by which the sample is prepared for injection into the GC/MS column. A sample is placed into a vial, the vial is sealed with a septum, and the sample is allowed to equilibrate at the desired analysis temperature. A portion of the vial ambient—the *headspace*—is then withdrawn through the septum with a syringe and is injected into the column and analyzed. This method thus analyzes the volatile species in an organic mixture, whereas the direct chromatographic techniques measure the actual mixture itself.

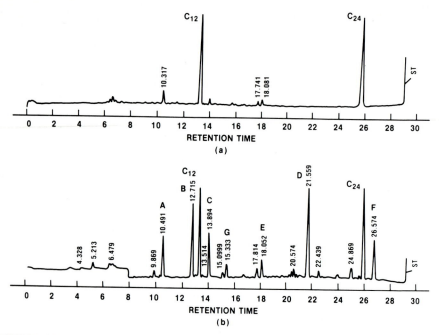

FIGURE 23

Gas chromatograms taken from heated silicon wafers after (a) a reverse-osmosis filtering and (b) exposure to a cleanroom atmosphere for two hours. Note the larger number of peaks in the exposed sample. Each peak corresponds to a different contaminant. (*After Fergason, Ref. 42.*)

This technique, in a modified fashion, was applied to determine the cause of lid-seal failures in a hermetic package. The IC package was first chilled to liquid nitrogen temperatures, fractured, and then quickly placed in the vial. The vial was then heated to the temperature required for the analysis (in this case, room temperature), and the headspace procedure was followed. The chromatogram of Fig. 24 and the indicated mass-spectral identifications disclosed that the cause of the failure was insufficient curing of the RTV rubber; most of the chromatographic peaks could be associated with RTV outgassing products during cure. The pressure increase due to RTV outgassing during the lid-sealing operation was high enough to break the lid seal.

Polymeric materials (photoresist is a good example) can be examined by a form of liquid chromatography known as Gel Permeation Chromatography (GPC) or Size Exclusion Chromatography (SEC).[43] This method determines the molecular size (or molecular weight) distribution of a liquid. The liquid is injected onto a porous column; a simple model for such a column is shown in Fig. 25. The column is assumed to contain an array of both large and small pores along its length. In the figure, three molecular sizes are shown as 1, 2, and 3 in order of decreasing size. A "large" molecule can proceed only down the central part of the column. It cannot fit into the smaller tunnels and consequently

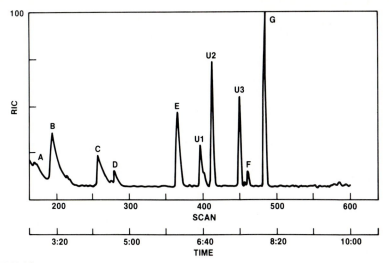

FIGURE 24

An example of a headspace analysis from a hermetic package with poor lid-sealing characteristics. The sample was cooled to liquid nitrogen temperature, fractured, and placed in a vial, after which it was allowed to warm to room temperature. The headspace gas chromatogram contained a number of peaks which were identified. Most of the peaks were associated with RTV curing. (A) Methoxy trimethyl silane, (B) trimethyl silanol, (C) 2-methoxy ethanol, (D) n-butanol, (E) methyl isobutyl ketone, (F) toluene, (G) 4-ethenyl cyclohexene. Peaks labeled by U were not identified. (*After D. L. Angst, AT&T Bell Laboratories, unpublished.*)

will progress through the column rapidly. The small molecules can interact with the smaller pores as well, and therefore spend time in the side tunnels before emerging from the column. These smaller molecules come out last. The intermediate molecules exit between the others and consequently the chromatographic components exit in order of size. For polymers, the higher-molecular-weight species will exit first. In addition to the analysis of the higher-molecular-weight fragments by SEC, normal- and reversed-phase liquid chromatographic techniques are useful for characterization of the lower-molecular-weight components. Both high- and low-molecular-weight

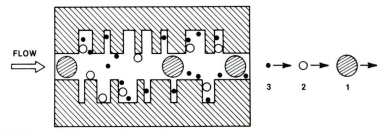

FIGURE 25

A simplified model for size exclusion chromatographic separation. This technique is useful for polymer characterization and provides important information about photoresists.

components have effects on resist behavior, and chromatography is emerging as an important method for the quality control of these materials.

Ion exchange resins can also be used for chromatographic applications. In this case, a solution containing cations and anions is passed through the resin column. The ions interact with the resin and are exchanged with ions on the column to various extents depending upon the ionic species. Eventually the ions leave the column and are passed through a suppressor column that selectively removes all of the ions lost by the original column as a result of the exchange. What is left are the sample ions themselves, which are detected by a conductivity detector. The ions are separated both by identity (F, Cl) and by charge state (+1, +2, −1, etc.). Ion chromatography is often used to determine the ionic contamination in reagents that will be used to process device wafers. Packaging materials can also be analyzed for ionic contamination after aqueous extraction. The method is very sensitive and can, with preconcentration of the sample, determine parts per billion (ppb) levels of some ionic impurities.

Phosphosilicate glass and BPSG films are often analyzed with ion chromatography. The wafer is first carefully weighed. The film is then stripped in a small volume of HF and the wafer is reweighed to obtain the weight of the film. The HF solution is then diluted, a standard is added, and the chromatogram is obtained. Figure 26 shows an example of a BPSG analysis. Four major peaks are observed. The first are phosphite and phosphate peaks, the ratio of which has been linked to physical properties of the glass. This ratio is also a strong function of the method used to deposit the glass.[44] The fourth peak is boron, which elutes as BF_4^-. The remaining peak is an internal standard which was added to allow accurate quantitation. The areas under the curves permit a quantitative determination of the boron and phosphorous concentrations in the original film.

12.4.3 Infrared Spectroscopy

Many molecular species have vibrational modes with characteristic frequencies in the infrared region. These vibrational modes can be excited by radiation from the infrared portion of the spectrum. The absorption of an incident IR beam by a sample is measured by various IR techniques. Varying the wavelength of the incident beam permits recording the absorption spectrum of a compound, which is often a "fingerprint" of the species present. The technique can be made quantitative, and is applicable to mixtures, since absorptivities are additive.

IR spectroscopy has two major applications: the identification of organic and inorganic components via their absorption spectrum "signatures" and quantitation of selected components. Extensive libraries of absorption spectra are available for comparison with unknowns.[45] IR spectroscopy is particularly valuable in the examination of the many organic solvents used in VLSI processing. The technique is relatively rapid and inexpensive. Digital processing of the spectra permits subtraction analyses, which are particularly useful for trace-level determination. IR methods can also be used for trace-gas analysis.

In dispersive IR instruments, a broad spectrum source of IR radiation is

ION CHROMATOGRAPHY

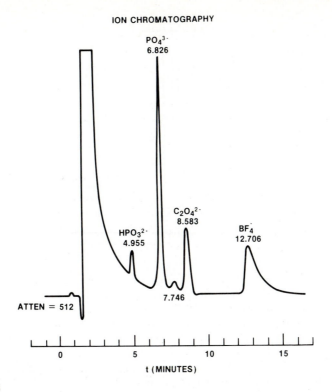

FIGURE 26
An ion chromatogram from a dissolved BPSG sample. The peak at 8.58 minutes is from a calibrating material which was intentionally added. (*After M. C. Hughes, AT&T Bell Laboratories, unpublished.*)

split into two beams of equal intensity. The first proceeds directly to the detector, while the second passes through the sample. The ratio of the two signals is used to determine the absorption. A more recent technique is based upon the Fourier Transform (FTIR) method. Here an interferometer, as shown in Fig. 27, is used to determine the absorption spectrum.[46] A beam of incident IR radiation is split by a semitransparent mirror into two spatially separated beams (A, B) which follow their respective dashed and dotted paths. Since the path traveled by the two beams is different, they will interfere with each other if the path difference is not a multiple of the wavelength. If one of the mirrors is uniformly translated, the path difference will change continuously. If the source is monochromatic, then the combined beams will go in and out of phase with each other, resulting in a sinusoidal modulation at the detector. The Fourier transform of such a signal will be a delta function proportional in strength to the incident intensity at the source frequency. If the source is polychromatic, the Fourier transform of the signal at the detector will produce its intensity spectrum. These transforms are calculated by computer. If a sample is now inserted, an absorption signal is obtained. Using

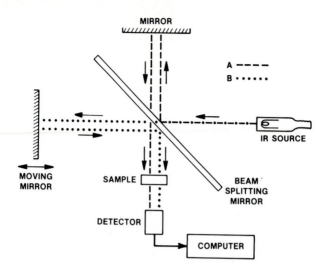

FIGURE 27
The Fourier Transform Infrared Spectrophotometer (FTIR). This instrument provides very rapid analysis of solid films such as BPSG. Recent instrumentation permits the use of a microscope to acquire spectra from features as small as 20 μm.

Fourier methods, the absorption spectrum is easily obtained from the transforms of the sample and the instrument background.

The advantage of the FTIR method is that it is a rapid one and allows signal integration methods, which increase signal-to-noise ratios and consequently improve the detection limits. Trace-level measurements are possible, and FTIR has been used extensively to measure low levels of oxygen and carbon in silicon starting material.[47]

BPSG films have been also analyzed with FTIR methods.[44] Figure 28 shows FTIR spectra from a variety of TEOS (tetraethylorthosilicate) oxides. A great deal of information is available in these spectra, but care has to be used in the interpretation. FTIR is sensitive to the structural configuration of the glass, and this structure is a function of the deposition chemistry and the annealing history. Annealing tends to sharpen the spectrum somewhat, and quantitation requires at least a minimal anneal to stabilize the structure. For a given process, however, simultaneous quantitation of B and P is possible, and the method therefore has important process-control potential.

12.4.4 Mass Spectrometry

A mass spectrometer separates a beam of ions according to the mass or mass-to-charge ratio of the ions. This technique requires low pressures, and the method was originally applied to the analysis of gases. In more modern instruments, solids, liquids, and gases can be analyzed, although the former two must be converted to gas first. There are several types of mass spectrometer; the two most

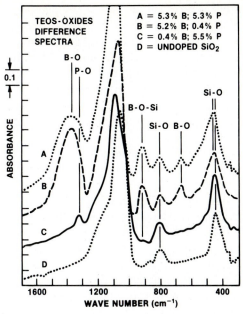

FIGURE 28
FTIR spectra taken from a variety of BPSG films. (*After Becker, et. al., Ref. 44.*)

commonly used versions are the magnetic-sector and the quadrupole.[48] The first of these uses a tandem electric and magnetic field geometry to create the mass separation, and it is characterized by its (adjustable) very high mass resolution. The quadrupole mass spectrometer has a lower mass resolution (by as much as a factor of 100) but has the distinct advantage of being able to accumulate a full mass spectrum in only a few milliseconds. It can therefore be used in dynamic situations such as GC/MS.

An early use of mass spectrometers for the analysis of solids was the Spark Source Mass Spectrometer (SSMS).[49] This instrument uses a solid sample as one (or both) of a pair of electrodes. A spark is generated which vaporizes and ionizes the sample material. This vapor is used as the source for the spectrometer which produces the mass spectrum. Sensitivities of ppm to ppb can routinely be achieved with this method, but it is not directly quantitative because of difficulties in stabilizing and reproducing the sparking condition. An important advantage of the method is, however, that in spite of quantitation difficulties, it is capable of identifying and measuring elements spanning the entire periodic table.

Semiconductor materials often need to be of extreme purity, and if enough of a bulk solid is available, the SSMS can be utilized. In the early days of the technology, when pure-silicon-crystal growth methods were first being developed, SSMS was a major analytical tool. A modern descendant of SSMS is the glow discharge mass spectrometer (GDMS). In this relatively new instrument, a solid rod of the sample about 1 mm × 2 cm is inserted into an argon glow discharge and acts as the cathode, with a potential of about a few hundred volts. The sputtered atoms from the sample are ionized by the plasma, and the positive ions are extracted through a slit and accelerated into a high-resolution mass spectrometer. The detected ions are usually singly charged and representative of the composition of the sample. Within a factor of two to three, the direct ion currents, corrected for instrumental factors and normalized to the matrix signal, can be used directly for determining elemental concentrations. Sensitivity factors are currently being determined by workers in the field. Elemental detection limits in the low parts per billion appear to be practical.

Recent problems with soft errors in large-scale memory devices have been connected with trace-level uranium and thorium concentrations in the materials used for the device fabrication.[50] The major materials implicated are packaging materials (ceramics, lids, etc.) and internal metallization. GDMS is capable of detecting 30 ppb of uranium in Al and can probably be applied to the other critical materials as well.[51] These high-sensitivity data can be accumulated in less than one hour (for a complete analysis), although the sample preparation for such ultra-trace methods is difficult. This type of measurement also has direct application for the evaluation of high-purity sputtering targets if core samples from the targets are supplied by the manufacturer for testing purposes.

Mass spectrometers are also used extensively for gas analysis. A special application of gas analysis is the determination of the moisture content of hermetically sealed ceramic packages.[52] It is believed that moisture in these packages can contribute to metallization corrosion failures, and a requirement that most manufacturers try to meet is a maximum of 5000 ppm moisture after a prescribed bake-out at 100°C. The challenge of determining this level of moisture is that the package volume is usually only a few tens of microliters. A difficult problem associated with this technique is that as soon as such a small amount of moisture is injected into an analytical instrument, especially a vacuum instrument such as a mass spectrometer, a portion of the moisture adsorbs onto the walls of the system and can never reach the detector. Various methods of circumventing this problem have been developed. The most popular is to make the measurement so rapidly that adsorption has not had time to occur. Another is to precondition the system with moisture so that the walls are in equilibrium before the analytical sample is admitted. Operation at high temperatures also tends to keep the water from adsorbing. Successful measurements have been made down to 500 ppm of moisture in such systems; calibration procedures for these measurements depend upon careful injection of controlled small volumes of gases of known (and measured) moisture content. An advantage of such a mass-spectrometric determination of moisture is that after the water signal (mass 18) has been measured, the nonreactive gases can be determined as well. Table 6 shows a typical analysis.

12.4.5   Other Methods of Analytical Chemistry

Other analytical methods that find extensive utilization for VLSI materials characterization include thermal methods of analysis and various forms of optical

TABLE 6
Typical hermetic package gas analysis by mass spectrometry

Gas	Concentration
Water	500 ppm
Hydrogen	0.66%
Helium	<100 ppm
Nitrogen	Balance
Methane	100 ppm
Oxygen	<100 ppm
Argon	<100 ppm
Carbon dioxide	400 ppm

spectroscopy. Flame atomic-absorption spectroscopy (AA) is a common method for looking at impurities in liquids or solutions. In this technique, a cathode lamp which emits light from the element of interest is used. (For example, if sodium is to be determined in water, a sodium lamp would be used.) This beam of light is collimated, focused, and passed through a hot flame into a detector. If the liquid is now quickly injected into the flame, it is atomized and sodium atoms will absorb and then re-emit the sodium wavelength photons. Since the re-emission will be in random directions, most of the absorbed radiation will not arrive at the detector. The absorbance of the sample solution is determined from the decrease in the detected signal. Quantitation is effected through carefully prepared standards in a matrix similar to the sample matrix. Parts-per-million sensitivity is common for this method, and ppb sensitivity is attained for a few elements. The method does not include all elements of the periodic table and is rather slow, since a separate lamp must be mounted in the instrument for each element to be determined.

A related method is Inductively Coupled Plasma Atomic Emission Spectroscopy (ICP), in which the liquid is injected into a very hot plasma torch, where it is immediately atomized and ionized. The emitted radiation is collected by a grating spectrometer, and the intensity, measured at a particular wavelength, is proportional to the concentration of the element, which is known to emit in that particular spectral range. ICP instruments often have both fixed and adjustable spectrometers. The fixed spectrometer can be set to determine as many as 20 elements simultaneously, and the adjustable spectrometer can detect additional elements in a sequential fashion. The sensitivity of ICP is often better than AA, with ppb sensitivities achievable in many cases. The sensitivity advantage of ICP is especially important for the refractory elements. Another advantage of the method is speed, and the technique often is used to ensure that incoming chemicals meet required specifications.

An even more sensitive instrument is the combination of an ICP with a mass spectrometer; the ICP–MS.[53] Here, the ions produced by the ICP torch are directed into a mass spectrometer. Detection limits for this method approach tens of parts per trillion for many elements in the periodic table. It is expected that this instrument will have a major impact on ultra-trace analysis of VLSI-grade chemicals.

12.4.6   Gas Analysis

The analysis of gases is an entire subject unto itself. For those molecules that have absorption bands in the infrared region, IR spectrophotometry can be used for quantitative analysis with detection limits as low as ppm. Levels of ppb and ppt have been determined with laser IR methods in which a monochromatic IR laser is tuned to a particular vibration frequency of the molecule being detected. [54] Because only one species can usually be detected with a particular laser (they can be tuned over a very limited range), such instruments are practical only for those gases for which very low detection limits are required. The technique has found extensive application in environmental studies.

GC techniques are also applicable to gas analysis when appropriate detectors are utilized. Mass spectrometry, however, dominates the application. Both high-resolution and low-resolution mass spectrometers are useful, and new instruments are now appearing which are designed specifically for the highly corrosive and/or toxic gases used in VLSI technology. Finally, sensors for specific gases have been designed with very high sensitivity and specificity. Portable oxygen and moisture monitors are the most popular of the genre.

12.5   SUMMARY AND FUTURE TRENDS

This chapter has discussed a large number of analytical techniques commonly applied to VLSI materials, and it is often difficult to choose the right one. Table 7 presents a brief guide to some of the various material properties which often need to be examined, and the analytical techniques which are available to provide the examination. The table is clearly not complete, but it should help guide the reader to the proper method. Often, more than one method is available, but it should be stressed that many of the choices are not mutually exclusive. A material might be examined by X-ray methods for the heavier elements, and then by Auger spectroscopy to gather light-element information. The field is rich and the quality of the harvest is often determined by the judiciousness of the choices.

As mentioned at the beginning of this chapter, analytical techniques will probably continue to advance in response to the demands of VLSI programs. Some of these new trends are already apparent. Analytical-beam techniques are progressing towards smaller beam sizes. Small-beam SIMS systems will be developed, and the same technology will lead to the introduction and utilization of the Scanning Ion Microscope. [55] Resolution of these instruments will approach 100 Å and below.

RBS beams with a diameter of 1.0 μm are already available, and raster scanning of these instruments will lead to the routine use of three-dimensional depth profiling. [56] SIMS methods will become more quantitative, and mapping

TABLE 7
A short guide to available analytical techniques

Phase	Characteristic	Sensitivity	Method
Solid			
Bulk			
	Composition	Major to 0.1%	EMP, SEM/EDS*, XRF
	Crystal structure	—	X-ray, electron diffraction
	Trace impurities	ppm, ppb	SIMS, GDMS, NAA, SNMS, PIXE, LEM, SSMS
	Localized defects, impurities	—	TEM, SEM, STEM
Surface			
	Composition	Major to 0.1%	ESCA, AES
		Trace	Surface SIMS, ISS
	Structure	—	TEM, LEED
Films			
	Composition versus depth	Major to 0.1%	AES, ESCA, RBS, NDP†
		Trace	SIMS, SNMS, NDP
	Structure	—	X-ray, TEM
	Thickness	$<1\ \mu m$	X-ray, RBS, TEM
Devices			
	Morphology	—	SEM, TEM
Liquids			
	Composition, inorganic	Trace	AA, ICP
		Ultra-trace	ICP–MS
	Composition, organic	Major to trace	GC, GC/MS
		Major to trace	IR, FTIR
	Thermal properties	—	Thermal analyzers
Gases			
	Composition	Major to trace	GC, GC/MS, IR, sensors, Mass spectrometry

*Scanning Electron Measure/Energy Dispersive (X-ray) Spectroscopy.
†Neutron Depth Profiling.

will become quantitative as well. Quantitative electron-microprobe mapping will become commonplace.

Lattice imaging with high-resolution and high-energy TEMs will also emerge as important techniques when atomic-scale features begin to dominate small-device performance. Finally, although SEM resolution will probably not improve significantly over what is currently available, the introduction of image-processing computers will enhance micrographs and assure the technologist that all of the available information has been extracted. Voltage-contrast measurements in the SEM will also become ubiquitous as other voltage-probing techniques become unusable.

The field of analytical chemistry has been lowering its detection limits more from the push of environmental and medical needs than from VLSI pressures, but VLSI will continue to benefit from these developments. Ultimately, analytical chemistry will approach the ability to detect the presence of tens of atoms routinely. When that day arrives, the VLSI engineer will probably still remain unsatisfied!

PROBLEMS

1 An integrated circuit is placed in a scanning electron microscope and is DC-biased such that one metal line on the structure is at +5 volts with an adjacent line held at ground potential. If a 10 keV electron beam is used to produce a scanning image, how will the bias affect (a) the secondary image and (b) the back-scattered electron image? What kind of an application does this suggest for the SEM?

2 Explain why point A in Fig. 6 is *not* a stable operating point for observing an insulating sample.

3 An SEM is being used for the measurement of the width of a photoresist feature. The operator notes that the resist appears dark in secondary mode. What does this tell the operator about charging, and what effect could this type of charging, if sufficiently high, have on the width measurement?

4 Why can hydrogen not be detected by Auger spectroscopy? Helium?

5 An investigator is examining a thin film of aluminum in an Auger instrument which is also provided with an X-ray spectrometer. The incident 10 keV electron beam produces an Auger spectrum which contains strong signals from Al and from oxygen. The X-ray detector detects only aluminum, even though it is capable of detecting the oxygen signal as well. Explain this observation.

6 Using Eq. 8, estimate the thickness of the aluminum layer of Fig. 9 if the Si X-ray line is just produced for a primary beam energy of 5, 10, 15, 20, and 30 keV. Ignore any absorption effects on the X-ray signals. Plot your results and suggest a possible practical application.

7 An RBS spectrum is collected from a tantalum silicide film deposited on a pure graphite substrate. What do you think the spectrum would look like? Compare it with Fig. 21.

8 If an observer dealing with a crystalline material whose lattice spacings are in the 1.0 Å range wishes to perform diffraction studies on this material with electrons, protons, or $^4He^+$ ions, approximately what should the energies of these particles be?

9 The mass resolution of a mass spectrometer is defined by $M/\Delta M$ where M is the mass number at which an analysis is being performed and ΔM is the mass difference that can be resolved at that mass number. Silicon has isotopic masses at 28, 29, and 30, so that a signal from P at mass 31 would be overlapped by SiH, which can be produced in most vacuum systems from the hydrogen background gas reacting with the Si ions from the sample. What mass resolution is required for a SIMS determination of a P concentration in Si?

10 A hermetically sealed package ambient is often found to have 1–3% hydrogen in it. Suggest a possible source for this. The sealing atmosphere may be assumed to be hydrogen-free.

REFERENCES

1 R. B. Markus and T. T. Sheng, *Transmission Electron Microscopy of Silicon VLSI Devices and Structures*, Wiley, New York, 1983.

2 L. A. Casper, Ed., *Microelectronic Processing: Inorganic Materials Characterization, ACS Symposium Series* No. 295, American Chemical Society, Washington, DC, 1986.

3 K. F. J. Heinrich, *Electron Beam X-Ray Microanalysis*, Van Nostrand Reinholt, New York, 1981.

4 L. C. Feldman and J. W. Mayer, *Fundamentals of Surface and Thin Film Analysis*, North-Holland, New York, 1986.

5 J. J. Hren, J. I. Goldstein, and D. C. Joy, Eds., *Introduction to Analytical Electron Microscopy*, Plenum, New York, 1979. D. E. Newbury, D. C. Joy, P. Echlin, C. E. Fiori, and J. I. Goldstein, *Advanced Scanning Electron Microscopy and X-Ray Microanalysis*, Plenum, New York, 1986.

6 H. A. Bethe, *Handbook of Physics*, Vol. 24, Springer, Berlin, 1933.

7 J. I. Goldstein, D. E. Newbury, P. Echlin, D. C. Joy, and E. Lifshin, *Scanning Electron Microscopy and X-Ray Microanalysis*, Plenum, New York, 1981.

8 M. J. Berger and S. M. Seltzer, *Nat. Acad. Sci./Nat. Res. Council Publication No. 1133*, Washington, DC, 1964.

9 G. Carter and J. S. Colligon, *Ion Bombardment of Solids*, American Elsevier, New York, 1968.

10 K. Kanaya and S. Okayama, *J. Phys. D. Appl. Phys.*, **5**, 43 (1972).

11 R. F. Egerton, *Electron Energy Loss Spectroscopy*, Plenum, New York, 1986.

12 L. Reimer, *Scanning Electron Microscopy*, Springer-Verlag, New York, 1985. See also the many volumes of *Scanning Electron Microscopy* published by SEM Inc. (1968–present), AMF O'Hare (Chicago).

13 B. R. Hammond and T. R. Vogel, "Non-encapsulated Microsectioning as a Construction and Failure Analysis Technique," *IEEE International Reliability Physics Symposium*, 1982, pp. 221–223.

14 A. Messiah, *Quantum Mechanics*, Vol. 1, North-Holland, Amsterdam, 1961. L. deBroglie, *Phil. Mag.*, **47**, 446 (1924).

15 L. E. Murr, *Electron and Ion Microscopy and Microanalysis*, Marcel Dekker, New York, 1982, p. 387.

16 B. P. Cullity, *Elements of X-Ray Diffraction*, Addison-Wesley, Reading, MA, 1978.

17 L. Reimer, *Transmission Electron Microscopy*, Springer-Verlag, New York, 1984.

18 R. W. Ditchburn, *Light*, Interscience, New York, 1963.

19 H. Windawi and Floyd F.-L. Ho, Eds., *Applied Electron Spectroscopy for Chemical Analysis*, Wiley, New York, 1982.

20 R. Jenkins, R. W. Gould, and D. Gedcke, *Quantitative X-Ray Spectrometry*, Marcel Dekker, New York, 1981.

21 A. C. Adams and S. P. Murarka, "Measuring the Phosphorus Concentration in Deposited Phosphosilicate Films," *J. Electrochem. Soc.*, **126**, 334 (1979).

22 E. P. Bertin, *Principles and Practice of X-Ray Spectrographic Analysis*, Plenum, New York, 1970.

23 W. K. Chu, J. W. Mayer and M. A. Nicolet, *Backscattering Spectrometry*, Academic Press, New York, 1978.

24 T. M. Buck, "Low Energy Ion Scattering Spectrometry," in *Methods of Surface Analysis*, A. W. Czanderna, Ed., Elsevier, New York, 1975.

25 G. Blaise, "Fundamental Aspects of Ion Microanalysis," in *Materials Characterization Using Ion Beams*, J. P. Thomas and A. Cachard, Eds., Plenum, New York, 1976.

26 R. E. Honig, "Surface and Thin Film Analysis of Semiconductor Materials," *Thin Solid Films*, **31**, 89 (1976).

27 C. W. Magee, "On the Use of Secondary Ion Mass Spectrometry in Semiconductor Device Materials and Process Development," *Ultramicroscopy*, **14**, 55 (1984).

28 R. J. Blattner and C. A. Evans, Jr., "High Performance Secondary Ion Mass Spectrometry," *Scanning Electron Microscopy*, **IV**, 55 (1980), SEM Inc., AMF O'Hare (Chicago).

29 S. R. Bryan, W. S. Woodward, R. W. Linten, and D. P. Griffis, "Secondary Ion Mass Spectrometry/Digital Imaging for the Three Dimensional Chemical Characterization of Solid State Devices," *J. Vac. Sci. Technol.*, **A3**, 2102 (1985).

30 A. Benninghoven, F. G. Rudenauer, and H. W. Werner, *Secondary Ion Mass Spectrometry*,

Wiley, New York, 1987.

31 H. W. Werner, "Analytical Assistance in Semiconductor and Electronic Material Technology," *Fresenius Z. Anal. Chem.*, **314**, 274 (1983).

32 W. Vandervorst, H. E. Maes, and R. F. DeKeersmaecker, "Secondary Ion Mass Spectrometry: Depth Profiling of Shallow As Implants in Silicon and Silicon Dioxide," *J. Appl Phys.* **56**, 1425 (1984).

33 P. K. Chu and S. L. Grube, "Quantitative Determination of Boron and Phosphorus in Borophosphosilicate Glass by Secondary Ion Mass Spectrometry," *Anal. Chem.* **57**, 1071 (1985).

34 B. F. Phillips, D. C. Burkman, W. R. Schmidt, and C. A. Peterson, "The Impact of Surface Analysis Technology on the Development of Semiconductor Cleaning Processes," *J. Vac. Sci. Technol.* **A1**, 646 (1983).

35 K.-H. Muller, K. Seifert, and M. Wilmers, "Quantitative Chemical Surface, In-depth, and Bulk Analysis by Secondary Neutrals Mass Spectrometry (SNMS)," *J. Vac. Sci. Technol.* **A3**, 1367 (1985).

36 Figure courtesy of C. A. Evans, Jr., Charles Evans and Assoc., 301 Chesapeake Drive, Redwood City, CA. The instrument used to produce these data was an early prototype.

37 W. G. Morris, W. Katz, H. Bakhru, and A. W. Habert, "RBS Analysis with a 1 μm Beam," *J. Vac. Sci. Technol.* **B3**, 392 (1985).

38 *Book of SEMI Standards*, Vol. 1, Semiconductor Equipment and Materials Institute, Mountain View, CA. Most recent annual edition.

39 P. Kruger, *Principles of Activation Analysis*, Wiley-Interscience, New York, 1971. D. Brune, B. Forkman, and B. Persson, *Nuclear Analytical Chemistry*, Verlag Chemie International Inc., Deerfield Beach, FL, 1984.

40 P. F. Schmidt and C. W. Pearce, "A Neutron Activation Analysis Study of the Sources of Transition Group Metal Contamination," *J. Electrochem. Soc.* **128**, 630 (1981).

41 H. Willard, L. L. Merritt, Jr., J. A. Dean, and F. A. Settle, Jr., *Instrumental Methods of Analysis*, Van Nostrand, New York, 1981.

42 L. A. Fergason, "Analysis of Organic Impurities on Silicon Wafer Surfaces," *Microcontamination*, April 1986, p. 33.

43 P. Paniez and A. Weill, "The Novolak Calibration Method Applied to the GPC Analysis of the UV Photoresists," *Microelectronic Eng.* **4**, 57 (1986).

44 F. S. Becker, D. Parlik, H. Schafer, and G. Staudigl, "Process and Film Characterization of Low Pressure Tetraethylorthosilicate-Borophosphosilicate Glass," *J. Vac. Sci. Technol.* **B4**, 732 (1986).

45 *Catalog of Infrared Spectrograms*, Sadtler Research Laboratories, Philadelphia, Pa. A continuously updated subscription service.

46 P. R. Griffiths, "Fourier Transform Infrared Spectrometry," *Science*, **222**, 297 (1983).

47 D. G. Mead, R. M. Gummer, and C. R. Anderson, "Temperature Dependent Infrared Characterization of Silicon Wafers" in ASTM Pub. No. STP 804, *Silicon Processing* by D. C. Gupta, Ed., American Society for Testing and Materials, Philadelphia, 1983.

48 B. S. Middleditch, Ed., *Practical Mass Spectrometry*, Plenum, New York, 1979. See also Ref. 41, Chapter 19.

49 F. D. Leipziger and R. J. Guidoboni, "Trace Element Survey by Spark Source Mass Spectrograph," in Ref. 2.

50 T. C. May and M. H. Woods, "Alpha Particle Induced Soft Errors in Dynamic Memories," *IEEE Trans. Electron. Devices* **ED26**, 2 (1979).

51 *The Determination of Trace Elements in Aluminum.* Applications Note No. 02.641, VG Isotopes Limited, Ion Path, Road Three, Winsford, Cheshire, England.

52 *RADC/NBS Workshop on Moisture Measurement and Control for Semiconductor Devices—III*, U. S. Dept of Commerce, Publication No. NBSIR 84-2852 (1984).

53 R. S. Houk, "Mass Spectrometry of Inductively Coupled Plasmas," *Anal. Chem.* **58**, 97A (1986).

54 J. Reid, R. L. Sinclair, W. B. Grant, and R. T. Menzies, "High Sensitivity Detection of Trace

Gases at Atomospheric Pressure Using Tunable Diode Lasers," *Opt. Quantum Electron.* **17**, 31 (1985).

55 R. Levi-Setti, P. H. LaMarche, K. Lam, and Y. L. Wang, "Initial Operation of a New High-Resolution Scanning Ion Microscope," *Proc. SPIE-Int. Soc. Opt. Eng.* **471**, 75 (1984).

56 B. L. Doyle and N. D. Wing, "Three Dimensional Profiling with the Sandia Nuclear Microprobe." *IEEE Trans. Nuc. Sci.* **NS30**, 1214 (1983).

CHAPTER
13

ASSEMBLY TECHNIQUES AND PACKAGING OF VLSI DEVICES

K. M. STRINY

13.1 INTRODUCTION

The packaging of the VLSI die (also called chip or bar) is a broad subject that ranges from pre-assembly wafer preparation to fabrication technologies for the packages that provide electrical connection and mechanical and environmental protection. Cost and/or performance considerations of the packaged device usually dictate the assembly and packaging details. Packaging significantly affects, and in many instances dominates, the overall cost, performance, and reliability of the packaged die. Appropriately, packaging is now receiving more attention by both packaged-device vendors and system builders.

The rapid increase in the number of devices per chip in VLSI and the performance of these devices create the major challenges that face packaging designers. Memories have been the driving force in the advancement of leading-edge wafer fabrication technology. However, they do not challenge packaging to the same extent as logic chips that dissipate more power and require more input/output (I/O) terminals. Although memory-chip complexity (in terms of bits per chip) and chip size have significantly increased over the last four design generations, the packaging challenge in terms of I/O requirements (Fig. 1) remains essentially constant, due to multiplexing techniques on the chip. In contrast, the number of I/O terminals required for logic (gate array, for example) and

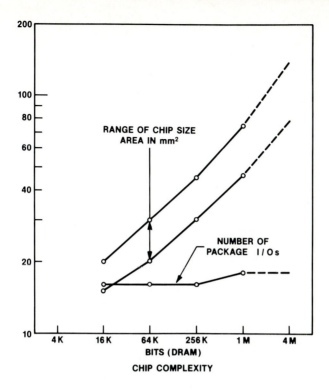

FIGURE 1
Continuous growth in DRAM complexity and size places little demand on package size and number of I/Os.

microprocessor devices continues to increase in proportion to the number of gates on the chip (Fig. 2). The number of signal terminals or package I/Os required for logic devices can be estimated using an empirical relationship known as Rent's Rule which is[1]

$$\text{Number I/O} = \alpha(\text{Gate Count})^{\beta} \qquad (1)$$

where α and β are constants determined by the device design.

Problem
Estimate the number of gates that can be included on a logic-gate-array chip which is to be assembled in a 100 I/O package. Assume $\alpha = 4.5$ and $\beta = 0.5$ in Eq. 1.

Solution
Using Rent's Rule, Eq. 1,

$$\text{Number I/O} = \alpha(\text{Gate Count})^{\beta}$$

Rearranging, we get

$$(\text{Gate Count})^{\beta} = \frac{\text{Number I/O}}{\alpha}$$

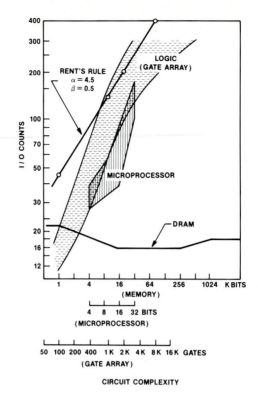

FIGURE 2
Comparison of I/O requirements for logic, microprocessor, and memory devices (DRAM) as a function of circuit complexity (*Courtesy of Kyocera International, Inc.*)

raising both sides of the equation to $1/\beta$ power

$$\text{Gate Count} = \left(\frac{\text{Number I/O}}{\alpha}\right)^{1/\beta}$$

Substituting knowns, we have

$$\text{Gate Count} = \left(\frac{100}{4.5}\right)^{\frac{1}{0.5}}$$

$$= (22.22)^2$$

$$= 493 \text{ gates}$$

Equation 1 is plotted in Fig. 2 to show the extent of agreement with the range of data represented in the curve. Figure 2 compares the extreme I/O demands placed on packaging by logic and microprocessors to the demands placed by memory, where the I/O count has been essentially constant.

The continual decrease in silicon feature size naturally leads to chips with higher I/O count, and hence, more interconnections to the silicon and more package pins. Economics, however, dictates that the chip be made as small as possible and placed into high-density packages requiring smaller lead spacing, hence placing new demands on the next level of assembly: for example, printed wiring boards (PWBs). The smallest possible VLSI chip, however, is probably large in size when compared to SSI/MSI and will challenge the mechanical integrity of the package. The package design will need to provide good heat dissipation, good electrical performance, and high reliability. Finally, the package must be easy to inspect after assembly to the next packaging level and be compatible with a variety of assembly, test, and handling systems.

This chapter discusses all of these packaging areas. It focuses on single-chip package types, assembly techniques, and thermal, electrical, and interconnection issues associated with the use of single-chip packages in a system's environment, particularly at the PWB level.

13.2   PACKAGE TYPES

There is a wide variety of package types that can be selected for VLSI devices. Recently there has been a proliferation of package types both for through-hole mounting and surfacing mounting to PWBs. Figure 3 illustrates the different methods of mounting to PWBs in use today. Some of the more popular package types and their availability as a function of the number of package I/Os are shown in Fig. 4. The through-hole (TH) types include the venerable dual-in-line package (DIP) and the newer pin-grid array (PGA), covering the range of I/Os from 8 through 300. Both types are available in hermetic ceramic and plastic types. In a

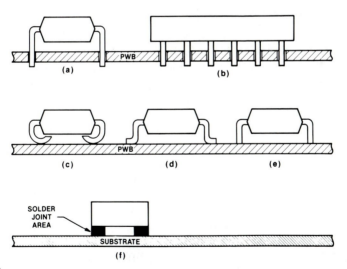

FIGURE 3

Examples of packages and mounting method to PWB in use today. (a) and (b) show typical through-hole packages; (c) to (f) show typical surface-mounted packages. (a) - Dual-in-line package (b) - Pin-grid-array package (c) - "J" leaded packages, leaded-chip-carrier or small-outline (d) - Gull-wing-leaded packages, chip-carrier or small-outline (e) - Butt-leaded package, small-outline dual-in-line type (f) - Leadless type, ceramic chip carrier mounted to a matching ceramic substrate.

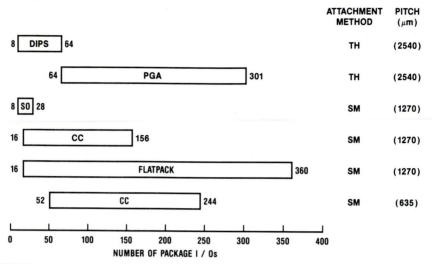

FIGURE 4
IC package types as a function of I/Os and method of attachment to PWBs.

hermetic package, the chip environment is decoupled from the external environment by a vacuum-tight enclosure. Package leads are typically on 2.54 mm pitch, although new designs are being introduced with 1.78 and 1.27 mm pitches. The finer-pitch through-hole packages present a challenge to PWB manufacturers to improve their process capability. The availability of new surface-mount (SM) packages has expanded greatly, although their acceptance has not advanced as quickly as foreseen.[2,3,4] Reasons for the slow growth are the lack of availability of devices at a competitive price, lack of confidence in placing and soldering surface-mount packages, and unforeseen technical difficulties such as lead-coplanarity maintenance and differences in thermal coefficients of expansion (TCE) between packages (especially ceramic types) and PWB materials. Nevertheless, the availability of the surface-mount packages shown in Fig. 4 (based on standards and registrations of the Joint Electron Device Engineering Council (JEDEC)) covers the entire range available also in through-hole types.[5] For SM package I/Os up to 28 terminals, the designer has a choice of a dual-in line type known as small outline (SO), available in plastic only, and the quad types known as chip carriers (CC) and flatpacks, both available in hermetic ceramic and plastic. Above 28 I/O terminals, all SM packages are quad types only (hermetic ceramic and plastic). CC packages are available in leaded plastic (PLCC) and leaded ceramic (LDCC) for mounting to PWBs, or leadless ceramic (LLCC) for mounting to boards with matching TCE or socketing. Leads may be of different geometric form ("J" versus gull wing), and located either on 1.27 or 0.635 mm pitches.

Packages designed for high performance are ceramic-based and usually hermetically sealed. New materials such as aluminum nitride[6] and silicon carbide[7] have been developed to enhance thermal performance for high-power applications. Low cost drives the need for plastic packages, which are typically batch-handled on automated assembly equipment to minimize labor costs. Performance features can be built into the plastic technology at little additional cost and with equivalent assembly yields and reliability, compared to the standard plastic package. An example of a performance plastic package is shown in Fig. 5. This photograph shows a novel packaging approach for shrinking a 1 megabit DRAM to accommodate it in a standard plastic DIP.[8] The design provides a second interconnection level inside the molded package to distribute power and ground and to provide under-chip fan-in capability for address I/Os. This package design avoids the need for on-chip metal buses and bonding pads along the long side of the chip. Both of these chip features would significantly increase the width of the chip and prevent its packaging in the industry-standard 18 I/O DIP.

Many other variations of IC packages exist in the industry.[9] These include variations on DIPs (skinny and shrink), single-in-line packages (SIP), zigzag-in-line (ZIP), chip-in-tape or tape-automated bonding (TAB), and bare chips (unpackaged). Some of these are illustrated in Fig. 6.

The major issues that must be addressed along with package selection concern various design aspects and the selection of the assembly interconnection

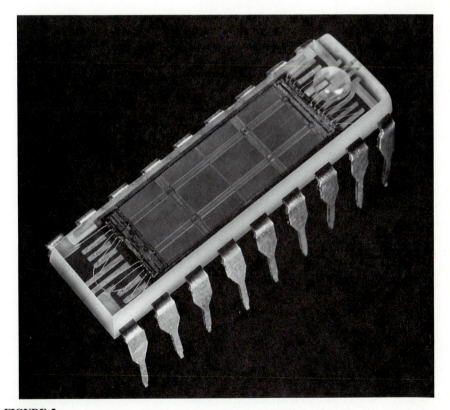

FIGURE 5
A 1 megabit DRAM in an 18 I/O plastic DIP (*Courtesy of AT&T Bell Laboratories.*)

MOUNTING METHOD	PACKAGE NAME	OUTLINE	LEAD SPACING
THROUGH-HOLE MOUNTING	S-DIP (SHRINK DIP)	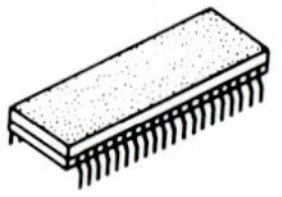	1.78 mm
	SKINNY DIP (OR SLIM DIP)	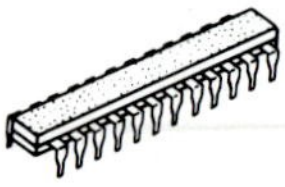	2.54 mm CERAMIC WIDTH 2/3 OR 1/2
	SIP (SINGLE INLINE)		2.54 mm
	ZIP (ZIGZAG INLINE)		2.54 mm
	QUIP (QUAD INLINE)		1.27 mm

FIGURE 6
A variety of package designs (*Courtesy of Kyocera International, Inc.*).

processes that best solve the needs of VLSI. These will be covered in Sections 13.3, 13.4, and 13.5.

13.3 PACKAGING DESIGN CONSIDERATIONS

13.3.1 VLSI Design Rules

The establishment of, and adherence to, good chip design rules, to achieve high yields in VLSI package assembly, are absolutely essential. Rules must be generated for the particular package type used and must be compatible with the assembly equipment to be used. These decisions should be made early, preferably before the chip layout is started and definitely before the layout is completed. As the I/O count grows and the active VLSI device-size shrinks, the space required for interconnection could represent a major fraction of the chip area. To avoid this problem, the effective bonding-pad sizes and spacings (also called pitch) should be reduced. Figure 7 shows the consequences of increasing I/O on the bonding-pad pitch for several chip sizes.[10] Not only must the pitch decrease, but the tolerances required to produce quality bonds must also decrease. Reliability and the ability to assemble VLSI chips automatically is affected directly by the chip layout. In the case of VLSI, where chip costs range from $5 to $50, the chip-to-package interface design must assure high assembly yields and avoid reliability problems associated with poor design rules.

An alternative to smaller bonding pads and pitch is the staggered-row bond

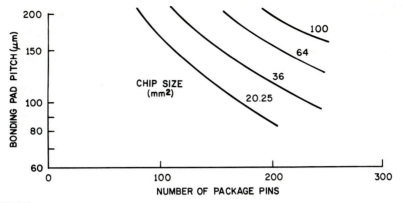

FIGURE 7
Bonding-pad pitch versus chip lead count for several chip sizes. For each number of package pins, the pitch is the maximum that can be accommodated on each chip with minimum increase of silicon area. (*After Otsuka and Usami, Ref. 10.*)

pads[11] shown in Fig. 8. Also illustrated in Fig. 8 are typical design rules for wire bonding.[11] Design rules cover recommended wire bond-pad sizes and pitch, clearances between pads and the edge of the separated chip, and clearances to internal adjacent metal conductors of other critical design features. Also specified is the maximum allowable angle of the wire from the normal direction off the chip. This rule attempts to minimize wire sweep in plastic molding, which could lead to potential edge shorts to silicon and shorts to adjacent conductors on the leadframes or metallized conductors of the package. Design rules[11] for wire bond spans (from the die bond pad to the substrate or to the leadframe bond pad) are shown in Table 1.

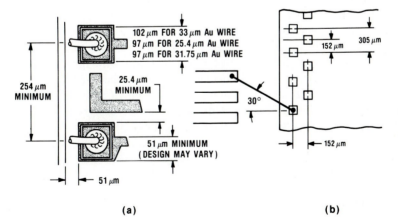

(a)   **(b)**

FIGURE 8
An arrangement of staggered bonding pads that results in a lower pitch than that obtainable with a single line of bonding pads. (a) Bonding pads size and spacing. (b) Maximum wire angle with respect to die edge. (*After Braden, Ref. 11.*)

TABLE 1

Au wire diameter versus recommended wire length for two package types [11]

Wire diameter (μm)	Package type	Minimum length (μm)	Maximum length (μm)
33	Plastic	1270	2540
33	Ceramic	1016	3175
25.4	Ceramic	1016	2540

Problem

How many bond pads placed on a 100 μm pitch can one place along a bond-pad locus that is 1000 μm long?

Solution

Let N equal the number of bond pads placed over the 1000 μm locus.
Therefore, we have (N − 1) spaces between the N bonding pads, each being 100 μm in length (pitch).
We can set up the following equation:

$$(\text{number spaces})\,(\text{pitch}) = (\text{length of locus})$$

or

$$(N-1)(100) = 1000$$

and

$$N = \frac{1000}{100} + 1 = 11$$

Design rules can be integrated into die design CAD systems as shown in Figs. 9, 10, and 11. Figure 9 is a typical plastic package wire-bond design template that shows the locations of all leadframe wedge-bond targets and the optimized location for the ball bond on the chip. The chip designer superimposes or merges this template with the proposed chip layout and can immediately see where bonding pads should be placed to achieve a manufacturable and reliable design. Figure 9 also illustrates two types of CAD wire-span templates used to verify design rules and to help decide cases where bond-pad placement falls outside the optimum zone. These templates are shown in more detail in Figs. 10 and 11. The use of a CAD system by the device designer assures the best possible design, particularly for high-volume, low-cost designs. In the case of high-performance custom designs, the same design rules apply. CAD tools may be too restrictive; hence, the chip and package designers must work together as a team to assure a manufacturable design.

13.3.2 Thermal Design Considerations

The objective of thermal design is to keep the operating junction temperature of a silicon die low enough to prevent the failure rate due to temperature-activated fail-

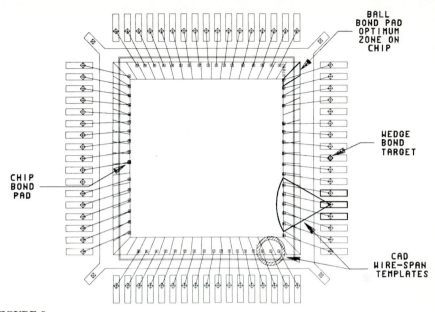

FIGURE 9
CAD template for positioning bonding pads on the die in their optimum location to achieve high
assembly yields and high reliability performance. Use of the template assures the device designer
that wire-span lengths meet the design rules (*Courtesy of AT&T Bell Laboratories.*)

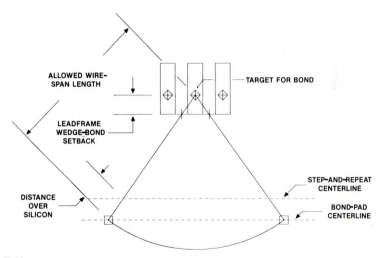

FIGURE 10
CAD template for checking adherence to wire-span guidelines. The template also provides an
extended zone (beyond the optimum shown in Fig. 9) for those cases where location in the optimum
zone is not compatible with the device layout. The maximum distance to the step-and-repeat centerline is controlled to minimize wire shorts to the die edge (*Courtesy of AT&T Bell Laboratories.*)

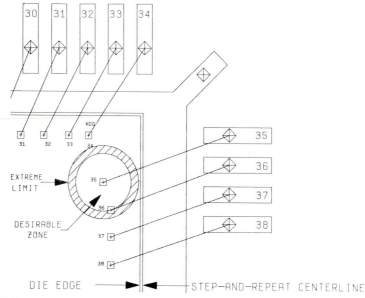

FIGURE 11
CAD template for checking the maximum distance that a wire spans over silicon, to minimize edge shorting. This example shows a violation of the guidelines. A correction would be to move the bonding pad to the right until the circle template became tangent to the step-and-repeat centerline (*Courtesy of AT&T Bell Laboratories.*)

ure mechanisms from exceeding the acceptable limit for a particular application. Only in the simplest possible applications can this objective be met by considering the packaged silicon device alone. Usually the packaged-device environment must be established for most of the following variables: PWB temperature, total power dissipation on the board, local-neighbor power dissipation, degree of forced-air cooling, space usable by the package both laterally and vertically (between boards), conductivity of the PWB, and ideal performance of the isolated package. The following paragraphs will discuss the thermal modeling of a single chip in a package.

In a simplistic heat-transfer model of a packaged die, the heat is transferred from the die to the surface of the package by conduction and from the package surface to the ambient by convection and radiation. Typically, in most applications, the temperature difference between the case* and ambient is small and radiation can be neglected. This model also neglects conduction heat transfer through the package terminals, which can be significant, particularly in high-I/O VLSI packages. The overall thermal resistance in this model can be considered as the sum of two thermal resistance components θ_{jc} and θ_{ca} and defined as

---

Case and *package* are synonymous terms when discussing heat-transfer models.

$$\theta_{ja} = \theta_{jc} + \theta_{ca}(°C/Watt) \tag{2}$$

$$\theta_{jc} = [T_j - T_c]/P \tag{3}$$

$$\theta_{ca} = [T_c - T_a]/P \tag{4}$$

where

θ_{ja} = Junction-to-Ambient Thermal Resistance

θ_{jc} = Junction-to-Case Thermal Resistance

θ_{ca} = Case-to-Ambient Thermal Resistance

T_j = Average Die or Junction Temperature (°C)

T_c = Average Case Temperature (°C)

T_a = Ambient Temperature (°C)

P = Power (Watts)

Problem
What is the junction-to-ambient thermal resistance for a device dissipating 550 mW into an ambient of 70°C and operating at a junction temperature of 125°C?

Solution
Use Eq. 2,

$$\theta_{ja} = \theta_{jc} + \theta_{ca}$$

Substitute Eq. 3 and Eq. 4 into Eq. 2.

$$\theta_{ja} = \frac{Tj - Tc}{P} + \frac{Tc - Ta}{P}$$

Combine terms

$$\theta_{ja} = \frac{Tj - Tc + Tc - Ta}{P} = \frac{Tj - Ta}{P}$$

Substitute known values

$$\theta_{ja} = \frac{125 - 70}{0.550} = 100°C/W$$

θ_{jc}, the conductive thermal resistance for low-power applications, is relatively insensitive to the ambient and is mainly a function of package materials and geometry. With the higher power requirements of most VLSI applications, one must also consider the temperature dependence of the materials selected in the design. For example, the thermal conductivity of common ceramic materials used in packaging is very temperature dependent as shown[12] in Fig. 12.

θ_{ca} depends on package geometry, package orientation in the application, and the conditions of the ambient in the operating environment (free- or forced-

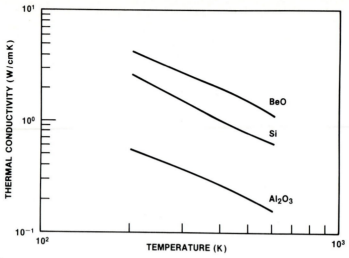

FIGURE 12
Influence of temperature on the thermal conductivity of integrated-circuit-related materials. (*After Oettinger, Ref. 12.*)

convection heat transfer). In actual practice, most designers use experimental methods to determine the thermal resistance of a package configuration. Thermal characterizations are done by mounting the package into one of the mounting configurations shown in Fig. 13 and monitoring both case and ambient temperatures. Results show that packaged devices with a preferred heat-flow path (most ceramic packages) may be thermally characterized independent of external heat-sinking,

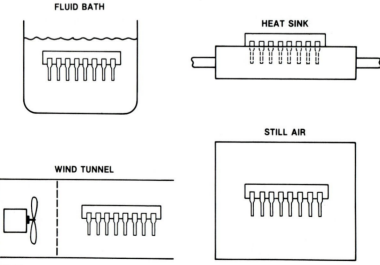

FIGURE 13
Typical package-mounting arrangements for thermal characterization. (*After Oettinger, Ref. 12.*)

while those with no preferred heat-flow path (most plastic packages) should be thermally characterized with a specific mounting arrangement and a convection heat-flow environment.[13,14] These results show that the use of simple modeling, supported by experimentally derived values of thermal resistance, is probably adequate for first-order determinations but should be followed up by a more sophisticated program using specially designed test chips and computer modeling.

13.3.3 Electrical Considerations

Electrical performance at the IC-package level has only recently been of general interest for silicon devices.[15] However, with the increasing speed of today's circuits and their potentially reduced noise margins, package design must be considered more carefully. Several electrical performance criteria are of interest, namely: low ground resistance (minimum power-supply voltage drop), short signal leads (minimum self-inductance), minimum power-supply spiking due to signal lines simultaneously switching, short paralleled-signal runs (minimum mutual inductance and cross talk), short-length signal runs near a ground plane (minimum capacitive loading), and the maximum use of matched impedances to avoid signal reflection. These criteria are, of course, not all mutually independent. They may be related through simple geometric variables, such as conductor cross section and length, dielectric thickness, and dielectric constant of the packaging body. These problems are usually handled with transmission-line theory in printed circuit boards, where the lengths of circuit paths are more obviously a cause for concern. A wealth of papers and books covers this topic, and all of the techniques described are generally applicable at the package level.[16-18]

Lead inductances on the new SM packages as a function of package size (I/O) are significantly lower than their TH counterparts, as shown in Fig. 14.

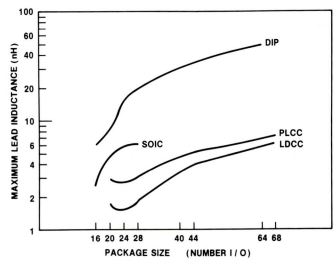

FIGURE 14
Lead inductances for various DIP, SOIC, PLCC, and LDCC package sizes. (*After Korf, Ref. 3.*)

This is due to the inherently shorter leads on leaded SM packages, no leads on leadless SM packages, and to shorter lead traces within the package because of the smaller size of SM packages compared to the I/O-equivalent TH package.

The most important practical electrical design problem in IC packages is noise reduction. Basically, when a line switches, the voltage induced in the ground line is given by

$$V_i = L_g di/dt \tag{5}$$

where V_i is the induced voltage, L_g is the inductance of the ground lead, and di/dt is the derivative of current with respect to time. If j lines are switching, then V is given by

$$V_i = L_g \sum_j di_j/dt \tag{6}$$

To reduce V_i, multiple grounds must be used to lower L_g. If m ground leads are used, the total inductance is approximately L_g/m. The practical result is that up to 25% of the leads may have to be grounded to control noise. This high percentage has a large impact on packaging density and creates an incentive to reduce L_g. Inductance can be reduced significantly through the use of large-area power and ground planes within the package.[19]

13.3.4   Mechanical Design Considerations

As the size of a VLSI die increases, it presents new challenges to the package designer. Ideally, one would prefer to use materials in package construction that are matched in physical properties, in particular the TCE. In the real world, however, we see that the materials currently in use have TCE that vary by orders of magnitude, as shown[20] in Fig. 15. In the case of SSI and MSI,

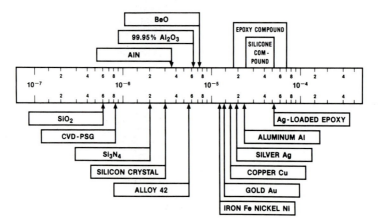

FIGURE 15
Thermal coefficients of expansion (TCE) of materials for semiconductor devices, $(^{\circ}C)^{-1}$. (*After Kakei, Ikeda, and Koshibe, Ref. 20.*)

the designer could achieve better matches by making engineering tradeoffs that did not degrade the packaged-device yields, performance, or reliability. For example, in SSI plastic molded packages the die (silicon) is attached directly to the lead frame (typically Alloy 42). Alloy 42 has poor thermal conductivity but closely matches the TCE of silicon. The use of Alloy 42 minimizes thermal stresses and has negligible impact on the thermal design because of the low power levels encountered in SSI. In VLSI, however, the use of a die with large Alloy 42 leadframe could bring on higher thermal stresses, resulting in die cracking and/or "popoff," or in degradation of performance and reliability due to high junction temperatures during normal operating conditions.[21]

In the previous discussion, only base materials at the die-attach interface were considered. In the actual design, the die is attached using materials such as solder, brazing alloy, or polymer adhesive, and using a process step introducing heat to melt solders (or brazing alloy) or to thermally cure the polymer adhesives. Physical properties of these materials are shown[22-24] in Table 2. The die-attach bond line, therefore, is a third material in the interface with its characteristic

TABLE 2
Elastic moduli and thermal conductivities of materials used in packaging[22-24]

Material	Elastic modulus E (GPa)	Thermal conductivity k (W/cm-$°$C)	Application
Alumina (Al_2O_3)	262	0.17	Substrate
Beryllia (BeO)	345	2.18	Substrate
Common Cu Alloys	119	2.64	Leadframe
Ni–Fe Alloys (Alloy 42)	147	0.15	Leadframe
Au–20%Sn	59.2	0.57	Die-bond adhesive and lid sealant
Au–3%Si	83.0	0.27	Die-bond adhesive
Pb–5%Sn	7.4	0.63	Flip-chip
Silicon	13.03	1.47	Electronic circuit
Au	78	3.45	Wire metallurgy
Ag-loaded epoxy	3.5	0.008	Die-bond adhesive
Epoxy (fused silica filler)	13.8	0.007	Molding compound

TCE, and with another physical property, stiffness, characterized by the tensile modulus (E), which significantly affects the thermal stress in the die-attach interface, as shown in Fig. 16. In addition, the thickness of the joint also directly impacts the thermal stress in the interface, as shown in Fig. 17. Thermal stress increases with increasing E and decreasing thickness. Eutectic die bonding has very high E and a very thin bond line, hence limiting its usefulness in VLSI applications to packages where the attachment is made to a substrate with a closely matched TCE.

Another major mismatch in the TCE of materials used in package construction is between the molding compound and die in a plastic package. This mismatch does not exist in ceramic hermetic high-performance VLSI, since no package materials come into direct contact with the IC surface. The only exception is when a die coat[25] is used on the device surface where the coating, either room temperature vulcanized (RTV) silicone rubber or polyimide, has a high TCE but compensates by being relatively flexible. The die itself is constructed of materials with over an order of magnitude difference in TCE, which compounds the thermal-stress problem in plastic packages.

Analysts have developed sophisticated stress models and introduced computer aids such as finite-element analysis (FEA), to increase their capability to characterize the thermal stresses and deformations throughout the entire plastic package, with the goal of optimizing the reliability of the package.[26] The plastic-encapsulated IC package is basically a composite structure consisting of silicon die, metal lead frame, and plastic molding compound, as shown in Fig. 18. A two-dimensional finite-element model of the cross section is shown in Fig. 19.

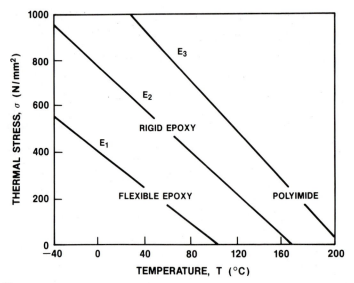

FIGURE 16
Thermal stress predicted for flexible epoxy (tensile modulus E_1), rigid epoxy (E_2), and polyimide (E_3), as a function of temperature where $E_3 > E_2 > E_1$. (*After Bolger and Mooney, Ref. 21.*)

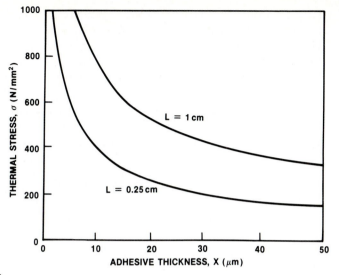

FIGURE 17
Effect of die length and adhesive thickness on thermal stress.(*After Bolger and Mooney, Ref. 21.*)

The model is divided into small elements interconnected at a discrete number of nodal points, creating a finite-element mesh. Material properties are assumed to be linearly related to the thermal load. Material interfaces are assumed to have perfect adhesion to each other. The model further assumes an isothermal initial condition of 170°C corresponding to the molding temperature. The thermal stresses are then calculated at the desired isothermal temperature, which is typically set at −65°C,

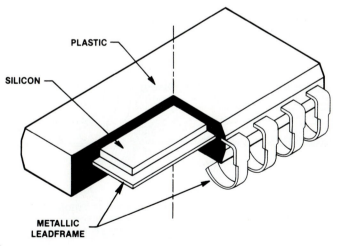

FIGURE 18
The plastic package is basically a composite structure consisting of silicon chip, metal leadframe, and plastic molding compound (*Courtesy of AT&T Bell Laboratories.*)

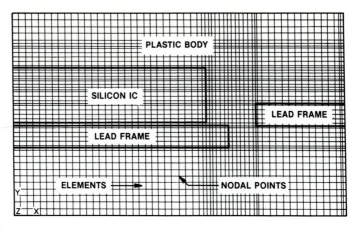

FIGURE 19
Plane stress analysis is assumed for the two-dimensional model. The model is divided into a large number of small elements interconnected at a discrete number of nodal points. Because of symmetry the model covers only half the IC package (*Courtesy of AT&T Bell Laboratories.*)

the lowest temperature expected during service or in transit. Figure 20 is a plot of the minimum principle stresses (compression) in both the plastic molded body and the silicon die. The stress contours represent isobars of equal stress within the package and die. Severe stress concentration is evident at the lower edge of the die and could lead to die cracking.

The use of finite-element analysis has become a powerful analytical tool for quickly predicting the impact of design and process changes on stress-related failure mechanisms. Such analyses were not possible with classical analytical techniques.

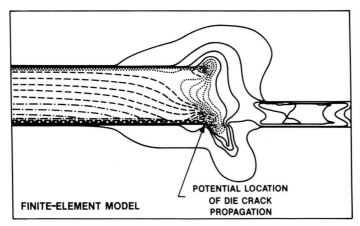

FIGURE 20
FEA showing contours of constant stress (minimum principal stress in compression) both in the plastic body and in the silicon chip (*Courtesy of AT&T Bell Laboratories.*)

13.4   VLSI ASSEMBLY TECHNOLOGIES

This section covers the basic assembly operations in use today for VLSI devices. A generic assembly flow chart applying equally to plastic or ceramic packages is shown in Fig. 21. In this section we will consider the steps from wafer preparation through wire bonding. Section 13.5 will deal with the various package technologies in use today.

13.4.1   Wafer Preparation

Significant advancements have been made in the area of wafer backgrinding, diamond-blade dicing saw technology, and wafer handling systems.[27-29] Larger-diameter wafers (hence thicker wafers), denser circuitry, narrower streets between circuits, and the placement of test patterns in streets have driven the development of these advancements. Thicker wafers place more demand on the separation process and on VLSI package design and assembly processes. Thermal stresses resulting from TCE mismatches between a large VLSI die and plastic molding materials (described in the last section) are very dependent on the thickness of the die, particularly in the small SM types. SM overall package dimensions are significantly smaller than their DIP counterparts, due to the finer pitch (1.27 mm) of the leads needed to achieve higher densities at the PWB level. The package thickness, in addition, is bound by the need for lower profiles, which are needed to achieve closer spacing between PWBs. To avoid these problems, wafers are thinned down using highly automated backgrinding processes.

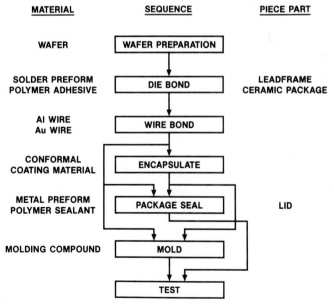

FIGURE 21
Generic assembly sequence for plastic and ceramic packages.

The once-traditional scribe-and-break method is now obsolete and has been replaced by diamond-wheel dicing. Advances in diamond blade technology[28] have led to better-quality cuts, allowing the narrowing of streets and improved capability for cutting through dissimilar materials such as passivation, metallization, and gold, which is used on the back of the wafer to optimize electrical and thermal performance. Wafers are, typically, adhesive-mounted to a tape that has been preassembled to a frame.[29] The use of tape ensures that the die is held in place after 100% saw-through and that die damage does not occur during subsequent transport to die bonding or storage. Wafer dicing machines are completely automatic and contain features such as automatic alignment systems, an integral wafer cleaning station, and quality monitors.[29] Die sorting is typically done in situ on an automatic die-bonding machine. The sawed-apart wafer still mounted in the tape-frame fixture is loaded into an automatic die bonder that picks the good chips from the array and places them into the packages or onto leadframes. This topic will be discussed in the next section.

13.4.2 Die Interconnection

Die interconnection consists of two steps. In the first step, the back of the die is mechanically attached to an appropriate mount medium, such as a ceramic substrate, multilayer-ceramic-package-piece part, or metal leadframe. This attachment sometimes enables electrical connections to be made to the back of the die. The two common die-bonding methods are hard solders (or eutectic) and polymers. In the second step, the bond pads on the circuit side of the die are electrically interconnected to the package. The three common schemes of interconnection to the chip bond pads are wire bonding, TAB, and flip-chip solder bonding, also called controlled-collapse bonding (CCB). All three processes are extensively used for SSI and MSI packaging. Wire bonding is further split into different processes, such as thermosonic and thermocompression ball-and-wedge using Au wire, and ultrasonic wedge–wedge using both Au and Al wires. All of these processes are directly applicable to the VLSI die. The most popular bonding process in general use for VLSI die up to 132 I/Os is the thermosonic ball-to-wedge process. This process will be described in some detail in Section 13.4.4. TAB and area-array solder interconnections are also in use today, but will not be described.[30,31]

13.4.3 Die Bonding

Die size is steadily growing, in proportion to the emergence of improved VLSI technology capabilities. Increased die size places more stringent demands on the processing and reliability of die bonding.[32] Die-bonding technology for a large VLSI die (greater than 5 mm on a side) requires special attention to thermal and stress management as discussed in Sections 13.3.2 and 13.3.4. In addition, similar care must be exercised in the selection of a die-bond process for the particular application and in the implementation of stringent process controls to assure high-quality bonds. VLSI die size also impacts wafer fabrication guidelines and assembly automation. These issues influence equipment selection and determine how easily a controllable process can be achieved.

The major choices for a die-attach process include hard solder (or eutectic), polymers (epoxies and polyimides), and Ag-filled glasses. A comparison of the die-attach processes for hermetic ceramic and plastic packages shows two emerging technologies for VLSI, eutectic for hermetic ceramic and polymers for plastic.[11] The eutectic process is essentially contamination-free, has excellent shear strength, and assures low-moisture packages.[33] The major disadvantages are that it is difficult to automate and that it places high thermal stresses on the die due to the high process temperature. New ceramic substrate materials such as AlN, whose TCE closely matches that of silicon, make this process more viable for large VLSI die in hermetic ceramic packages.[6] The polymer die-attach process has two materials that are contenders for VLSI in plastic. Epoxies (with and without Ag flakes for electrical and thermal conduction) have the advantage of being easy to automate, and require low-temperature curing which minimizes thermal stress in large die. Polyimides (with and without Ag) are also easy to automate. They require higher curing temperatures, leading to high thermal stress, especially when used with high TCE copper leadframes. Copper is the preferred material for handling the power dissipation requirements of VLSI. On the other hand, epoxies are higher in hydrolyzable ions that could have a direct impact on reliability. Epoxy suppliers have improved their materials to remove the major objections and these materials are being widely used today for VLSI. Polyimides are difficult to process without voids, which affect thermal performance and have contributed to die cracking.[32] The use of polyimides should be limited to packages using bonding media with a closely matched TCE, such as Alloy 42 leadframes and ceramic substrates. The following sections will describe the eutectic and epoxy die-attach processes as they are being used today in VLSI assembly.

Eutectic die bonding. Figure 22 illustrates the fundamental aspects of a die bond. Eutectic die bonding metallurgically attaches the die to a substrate material (typically a metal leadframe made of Alloy 42) or to a ceramic substrate (usually, 90 to 99.5% Al_2O_3). Metallization is often required on the back of the die to make it wettable by the die-bonding preform, which is a thin sheet (usually

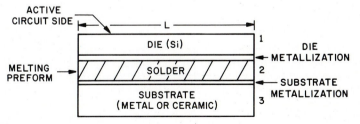

FIGURE 22
The basic structure of a silicon device die bonded with a metal preform.

< 0.05 mm) of the appropriate solder-bonding alloy.[32] The substrate material is usually metallized with plated Ag (leadframes) or Au (leadframes or ceramic). Table 3 lists the compositions and melting points for solder-preform materials.[34]

Solder die bonding to ceramic packages, which are to be hermetically sealed, is usually performed with a Au or Au–2%-Si preform. The TCE mismatch between silicon and the ceramic substrate is relatively small, so that thermal stress due to a TCE mismatch is not an issue. In the presence of some mechanical scrubbing and temperature of about 370°C, the preform reacts to dissolve the silicon. The Au–Si eutectic composition is reached and then exceeded, because the bonder temperature is typically set to more than 400°C. As the composition of the composite structure becomes more silicon-rich, it freezes and the die bond is completed. Many applications require lower temperatures or more ductile die-bonding solders, to be compatible with other substrate materials or process steps.

Epoxy die bonding. Silver-filled epoxy adhesives are of major interest today as die-bond materials. The use of silver fillers (typically flakes) makes the epoxy both electrically conductive, to provide low resistance between the die and the substrate, and thermally conductive, to allow a good thermal path between the die and the rest of the package.

Epoxy die-bond adhesives are preferred for VLSI for the many reasons discussed earlier. In addition to being less expensive than the high-gold-content hard solders, they are more flexible and lend themselves to automation. For example, in the manufacture of plastic DIPs, epoxy can be transfer-printed onto die paddles (pads) at very high rates, and the die can be placed with high-speed pick-and-place tools. The process is schematically shown in Fig. 23. Automatic die bonders have the capability of handling a wide range of die sizes, picking up

TABLE 3
Compositions and melting points for die-attach preforms[34]

Composition	Temperature (°C)	
	Liquidus	Solidus
80% Au, 20% Sn	280	280
92.5% Pb, 2.5% Ag, 5% In	300	—
97.5% Pb, 1.5% Ag, 1% Sn	309	309
95% Pb, 5% Sn	314	310
88% Au, 12% Ge	356	356
98% Au, 2% Si	800	370
100% Au	1063	1063

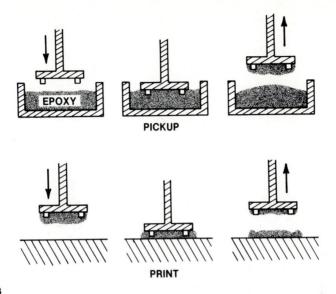

FIGURE 23
In transfer printing, a special tool picks up epoxy from a preleveled pot and prints a pattern on to
the die-attach surface. Dimples in the transfer tool control the thickness of the transferred epoxy.

either from expanded wafer presentation in tape frames, or from waffle packs
(die sorted off-stream from the die bonder, either manually or automatically).[35]
Accurate die placement is attainable and affects automatic wire bonding by yield-
ing greater consistency of wire lengths and improved looping characteristics. In
addition, die-placement accuracy enhances the wire bonder's pattern-recognition
performance and efficiency. Die collets are preferred over surface pickup to minimize damage to the fine features on the VLSI die circuitry. Because epoxies are
thermosetting polymers (cross-link when heated), they must be cured at elevated
temperatures to complete the die bond. Typical cure temperatures range from 125
to 175°C.

In general, epoxy die bonds are as good as or better than their metal
counterparts, except in the most demanding applications; that is, applications that
require high temperatures, high current through the die bond, critical thermal-
performance requirements, or those with very surface-sensitive devices bonded
in a ceramic hermetic package.

13.4.4 Wire Bonding

Wire bonding is always performed on ICs after they have been die-bonded to
the appropriate piece part, be it a leadframe paddle or the cavity of a ceramic
package or substrate. Typically, gold wire is ball–wedge bonded (thermosonic
or thermocompression); that is, ball-bonded to the chip bond pad (typically
aluminum) and wedge-bonded to the package substrate (typically Au or Ag),
as shown[36] in Fig. 24. One advantage of ball–wedge bonding comes from the
symmetrical geometry of the capillary tip. The ball bond is formed by the inner

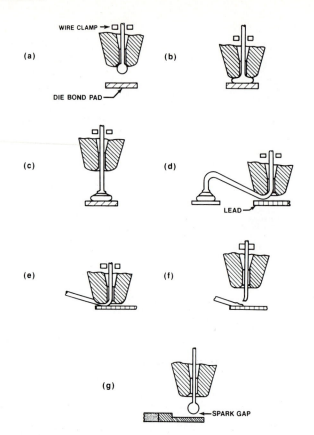

FIGURE 24

Tailless ball-and-wedge bonding cycle. (a) The capillary is targeted on the die's bond pad and positioned above the die with ball formed on the end of the wire and pressed against the face of the capillary. (b) The capillary descends, bringing the ball in contact with the die. The inside cone, or radius, grips the ball in forming the bond. In a thermosonic system, ultrasonic vibration is then applied. (c) After the ball is bonded to the die, the capillary rises to the loop-height position. The clamps are open and wire is free to feed out the end of the capillary. (d) The lead of the device is positioned under the capillary, which is then lowered to the lead. Wire is fed out the end of the capillary, forming a loop. (e) The capillary deforms the wire against the lead, producing a wedge-shaped bond which has a gradual transition into the wire. In a thermosonic machine, ultrasonic vibration is then applied. (f) The capillary rises off the lead, leaving the stitch bond. At a preset height, the clamps are closed while the capillary is still rising with the bonding head. This prevents the wire from feeding out the capillary, and pulls at the bond. The wire will break at the thinnest cross section of the bond. (g) A new ball is formed on the tail of wire which extends from the end of the capillary. A hydrogen flame or an electronic spark may be used to form the ball. The cycle is completed and ready for the next ball bond. (*After Kulicke and Soffa Industries, Ref. 36.*)

portion of the tip (Fig. 24b), and then the wedge bond can be performed anywhere on a 360° arc around the ball bond, using the outer portion of the tip (Figs. 24d and 24e). This capability to dress the wire in any direction from the ball is the key factor that makes this process attractive for high-speed automated bonding; that is, the bonding head or package table does not have to rotate to form the wedge bond. The versatility of ball–wedge bonding, requiring only *X*-, *Y*-, and

Z-axis head motion, leads to greater automation potential over those requiring Z head and X, Y, and θ table motion (Fig. 25).[37] Figure 26 shows a high-I/O VLSI device that is successfully ball-and-wedge bonded on today's state-of-the-art automatic wire bonders running in excess of 6 wires per second.

The major issue in establishing a quality and reliable wire-bonding process is process control. Two types of process tests are in use today. These are the wire-pull test[38] and the ball-shear test.[39] Wire-bond pull tests have been around a long time for evaluating both ball and wedge bonds. A typical setup is shown in Fig. 27a. The test is done by engaging the wire with a hook near the center of the wire span and pulling to destruction. The magnitude of the pull force at failure and the modes of failure (Fig. 28) are noted. The magnitudes of tension force in the wire (Fig. 27b) at break can be expressed as

$$T_1 = \frac{F[(ad)^2 + h^2]^{1/2}[(1-a)\cos\alpha + (\frac{h+H}{d})\sin\alpha]}{h + aH} \tag{7}$$

$$T_2 = \frac{F[1 + \frac{(1-a)^2 d^2}{(h+H)^2}]^{1/2}(h+H)(a\cos\alpha - \frac{h}{d}\sin\alpha)}{h + aH} \tag{8}$$

where

T_1 and T_2 = Tension in wire at the ball (1) and wedge (2)

a, d, h, H, α = Defined in Fig. 27

F = Pull force at wire break

If both bonds are on the same level ($H = 0$) and the loop is pulled in the

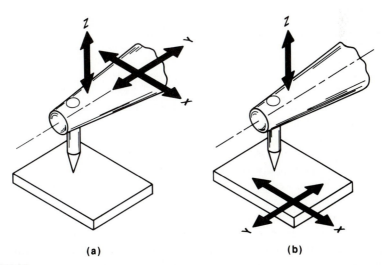

(a)   (b)

FIGURE 25
Comparison between head motions of a ball-to-wedge bonder (a) and a wedge-to-wedge bonder (b). (*After Nachon and Tai, Ref. 37.*)

FIGURE 26
Thermosonic ball wire bonds on a 72 I/O gate-array VLSI chip (*Courtesy of AT&T Bell Laboratories.*)

center ($a = 0.5$) and perpendicular to the substrate ($\alpha = 0$), the following familiar form is obtained

$$T = (F/2)[1 + (d/2h)^2]^{1/2} \tag{9}$$

where

$$T = \text{Tensile force in the wire}$$

In this simplified form, both ball- and wedge-bonds are under equal loads. However, the magnitude of the force actually transmitted is very dependent on the d/h ratio. This makes it difficult to judge the quality of the bond based on the magnitude of the pull force at failure. As a consequence, more emphasis was placed on failure-mode analysis (FMA) to determine the quality of the bonds (Fig. 28). FMA results showed that wire break above the ball was the major failure mode, and that the pull test was not really evaluating the quality of metallurgical bonds to the device bonding pad.

The ball-shear test has also been around a long time, but its use was very limited until recently. Today, after extensive development and evaluation, it has become the accepted method by many manufacturers for evaluating quality of ball bonds. The basic setup for shear testing is shown in Fig. 29. A shear load is applied to the ball bond as shown and loaded to destruction. The magnitude of

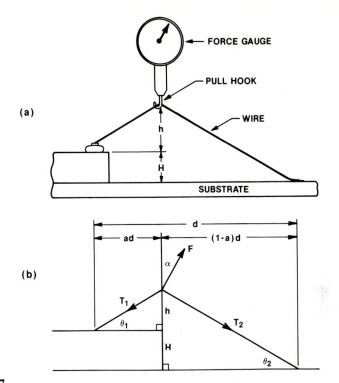

FIGURE 27
Setup and force-equilibrium diagram for the wire-pull test.

the force at failure and the failure mode (Fig. 30) are noted. A bonding process that is under control will exhibit high shear forces with a tight distribution, and, typically, fail in the ball bond. The test is also useful to study the effects of contamination on bondability (Fig. 31) and to characterize the reliability of Au–Al

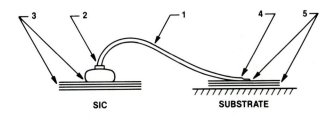

FIGURE 28
Failure modes in wire-bond pull tests.

Mode	Description
1	Break in the wire
2	Wire break above the ball.
3	The ball-bond failure location should be determined: gold-to-aluminum or in the SIC buried layers.
4	Break in wire at the wedge-bond heel.
5	Wedge-bond interface failure. The location determined: gold-to-gold or in the metallized layers.

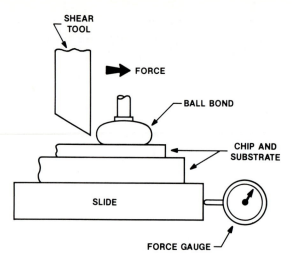

FIGURE 29
Setup of the shear test. Force is applied by the shear tool, and the force needed to shear is recorded on the force gauge.

intermetallic formation as a function of thermal aging.[40] An example of thermal-aging effects monitored by shear testing is shown in Fig. 32. In this plot, the difference in sensitivity of the two tests is observed where the pull-test data show no change over a time period where the shear force dropped by almost a factor of three.

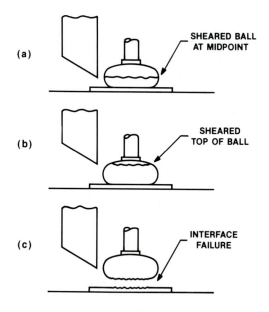

FIGURE 30
Failure modes found in ball bonds during shear testing. A good bond shears through the ball and exhibits very little interface failure. Low shear values can occur if the tool height is improperly set. FMA is essential in interpreting force values determined by the shear test.

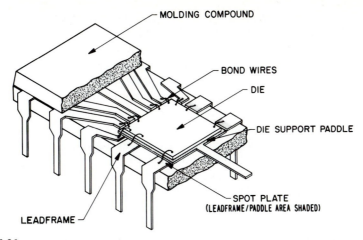

FIGURE 36
Ball-and-wedge-bonded silicon die in a plastic DIP. The die support paddle may be connected to one of the external leads. For most commercial products, only the die paddle and the wedge-bond pads are selectively plated (Au or Ag). The external leads are solder-plated or -dipped after package molding. (*After Howell, Ref. 42.*)

The two basic types of molded devices are postmolded and premolded devices. The postmolded package is the lower-cost of the two and is the predominant package technology in use today. This technology uses thermosetting (cross-linking) epoxy resins and is molded around the leadframe–chip subassembly after the chip is wire-bonded to the leadframe. The postmolding process is relatively harsh. To avoid the exposure of the die and its wire bonds to viscous molding material, the premolded-package concept was developed. Here the package is molded first, and then the chip and interconnects are added. The molding may be performed with the thermoset mentioned above or with thermoplastic (melting) polymers, such as polyphenylene sulfide. The premolded package is the plastic equivalent of the refractory-ceramic-cavity package and has great future potential for VLSI devices. Only the postmolded device and its material, and construction will be discussed here in detail.

The polymers most commonly used for IC packaging are epoxies. The original epoxy resin, used to mold ICs, was made by condensing epichlorohydrin with bisphenol-A to produce a material called Epoxy-A. An excess of epichlorohydrin was used to leave epoxy groups on each end of the low-molecular-weight polymer.

Today, NOVOLAC epoxies are generally preferred because their higher functionality gives them heat resistance, derived from having each repeating group contain an epoxy group. These resins, called Epoxy-B, are made by reacting epichlorohydrin with NOVOLAC phenolic resin and a base. The synthesis of the resins produces sodium chloride as a by-product. Both sodium and chloride ions are deleterious to device reliability, and, therefore, these by-products must be carefully washed from the resins before they are compounded into molding compounds.

Fillers (usually amorphous or crystalline SiO_2 or glass fibers, and sometimes Al_2O_3) are added to the resin so that the resultant mixture is 65 to 73% filler, by weight. Amorphous SiO_2 is used when a minimum expansion coefficient is desired at some sacrifice of thermal conductivity, while crystalline SiO_2 improves thermal conductivity at the expense of the TCE. Al_2O_3 has high thermal conductivity (as a filler material) but is very abrasive to molds. Fillers greatly improve the mechanical strength of the resin, reduce its TCE, and, therefore, reduce its shrinkage after molding. Small amounts of pigments, mold release agents, antioxidants, water getters, plasticizers, and flame retarders must also be added to complete and optimize the molding resin.

The rheological, chemical, and thermophysical properties of molding compounds are important to understand, both for the molding process and for their interrelation with the reliability of the finished device. The two most important characteristics that must be evaluated for a molding compound are its moldability and expansion coefficient. The molding process is not cost-effective unless it has high yield, and the most common reliability problems can be related to the mechanical quality of the molded part and its ability to withstand thermal stresses, as discussed in Section 13.3.4. Epoxy molding compound vendors have made significant changes over the past five years to meet the challenges of VLSI devices.[20] The primary driving force has been memory-chip sizes where both the 256K and 1M DRAM have established new criteria for mold-compound properties.

13.5.4   Molding Process

Thermoset molding materials are usually transfer-molded in large multicavity molds (Fig. 37). After entering the pot, the preheated molding compound, under

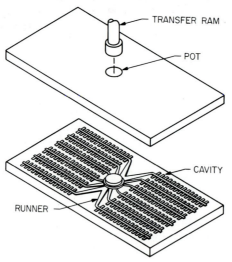

FIGURE 37
Schematic of a 168 cavity transfer mold showing ram, pot, and runner system.

pressure and heat, melts and flows to fill the mold cavities containing leadframe strips with their attached ICs. The IC leadframe often has long, fragile lead fingers, and the die is interconnected to these thin, fragile leads with 25 to 30 μm diameter Au wire. To avoid damaging this fragile structure, the viscosity and velocity of the molding compound must fall within certain ranges. Commercial molding compounds are designed to meet these requirements when molded at approximately 175°C and at pressures of about 6 MPa. The mold cycle requires from 1 to 5 minutes. Figure 38 shows an 80 cavity mold used for a 68 I/O plastic chip carrier.

To control the velocity of the molten molding compound, each device cavity has a gate or restriction to slow the material flow. The fluid mechanics of molding are relatively complex, because the materials are non-Newtonian. In addition, partial cross-linking can occur during the molding process, affecting the material viscosity. Elaborate mold designs have been proposed to compensate for these variables. An automated multipot approach as shown in Fig. 39 can also provide improved control of these molding variables.

13.5.5 Plastic Composite Package

No one has yet been able to produce a postmolded or premolded PGA package as a viable alternative to expensive multilayer ceramic PGA packages. However, over the past few years the industry has developed a low-cost plastic PGA using the existing PWB hybrid technology. These packages offer significant cost

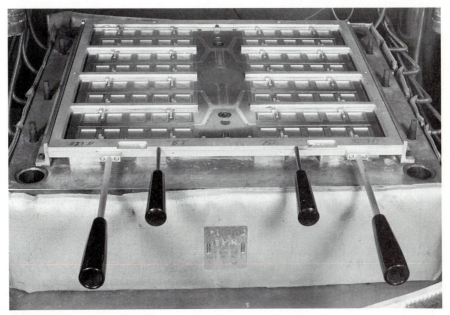

FIGURE 38
80 cavity transfer mold for a 68 I/O plastic chip carrier. Each lead frame accommodates 5 chips, and the cavity gate is located in one of the corners of the package. (*Courtesy of AT&T Bell Laboratories.*)

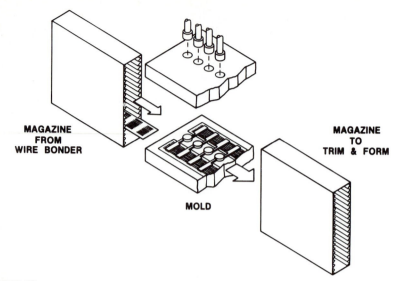

FIGURE 39
Schematic of a two-strip multipot automatic molding system. Leadframes are loaded and unloaded directly from and into magazines that are compatible with assembly equipment at locations other than the molding operations.

advantages over ceramic packages and can provide higher electrical conductivity on metal conductors, thus permitting more efficient designs and elimination of costly multilayer constructions.[43,44]

13.5.6 Special Package Considerations

This section looks at other aspects of packaging that must be addressed to have a complete package-assembly methodology. Among them are cleaning, package or lid sealing, encapsulation, α-particle protection, assembly environment, and miscellaneous assembly operations.

Cleaning. The most critical cleaning step the die undergoes is performed before bonding, encapsulation, or final lid seal. The cleaning process must be chemically compatible with the die metallurgy. Aluminum has a very narrow range of pH values in which its oxide protects it from corrosion in aqueous solution. Corrosion reactions, with metal dissolution, can occur in both basic and acidic solutions. Two objectives are usually sought in cleaning. One is the removal of organic species that can affect bondability; an organic solvent is usually required for this. The second is the removal of ionic species that can cause corrosion during the life of the device, or in an unusual instance, contribute to surface-charge accumulation. Water is a very good solvent for ionic species. Because environmental considerations have curtailed the use of powerful organic solvents such as trichloroethylene, the Freons are often used for organic and ion removal. Studies have been made on the efficacy of several combinations of water-, Freon

TA (contains 11% acetone), and oxygen-plasma cleaning, both with respect to the initial corrosion of aluminum and to the subsequent reliability of aluminum test devices in a humid environment.[45,46]

Although a preference for oxygen-plasma and water-rinse over a single Freon cleaning can be stated, a Freon cleaning alone is probably adequate to meet most reliability objectives for both hermetic- and plastic-packaged VLSI devices.

Ceramic package lid sealing. The major objective of package sealing is to protect the device from external contaminants during its lifetime. Further, any contaminants present before sealing must be removed to an acceptable level before or during sealing. Packages may be hermetically sealed with glass or metals, or lidded with polymers. The definition of hermeticity used here includes not only the ability to pass a fine-vacuum leak test, but also to exclude environmental contaminants for a long time period.

Figure 40 shows the relative capabilities of several materials to exclude moisture over long periods of time.[47] Clearly, organic sealants are not good candidates for hermetic packages. However, in some cases, organic sealants, properly integrated into the package design, do meet the operational definition of a hermetic seal given above.[48] For almost all high-reliability applications the hermetic seal is made with glass or metal. Glass sealing was mentioned previously for CERDIP and the process is essentially the same for lid sealing. Many of the metal alloys used for die bonding are suitable for lid sealing. A leak-tight metal

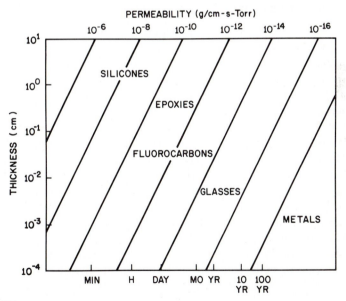

FIGURE 40
The calculated time for moisture to permeate various sealant materials (to 50% of the exterior humidity) in one defined geometry. Organics are orders of magnitude more permeable than materials typically used for hermetic seals. (*After Traeger, Ref. 47.*)

seal that excludes the external environment can be made without difficulty. The real difficulty has been freeing the package of contaminants, especially water, before sealing.[33,49]

Particular problems can arise in a device in a low operating temperature, where water can condense inside the package. The following reactions can then lead to corrosion in the presence of small mounts of halide contamination:

$$6HCl + 2Al \rightarrow 2AlCl_3 + 3H_2 \tag{10}$$

$$AlCl_3 + 3HOH \rightarrow Al(OH)_3 + 3HCl \tag{11}$$

$$2Al(OH)_3 + Aging \rightarrow Al_2O_3 + 3H_2O \tag{12}$$

Note that HCl is regenerated in these reactions until the aluminum is completely consumed. A relative humidity of 65% is apparently necessary to sustain these reactions. The level of moisture depends on the particular case, but no package should be sealed with greater than 6000 ppm water by volume (dew point = 0°C).[49]

The technology to measure such moisture levels has been thoroughly studied.[50] Yet, today, the military specification limit is still 5000 ppm by volume, mainly because of the technical difficulties in measuring moisture in small packages. Although 5000 ppm is the specification limit, the package assembler wants to achieve as low a moisture level as possible. A particularly effective method to achieve low moisture levels has been suggested.[33] This method makes use of the ability of atomic silicon to react with water to form SiO_2. The atomic silicon is formed during the lid seal, through the melting of the die-bond material or an added gettering preform. The proposed reaction is

$$Si + 2H_2O \rightarrow SiO_2 + 2H_2 \tag{13}$$

Encapsulation. Often the cost or difficulty of making a hermetic seal in a particular package type is prohibitive. In these cases, in addition to polymer lid sealing, surface die coatings can be used to protect aluminum metallurgy from atmospheric contaminants such as water. Silicones are very effective for this purpose.[51,52] A thin layer of silicone (0.25 mm) is not a diffusion barrier to water, based on the data in Fig. 40; the protection cannot be happening by simple exclusion of water. Qualitatively, the polymer must be preventing moisture condensation at the surface, and thus inhibiting leakage between adjacent metal lines. One proposed mechanism suggests that partially oriented siloxane dipoles at the substrate surface, working in concert with polar groups in the substrate, would sufficiently increase the work-of-adhesion of water molecules to prevent water from entering the interfacial (die surface to encapsulant) region.[53]

Silicone encapsulants are used by several electronic systems companies with excellent success for both aluminum and gold metallurgy devices. Still under study is a lower limit on line spacing, below which these materials are no longer effective.

α-Particle protection. Soft errors in memory circuits due to α-particles emanating from packaging materials were first reported[54] in 1978; there have been many papers on the subject since then. The α-particles are emitted by the decay of uranium and thorium atoms contained as impurities in the packaging materials.

Decreasing design rules of VLSI devices make them more sensitive to this problem.[20,54] Packaging materials (particularly ceramics) will probably not be pushed below the 0.001 to 0.01 α-particles/cm^2-h level. Because α-particles have low penetrating power in solids, low-α materials have been suggested as α-absorbing coatings on silicon die.[25,55] A 0.001 α-particle/cm^2-h emission rate, the lowest level anticipated, has been reported, using silicone coatings.[25] Although great progress can be made by reducing α-particle emission in current packaging materials, error correction is required to control soft errors caused by α-particles.

Assembly environment. Not much attention was given to the control of the environment in the assembly room for SSI/MSI devices. The ever-decreasing feature sizes on the VLSI die, down into the submicron range, makes environmental control a new issue that must be addressed. As shown in Fig. 41, the particle density for particles of one micron or greater in a typical room environment is astronomical. These particles, if not removed before molding in plastic packages, or lid sealing, in ceramic packages, can degrade the reliability of the packaged device by soft-error failures due to α emission from the particles, and hard failures due to metal corrosion from particles with high levels of contaminants.[56]

Miscellaneous assembly operations. There are a number of assembly operations that are seldom covered in technical publications or presented at technical conferences but give the packaging engineer trouble in day-to-day activities. Some

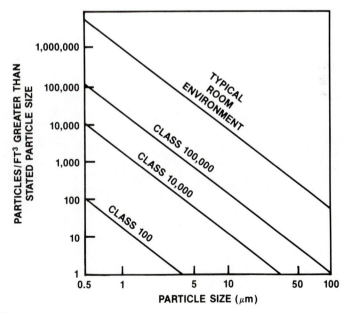

FIGURE 41
Particle-size distribution in a typical room atmosphere and in three classes of clean environments. (*After Lewis and Berg, Ref. 56.*)

of these have recently been addressed in the trade journals and will be mentioned here for completeness. In the fabrication of molded-plastic packages, a simple operation called trim and form has become a major issue when tooling for chip carriers with "J" leads.[57] The major problems have been in achieving and maintaining coplanarity of the leads to within 100 μm over the four sides of the quad-type package and providing sufficient compliance in the leads to achieve a reliable surface-mount solder joint to the next level of assembly on PWBs. Other aspects of assembly describe leak-testing methods,[58] component-marking systems,[59] lead scanning for quality,[60] handling of IC packages through electrical testing,[61] and new moisture-measurement techniques for hermetic packages.[62]

13.6   SUMMARY AND FUTURE TRENDS

Package pin counts will undoubtedly continue to increase with increasing silicon capability. Alumina ceramic packages will dominate the high-performance VLSI packaging technology until factors such as its high dielectric constant, modest thermal conductivity, or cost force a change to other contending materials. Higher packing density on the PWB level will drive the package design toward smaller lead pitches.[63,64] An example of a fine-pitch plastic-chip-carrier design that could be the predominant low-cost VLSI plastic package for the next decade is shown in Fig. 42.

If chip integration keeps pace with decreasing feature size (keeping the relative proportions of chip size and interconnection area), today's chip-interconnection technologies, especially wire bonding, will be challenged, and could be unusable[65] at I/O > 200. If wire bonding is to survive, it will need improvements in placement accuracy to meet demands of tighter bond-pad pitches.

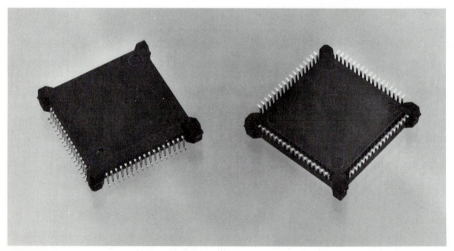

FIGURE 42
New design concept for fine-pitch (635 μm) high-density plastic chip carrier featuring gull-wing leads and protective bumpers (*Courtesy of AT&T Bell Laboratories.*)

Advancements in substrate metallization to finer pitches, and in lead-frame fabrication to permit features (wedge-bond fingers) narrower than the thickness of lead frames, will be needed to keep wire spans within reasonable limits as the chip sizes continue to decrease. Wire bonding of high-I/O die will have to be done in class-100 clean rooms, to optimize automatic wire-bonder performance and achieve acceptable bonding yields.[65] Some limited success has already been demonstrated in the laboratory, using the 320 I/O prototype PGA package shown in Fig. 43.

TAB and flip-chip will be revisited, to assess their future in VLSI packaging.[30,31,66] Both technologies lend themselves to hybrids, especially TAB with its pretestability features. Flip-chip has the advantage of a very-low-inductance connection to the package, and the best capability in terms of the density of interconnection (using an area array) to the device, compared to wire or TAB.

The design of systems will depend more and more on a systematic optimization of the entire interconnection scheme to achieve the potential benefits of improved silicon capability. This optimization may lead to completely new requirements for assembly and packaging, but more than likely will lead to a sorting-out of the various existing assembly technologies and packages.

A recent book written by Sinnadurai[67] is recommended as suggested reading for those individuals desiring a broader and more in-depth discussion on many of the topics covered in this chapter.

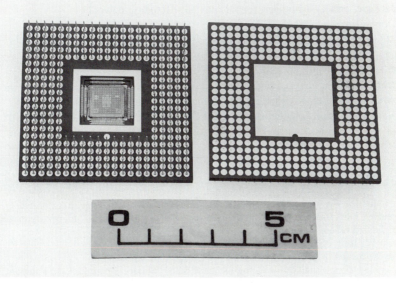

FIGURE 43
Multilayer ceramic PGA with a 320 I/O cavity on the pin side, two-tiered wedge-bond pads, and scratch pads for electrical probing (*Courtesy of AT&T Bell Laboratories.*)

PROBLEMS

1 What bonding-pad pitch is required to place 244 I/Os on a pad-limited VLSI chip that is 6 mm square (36 mm^2 in area)? Assume that the chip size is defined by the loci of the bonding-pad centerlines. Check answer from Fig. 7.

2 The minimum practical lower limit on bond-pad pitch today for thermosonic ball–wedge wire bonding is 178 μm. What is the maximum percent reduction in silicon area that can be achieved using a staggered bond-pad pattern with design rules shown in Fig. 8, as compared to a uniform single-row pattern at 178 μm pitch for a VLSI chip with 244 I/Os? Use same assumption as in Problem 1.

3 Figure 8 shows design rules for interfacing with patterned substrates or leadframes. When using a package with a stamped leadframe, the minimum pitch on the leadframe is limited by the leadframe thickness. The capability of leadframe stamping today is to provide feature widths and spaces equal to the thickness of the leadframe. Under these restrictions, the minimum die size is controlled by the fan-in design rules of Fig. 8 and the maximum wire span requirements of Table 1. Calculate the minimum pitch for a 244 I/O quad leadframe (square) using a 152 μm thick leadframe. Assume gold wire is used for bonding, with a diameter of 33 μm. Compare results with Problem 2 and discuss implications.

4 A 68 I/O PLCC has a thermal resistance of 37°C/watt in natural free convection. Calculate the maximum allowable junction temperature for a device dissipating 0.5 watt in an ambient of 70°C. Find the maximum power dissipation for a device operating at a 125°C junction temperature in a 70°C ambient.

5 Suppose you have a 5 volt, 1 ns output-rise-time silicon chip with 68 signal leads and you want to package it so that inductive noise in the ground line is kept to 0.2 volt. You want to switch 18 lines simultaneously, and you want each package lead to have approximately 7 nH inductance. The output buffers draw 25 mA. How many package leads do you need for this device, excluding power?

6 Derive Eqs. 7 and 8.

7 Show that the above reduces to the simple form in Eq. 9. Assume $\alpha = 0$.

8 The permeability of polymers to water is very low. According to Fig. 40, moisture will permeate through polymers in the order of 10 to 100 days. Yet, when used with plastic molded packages these same polymers will provide adequate moisture protection and meet reliability requirements. Explain why this happens.

9 How many grams of silicon are required to react with 5000 ppm by volume of water in a 0.1 cm^3 die cavity? Assume room temperature and atmospheric pressure in the cavity.

REFERENCES

1 C. J. Bartlett, "Advanced Packaging for VLSI," *Solid State Technology*, p. 119, June (1986).

2 Spencer Chin, Associate Ed., "The Chip Carrier's Checkered Progress," *Electronic Products*, p. 43, July 15 (1986).

3 Dana W. Korf, "SMT Squeezes Computer Power Into Tighter Spaces," *Electronic Packaging and Production*, p. 64, Jan. (1985).

4 Integrated Circuit Engineering Corp. Technical Staff, "Surface Mount Packaging Report," *Semiconductor International*, p. 72, June (1986).

5 *JEDEC Registered and Standard Outlines for Semiconductor Devices*, Publication 95, Electronic Industries Association, Washington, D.C., 1985.

6 Waltraud Werdecker and Fritz Aldinger, "Aluminum Nitride—An Alternate Ceramic Substrate for High Power Applications in Microcircuits," *IEEE Trans. Components, Hybrids, Manuf. Technol.*, **4**, 399 (1984).

7 R. N. Merryweather and J. A. Solansky, "Subrate Developments for Hybrids," *Hybrid Circuit Technology*, p. 31, Nov. (1986).

8 Carole Patton, "Novel Arrangement of Contacts and Buses Shrinks 1-Mbit RAM," *Electronic Design*, p. 42, June 27 (1985).

9 Charles L. Cohen, "Japan's Packaging Goes World Class," *Electronics*, p. 26, Nov. 11 (1985).

10 Kanji Otsuka and Tamotsu Usami, "Ultrasonic Wire Bonding Technology for Custom LSI's with Large Numbers of Pins," *Proc. 31st Electronic Components Conf.*, 1981, p. 350.

11 Jeff Braden, "VLSI Design Rules," *Semiconductor International*, p. 178, May (1986).

12 Frank F. Oettinger, "Thermal Evaluation of VLSI Packages Using Test Chips—A Critical Review," *Solid State Technology*, p. 169, Feb. (1984).

13 G. K. Baxter, "A Recommendation of Thermal Measurement Techniques for IC Chips and Packages," *Proc. Intl. Rel. Phys. Symp.*, 1977, p. 204.

14 R. J. Hannemann, "Microelectronic Device Thermal Resistance: A Format for Standardization," in *Heat Transfer Digest*: vol. 20, *Heat Transfer in Electronic Equipment*, American Society of Mechanical Engineers, p. 39, Nov. (1981).

15 Leonard W. Schaper, "The Impact of Inductance on Semiconductor Packaging," *Proc. Intl. Rel. Phys. Symp.*, p. 38 (1981).

16 A. J. Rainal, "Transmission Properties of Various Styles of Printed Wiring Boards," *Bell System Tech. J.*, **56**, p. 995 (1979).

17 John A. DeFalco, "Reflections and Crosstalk in Logic Circuit Interconnections," *IEEE Spectrum*, p. 44, July (1970).

18 Ryotaro Kamikawai, Masaaki Nishi, Keiichiro Nakanishi, and Akira Masaki, "Electrical Parameter Analysis from Three-Dimensional Interconnection Geometry," *IEEE Trans. Components, Hybrids, Manuf. Technol.*, **2**, 269 (1985).

19 Cecil W. Deisch, John F. Gogal, and John W. Stafford, "Design of a High Performance DIP-Like Pin Array Package for Logic Devices," *IEEE Trans. Components, Hybrids, Manuf. Technol.*, **3**, 305 (1983).

20 Masao Kakei, Yasuhisa Ikeda, and Shigeru Koshibe, "Low Stress Molding Compounds for VLSI Devices," *Nikkei Electronics Microdevices*, p. 82 (1984).

21 Justin C. Bolger and Charles T. Mooney, "Die Attach in Hi-Rel P-DIPS: Polyimides or Low Chloride Epoxies?," *IEEE Trans. Components, Hybrids, Manuf. Technol.*, **4**, 394 (1984).

22 Dennis R. Olsen and Howard M. Berg, "Properties of Die Bond Alloys Relating to Thermal Fatigue," *Proc. 27th Electronics Components Conf.*, 1977, p. 193.

23 Kyocera International, *Design Guidelines for Multilayer Ceramics*, Publication Number A-125E-2.

24 Peter J. Planting, "An Approach for Evaluating Epoxy Adhesives for Use in Hybrid Microelectronic Assembly," *IEEE Trans. Parts, Hybrids, Packag.*, **11**, 305 (1975).

25 Malcolm M. White, Joseph W. Serpiello, Kurt M. Striny, and W. Rosensweig, "The Use of Silicone RTV Rubber for Alpha Particle Protection on Silicon Integrated Circuits," *Proc. Intl. Rel. Phys. Symp.*, 1981, p. 43.

26 Steven Groothuis, Walter Schroen, and Masood Murtuza, "Computer Aided Stress Modeling for Optimizing Plastic Package Reliability," *Proc. 23rd Intl. Rel. Phys. Symp.*, p. 184 (1985).

27 Chuck Murray, Assoc. Ed., "Die Separation: Changing to Meet Industry Needs," *Semiconductor International*, p. 49, June (1986).

28 Gerry B. Gariepy, "Diamond Blade Technology in Die Separation," *Solid State Technology*, p. 95, July (1985).

29 H. Factor and K. Kaufman, "Water Dicing: On The Threshold of Automation," *Solid State Technology*, p. 81, July (1985).

30 James F. Marshall, "New Applications of Tape Bonding for High Lead Count Devices," *Solid State Technology*, p. 175, Aug. (1984).

31 L. S. Goldmann and P. A. Totta, "Area Array Solder Interconnections for VLSI," *Solid State Technology*, p. 91, June (1983).

32 R. K. Shukla and N. P. Mencinger, "A Critical Review of VLSI Die-Attachment in High Reliability Applications," *Solid State Technology*, p. 67, July (1985).

33 M. L. White, K. M. Striny, and R. E. Sammons, "Attaining Low Moisture Levels in Hermetic Packages," *Proc. 20th. Intl. Rel. Phys. Symp.*, 1982, p. 253.

34 C. E. T. White and J. Slatery, "An Update on Preforms," *Circuit Manufacturing*, p. 78, March (1978).

35 Bruce W. Hueners, "Advances in Automatic Bonding," *Solid State Technology*, p. 189, Feb. (1985).

36 Kulicke and Soffa Industries, *Bonding Tools and Production Accessories*, Bonding Handbook and General Catalog (1980).

37 B. Nachon and Y. Tai, "Capillary in High-Speed-Bonding Process," KS Microswiss, April (1984).

38 G. G. Harmon and C. A. Cannon, "The Microelectronic Wire Bond Pull Test—How to Use It, How to Abuse It," *IEEE Trans. Components, Hybrids, Manuf. Technol.*, **3**, 203 (1978).

39 George G. Harmon, "Microelectronic Ball-Bond Shear Test—A Critical Review and Comprehensive Guide to its Uses," *Solid State Technology,* p. 186, May (1984).

40 M. L. White, "The Removal of Die Bond Epoxy Bleed Material by Oxygen Plasma," *Proc. 32nd Electronics Components Conf.*, May 1982, p. 262.

41 R. A. Gardner and R. W. Nufer, "Properties of Multilayer Ceramic Green Sheets," *Solid State Technology*, p. 38, May (1974).

42 J. R. Howell, "Reliability Study of Plastic Encapsulated Copper Lead Frame Epoxy Die Attach Packaging System," *Proc. Intl. Rel. Phys. Symp.*, p. 104 (1981).

43 Candice H. Brown, "Low Cost Pin Grid Array Packages," *Solid State Technology*, p. 239, May (1985).

44 Martin F. Blackshaw and Francis J. Dance, "Design, Manufacture, and Assembly of High Pin Count Plastic Pin Grid Array Packages," *Solid State Technology*, p. 141, Aug. (1986).

45 Melanie Iannuzzi, "Development and Evaluation of a Preencapsulation Cleaning Process to Improve Reliability of HIC's with Aluminum Metallized Chips," *IEEE Trans. Components, Hybrids, Manuf. Technol.*, **4**, 429 (1981).

46 Kurt M. Striny and Arthur W. Schelling, "Reliability Evaluation of Aluminum-Metallized MOS Dynamic RAM's in Plastic Packages in High Humidity and Temperature Environments," *IEEE Trans. Components, Hybrids, Manuf. Technol.*, **4**, 476 (1981).

47 R. K. Traeger, "Hermeticity of Polymeric Lid Sealants," *Proc. 25th Electronics Components Conf.*, 1976, p. 361.

48 Irving Memis, "Quasi-Hermetic Seal for IC Modules," *Proc. 30th Electronic Components Conf.*, 1980, p. 121.

49 Robert W. Thomas, "Moisture, Myths, and Microcircuits," *IEEE Trans. Parts, Hybrids, Packag.*, **12**, 167 (1976).

50 "Moisture Measurement Technology for Hermetic Semiconductor Devices," March 22–23, 1978, NBS (National Bureau of Standards) Publ., p. 400 (1981).

51 Ching-Ping Wong, "Room Temperature Vulcanized (RTV) Silicone as Integrated Circuit (IC) Coating," *Proc. ISHM*, p. 315 (1981).

52 R. G. Manke, "A Moisture Protection Screening Test for Hybrid Circuit Encapsulants," *IEEE Trans. Components, Hybrids, Manuf. Technol.*, **4**, 492 (1981).

53 George Cvijanovich, "Active Protection of IC Surfaces," *Semiconductor International*, p. 57, May (1979).

54 T. C. May and M. H. Woods, "Alpha-Particle-Induced Soft Errors in Dynamic Memories," *IEEE Trans. Electron Devices*, **26**, p. 2 (1979).

55 Katsuhiko Yamaguchi and Mazuo Igaroshi, "Screen Printing Grade Polyimide Paste for Alpha-Particle Protection," *IEEE, Proc. 36th Elect. Components Conf.*, May 1986, p. 340.

56 Gary L. Lewis and Howard M. Berg, "Particulates in Assembly: Effect on Device Reliability," *IEEE, Proc. 36th Elect. Components Conf.*, May 1986, p. 100.

57 Peter Burggraaf, Senior Ed., "Trim and Form Equipment Review," *Semiconductor International*, p. 62, June (1986).

58 Peter H. Singer, Assoc. Ed., "Leak Detection in IC Packages," *Semiconductor International*, p. 91, Nov. (1984).

59 Ron Iscoff, West Coast Ed., "Component Marking Technology Trends," *Semiconductor International*, p. 38, Dec. (1985).

60 Eric J. Penn, "Lead Scanning—Luxury or Necessity?" *Electronic Packaging and Prod.*, p. 78, Sept. (1985).

61 Pieter Burggraaf, Senior Ed., "IC Handler Product Trends," *Semiconductor International*, p. 93, Sept. (1985).

62 Didier Kane and Michael Brizoux, "Recent Developments on Moisture Measurement by Surface Conductivity Sensors," *Proc. Intl. Rel. Phys. Symp.*, 1986, p. 69.

63 Tobias Naegele, "High Pin-Count ASICs May Get New JEDEC Package," *Electronics*, p. 36, Oct. 30 (1986).

64 Jerry Lyman, "Packaging," *Electronics*, p. 102, Oct. 16 (1986).

65 Gerard Dehaine, Karel Kurzweil, and Pierre Lewandowski, "Crossing the 200 Wire Barrier in VLSI Bonding," *Proc. Intl. Symp. Microelectronics*, 1984, p. 353.

66 Donald B. Brown and Martin G. Freedman, "Is There a Future for TAB?" *Solid State Technology*, p. 173, Sept. (1985).

67 F. N. Sinnadurai, *Handbook of Microelectronics Packaging and Interconnection Technologies*, Electrochemical Publications Ltd. (1985).

CHAPTER
14

YIELD
AND
RELIABILITY

W. J. BERTRAM

14.1 INTRODUCTION

Previous chapters discussed the science and technology of VLSI circuit fabrication. This chapter considers two conditions that must be satisfied in order for VLSI to be a useful, growing technology. First, the fabricated circuits must be capable of being produced in large quantities at costs that are competitive with alternative methods of achieving the circuit and systems function. Second, the circuits must be capable of performing their function throughout their intended life. To produce circuits that meet these two requirements we must understand the mechanisms that lead to high cost and unreliable devices. Once these mechanisms are understood and quantified, we can develop the technology required to meet our goals.

The optimum size of an IC, with respect to the number of circuit functions, is a compromise between several competing considerations: partitioning of the system (or subsystem), the expected yield of good circuits, packaging and system assembly cost, and the overall reliability of the complete system. If the system is divided into a large number of small ICs, then the yield of these circuits may be very high; however, the total cost of packaging a large number of these chips and assembling them on a circuit board may be greater than the cost of manufacturing a smaller number of larger circuits at a reduced yield, and packaging and assembling these circuits on a circuit board. To divide the complete system into an optimum number of ICs, we must be able to predict the yield and cost of an IC as a function of circuit size. This optimization is further complicated because the number of required circuit functions may increase as the system is divided into smaller and smaller ICs. Overall system performance may also suffer because the time to propagate signals between different parts of the system is greater if the different parts are on separate ICs than if all circuit functions are contained on one IC. Similarly, to make an optimum division of the system, we must be able to predict the total system reliability as a function of the number of ICs of varying size.

In this chapter we will discuss the various methods that have been developed to quantify the measurements of VLSI circuit yield and reliability. The various mechanisms that have been identified as seriously limiting the yield and reliability of VLSI circuits will be reviewed. Finally, we will discuss the techniques that are used to assure that new device technologies are not limited in their usefulness by yield- or reliability-reducing mechanisms.

14.2 MECHANISMS OF YIELD LOSS IN VLSI

Ideally, in a properly fabricated wafer of VLSI circuits, we would expect all of the circuits on the wafer to be good, functional circuits. In practice, the number of good circuits may range from close to 100% to only one or fewer good circuits per wafer. Usually, the causes for less than perfect yield fall into three basic categories: parametric processing problems, circuit design problems, or random point defects in the circuit. Each of these categories is discussed in this section.

14.2.1 Processing Effects

If one looks at a map or photograph of a tested wafer, one of the most obvious features is that the wafer may be divided into regions with a very high proportion of good chips and other regions where the yield of good chips is very low or even zero. A photograph of such a wafer is shown in Fig. 1. This section discusses processing effects that may lead to the existence of low yield regions. These effects include variations in the thickness of oxide or polysilicon layers, in the resistance of implanted layers, in the width of lithographically defined features, and in the registration of a photomask with respect to previous masking operations. Many of these variations depend upon one another. For example, regions where the polysilicon layer is thinner than average become overetched when the wafer is etched for the length of time needed to clear the polysilicon in regions where the layer is thicker than average. Polysilicon gate lengths are shorter in the thinner polysilicon regions than in the thicker polysilicon regions. This effect may result in regions in which the channel lengths are too short and the transistors cannot be turned off when the appropriate gate voltage is applied. Consequently, the circuits may not function or may have excessive leakage currents.

Variations in the doping of implanted layers can lead to variations in the contact resistance to the implanted layers. Variations in the thickness of the deposited dielectric can lead to variations in contact window size. Both of these effects can lead to nonoperative circuits if the circuit contains paths whose performance is dependent upon having a low value of contact resistance.

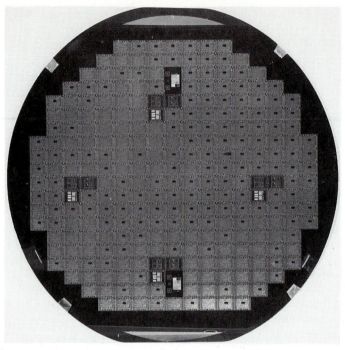

FIGURE 1
Photograph of an IC wafer showing regions of high and low yield. The chips with a black ink dab
are bad. The unique-looking chips in the four groups are test chips.

During the processing of a wafer various operations are carried out that
result in small but critical changes in the size of the wafer. For example, when
a wafer is oxidized, the SiO_2 formed has approximately twice the volume of
the silicon consumed in the oxidation process. The resultant composite wafer
consists of an interior silicon layer which is in tension, and oxide layers on the
two surfaces which are in compression. The resultant wafer is larger in diameter
than the original silicon wafer. If the degree of tension exceeds the elastic limit of
the material, deformation occurs. If the oxide is then removed from one surface,
the wafer will have a convex bow on the remaining oxide surface.

During processing, it is not uncommon for wafer size to vary in excess of
20 ppm. Thus, a 125 mm wafer changes size by 2.5 μm, which is far greater
than the desired realignment tolerance. If such variations in wafer size are not
compensated for, areas of the wafer may have inoperative circuits because of
misalignment. In addition, we may find areas of the wafer where improper cleaning has left chemical residue that can lead to excessive formation of oxidation-
induced stacking faults. These stacking faults may result in excessive leakage
currents and subsequent circuit failure.

As advanced processes and processing techniques are developed, many of
these yield-limiting effects will be reduced or eliminated; however, new ones
may be introduced!

14.2.2   Circuit Sensitivities

In addition to areas of a wafer where the yield is low because processing difficulties have led to device parameters outside the specified range, certain areas of
a wafer may have low device yield because the design of the circuit has failed to
take into account the expected variation in device parameters and the correlation
between variations in different parameters.

Threshold voltage (V_T) and channel length (L) of the MOS transistors are two
of the most important parameters in MOS circuit design. Variations in substrate
doping, ion-implantation dose, and gate-oxide thickness will cause variations in
the threshold voltage. Variations in gate length and source and drain junction
depth cause the channel length to vary. The threshold voltage and channel length
variations are generally not correlated with each other. However, the speed of a
circuit generally increases as the threshold voltage and channel length decrease.
Circuit performance is often simulated under high-speed (small V_T and L) and
low-speed (large V_T and L) conditions. It is important that circuit performance
should also be simulated for small V_T and large L, and large V_T and small L.
Circuit design must also consider variations in other device parameters, such
as the resistance of implanted regions, capacitance of conductors to substrate,
contact resistance, and leakage currents.

Often two circuits with nominally the same size and complexity, processed
in the same technology, have vastly different yields. The low yield in the one
case is due to a lack of understanding of the sensitivity of the circuit to device
parameters. Here, higher yield requires a cooperative effort between the circuit
designer, who identifies the specific device parameters to which the circuit is
sensitive, and the process engineer, who optimizes both the value and range of
variation of those parameters. Once the circuit sensitivities to specific process
parameters have been determined, redesign of the circuit to reduce these sensitivities produces high yield and low cost with minimal attention from the process
engineer.

14.2.3   Point Defects

We typically find that the yield of circuits is not 100%, even when all processing
parameters throughout the wafer are not only within the expected range but are
also within the range where the circuit performs satisfactorily. The cause of this
less-than-perfect yield is generally attributed to *point defects*. A point defect is a
region of the wafer where the processing is imperfect, and the size of this region
is small compared to the size of the chip. For example, consider a chip 2000 μm
square with features that have a size of 2 μm. A 3 μm diameter dust particle
on the wafer can cause a break in a metal conductor. Similarly, a 200 μm dust
particle could cause a large area of metal to be missing from the chip. Both these
particles could be considered point defects. A region of the wafer that is 1 cm in
diameter, covering 20 or 30 chips, where the metal is missing from the chips, is
not considered a point defect.

Many types of processing defects are considered to be point defects. One of the most common causes of this type of defect is dust or other particles in the environment. These particles may fall on wafers being transported in the processing facility, or they may be generated in a film-deposition operation and be incorporated into the film. They may also be present in photoresist solutions and deposited on the wafer during the photoresist operation, or they may be silicon particles, knocked out of the wafer during handling, which adhere to the wafer surface. Isolated oxidation-induced stacking faults that cause excessive junction leakage and circuit failure can also be considered point defects, as can isolated spikes in an epitaxial film or pinholes in a dielectric film.

Point defects can occur on lithographic masks as well as on silicon wafers. Dust or particles on the mask substrate during the mask-generation process can cause permanent defects in the mask. Particles which adhere to the mask during use cause a gradual increase in the density of point defects that are reproduced on the wafers. Periodic cleaning of the mask will remove these defects.

Successful IC fabrication requires continued monitoring of the density of point defects. Monitoring may be carried out by visual or Scanning Electron Microscope observation of circuit quality during all steps of fabrication. When any variation in the defect density of a particular operation is observed, the appropriate corrective action can be taken. Special operations may also be performed to monitor defects. Wafers can be etched to remove all films, and the silicon substrate treated to reveal stacking faults, whose density is then monitored. Other wafers may be patterned by using special masks which allow electrical measurement of the incidence of dielectric breakdown defects. As new types of defects are identified as the cause of circuit failure, methods of monitoring the density of these defects must be instituted.

The circuit yield and the cause of circuit failure must also be monitored. The use of a circuit, such as a memory chip, in which a failure can be related back to a specific area of the chip (for example, one of the six transistors of a static memory cell) has been most useful. The successful operation of an IC facility requires that the density of point defects be continuously monitored, controlled, and reduced.

14.3 MODELING OF YIELD-LOSS MECHANISMS

Modeling IC yield in terms of fundamental parameters that are independent of the particular IC and characteristic of the process and processing line is important for several reasons. By accurately modeling the yield, we can predict the cost and availability of future circuits, provided they are related in technology and design to the circuits used to develop the yield-modeling parameters. Once we know the yield-modeling parameters, we can compare the processing quality of different process lines and thus indicate where improvements in process facilities are required. Yield-modeling parameters, for a given process and process line, that show significant variation for a particular circuit may indicate a sensitivity to device parameters in that circuit. Further study of the circuit design may reveal a possible change in design or process that will result in a significant improvement in device yield.

Generally, IC yield is expressed in the form

$$Y = Y_0 Y_1 (D_o, A, \alpha_i) \tag{1}$$

where Y is the ratio of good chips per wafer to the number of chip sites per wafer, $(1 - Y_0)$ is the fraction of chip sites that yield bad chips either because of process-related effects or because of circuit sensitivities, and $(1 - Y_1)$ is the fraction of the remaining chip sites that yield bad chips. Y_1 is a function of D_o, A, and α_i, where D_o is the density of point defects per unit area, A is the area of the chip, and α_i are parameters unique to different models of the yield.

All of the models to be discussed in Sections 14.3.1 to 14.3.4 predict that the yield decreases monotonically as the area of the chip increases. The models can be quite useful in predicting IC yield, provided we stay within the range of parameters developed and verified in the models. IC yield modeling is notorious for underestimating the yield of future ICs; IC processing is an ever-developing art. Yield modeling can identify those processes and mechanisms that limit the yield of present ICs. As the yield-limiting features are identified, the processes are improved or eliminated as needed. For example, projection optics has replaced contact printing, resulting in lower defect density; dry etching has replaced wet chemical etching, resulting in better feature-size control; and ion implantation has replaced diffusion, resulting in better control of resistance, junction depth, and threshold voltage.

Yield modeling is not a predictor of the future; rather, it is a tool for improving the present generation of processes and designs. The following sections discuss the features of some of the more common methods for modeling the effect of point defects on IC yield.

14.3.1 Uniform Density of Point Defects

In the area of the wafer where the yield has not been degraded by either processing effects or circuit sensitivities, the remaining cause of loss of good chips is randomly distributed point defects. Figure 2 shows a grid of 24 chip sites with 10 defects randomly distributed over the area. In this example, 16 of the 24 sites have zero defects; that is, they are good chip sites. Of the remaining eight sites, six have one defect, and two have two defects; no sites have more than two defects. The problem of determining the yield of good chips is identical to the classical statistics problem of placing n balls in N cells, and then calculating the probability that a given cell contains k balls.[1] If n defects are distributed randomly among N chips, the probability P_k that a given chip contains k defects is given by the binomial distribution

$$P_k = \frac{n!}{k!(n-k)!} \cdot \frac{1}{N^n}(N-1)^{n-k} \tag{2}$$

In the limit that N and n are both large, and that the ratio $n/N = m$ remains

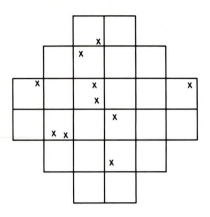

FIGURE 2
An array of 24 chip sites with 10 defects (indicated by an x) randomly distributed on the sites.

finite, the binomial distribution can be approximated by the more tractable Poisson distribution

$$P_k \cong e^{-m} \frac{m^k}{k!} \tag{3}$$

The probability that a chip contains no defects, which is the yield, is given by

$$Y_1 = P_0 = e^{-m} \tag{4}$$

and the probability that a chip contains one defect is given by

$$P_1 = me^{-m} \tag{5}$$

If the area of a chip is A, the total area of the chips in the usable part of the wafer is NA, and the density of defects is $n/NA = D_o$. The average number of defects per chip, m, is

$$m = \frac{n}{N} = D_o NA/N = D_o A \tag{6}$$

and

$$Y_1 = P_0 = e^{-D_o A} \tag{7}$$

The Poisson estimate of the yield of good chips was used to predict the yield during the early manufacture of ICs. However, the actual yield obtained on larger circuits was considerably greater than that estimated using the D_o values measured by fitting the yield of smaller circuits to the Poisson formula. In fact, the low yields estimated by the Poisson formula no doubt delayed early work in the development of ICs.

14.3.2 Simple Nonuniform Distributions of D

The discrepancy between the measured and predicted yields of early ICs led researchers to investigate the effect of nonuniform distributions of D_o across a wafer on the yield of that wafer. The yield of chips on a wafer, where D_o is nonuniform across the wafer, can be expressed as[3]

$$Y = \int_0^\infty e^{-DA} f(D)\,dD \tag{8}$$

where $f(D)$ is the normalized distribution function of defect density (pdf), that is,

$$\int_0^\infty f(D)\,dD \equiv 1$$

Three distributions of D_o are evaluated: the delta function, a triangular distribution, and a rectangular distribution (Fig. 3). The results are as follows:

$$Delta\ function : Y_1 = e^{-D_oA} \tag{9}$$

$$Triangular : Y_2 = \left\{ \frac{1 - e^{-D_oA}}{D_oA} \right\}^2 \tag{10}$$

$$Rectangular : Y_3 = \frac{1 - e^{-2D_oA}}{2D_oA} \tag{11}$$

If we look at the form of these expressions for values of $D_oA \gg 1$ we find that

$$\lim_{D_oA \gg 1} Y_1 = e^{-D_oA} \tag{12}$$

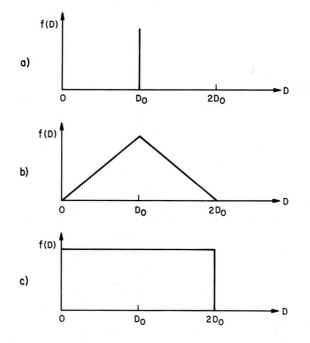

FIGURE 3
Distribution of defect density. (a) Delta function. (b) Triangular. (c) Rectangular. (*After Murphy, Ref. 3.*)

$$\lim_{D_oA \gg 1} Y_2 = \frac{1}{(D_oA)^2} \tag{13}$$

$$\lim_{D_oA \gg 1} Y_3 = \frac{1}{2D_oA} \tag{14}$$

Equations 13 and 14 give much less pessimistic estimates of the yield than the original Poisson estimate (Eq. 12). These expressions found considerable use, with Y_3 being the expression that most closely fit the observed yield of large ICs. The nonphysical characteristics of the above distributions led workers to investigate other, more physical distributions that could also predict the observed high yield of large ICs.

14.3.3   The Gamma Distribution of D

A more physical distribution for calculating IC yield is the Gamma distribution.[4,5,6] The probability density function (pdf) of this distribution is

$$f(D) = \frac{1}{\Gamma(\alpha)\beta^\alpha} D^{\alpha-1} e^{-D/\beta} \tag{15}$$

$$D, \alpha, \beta > 0$$

where α and β are the two distribution parameters and $\Gamma(\alpha)$ is the Gamma function. In this distribution the average density of defects is given by $D_o = \alpha\beta$, the variance of D is given by $\mathrm{var}(D) = \alpha\beta^2$, and the coefficient of variation is given by

$$\frac{\sqrt{\mathrm{var}(D)}}{D_o} = \frac{1}{\sqrt{\alpha}}$$

The probability of a chip having k defects is given by

$$P_k = \int_o^\infty e^{-m} \frac{m^k}{k!} f(D) dD$$

$$= \frac{\Gamma(k+\alpha)}{k!\Gamma(\alpha)} \frac{(A\beta)^k}{(A\beta + 1)^{k+\alpha}} \tag{16}$$

where $\int_o^\infty f(D)dD \equiv 1$. The probability of having no defects on a chip is

$$Y_4 = P_o = \frac{1}{(A\beta + 1)^\alpha}$$

$$= \frac{1}{(1 + SD_oA)^{1/S}} \tag{17}$$

where

$$S = \frac{\mathrm{var}(D)}{D_o^2} = \frac{1}{\alpha}$$

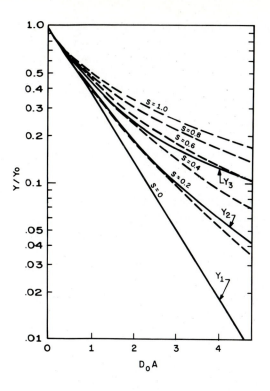

5

good chips as a function of D_oA for the Gamma (broken curves) and the delta function, *, and rectangular distributions (solid curves).

nsiderably among the different types of product manufactured in different ing facilities. Also, many different types of defects affect the yield, and imeters D_o and S vary considerably for different types of defects.

Yield of Chips with Redundant Circuitry

arge MOS memory chips are designed with redundant circuitry,[2] which switched in to replace defective circuit elements. This replacement is arly simple in the case of the MOS memory, which in large part is ed of reiterative circuitry. The replacement of a defective circuit is usually lished using fusible links, which can be fused as needed using laser or chniques.

a memory chip were designed with one redundant column, for example, p with a defect in a column could be repaired by electrically substituting undant column for the defective column. The yield of this chip would then n by

$$Y_1 = P_0 + \eta P_1 \qquad (20)$$

P_0 is the probability of a chip containing no defects, P_1 is the probability

In Eq. 17 the yield is a function of the average defe
A, and the square of the coefficient of variance S c
general, the Gamma distribution of D is a skewed
zero to infinity. Figure 4 shows the shape of the distr
values of the shape parameter S. In the limiting ca
distribution reduces to a delta function

$$f(D) = \delta(D - D_o)$$

where $S \cong 0$, $\beta = SD_o$, and the yield Y_4 is given by t

$$Y_4 = e^{-D_o A}, \quad S \to 0$$

Figure 5 is a plot of the Gamma yield function of
Y_2 and Y_3 are plotted for comparison. The yield fun
of D_o, Y_1, is identical to the Gamma yield function
yield function is clearly a good approximation to the
a wide range of $D_o A$ values. Furthermore, the Gam
potential to represent quite a large variation in the
yield versus area curve. For these reasons the Gamma
common function for representing IC yield. Values

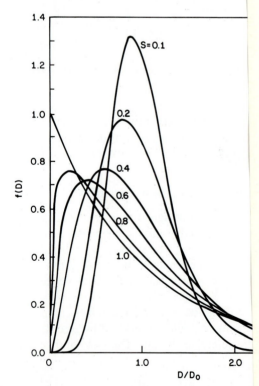

FIGURE 4
Gamma distribution of defect density.

FIGUR
Yield o
triangul

vary c
proces
the pa

14.3.

Many
can be
particu
compo
accom
other
I
any ch
the rec
be giv

where

of a chip containing one defect, and η is the probability that a chip containing one defect can be repaired by using the single redundant column. This yield model can be extended to chips containing more complex redundant circuitry.

14.3.5   Mutiple Types of Defects

IC yield is affected by many types of defects. Each type affects different circuits to a different extent. For example, defects in the gate oxide occur only in the gate region of the transistors. Junction leakage caused by oxidation-induced stacking faults can occur in both the gate region and the source and drain regions. Shorts between metal runners are important only in those regions of the chip with a high density of closely packed metal runners.

Since yield loss so strongly depends on the different defect mechanisms, each type should be considered independently. Each defect mechanism should be characterized by its mean defect density D_{no}, the shape factor of the distribution of defects S_n, and the portion of the total chip area A_n that is susceptible to that particular type of defect. Making use of the Gamma yield function, the yield for each type of defect is

$$Y_n = \frac{1}{(1 + S_n A_n D_{no})^{1/S_n}} \tag{21}$$

The overall yield is then the product of the yield for each known type of defect, that is,

$$Y = \prod_{n=1}^{N} Y_n = \prod_{n=1}^{N} (1 + S_n A_n D_{no})^{-1/S_n} \tag{22}$$

When chip yield is determined by several major types of defects, the parameters S_n, A_n, and D_{no} must be determined individually, and the yield of the overall process estimated[7,8,9] using Eq. 22.

For a mature product processed in a well-controlled, high-yield fabrication line, all of the major yield-limiting defects have probably been controlled or eliminated. Thus the yield (represented by Eq. 22) is the product of many terms, all close to unity, that is, $S_n A_n D_{no} \ll 1$. The yield is

$$Y = \prod_{n=1}^{N} (1 + S_n A_n D_{no})^{-1/S_n}$$

$$\ln Y = \sum_{n=1}^{N} \left[-\frac{1}{S_n} \ln(1 + S_n A_n D_{no}) \right] \tag{23}$$

Since $S_n A_n D_{no} \ll 1$,

$$\ln(1 + S_n A_n D_{no}) \approx S_n A_n D_{no}$$

and

$$\ln Y = \sum_{n=1}^{N} \left[-A_n D_{no} \right] \tag{24}$$

or

$$Y = \exp\left(-\sum_{n=1}^{N} A_n D_{no} \right) \tag{25}$$

$$= e^{-A\overline{D}}$$

where

$$\overline{D} = \frac{1}{A} \sum_{n=1}^{N} A_n D_{no}$$

Thus for a mature, high-yield chip, the yield is again represented by an exponential, independent of the shape parameters for each type of defect. The composite defect density $\overline{D}$, however, is a function of the type of circuit, and circuit layout, through the fractional areas A_n. The $\overline{D}$ obtained from one chip design is not applicable to another chip design. The yield expressed in Eq. 25 does, however, apply to the type of experiment that studies the yield of multiples of a single chip.[7,10,11] In that case we would expect the yield of M-multiple chip sites, for a mature, high-yield chip, to be

$$Y_M = e^{-MA\overline{D}} \tag{26}$$

as is observed in many experiments.

14.3.6   Radial Distribution of Defects

In the previous sections we assumed that the density of defects is not uniform, but varies across a wafer and from wafer to wafer. Several researchers have performed detailed studies on the origin of the defect-density variation for their own particular process.[8,9,12,13,31] They find that certain types of defects show a strong radial dependence in the frequency of their occurrence. This relationship holds especially for cosmetic defects, such as handling damage, misalignment errors, photoresist residue, and so forth. In one study,[31] it was found that not only was the yield in the outer third of the wafer only 60% of that in the center of the wafer, but that the reliability of chips from the outer third was significantly lower than the reliability of those from the center of the wafer.

The radial variation in defect density can be represented by the expression

$$D(r) = D_o + D_R e^{(r-R)/L} \tag{27}$$

where D_o is the defect density associated with the center of the wafer, D_R is the increase in defect density at the edge of the wafer, r is the radial coordinate, R is the radius of the wafer, and L is the characteristic length associated with the edge-related defects.

The yield of a wafer with a radial distribution of defects may be obtained by integrating the Poisson yield function over the area of the wafer:

$$Y_R = \frac{2}{R^2} \int_0^R e^{-D(r)A} r\, dr \qquad (28)$$

Equation 28 can be evaluated exactly, but not in a simple, closed form. Computer calculations of Y_R for specific cases have been performed.[9]

The radial distribution of defect density in Eq. 27 can also be transformed to a distribution function of $D, f(D)$. The yield can then be calculated by using Eq. 9, or $f(D)$ can be approximated by a Gamma distribution, and the results of Section 14.3.3 are then applicable.

14.3.7 Summary

The previous sections have examined some causes which lead to loss of yield in the manufacture of ICs. One characteristic of the yield-loss mechanisms is the dependency upon the technology, the type of circuit, the particular circuit design, and the maturity of the processing technology, facility, and circuit design. Yield-loss mechanisms change with time. As the technology matures, the major causes of yield loss are identified and eliminated. Similarly, as the sensitivities of a circuit to process variations are identified, the process control is improved, and the circuit will be redesigned to reduce the sensitivities. Finally, when both process and design have been optimized, the remaining yield-loss mechanism is a low level of random defects. Identification and elimination of yield-loss mechanisms will continue with the development of new technologies.

14.4 RELIABILITY REQUIREMENTS FOR VLSI

Before we begin a detailed study of reliability, it will be instructive to consider several examples of the effect of device failure on system performance. These examples will give the reader perspective as to the acceptable values of device-failure rates and point out the disastrous effects of excessive device failures on the system's economic viability.

A series of studies on semiconductor-device-failure rates and mechanisms began shortly after the development of the transistor, when serious consideration was given to the construction of large solid-state computer systems and electronic telephone switching systems (ESS). One of these early systems could contain on the order of 100,000 discrete transistors, plus other components. If we take one device failure per month as a goal, then the devices must have a failure rate, λ, of

$$\lambda < \frac{1 \text{ Failure}}{10^5 \text{ Devices} \times 720 \text{ Hours}}$$

$$= 14 \times 10^{-9} \text{ Failure/Device-hour} \qquad (29)$$

We will define the unit of failure rate to be 1 *Failure Unit* $\equiv$ 1 FIT $\equiv$ 1 Failure/10^9 Device-hour. Thus, we see that our goal for the device-failure rate of our hypothetical system is

$$\lambda < 14 \text{ FIT}$$

(Note that because the system will be designed with redundant elements, a single device failure most likely will not result in a system failure.) Over the expected 10 year life of the typical system, 120 devices ($\approx$ 0.1% of the devices in the system) will fail. A total of 120 service calls will have to be made. Table 1 summarizes the effect of different device-failure rates on the number of failures per month and the total percentage of devices that will fail in the system's 10 year life. It can be seen that a device failure rate of 10 FIT is very desirable, a failure rate of 100 FIT is probably acceptable, and a failure rate of 1000 FIT is unacceptable, because such a rate would require repairs approximately twice a day and, depending on the number of devices per circuit board, the replacement of close to the entire complement of system circuit boards during the life of the system.

The previous example was based on an original semiconductor device system using 100,000 discrete transistors. Table 2 summarizes similar considerations for two modern systems: a large data set containing 150 to 225 ICs, and a private telephone system containing 10,000 or more ICs. The acceptable device-failure rates for both of these systems are the same as for the first example (i.e., a device-failure rate of less than 10 FIT is desirable, and a failure rate of 1000 FIT is unacceptable).

The next consideration is how to demonstrate that the devices actually have the desired failure rate of less than 10 FIT. Table 3 shows the expected time for 1 failure to occur in 100 devices being tested. Also tabulated are the estimated device failure rates if we test 500 devices for 6 months and observe no failures. In both cases we see that if we are to prove our desired device-failure rate of 10 FIT, we must test large numbers of devices for totally unreasonable lengths of time, in terms of the development schedule of a device or system. Proof of adequate device reliability is time-consuming and expensive; however, the cost of inadequate device reliability is disastrously expensive.

The following sections discuss methods to quantitatively measure and predict device failure rates, and to identify and eliminate the failure mechanisms.

TABLE 1
Effect of transistor failure on system performance

Failure rate	Failure / Month*	Total % of devices to fail in 10 years
10 FIT	0.7	0.1
100 FIT	7	1
1000 FIT	70	10

*Based on 100,000 devices per system.

TABLE 2
Effect of device failure on modern system performance

	Data set 150 to 225 ICs		
Device failure rate (FIT)	Mean time to data set failure (year)	Percent of sets fail per month
10	51	0.16
100	5	1.6
1000	0.5	16.0

	Private telephone system Bulk of installations 5000 to 10,000 ICs Large installations 50,000 ICs		
Device failure rate (FIT)	System failure/month (for 10,000 IC system)	Percent of circuit packs fail in 10 years
10	0.07	1
100	0.7	10
1000	7.0	65

14.5 MATHEMATICS OF FAILURE DISTRIBUTIONS, RELIABILITY, AND FAILURE RATES

The term reliability has many popular connotations. For a useful, mathematical description of reliability, however, we must first define the term precisely. A generally accepted definition of reliability is the probability that an item will perform a required function under stated conditions for a stated period of time.

The "required function" must include a definition of satisfactory operation and unsatisfactory operation, or failure. For an IC the required function is generally defined by a test program for an automatic test set. The program's output

TABLE 3
Reliability prove-in

For 100 devices on test		If we run 500 devices for 6 months with no failures	
Failure rate (FIT)	Time for 1 failure (year)	Confidence level (%)	Failure rate (FIT)
10	114	Best estimate	325
100	11	60	430
1000	1	90	1100
		95	1400
		99	2100

simply states "good" or "bad." In many cases, however, initial test programs are not complete; they do not test the circuit under all the required conditions. As new device-failure modes are identified, the appropriate tests are included in later generations of the test program.

The "stated conditions" in the definition comprise the total physical environment, including the mechanical, thermal, and electrical conditions of expected use (and also of periods of disuse, such as storage, etc.).

The "stated period of time" is the time during which satisfactory operation is required. This will vary depending upon the usage of the system. In some cases the time can be relatively short, as in the case of an emergency beacon transmitter on an aircraft. Often the period of use is preceded by a long period of disuse. Other systems are in continuous usage. For such continuously operating systems another concept, *availability*, is defined. Availability is the probability that an item will operate when needed, or the average fraction of time that a system is expected to be in an operating condition.

The next sections present the basic reliability concept and related concepts in terms of mathematical functions. A more rigorous treatment can be found in Reference 5.

14.5.1   Cumulative Distribution Function

Assume that a device or system is operating at time $t = 0$. The probability that the device will fail at or before time t is given by the function $F(t)$. This is a cumulative distribution function (or cdf) with the properties

$$F(t) = 0 \qquad\qquad t < 0$$

$$0 \le F(t) \le F(t') \qquad 0 \le t \le t' \tag{30}$$

$$F(t) \to 1 \qquad\qquad t \to \infty$$

14.5.2   Reliability Function

The reliability function $R(t)$ is the probability that the device will survive to time t without failure. The function is related to $F(t)$ and is given by

$$R(t) = 1 - F(t) \tag{31}$$

14.5.3   Probability Density Function

The derivative of $F(t)$ with respect to time is known as the probability density function (or pdf) and is represented by $f(t)$. The pdf is related to the cdf by

$$f(t) = \frac{d}{dt}F(t) \tag{32}$$

or

$$F(t) = \int_0^t f(x)dx \tag{33}$$

Similarly,

$$R(t) = \int_t^\infty f(x)dx \tag{34}$$

and

$$f(t) = -\frac{d}{dt}R(t) \tag{35}$$

14.5.4 Failure Rate

In most applications the quantity that is of most concern is the instantaneous failure rate, often referred to as the hazard rate. By the term failure rate, we will always mean the instantaneous failure rate, not the average failure rate.

The fraction of devices that were good at time t and that fail by time $t + \Delta$ is given by

$$F(t + \Delta) - F(t) = R(t) - R(t + \Delta) \tag{36}$$

The average failure rate during the time interval Δ is given by

$$\text{average failure rate} = \frac{1}{\Delta}\frac{R(t) - R(t + \Delta)}{R(t)} \tag{37}$$

In the limit as Δ approaches zero, this becomes the instantaneous failure rate $\lambda(t)$, given by

$$\lambda(t) = -\frac{1}{R(t)}\frac{dR(t)}{dt} \tag{38a}$$

$$= \frac{f(t)}{R(t)} \tag{38b}$$

$$= \frac{f(t)}{1 - F(t)} \tag{38c}$$

$$= -\frac{d}{dt}\ln R(t) \tag{38d}$$

Using Eq. 38d, we can express the reliability function as

$$R(t) = \exp\left[-\int_0^t \lambda(x)dx\right] \tag{39}$$

14.5.5 Mean Time to Failure

A common measure of reliability is the mean time to failure (MTTF) of the device or system, which can be expressed as

$$\text{MTTF} = \int_0^\infty tf(t)dt \tag{40}$$

MTTF is the device's average age at failure for a population whose reliability is represented by the function $R(t)$ with a pdf of $f(t)$.

14.6 COMMON DISTRIBUTION FUNCTIONS

It is desirable to have a single mathematical model that represents the failure rate of devices over their entire life. The failure rate of an IC, and of most common items as well, generally varies as a function of time in the manner illustrated in Fig. 6. During the early life of the device the failure rate is high, but it decreases as a function of time. The failures during this period are called "early failures" or "infant mortality." The causes of early failure are generally manufacturing defects. In many devices the early failure period is the most important because of the very high failure rates that can be observed during this period. Section 14.7.4 discusses methods of eliminating or shortening this early failure period.

During the midlife or steady-state period, the failure rate is generally low and fairly constant. Device failures are a result of a large number of unrelated causes.

Eventually the device enters the final, or wearout, period of its life. This period is more commonly observed in other items, such as electric light bulbs, than in ICs. In most ICs no wearout mechanisms are observed during the useful life of the device. However, some failure mechanisms (e.g., mobile ion shift, corrosion, and electromigration) can be observed in low-quality ICs.

Of the common distribution functions we will be discussing, no single one adequately represents all three periods of a device's life. At most, one distribution function can be used to represent two of the periods. Note that the distribution functions discussed in the next sections are treated in more detail in References 5 and 6.

14.6.1 Exponential Distribution Function

The simplest distribution function, the exponential function, is characterized by a constant failure rate over the lifetime of the device. This function is useful for

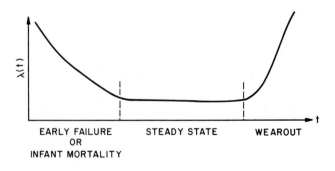

FIGURE 6
Failure rate versus time for typical ICs.

representing a device in which all early failure and wearout mechanisms have been eliminated. The exponential distribution is characterized by the following functions:

$$\lambda(t) = \lambda_0 \equiv \text{constant}$$

$$R(t) = e^{-\lambda_o t}$$

$$F(t) = 1 - e^{-\lambda_o t} \tag{41}$$

$$f(t) = \lambda e^{-\lambda_o t}$$

$$MTTF = \int_0^\infty t\lambda_o e^{-\lambda_o t} dt = \frac{1}{\lambda_o}$$

14.6.2 Weibull Distribution Function

In the Weibull distribution function the failure rate varies as a power of the age of the device. The failure rate is represented as

$$\lambda(t) = \frac{\beta}{\alpha} t^{\beta - 1} \tag{42}$$

where α and β are constants. For $\beta < 1$ the failure rate decreases with time, and the Weibull distribution may be used to represent the early failure period of a device. For $\beta > 1$ the failure rate increases with time, and the Weibull distribution may be used to represent the wearout period of a device. If $\beta = 1$ the failure rate is constant, and the exponential distribution is a special case of the Weibull distribution with $\beta = 1$. From Eq. 42 we can also calculate that

$$R(t) = \exp\left\{ -\frac{1}{\alpha} t^\beta \right\}$$

$$F(t) = 1 - \exp\left\{ -\frac{1}{\alpha} t^\beta \right\} \tag{43}$$

$$f(t) = \frac{\beta}{\alpha} t^{\beta - 1} \exp\left\{ -\frac{1}{\alpha} t^\beta \right\}$$

$$MTTF = \alpha^{1/\beta} \Gamma(1 + 1/\beta)$$

In some applications of the Weibull distribution function, we can obtain a better fit to the experimental failure rate by introducing a third parameter. In Eqs. 42 and 43 the time t is replaced by

$$X = t - \gamma \tag{44}$$

where the parameter γ corresponds to some portion of the life of the device that has been used up during device manufacture, device burn-in (Section 14.7.4), or device testing.

In fitting experimental data to an assumed distribution function to determine the quality of fit and the distribution parameters, we need to design plotting paper on which the experimental data lie on a straight line if the data can be represented by the assumed distribution. Such plotting paper is available for use with the Weibull distribution. For the Weibull distribution

$$1 - F(t) = \exp\left\{ -\frac{1}{\alpha} t^{\beta} \right\} \qquad (45)$$

and

$$\ln\left\{ \ln\left[\frac{1}{1 - F(t)} \right] \right\} = \beta \ln t - \ln \alpha \qquad (46)$$

which is linear in the form $y = mx + b$. Figure 7 shows an example of Weibull plotting paper; the ordinate is marked in cumulative percent failed devices $F(t)$, and the abscissa is $\ln t$. If the experimental data are represented by the two-parameter Weibull distribution, a straight line is obtained when the data are plotted (Fig. 7). The slope of the line is the parameter β.

    The early failure rate of many ICs can be represented by a Weibull distribution, and Weibull plotting paper is commonly used in IC reliability studies.

14.6.3   Duane Plotting

In some areas, especially those relating to system design, we can analyze failure data by using a technique known as a Duane plot.[14] This graphical analysis technique allows a quick prediction of device-failure rates when the number of device failures is small, and is most often used for evaluating prototype systems.

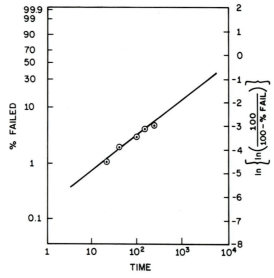

FIGURE 7
An example of Weibull plotting paper showing the failure distribution for a typical device.

The technique consists of plotting the log of the average failure rate (AFR) versus log time. The average failure rate is defined as the fraction of failed devices at time t divided by the time:

$$\text{average failure rate} = \text{AFR} = \frac{F(t)}{t} \tag{47}$$

Often on such a plot, the data plot as a straight line with negative slope S, that is,

$$\ln \text{AFR} = -S \ln t + \ln K$$

or

$$\text{AFR} = \frac{F(t)}{t} = Kt^{-S} \tag{48}$$

where K is a constant. If that is the case, then

$$F(t) = Kt^{1-S}$$

$$f(t) = K(1 - S)t^{-S}$$

and

$$\lambda(t) = \frac{f(t)}{1 - F(t)} = \frac{K(1 - S)t^{-S}}{1 - Kt^{(1-S)}} \tag{49}$$

If only a small fraction of devices have failed, then $F(t) << 1$ and

$$\lambda(t) \cong K(1 - S)t^{-S} \tag{50}$$

or

$$\lambda(t) = (1 - S)\text{AFR} \tag{51}$$

Equation 50 corresponds to a Weibull distribution with $\beta = (1 - S)$ and $\alpha = 1/K$. Thus, a Duane plot is an alternative to a Weibull plot to analyze data in cases where $F(t) << 1$.

The danger in using a Duane plot is that the limitation $F(t) << 1$ is not always recognized, and Eq. 50 or 51 rather than Eq. 49 is used to extrapolate $\lambda(t)$ into the future. Also, extrapolating $\lambda(t)$ during the early failure period into the steady-state period may result in gross underestimation of the long-term failure rate.

14.6.4 Log-Normal Distribution Function

The log-normal distribution function has been used quite successfully to describe the failure statistics of semiconductor devices over long periods of time. Depending on the values of the distribution parameters, this function can represent any one of the three periods in the life of a device.

The probability density function (pdf) of the log-normal distribution is given by

$$f(t) = \frac{1}{\sigma t \sqrt{2\pi}} \exp\left[-\frac{1}{2}\left(\frac{\ln t - \mu}{\sigma} \right)^2 \right] \tag{52}$$

$$= \frac{1}{\sigma t \sqrt{2\pi}} \exp\left[-\frac{1}{2}\left(\frac{1}{\sigma}\ln\frac{t}{t_{50}} \right)^2 \right]$$

where the median time to failure (the time when 50% of the devices have failed) is given by

$$t_{50} = e^{\mu} \tag{53}$$

The average time to failure is

$$\bar{t} = \exp\left(\mu + \frac{\sigma^2}{2} \right) \tag{54}$$

and the scale parameter σ is approximately

$$\sigma \approx \ln\left(\frac{t_{50}}{t_{16}} \right) \tag{55}$$

where t_{16} is the time when 16% of the devices have failed (more precisely, 15.866%).

The cdf is given by

$$F(t) = \frac{1}{\sigma\sqrt{2\pi}} \int_0^t \frac{dx}{x}\exp\left[-\frac{1}{2}\left(\frac{\ln x - \mu}{\sigma} \right)^2 \right] \tag{56}$$

and the failure rate is

$$\lambda(t) = \frac{f(t)}{1 - F(t)} \tag{57}$$

The log-normal distribution does not lend itself to easy numerical calculation. Most data analyses using the log-normal distribution are done either graphically or on a digital computer. Figure 8 shows the log-normal failure rate as a function of time and the cdf for different values of the scale parameter σ. The log-normal distribution may represent the early failure, the steady state, or the wearout period of a device.

Figure 9 shows a normalized plot of log-normal failure rate versus time. The normalized failure rate $\lambda(t) \cdot t_{50}$ is a function only of t/t_{50} and the scale parameter σ. This figure also plots the locus of constant failure rate at a given time. If the constant failure rate is λ_0 and the given time is t_0, then

$$\lambda t_{50} = (\lambda_0 t_{50}) \cdot (t_0/t_0) = (\lambda_0 t_0)/(t_0/t_{50})$$

This equation plots as a straight line with slope of -1 in the log-log plot of Figure 9.

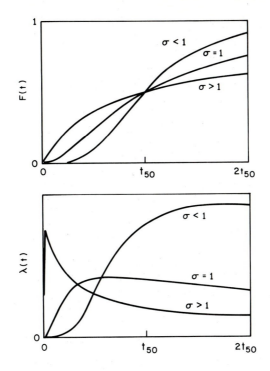

FIGURE 8

Cumulative distribution function $F(t)$ and failure rate $\lambda(t)$ for the log-normal failure distribution.

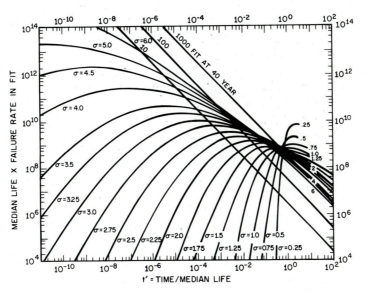

FIGURE 9

Normalized failure rate versus normalized time for the log-normal failure distribution. In this plot, a line with slope -1 is the locus of a given failure rate at a given time with σ as the parameter. Loci are shown for three failure rate values at 40 years. (*After Goldthwaite, Ref. 15.*)

Consider, for example, the case of a failure distribution with $\sigma = 2$. For short times the failure rate is low but increasing with time. When the normalized time ($t' = t/t_{50}$) is approximately 3×10^{-3}, the failure rate equals 10 FIT, and the time equals 40 years. The median life of the failure distribution required for this to happen is 1×10^8 h. If the median life is greater than 1×10^8 h, the failure rate for $\sigma = 2$ will be less than 10 FIT at 40 years. If the median life is less than 1×10^8 h, the failure rate will be greater than 10 FIT at 40 years.

For the case $\sigma = 4.5$, the normalized failure rate has a maximum of 2×10^{12} at $t' \cong 2 \times 10^{-9}$. If this maximum failure rate is 10 FIT, then the median life is 2×10^{11} h, and the maximum failure rate of 10 FIT occurs at $t = 400$ h. If the median life, for $\sigma = 4.5$, is less than 2×10^{11} h, then the maximum failure rate is greater than 10 FIT.

Table 4 summarizes the minimum median life required to meet a maximum failure rate at 40 years. The boldface values in the table correspond to those cases where the maximum failure rate occurs in less than 40 years.

Experimental data can be fitted to the log-normal distribution function by plotting the data on standard log-normal plotting paper (Fig. 10). The σ of the distribution is estimated by the intersection of a line, parallel to the experimental data and passing through the index point $\oplus$, with the σ scale on the right-hand side of the plot.

14.6.5 Two Failure Populations

The study of a device's failure rate is often complicated by the existence of more than one failure mechanism. The failure statistics of each individual mechanism may be governed by a log-normal distribution, but the parameters of the distribution, σ and μ, are different for each mechanism. Quite often one of the failure modes is characterized by a short median life and represents a small percentage of the total population that is usually referred to as the sport population. This mode represents an early failure mechanism. The remainder of the population has a

TABLE 4
Log-normal distribution: minimum median life required to meet a maximum failure rate specification at 40 years*

σ	10 FIT (h)	100 FIT (h)	1000 FIT (h)
0	$t_{50} > 3.5 \times 10^5$	3.5×10^5	3.5×10^5
0.5	2×10^6	1×10^6	7×10^5
1	7×10^6	3×10^6	8×10^5
2	1×10^8	1.3×10^7	**1.3×10^6**
3	1×10^9	**1×10^8**	**1×10^7**
4	**3×10^{10}**	**3×10^9**	**3×10^8**
5	**2×10^{12}**	**2×10^{11}**	**2×10^{10}**

*Boldface values correspond to those cases where the maximum failure rate occurs in less that 40 years.

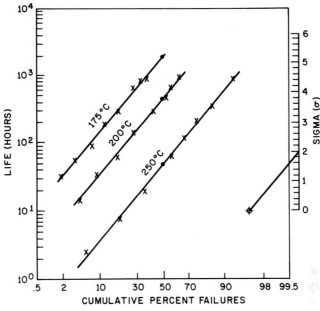

FIGURE 10
An example of log-normal plotting paper showing device failure distribution at three temperatures.

considerably longer median life and represents the steady-state failure mechanism. In actual use, the sport population may be the predominant failure mechanism seen in the field.

Figure 11 shows bimodal failure data plotted on log-normal paper. Using trial and error techniques, the data is resolved into the two component distributions. Using each distribution's parameters, the failure rate can be expressed as

$$\lambda_T(t) = \frac{\epsilon_s f_s(t) + (1 - \epsilon_s)f_M(t)}{1 - \epsilon_s F_s(t) - (1 - \epsilon_s)F_M(t)} \tag{58}$$

where ϵ_s is the fraction of sports in the population, and $f_s(t)$, $F_s(t)$ and $f_M(t)$, $F_M(t)$ are the distribution parameters of the sport and main populations, respectively.

14.7 ACCELERATED TESTING

In the previous sections we have shown that the required failure rate of an IC is on the order of 100 FIT or less. From Tables 3 and 4 we see that, under normal operating conditions, with a failure rate of 100 FIT, the time required to observe one failure in 100 devices is about 100,000 h (i.e., 11.4 years!). The required median life is on the order of 10^5 to 10^{11} h, depending on the scale parameter σ. Clearly, it is impractical to study the failure characteristics of devices with the

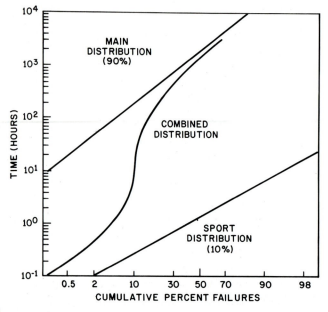

FIGURE 11
A plot of a bimodal failure distribution with the two constituent log-normal distributions. (*After Peck, Ref. 29.*)

requisite reliability under normal operating conditions. If we are to study failure characteristics, some means must be found to accelerate the mechanisms that cause devices to fail.

There are five common stresses used to accelerate device failure mechanisms: temperature, voltage, current, humidity, and temperature cycling. Temperature cycling is used to accelerate mechanical failure of the chip or assembled package (see Chapter 13).

In any study performed under accelerated aging conditions, different failure mechanisms may be accelerated by different amounts for the same applied stress. A device may fail at normal operating conditions because of two completely different failure mechanisms. Under the applied stress, one of these failure modes may be accelerated much more than the other. Thus, in these accelerated aging studies, we would see only one failure mode, and we may successfully eliminate that one mode. However, under normal operating conditions, we would have reduced the device failure rate by only a factor of two.

Accelerated aging is a useful tool only if we know the device failure mechanisms and the acceleration of these mechanisms as a function of the applied stress. Adequate studies must be done under normal operating conditions to satisfy ourselves that no failure mechanisms remain that were not accelerated by the applied stress. The ultimate test of device reliability is long hours of operation under expected conditions of normal use.

14.7.1 Temperature Acceleration

Many of the mechanisms that cause failure are chemical or physical processes that can be accelerated by temperature. The reaction rate R at which these processes proceed is governed, in many cases, by the Arrhenius equation

$$R = R_0\exp(-E_a/kT) \tag{59}$$

where E_a is the activation energy (in electron volts) of the process, k is the Boltzmann constant equal to 8.6×10^{-5} eV/K, and T is the temperature in Kelvin.

We can visualize some parameter of the IC that changes as a function of time. This parameter has some initial value, and IC failure takes place when the parameter exceeds some value which we call the failure criterion. Figure 12 shows the parameter increasing at different rates at the two temperatures T_1 and T_2, where $T_2 > T_1$. For the two temperatures, failure takes place at time t_1 and t_2, respectively, where $t_2 < t_1$. We have arbitrarily shown the parameter value to change linearly with time.

Assuming that the reaction causing the parameter value to change is governed by the Arrhenius equation, the ratio of the two times to failure is given by

$$\frac{t_1}{t_2} = \frac{R_2}{R_1} = \exp\left[\frac{E_a}{k}\left(\frac{1}{T_1} - \frac{1}{T_2}\right)\right] \tag{60}$$

The acceleration factor due to the increased temperature of operation is the ratio of the two times to failure:

$$\text{Acceleration Factor} = \frac{t_1}{t_2} = \exp\left[\frac{E_a}{k}\left(\frac{1}{T_1} - \frac{1}{T_2}\right)\right] \tag{61}$$

IC failure occurs when the destructive reaction has proceeded to some value equal to the failure criterion. Thus the product of the reaction rate R and the time to failure t_F is a constant. The time to failure at different temperatures is

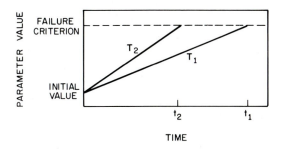

FIGURE 12
Change of parameter value with time at different temperatures.

$$t_F = \frac{\text{constant}}{R} = \text{constant} \times \exp\!\left(E_a/kT\right) \tag{62}$$

or

$$\ln t_F = \text{constant} + E_a/kT \tag{63}$$

The validity of the Arrhenius assumption and the value of the activation energy for a given failure mechanism can be determined by plotting the natural logarithm of the time to failure versus the inverse of the absolute temperature. If a linear plot over a reasonable range of temperature is observed, then extrapolating the failure time back to the operating temperature is valid. Note that the silicon chip temperature is higher than the package temperature; the difference is determined by the power dissipated by the chip and the thermal resistance of the package.

In this discussion we have assumed that the failure reaction is linear in time. Some reactions may have a different functional dependence. Generally the same treatment applies.[16]

To illustrate the acceleration of failure by temperature, we show in Fig. 10 the failure distribution of a group of discrete transistors operated at three different temperatures. At all three temperatures the failure distribution is log-normal. The scale parameter at all three temperatures has a value of 2.0, supporting the assumption that the same failure mechanism is dominant at all three temperatures. This assumption should be verified by detailed analysis of the failed devices.

Figure 13 shows a plot of the median time to failure at the three temperatures versus the inverse absolute temperature. The three data points fall on a straight line; the slope of the line corresponds to an activation energy of 1.0 eV. The

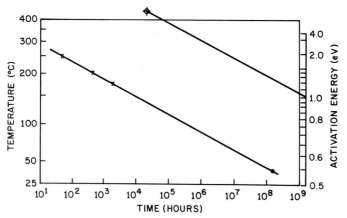

FIGURE 13

Arrhenius plot of the failure data from Fig. 10. The median life at each temperature is plotted versus inverse absolute temperature ($1/T$). For convenience, the ordinate is marked in as a Celsius temperature scale.

median life extrapolated to an operating temperature of 40°C is approximately 2 $\times$ 10^8 h. The extrapolation in median time to failure from the data point at 175°C to the operating temperature is a factor of 10^5. Referring to Table 4, we see that a median time to failure of 2×10^8 h, with a scale factor $\sigma = 2.0$, corresponds to a failure rate at the operating temperature which reaches a value of 10 FIT after 40 years of use.

Table 5 summarizes the acceleration factor due to operating at an increased temperature for two values of the activation energy. The table also shows the time at the increased temperature that is equivalent to 40 years of operation at a 60°C chip-operating temperature. For a failure mechanism with a 1.0 eV activation energy, 11 hours of operation at 200°C is equivalent to 40 years of operation at 60°C.

The above discussion assumes that the activation energy is positive, that is, that devices fail faster at higher temperatures. For several known failure mechanisms, however, the failure rate decreases at elevated temperatures. In one of these mechanisms, hot electrons generated in the silicon are captured in the gate oxide, causing a shift with time in the threshold voltage.[17] These captured electrons are gradually released back to the silicon; the higher the temperature, the greater the release rate. At high temperatures no shift in threshold voltage is seen, because the release rate is equal to or greater than the capture rate. At low temperatures very large shifts in threshold voltage are seen, because the release rate becomes much smaller than the capture rate. It is imperative that the device failure rate be studied at *all* extremes of temperature.

14.7.2 Voltage and Current Acceleration

Voltage and current are effective accelerating stresses for many of the common failure mechanisms observed in ICs. Voltage (or in some cases electric field) causes acceleration of failure caused by dielectric breakdown, interface charge accumulation, charge injection, and corrosion. With most ICs, varying the applied voltage over a very large range is not possible. A device designed to operate at

TABLE 5
Acceleration factor and time equivalent to 40 years

T_{high}* (°C)	Acceleration factor		Time (h) equivalent to 40 year	
	$E_a = 1.0$ eV	0.5 eV	1.0 eV	0.5 eV
300	2.2×10^6	1500	0.2	233
250	3.2×10^5	570	1.1	616
200	3.1×10^4	176	11	2000
150	1700	41	200	8526
125	300	17	1200	20,200
85	11.5	3.4	30,000	103,000

*$T_{\text{amb}} = 60$°C

an applied voltage of 5 V usually will not function properly if the applied voltage is outside the 4 to 7 V range. Of course, some devices operate over a much wider range of applied voltage.

Most studies indicate that the reaction rate R of the failure mechanism is proportional to a power of the applied voltage. The power is usually a function of temperature, that is,

$$R(T, V) = R_0(T)V^{\gamma(T)} \tag{64}$$

The coefficient $R_0(T)$ is usually an Arrhenius function of T, and the parameter $\gamma(T)$ varies between about 1 to 4.5. Thus, if we operate a given device at 7 V rather than the nominal 5 V, we can accelerate the failure anywhere from 40% to as much as a factor of 4, not a very large acceleration.

In the case of dielectric breakdown, a different type of acceleration occurs.[18] For a given applied field a certain fraction of the devices fail in a very short time (on the order of seconds). Very few additional failures occur as the field is maintained. If the applied field is then increased, an additional fraction of failures occur, again in a relatively short time. In a case such as this, operation of the device at an increased voltage is more in the nature of a screening or burn-in rather than a method of accelerating the failure mechanism.

Device operation at increased current levels is used primarily as a method of accelerating failures caused by electromigration in the metallic conductors. Studies show that the reaction rate R of electromigration is a function of temperature and current density, J, of the form

$$R(T, J) = R_0(T)J^{\gamma(T)} \tag{65}$$

Again the coefficient $R_0(T)$ is an Arrhenius function whereas the parameter $\gamma(T)$ varies between about 1 and 4.

It is usually not possible to independently vary the current in an IC, as it is determined by the circuit design. Most studies of the acceleration of electromigration failures use special test structures.[19] These structures are used to determine the maximum allowable current density for which the failure rate of conductors will be acceptable. These values of maximum current density are then used as a design guide in IC layout.

14.7.3 Stress-Dependent Activation Energy

In Section 14.7.1 the acceleration by temperature of chemical or physical processes which can lead to device failure was discussed. In many cases, for example the situation illustrated in Figs. 10 and 13, the process is adequately described through the application of the Arrhenius equation (Eq. 59). In Section 14.7.2 the acceleration of failure by other stresses, such as voltage or current, was described. The parameter describing the dependence of the reaction rate of the failure mechanism, $\gamma(T)$ generally varies over a relatively large range (1 to 4 or more). Studies have also shown that for failure accelerated by voltage (dielectric breakdown) and current (electromigration) the activation energy is dependent upon the applied stress.[33,34] Some of these results are illustrated in Figs. 14 and 15.

A generalized Eyring model has been developed in order to better understand thermally and stress activated failure mechanisms.[32,33] The Arrhenius model is based on the assumption that a definite amount of energy is required for the reaction involved in the failure mechanism to take place. The activation energy, E_a, is a measure of this energy. This energy is supplied by the thermal energy of the reactants. The Eyring model assumes that the energy required for the reaction to take place is affected by the applied stress—in the case of VLSI circuits, generally voltage or current. Under this assumption, the reaction rate of the failure mechanism, R, is

$$R \alpha \sinh\left[\frac{a(T)S}{kT}\right] \exp\left(-\frac{Q}{kT}\right) \tag{66}$$

where S is the applied stress, and $a(T)$ is a temperature-dependent parameter given by

$$a(T) = kT\gamma(T) = kT\left(\gamma_0 + \frac{\gamma_1}{kT}\right) \tag{67}$$

Q is related to the Arrhenius activation energy and is given by

$$Q = E_{ao} + a(T)S_B \tag{68}$$

where S_B is the breakdown stress, the value of applied stress where failure of the device occurs essentially instantaneously.

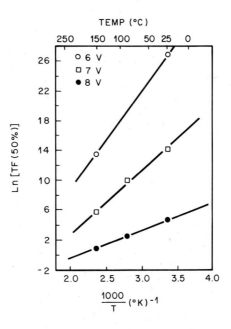

FIGURE 14
Arrhenius plots of time-to-failure data for SiO_2 dielectric breakdown. Three stress levels are shown. (*After McPherson and Baglee, Ref. 33.*)

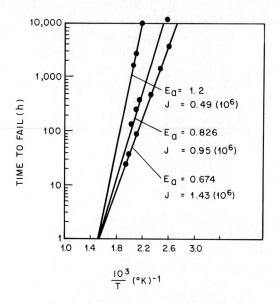

FIGURE 15
Arrhenius plots of time-to-failure data for electromigration in Al–Si metallization. Three stress levels are shown. (*After Partridge and Littlefield, Ref. 34.*)

The reaction rate can also be expressed as being

$$R \propto \sinh[\gamma(T)S]\exp\left(-\frac{E_{ao} + \gamma_1 S_B}{kT}\right)\exp(-\gamma_0 S_B) \tag{69}$$

Under the conditions of low stress, Eq. 69 reduces to

$$R \propto \gamma S \exp\left(-\frac{E_{ao} + \gamma_1 S_B}{kT}\right) \tag{70}$$

while under the condition of high stress, Eq. 69 reduces to

$$R \propto e^{\gamma_0 S}\exp\left(-\frac{E_{ao} + \gamma_1 S_B - \gamma_1 S}{kT}\right) \tag{71}$$

From Eqs. 70 and 71, we see that the reaction rate is a function of the applied stress, S, and that in the case of high stress, Eq. 71, the effective activation energy will also decrease with increased stress.

The application of Eyring modeling of VLSI failure mechanisms is a new field in which we expect to see important advances in the future.

14.7.4 Humidity–Temperature Acceleration

Usually, state-of-the-art ICs are packaged in hermetic packages, allowing evaluation and initial reliability studies to be carried out under optimum conditions.

As the devices mature, competitive pressure to reduce costs results in the use of lower-cost, nonhermetic packages. The presence of water vapor in the chip environment introduces a new variety of possible failure mechanisms.

Studies show that water vapor quickly permeates plastic packaging material.[20] In the first step of the permeation process, water vapor transports contaminants from the surface of the package through the plastic and leaches impurities from the plastic packaging material itself. The surface of the chip is very quickly exposed to the water vapor and various contaminants.

The second step is the diffusion of the contaminated water vapor through the passivation layer of the chip, usually a relatively slow process. However, if the passivation layer contains defects or cracks, the water vapor penetrates through the passivation layer much more quickly. In these studies the penetration of water vapor through the passivation layer determined the reaction rate of the failure mechanisms.[20]

Once the water vapor reaches the metallization level of the chip, electro-chemical corrosion can occur. The ions needed for this corrosion process can arise from two sources. Ions can diffuse as a contaminant through the passivation layer along with the water vapor. If the intermediate dielectric of the chip is a phosphorus-doped glass, the water vapor can leach phosphorus from the interme-diate dielectric. In either case, electrochemical corrosion is a very rapid process, which results in metallization failure. This type of failure mechanism can be accelerated by increasing the partial pressure of water vapor in the environment. Increasing the ambient temperature also increases the rate-limiting diffusion of the water vapor through the passivation layer. Table 6 summarizes the analysis for the acceleration of the failure of two types of devices.[20] The acceleration factor obtained for the two types of devices is significantly different.

As with other types of stress, the application of an acceleration factor obtained for one type of failure mechanism in a given device to any other device is extremely dangerous. For each device, and each failure mechanism in that device, the accelerating effect of temperature, humidity, and voltage must be redetermined.

TABLE 6
Humidity–induced failures[20]

Device	Bipolar	NMOS
Passivation layer	Sputtered SiO_2	SiN
Intermediate dielectric	Sputtered SiO_2	P-glass
Mechanism	Cl corrosion (contamination)	P corrosion (from P-glass)
σ	0.43	1.34
E_a	1.1 eV	0.3 eV
Accelerating factor ($125°C$–$60°C$)	300	5.5
40 yr/Acceleration	1200 hours	7.3 years

14.7.5 Burn-in

For mature products in high-volume manufacture, a considerable effort is made in determining and eliminating the predominant failure mechanisms identified in the initial reliability studies so that the steady-state failure rate of the device meets or exceeds the design goal. Possible wearout mechanisms will have been identified and eliminated, either through modifications in the process or in the design. However, the manufactured devices will normally show the existence of continuing early failure or infant mortality type of failures.

Generally, manufacturing defects cause the infant mortality type of failures. Examples of such defects include oxide pinholes, photoresist or etching defects resulting in near-opens or shorts, contamination on the chip or in the package, scratches, weak chip or wire bonds, and partially cracked chips or packages.

The contribution of the infant mortality failures to the overall failure rate can be modeled as a subpopulation of sport devices, using either the log-normal or some other distribution function. If the steady-state failure rate is extremely low, then the infant mortality failure rate is most easily modeled using the Weibull model, without making reference to any subpopulation. For this case the overall failure rate is expressed as

$$\lambda(t) = \frac{\beta}{\alpha} t^{\beta-1} + \lambda_{SS} \tag{72}$$

where $\lambda(t)$ is the device failure rate, β and α are the parameters of the infant mortality failure distribution, and λ_{SS} is the constant steady-state failure rate.[35]

The purpose of a burn-in procedure is to operate the devices for some period of time during which most of the devices that are subject to infant mortality failure actually fail. Failing the burn-in procedure is preferable to having devices fail after they are installed in a system and delivered to the customer. The conditions during burn-in presumably accelerate the failure mechanisms that contribute to the infant mortality failures. The burn-in may possibly consist of operating the devices under conditions of increased temperature, increased voltage, and high-current (if possible) load conditions.

Studies of the acceleration of infant mortality failures under conditions of increased operating temperature[21] show that infant mortality failure mechanisms have an activation energy E_a of 0.37 to 0.42 eV. The acceleration factors for infant mortality failures by increased operating voltage have not been reported.

If devices are operated under burn-in conditions for some time, t_{BI}, where the acceleration of the infant mortality failures is A_{BI}, then the failure rate of the devices when they are placed in service is

$$\lambda_{\text{EFF}}(t) = \lambda(t + t_0) = \lambda(t + t_{BI} \cdot A_{BI}) \tag{73}$$

Figure 16 shows the effect of burn-in on the device-failure rate. If the activation energy of the infant mortality failure is 0.4 eV, then a 168 hour burn-in at 150°C is equivalent to over 4 months of operation at 60°C. This can result in a significant improvement in the reliability of the installed system.

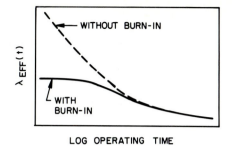

FIGURE 16
Failure rate versus time-in-use for devices with and without burn-in. The broken curve is the failure rate $\lambda(t)$ without burn-in. The solid curve is the failure rate $\lambda_{EFF}(t) = \lambda(t + t_0)$ with burn-in.

Other methods of eliminating potential infant mortality failures from the devices shipped to customers are also in use, such as cycling the packaged devices over an extended temperature range to eliminate weak chip or wire bonds and partially cracked chips or packages. As mentioned in Section 14.7.2, device operation at an applied voltage above the nominal operating voltage screens out devices with oxide defects that may fail when the device is subjected to an overvoltage condition in actual service.

14.8   FAILURE MECHANISMS

Previous sections have examined the techniques for evaluating and predicting IC reliability and failure rates. Equally as important as the evaluation of reliability is the improvement of reliability. The viability of VLSI circuits is critically dependent upon the identification and elimination of those failure mechanisms that are introduced as technology makes its inexorable push toward higher-density circuits of ever-increasing performance.

Table 7 summarizes the failure mechanisms that are commonly identified as degrading IC reliability. Many of these processes have been covered in previous chapters. More detailed information on specific failure processes is available in several review articles on the reliability of silicon ICs. [22,23]

Table 7 lists those factors that contribute to each failure process and those that may accelerate device degradation. In integrated circuit reliability studies, the accelerating factors are varied, or applied at an increased value during a burn-in procedure. Table 7 includes the activation energy E_a for the processes that are accelerated by temperature and the coefficient for the processes that are accelerated as a power γ of the applied field or current density. Of prime importance in Table 7 is not the particular value of E_a or γ for a particular process, but the wide range over which these parameters vary. The exact value of E_a or γ must be determined for each process and each device made in that process.

In addition to the failure processes listed in Table 7, other failure mechanisms, two of which we will discuss next, are observed in silicon ICs.

TABLE 7
Time-dependent failure mechanisms in silicon semiconductor devices[23,30]

Device association	Failure mechanism	Relevant factors	Accelerating factors	Acceleration (E_a = apparent activation energy for temp.)
Silicon oxide and silicon–silicon oxide interface	Surface charge accumulation	Mobile ions, V, T	T	E_a = 1.0 to 1.05 eV (depends upon ion density)
	Dielectric breakdown	$\mathscr{E}$, T	$\mathscr{E}$, T	E_a = 0.2 to 1.0 eV $\mathscr{E}^\gamma$, $\gamma(T)$ = 1 to 4.4
	Charge injection	$\mathscr{E}$, T, Q_{ss}	$\mathscr{E}$, T	E_a = 1.3 eV (slow trapping)
				$E_a \approx -1$ eV (hot electron injection)
Metallization	Electromigration	T, J, A, gradients of T and J, grain size	T, J	E_a = 0.5 to 1.2 eV J_γ, $\gamma(T)$ = 1 to 4
	Corrosion (chemical, galvanic, electrolytic)	Contamination, H, V, T	H, V, T	Strong H effect, $E_a \approx 0.3$ to 1.1 eV (for electrolysis) V may have thresholds
	Contact degradation	T, metals, impurities	Varied	
Bonds and other mechanical interfaces	Intermetallic growth	T, impurities, bondstrength	T	Al–Au: E_a = 1.0 to 1.05 eV
	Fatigue	Bond stength, temperature cycling	T extremes in cycling	
Hermeticity	Seal leaks	Pressure differential, atmosphere	Pressure	

Note: V is voltage, $\mathscr{E}$ is electric field, A is area, T is temperature, J is current density, and H is humidity.

14.8.1 Electrostatic Discharge Damage

The gate oxide in modern VLSI circuits is presently on the order of 250 Å thick and will be considerably thinner in future processes. The dielectric breakdown strength of SiO_2 is approximately 8×10^6 V/cm. Thus, a 250 Å gate oxide will not sustain voltages in excess of 20 V. This is well in excess of the normal operating voltage of VLSI circuits. However, during handling of the devices, voltages higher than the breakdown voltage of the gate oxide may be placed upon the circuits.

A major source of excessive voltage arises from triboelectricity (electricity produced by rubbing two materials together). A person walking across a room or simply removing an IC from its plastic packaging material can generate from 15,000 to 20,000 V. The discharge of the triboelectrically generated charge into the IC can either damage the circuit sufficiently to cause immediate failure or damage the device slightly, but enough to cause subsequent failure early in the operating life of the device.

Burn-in will not reduce device failure caused by electrostatic discharge (ESD). In fact, additional handling of the devices during the burn-in procedure may actually increase the early device-failure rate if adequate safeguards are not taken to prevent the buildup of static electricity. Electrostatic charging theory, ESD failure models, and methods for preventing ESD failures are available.[24]

Even with proper precautions the discharge of possibly several hundred volts of static electricity occurs during IC handling. The input and output circuit terminals are therefore designed with protection networks on the chip that provide a path for the discharge of the current and prevent the generation of excessive voltage across the gate oxide of the devices. Figure 17 illustrates several types of gate-protection networks. For the circuit in Figure 17a, if a positive voltage is discharged into the input terminal, the upper diode is forward biased, and the discharge current will flow through the diode to the V_{DD} terminal of the device. No excess voltage is generated across the gate oxide. Similarly, for the discharge of a negative voltage into the input terminal, the current will forward bias the lower diode and flow to the V_{SS} terminal. For the circuit in Figure 17b, when a discharge into the input terminal occurs, one of the Zener diodes will be forward biased and the other will be biased to its reverse breakdown potential, allowing the discharge of the current to the V_{SS} terminal. For the circuit in Figure 17c, the discharge into the input terminal either flows through the source diode to the V_{SS} terminal, or the potential turns on the high-threshold voltage transistors, allowing the current to flow to V_{SS}. The spark gap in Figure 17d provides an additional path for the discharge of current to V_{SS}. In all the circuits the resistor limits the peak current flow. The diodes and other current paths of these protection networks must be designed to handle the high currents and powers produced in a typical ESD event. A human body, initially charged to a potential of 2000 V, will cause a peak current flow of several amperes and a peak power dissipation of several kilowatts when the body is discharged into an IC.

14.8.2 Alpha-Particle-Induced Soft Errors

Radioactive elements, such as uranium or thorium, are naturally occurring impurities in IC packaging materials. The alpha particles emitted by these materials can cause soft errors in the ICs.[25,26] The term *soft error* refers to a random failure not related to a physically defective device. The penetration of an alpha particle into the silicon causes the generation of an electron-hole plasma along the path of the particle. The generated carriers can cause the loss of information stored in

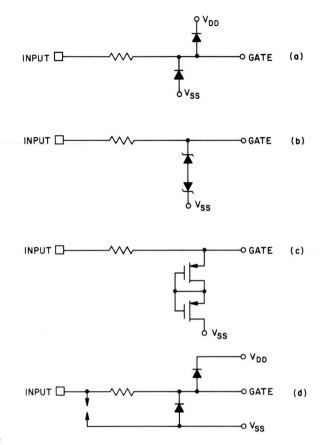

FIGURE 17
Schematic for several types of gate-protection networks. (a) Diode. (b) Zener Diode. (c) Transistor.
(d) Spark Gap. (*After Wood, Ref. 23.*)

the memory cells of a dynamic memory or stored in the depletion region of the
drains of devices making up circuits such as dynamic shift registers or other logic
units. The adsorption of a 4 MeV alpha particle can generate 10^6 electron-hole
pairs, which is equal to or greater than the charge stored in a dynamic memory
cell.[27] To illustrate the seriousness of the problem, a typical soft error rate in
a memory system containing one thousand 16K memory devices may be on the
order of 1 soft error per 1000 hours, corresponding to a device failure rate of
1000 FIT.[22]

The incidence of soft errors can be reduced by surrounding or coating the IC
chip with a material having a very low density of radioactive contamination.[28] For
example, alpha particles with an energy up to 8 MeV are completely absorbed in
50 μm of silicone rubber, and this material does not emit any significant amount
of alpha particles.

14.8.3 Continuously Variable Failure Population

In a previous section we considered the effect of two independent failure mechanisms on the characterization of the failure rate of a device. Recent studies indicate that the more complex situation of a continuously variable failure population may exist.[31] In this study the device wafers were partitioned into a center area and two annular areas, each containing an equal number of chip sites. The yield and reliability of chips from these three regions were then studied. It was found that the yields from chip sites in the central two regions were essentially equivalent, but the yield from chip sites in the outer region was reduced by 40%.

The failure rate of devices from the three regions was then studied under various conditions. The results were normalized to the failure rates observed for devices from the central region. It was found that the failure rate for devices from the outer region was greater than that for devices from the central region.

For all types of stresses and failure mechanisms, the device from the outer region had a failure rate more than 4 times greater. For certain specific failure mechanisms, the failure rate of devices from the outer region was more than 10 times greater.

These results pose several interesting questions which must be answered by future work in the area of yield and reliability. First, are the distribution functions previously described adequate to handle such a continuously variable failure population? Second, it appears that significant improvement in device reliability could be achieved by eliminating low-yield portions of the device population: What are the algorithms that properly relate device reliability and device yield (and cost) for products that show characteristics such as those described?

14.9 SUMMARY AND FUTURE TRENDS

The trend in modern VLSI technology, as exemplified in previous chapters, has the tendency to exacerbate the problems of maintaining high yield and reliability. Progress in VLSI technology is achieved at the cost of finer dimensions, larger chip area, more complex processes, and the introduction of new and more complex material systems. All of these inevitably lead to initially higher defect density. Identifying, characterizing, and ultimately eliminating the causes of circuit failure and low yield is a necessary part of the development of a new technology.

Similarly, the cost of progress will be devices operating with higher electric fields, higher current densities, and higher power dissipation. All of these factors will accelerate the failure of devices due to those failure mechanisms previously identified and characterized. In addition, the introduction of new material systems introduces new types of IC failure. The successful development of new technologies must include the identification and characterization of these failure processes. The combined efforts of technology development along with yield and failure-mode analysis will provide the technology needed to produce VLSI circuits usable in economically viable systems.

PROBLEMS

1 The wafer in Fig. 1 has a total of 266 chip sites and 99 good chips. If the wafer grid is divided into new grids containing double, triple, and quadruple chip sites, there are 126, 83 and 58 double, triple, and quadruple chip sites, respectively. The number of good double, triple, and quadruple chips is 29, 14, and 7, respectively. Examination of the wafer shows that no good chips are found over approximately 46.6% of the area of the wafer, corresponding to a Y_0 of 53.4%.

(a) From the yield of good chips, estimate the D_oA based upon Poisson, triangular, and rectangular distribution functions of D.

(b) Using these D_oA estimates, predict the yield of a chip with four times the chip area in Fig. 1. Compare your answer to the observed yield of good quadruple chips.

2 Using the data from Problem 1 for the observed yield of multiple chips, determine whether the data give a good fit to Eq. 26, or if a better fit would be obtained using the Gamma distribution yield of Eq. 17.

3 A group of 10 devices is placed on an accelerated aging test at 150°C and operated for a total time of 3000 h. The devices are removed from the test and tested for electrical characteristics every 100 h. A different device fails at each of the following test times: 200, 800, 1000, 1700, 2800, and 2900 h. At 3000 h four devices are still operating. Fit the data to both log-normal and Weibull distribution functions. (Note: When plotting data of this type it is accepted practice to use #Failed/(#On Test + 1) for the fraction of devices failed.)

(a) Does the data fit one distribution function better than the other?

(b) What are the median life and σ for the log-normal distribution?

(c) What is the median life for the Weibull distribution?

(d) At what time do the two distributions predict that 99% of the devices will fail?

(e) A second group of 10 devices from the same manufacturing lot is aged under identical conditions. Failures were found at 200, 1600, 1700, and 2400 h. At 3000 h six devices were still good. Analyze this data in the same manner as above.

(f) Combine the results from the two experiments, and estimate the median life and σ for these devices when aged at 150°C.

4 A third group of 20 devices from the same production lot as in Problem 3 is aged at 200°C. The devices are tested after every 10 h of aging. Devices fail at the following times: 10, 30, 40, 50 (two failures), 80, 100 (two failures), and 110 h. Testing was discontinued at this time.

(a) What is the best estimate of the log-normal median life and σ?

(b) Assuming that failure mode analysis indicates that the devices are failing by the same mechanism at both 150 and 200°C and that the Arrhenius relation thus applies, calculate the activation energy and the estimated mean life at operating temperatures of 70 and 50°C.

(c) Making reasonable estimates for the error in the determination of the median life at 200°C and 150°C, calculate a reasonable estimate for the lower limit of the median life at 70°C.

5 Use the results of Problems 3 and 4 to answer the following.

(a) What is the best estimate of the failure rate of the device when operated at a temperature of 70°C, at an operating time of 4 months, 1 year, and 10 years?

(b) Using the lower limit for the value of the mean life at 70°C (from Problem 4), what is the worst-case prediction of the failure rate at 4 months, 1 year, and 10 years?

(c) What are the best estimates of the failure rate if the operating temperature is reduced to 50°C?

6 For a median life of 4.0×10^6 h and a σ of 1.3,

(a) What fraction of devices will have failed after 10 years of use at 70°C?

(b) If 100 of these devices are used in a system, what fraction of the systems will have failed in 10 years? (Hint: Use the Poisson approximation to the binomial probability, Eq. 4, that no failed devices are found in a given system.)

(c) What is the expected fraction of failed devices after 10 years of operation using the lower limit for the median life of 1.8×10^6 h?

(d) What is the expected fraction of failed systems?

(e) What is the expected fraction of failed devices and systems after 10 years of operation if the system is operated at 50°C where the expected mean life is 4.0×10^7 h?

7 Twenty sample devices from another production lot of the same devices as in the previous problems were evaluated at an aging temperature of 150°C. Some of the devices in this lot were believed to have been improperly processed and contaminated with sodium. These 20 devices were tested after 2, 4, 8, . . .,4096 h of aging. Failures were found at the following times: 32, 64(three failures), 128, 512, 1024(two failures), 2048(three failures) and 4096(two failures) h. Plot the failure distributions on log-normal plotting paper.

(a) Estimate from the inflection point of the failure distribution the fraction of contaminated or sport devices in the population.

(b) Separate the devices into sport and normal devices. Replot the failure data to determine the median life and σ of each population.

8 Assume that the sodium-contaminated sport devices compose 25% of the population, that the median life at 150°C is 60 hours, the σ is 0.7, and the activation energy is 1.0 eV. Assume that the main population has a median life at 150°C of 4800 h, a σ of 1.3, and an activation energy of 1.10 eV.

(a) Plot the failure rate of the devices as a function of time-in-use at a temperature of 70°C for devices used as produced.

(b) Would a 150 h burn-in at 150°C provide an effective method of reducing the failure rate of these devices during normal operation at 70°C?

REFERENCES

1 W. Feller, *An Introduction to Probability Theory and Its Applications*, 3rd ed., Wiley, New York, 1968.

2 S. E. Schuster, "Multiple Word/Bit Line Redundancy for Semiconductor Memories," *IEEE J. Solid State Circuits*, **SC13**, 698 (1978).

3 B. T. Murphy, "Cost-Size Optima of Monolithic Integrated Circuits," *Proc. IEEE*, **52**, 1537 (1964).

4 C. H. Stapper, Jr., "On a Composite Model to the I.C. Yield Problem," *IEEE J. Solid State Circuits*, **SC10**, 537 (1975).

5 W. W. Hines and D. C. Montgomery, *Probability and Statistics in Engineering and Management Science*, Wiley, New York, 1972.

6 K. V. Bury, *Statistical Models in Applied Science*, Wiley, New York, 1975.

7 H. Murrman and D. Kranzer, "Yield Modeling of Integrated Circuits," *Siemens Forsch. Entwicklungs Ber.*, **9**, 38 (1981).

8 C. H. Stapper, "LSI Yield Modeling and Process Monitoring," *IBM J. Res. Dev.*, **20**, 228 (1976).

9 A. Gupta, W. A. Porter, and J. W. Lathrop, "Defect Analysis and Yield Degradation of Integrated Circuits," *IEEE J. Solid State Circuits*, **SC9**, 96 (1974).

10 C. H. Stapper, "Yield Model for 256K RAMs and Beyond," *IEEE Int. S.S.C.C. Digest of Tech. Papers*, 1982, p. 12.

11 W. G. Ansly, "Computation of Integrated Circuit Yields from the Distribution of Slice Yields for the Individual Devices," *IEEE Trans. Electron Devices*, **ED15**, 405 (1968).

12 A. Gupta and J. W. Lathrop, "Yield Analysis of Large Integrated Circuit Chips," *IEEE J. Solid State Circuits*, **SC7**, 389 (1972).

13 T. Yanagawa, "Yield Degradation of Integrated Circuits Due to Spot Defects," *IEEE Trans. Electron Devices*, **ED19**, 190 (1972).

14 J. T. Duane, "Learning Curve Approach to Reliability Monitoring," *IEEE Trans. Aerosp.*, **2**, 563 (1964).

15 L. R. Goldthwaite, "Failure Rate Study for the Log-normal Lifetime Model," *IEEE Proc. 7th Symposium on Reliability and Quality Control*, 1961, p. 208.

16 R. E. Weston and H. A. Schwartz, *Chemical Kinetics*, Prentice-Hall, Englewood Cliffs, NJ, 1972.

17 R. C. Sun, J. T. Clemens, and J. T. Nelson, "Effects of Silicon Nitride Encapsulation on MOS Device Stability," *Reliability Physics, 18th Annual Proceedings*, 1980, pp. 244–251.

18 E. S. Anolick and L-Y. Chen, "Application of Step Stress to Time Dependent Breakdown," *Reliability Physics, 19th Annual Proceedings*, 1981, pp. 23–27.

19 S. Vaidya, D. B. Fraser, and A. K. Sinha, "Electromigration Resistance of Fine-Line Al for VLSI Applications," *Reliability Physics, 18th Annual Proceedings*, 1980, pp. 165–170.

20 J. E. Gunn and S. K. Malik, "Highly Accelerated Temperature and Humidity Stress Test Technique (HAST)," *Reliability Physics, 19th Annual Proceedings*, 1981, pp. 48–51.

21 D. S. Peck, "New Concerns About Integrated Circuit Reliability," *Reliability Physics, 16th Annual Proceedings*, 1978, pp. 1–6.

22 C. M. Bailey, "Basic Integrated Circuit Failure Mechanisms," in L. Esaki, Ed., *Large Scale Integrated Circuits: State of the Art and Prospects*, Lange Voorkout, The Hague, 1982.

23 J. Wood, "Reliability and Degradation of Silicon Devices and Integrated Circuits," in M. J. Howes and D. V. Morgan, Eds., *Reliability and Degradation*, Wiley, New York, 1981.

24 B. A. Unger, "Electrostatic Discharge Failures of Semiconductor Devices," *Reliability Physics, 19th Annual Proceedings*, 1981, pp. 193–199.

25 T. C. May and M. H. Woods, "A New Physical Mechanism for Soft Errors in Dynamic Memories," *Reliability Physics, 16th Annual Proceedings*, 1978, pp. 33–40.

26 T. C. May and M. H. Woods, "Alpha-Particle-Induced Soft Errors in Dynamic Memories," *IEEE Trans. Electron Devices*, **ED26**, 2 (1979).

27 C. M. Hsieh, P. C. Murley, and R. R. O'Brien, "Dynamics of Charge Collection from Alpha Particle Tracks in Integrated Circuits," *Reliability Physics, 19th Annual Proceedings*, 1981, pp. 38–42.

28 M. L. White, J. W. Serpiello, K. M. Striny, and W. Rosenzweig, "The Use of Silicone RTV Rubber for Alpha Particle Protection of Silicon Integrated Circuits," *Reliability Physics, 19th Annual Proceedings*, 1981, pp. 43–47.

29 D. S. Peck, "The Analysis of Data from Accelerated Stress Tests," *Reliability Physics, 9th Annual Proceedings*, 1971, pp. 69–77.

30 D. S. Peck, "Practical Applications of Accelerated Testing—Introduction," *Reliability Physics, 13th Annual Proceedings,* 1975, pp. 253–254.

31 H. A. Bonges III, "Radial Dependency of Reliability Defects on Silicon Wafers," *Reliability Physics, 24th Annual Proceedings*, 1986, pp. 172–174.

32 J. W. McPherson, "Stress Dependent Activation Energy," *Reliability Physics, 24th Annual Proceedings*, 1986, pp. 12–18.

33 J. W. McPherson and D. A. Baglee, "Acceleration Factors for Thin Gate Oxide Stressing," *Reliability Physics, 23rd Annual Proceedings*, 1985, pp. 1–5.

34 J. Partridge and G. Littlefield, "Aluminum Electromigration Parameters," *Reliability Physics, 23rd Annual Proceedings*, 1985, pp. 119–125.

35 J. C. North and D. P. Holcomb, "An Infant Mortality and Long-Term Failure Rate Model for Electronic Equipment," *AT&T Technical Journal*, **64**, 15 (1985).

APPENDIX
A

PROPERTIES
OF SILICON
AT 300 K

Properties	Si
Atoms/cm^3	5.0×10^{22}
Atomic weight	28.09
Breakdown field (V/cm)	$\sim 3 \times 10^5$
Crystal structure	Diamond
Density (g/cm^3):	
Solid	2.33
Liquid (1412°C)	2.53
Dielectric constant	11.9
Effective density of states of conduction band, N_C (cm^{-3})	2.8×10^{19}
Effective density of states in valence band, N_V (cm^{-3})	1.04×10^{19}
Effective mass, m^*/m_0:	
Electrons	$m_l^* = 0.98,\ m_t^* = 0.19$
Holes	$m_{lh}^* = 0.16,\ m_{hh}^* = 0.49$
Electron affinity (V)	4.05
Energy gap (eV)	1.12
Heat capacity (cal/g-mol-°C):	
Solid	4.78
Liquid (1412°C)	6.76
Index of refraction	3.42

Properties	Si
Intrinsic carrier concentration (cm^{-3})	1.45×10^{10}
Intrinsic Debye length (μm)	24
Intrinsic resistivity (Ω-cm)	2.3×10^5
Lattice constant (Å)	5.43095
Linear coefficient of thermal expansion, $\Delta L/L\Delta T$(°C^{-1})	2.6×10^{-6}
Melting point (°C)	1412
Minority-carrier lifetime(s)	2.5×10^{-3}
Mobility (drift) (cm^2/V-s):	
μ_n(electrons)	1500
μ_p(holes)	475
Optical phonon energy (eV)	0.063
Phonon mean free path λ_0 (Å):	
Electrons	76
Holes	55
Poission's ratio	0.42
Specific heat (J/g–°C)	0.7
Thermal conductivity (W/cm-°C):	
Solid	1.5
Liquid (1412°C)	4.3
Thermal diffusivity (cm^2/s)	0.9
Torsion modulus (kg/mm^2)	4050
Vapor pressure	1 at 1650°C
(Pa)	10^{-6} at 900°C
Young's modulus (kg/mm^2)	10,890

APPENDIX
B

LIST OF SYMBOLS

Symbol	Description	Unit
a	Lattice constant	Å
B	Magnetic induction	Wb/m^2
c	Speed of light in a vacuum	cm/s
C	Capacitance	F
D	Electric displacement	C/cm^2
D	Diffusion coefficient	cm^2/s
E	Energy	eV
E_F	Fermi energy level	eV
E_g	Energy bandgap	eV
$\mathscr{E}$	Electric field	V/cm
$\mathscr{E}_m$	Maximum field	V/cm
f	Frequency	Hz
h	Planck's constant	J-s
$h\nu$	Photon energy	eV
I	Current	A
J	Current density	A/cm^2
k	Boltzmann constant	J/K
kT	Thermal energy	eV
L	Length	cm or μm
m_0	Electron rest mass	kg
m^*	Effective mass	kg

Symbol	Description	Unit
$\bar{n}$	Refractive index	
n	Density of free electrons	cm^{-3}
n_i	Intrinsic density	cm^{-3}
N	Doping concentration	cm^{-3}
N_A	Acceptor impurity density	cm^{-3}
N_D	Donor impurity density	cm^{-3}
p	Density of free holes	cm^{-3}
P	Pressure	N/m^2
q	Magnitude of electronic charge	C
Q_{it}	Interface-trap density	charges/cm^2
R	Resistance	Ω
t	Time	s
T	Absolute temperature	K
v	Carrier velocity	cm/s
v_s	Saturation velocity	cm/s
V	Voltage	V
V_{bi}	Built-in potential	V
V_B	Breakdown voltage	V
W	Thickness	cm or μm
x	x direction	
∇T	Temperature gradient	K/cm
ϵ_0	Permittivity in vacuum	F/cm
ϵ_s	Semiconductor permittivity	F/cm
ϵ_i	Insulator permittivity	F/cm
ϵ_s/ϵ_0 or ϵ_i/ϵ_0	Dielectric constant	
τ	Lifetime or decay time	s
θ	Angle	rad
λ	Wavelength	μm or Å
ν	Frequency of light	Hz
μ_0	permeability in vacuum	H/cm
μ_n	Electron mobility	cm^2/V-s
μ_p	Hole mobility	cm^2/V-s
ρ	Resistivity	Ω-cm
ϕ	Barrier height or imref	V
ϕ_m	Metal work function	V
ω	Angular frequency ($2\pi f$ or $2\pi\nu$)	Hz
Ω	Ohm	Ω

APPENDIX
C

INTERNATIONAL SYSTEM OF UNITS

Quantity	Unity	Symbol	Units
Length	meter	m	
Mass	kilogram	kg	
Time	second	s	
Temperature	kelvin	K	
Current	ampere	A	
Frequency	hertz	Hz	$1/s$
Force	newton	N	$kg\text{-}m/s^2$
Pressure	pascal	Pa	N/m^2
Energy	joule	J	$N\text{-}m$
Power	watt	W	J/s
Electric charge	coulomb	C	$A\text{-}s$
Potential	volt	V	J/C
Conductance	siemens	S	A/V
Resistance	ohm	Ω	V/A
Capacitance	farad	F	C/V
Magnetic flux	weber	Wb	$V\text{-}s$
Magnetic induction	telsa	T	Wb/m^2
Inductance	henry	H	Wb/A

Quantity	Symbol	Value
Angstrom unit	Å	$1\ \text{Å} = 10^{-1}\ \text{nm} = 10^{-4}$ $\mu\text{m} = 10^{-8}\ \text{cm}$
Avogadro constant	N_{AVO}	$6.02204 \times 10^{23}\ \text{mol}^{-1}$
Bohr radius	a_B	$0.52917\ \text{Å}$
Boltzmann constant	k	$1.38066 \times 10^{-23}\ \text{J/K}\ (R/N_{AVO})$
Elementary charge	q	$1.60218 \times 10^{-19}\ \text{C}$
Electron rest mass	m_0	$0.91095 \times 10^{-30}\ \text{kg}$
Electron volt	eV	$1\ \text{eV} = 1.60218 \times 10^{-19}$ $\text{J} = 23.053\ \text{kcal/mol}$
Gas constant	R	$1.98719\ \text{cal-mol}^{-1}\text{K}^{-1}$
Permeability in vacuum	μ_0	$1.25663 \times 10^{-8}\ \text{H/cm}(4\pi \times 10^{-9})$
Permittivity in vacuum	ϵ_0	$8.85418 \times 10^{-14}\ \text{F/cm}(1/\mu_0 c^2)$
Planck constant	h	$6.62617 \times 10^{-34}\ \text{J-s}$
Reduced Planck constant	$\hbar$	$1.05458 \times 10^{-34}\ \text{J-s}\ (h/2\pi)$
Proton rest mass	M_p	$1.67264 \times 10^{-27}\ \text{kg}$
Speed of light in vacuum	c	$2.99792 \times 10^{10}\ \text{cm/s}$
Standard atmosphere		$1.01325 \times 10^5\ \text{N/m}^2$
Thermal voltage at 300 K	kT/q	$0.0259\ \text{V}$
Wavelength of 1 eV quantum	λ	$1.23977\ \mu\text{m}$

INDEX